Modern Drug Synthesis

Modern Drug Synthesis

Edited by

Jie Jack Li
Bristol-Myers Squibb Company

and

Douglas S. Johnson
Pfizer Global Research and Development

A JOHN WILEY & SONS, INC., PUBLICATION

For general information on our other products and services or for technical support, please contact our Customer Care Department within the United States at (800) 762-2974, outside the United States at (317) 572-3993 or fax (317) 572-4002.

Wiley also publishes its books in a variety of electronic formats. Some content that appears in print may not be available in electronic formats. For more information about Wiley products, visit our web site at www.wiley.com.

Library of Congress Cataloging-in-Publication Data:

Modern drug synthesis / edited by Jie Jack Li, Douglas S. Johnson.
    p. cm.
  Includes index.
  ISBN 978-0-470-52583-8 (cloth)
  1. Drugs—Design. 2. Pharmaceutical chemistry. I. Li, Jie Jack. II. Johnson,
Douglas S. (Douglas Scott), 1968–
   RS420.M6325 2010
   615'.19—dc22   2010007983

Printed in the United States of America.

10 9 8 7 6 5 4 3 2 1

Contents

Preface

Our first two books on drug synthesis, *Contemporary Drug Synthesis* and *The Art of Drug Synthesis,* were published in 2004 and 2007, respectively. They have been warmly received by the chemistry community. Here is our third installment in the *Drug Synthesis* series.

   This book has five sections. Section I, "Infectious Diseases" covers three drugs; Section II, "Cancer" reviews five drugs, three of which are kinase inhibitors; Section III covers eight drugs that target cardiovascular and metabolic diseases; Section IV on central nervous system diseases concerns four classes of recent drugs; and Section V summarizes two drugs: raloxifene hydrochloride (Evista) for the treatment of osteoporosis and latanoprost (Xalatan) for ophthalmological indications.

   In this installment of the *drug synthesis series*, we have placed more emphasis on other aspects of medicinal chemistry in addition to the synthesis. To that end, each chapter is divided into seven sections:

1.   Background
2.   Pharmacology
3.   Structure–activity relationship
4.   Pharmacokinetics and drug metabolism
5.   Efficacy and safety
6.   Syntheses
7.   References

We are indebted to the contributing authors from both industry and academia. Many of them are veterans in medicinal chemistry. Some of them discovered the drugs that they reviewed. As a consequence, their work tremendously elevated the quality of this book. Meanwhile, we welcome your critique and suggestions so we can make this *drug synthesis series* even more useful to the medicinal/organic chemistry community.

Jack Li and Doug Johnson
February 1, 2010

Contributors

Dr. Joseph D. Armstrong III
Department of Process Research
Merck & Co. Inc.
PO Box 2000
Rahway, NJ 07065

Dr. Frank R. Busch
Pharmaceutical Sciences
Pfizer Global Research and Development
Eastern Point Rd.
Groton, CT 06340

Dr. Victor J. Cee
Amgen, Inc.
Mailstop 29-1-B
1 Amgen Center Dr.
Thousand Oaks, CA 91320

Dr. Jotham W. Coe
Pfizer Global Research and Development
Eastern Point Rd.
Groton, CT 06340

Dr. Jason Crawford
Centre for Drug Research &
Development
364-2259 Lower Mall
University of British Columbia
Vancouver, BC, Canada V6T 1Z4

Dr. David J. Edmonds
Pfizer Global Research and Development
Eastern Point Rd.
Groton, CT 06340

Dr. Scott D. Edmondson
Medicinal Chemistry
Merck & Co. Inc.
PO Box 2000
Rahway, NJ 07065

Prof. Arun K. Ghosh
Departments of Chemistry and
Medicinal Chemistry
Purdue University
560 Oval Drive
West Lafayette, IN 47907

Dr. David L. Gray
CNS Medicinal Chemistry
Pfizer Global Research and Development
Eastern Point Rd.
Groton, CT 06340

Benjamin S. Greener
Pfizer Global Research and Development
Sandwich Laboratories
Ramsgate Rd.
Sandwich, Kent, UK, CT13 9NJ

Dr. Kevin E. Henegar
Pharmaceutical Sciences
Pfizer Global Research and Development
Eastern Point Rd.
Groton, CT 06340

Dr. R. Jason Herr
Medicinal Chemistry
Albany Molecular Research, Inc.
26 Corporate Cir.
PO Box 15098
Albany, NY 12212-5098

Dr. Shuanghua Hu
Discovery Chemistry
Bristol-Myers Squibb Co.
5 Research Parkway
Wallingford, CT 06492

Yazhong Huang
Discovery Chemistry
Bristol-Myers Squibb Company
5 Research Parkway
Wallingford, CT 06492

Dr. Julianne A. Hunt
Franchise and Portfolio Management
Merck & Co., Inc.
Rahway, NJ 07065

Kapil Karki
Pfizer Global Research and Development
Eastern Point Rd.
Groton, CT 06340

Dr. Brian A. Lanman
Amgen, Inc.
Mailstop 29-1-B
1 Amgen Center Drive
Thousand Oaks, CA 91320

Dr. Jie Jack Li
Discovery Chemistry
Bristol-Myers Squibb Co.
5 Research Parkway
Wallingford, CT 06492

Dr. John A. Lowe, III
JL3Pharma LLC
28 Coveside Ln.
Stonington CT 06378

Cuthbert D. Martyr
Department of Chemistry
Purdue University
560 Oval Drive
West Lafayette, IN 47907

Dr. David S. Millan
Pfizer Global Research and Development
Sandwich Laboratories
Ramsgate Rd.
Sandwich, Kent, UK, CT13 9NJ

Dr. Sajiv K. Nair
Pfizer Global Research and Development
10770 Science Center Dr.
La Jolla, CA92121

Dr. Martin Pettersson
Pfizer Global Research and Development
Eastern Point Rd.
Groton, CT 06340

Dr. Marta Piñeiro-Núñez
Eli Lilly and Co.
Lilly Corporate Center
Indianapolis, IN 46285

Dr. David Price
Pfizer Global Research and Development
Eastern Point Rd.
Groton, CT 06340

Dr. Subas Sakya
Pfizer Global Research and Development
Eastern Point Rd.
Groton, CT 06340

Dr. Robert A. Singer
Pharmaceutical Sciences
Pfizer Global Research and Development
Eastern Point Rd.
Groton, CT 06340

Dr. Jennifer A. Van camp
Hit to Lead Chemistry
Global Pharmaceutical R&D
Abbott Laboratories
200 Abbott Park Rd.
Dept. R4CW, Bldg. AP52N/1177
Abbott Park, IL 60064-6217

Dr. Feng Xu
Department of Process Research
Merck & Co. Inc.
PO Box 2000
Rahway, NJ 07065

Dr. Ji Zhang
Process Research and Development
Bristol-Myers Squibb Co.
1 Squibb Dr.
New Brunswick, NJ 08903

I

INFECTIOUS DISEASES

1

Raltegravir (Isentress): The First-in-Class HIV-1 Integrase Inhibitor

Julianne A. Hunt

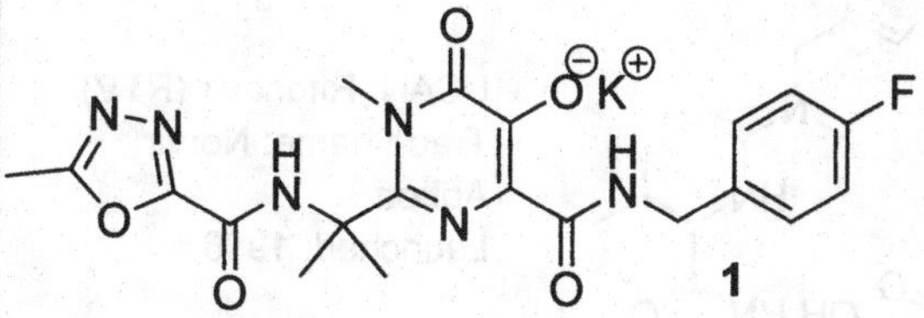

USAN: Raltegravir (RAL)
Trade name: Isentress®
Merck & Co., Inc.
Launched: 2007

1.1 Background

HIV/AIDS is a global pandemic. Nearly three decades of HIV/AIDS research has resulted in the development of more than 20 antiretroviral drugs approved by the U.S. Food and Drug Administration (FDA) for treatment of the disease; combinations of these drugs, known as highly active antiretroviral therapies (HAART), have dramatically decreased morbidity and increased life expectancy.[1] Nevertheless, HIV/AIDS remains a significant cause of morbidity and mortality worldwide. The Joint United Nations Program on HIV/AIDS estimated in 2007 that more than 33 million people lived with the disease worldwide and that AIDS killed more than 2 million people, including 330,000 children. In the United States alone, the Centers for Disease Control and Prevention (CDC) estimated in 2006 that more than one million people were living with HIV infection.[2-4] Although antiretroviral drugs have undeniably changed the lives of many HIV-positive individuals, an unmet medical need clearly exists.

Raltegravir, or Isentress (1), is the first FDA-approved inhibitor of HIV integrase. HIV/AIDS drugs are categorized according to their mode of action as nucleoside and nucleotide reverse transcriptase inhibitors [NRTIs, e.g., tenofovir (2)], nonnucleotide reverse transcriptase inhibitors [NNRTIs, e.g., efavirenz (3)] protease inhibitors [PIs, e.g., ritonavir (4)], fusion inhibitors [e.g., enfuvirtide (5)], entry inhibitors

3

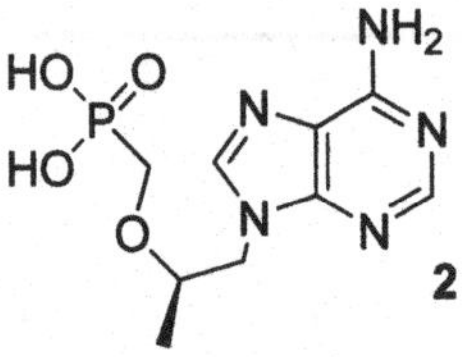

USAN: Tenofovir (TDF)
Trade name: Viread®
Gilead Sciences
Launched: 2001

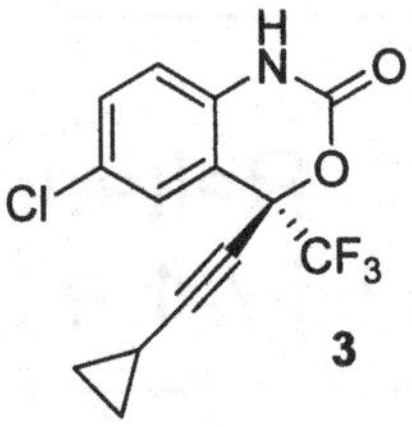

2

USAN: Efavirenz (EFV)
Trade name: Sustiva®
Bristol-Myers Squibb
Launched: 1998

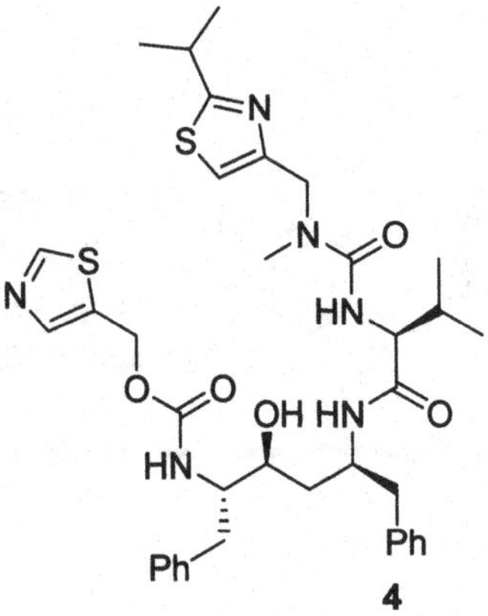

3

USAN: Ritonavir (RTV)
Trade name: Norvir®
Abbott
Launched: 1996

4

Ac-YTSLIHSLIEESQNQQEKN
EQELLELDKWASLWNWF-CONH$_2$

5

USAN: Enfuvirtide (T20)
Trade name: Fuzeon®
Roche
Launched: 2003

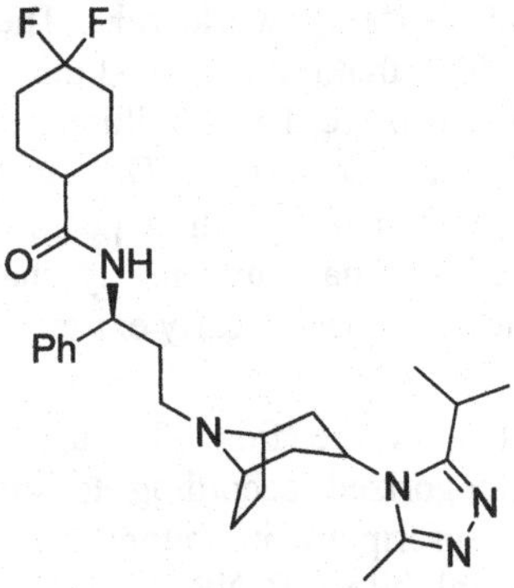

USAN: Maraviroc (MVC)
Trade name: Selzentry®
Pfizer
Launched: 2007

6

[e.g., the CCR5 antagonist maraviroc (6)], and integrase inhibitors (INSTIs, e.g., raltegravir).

Until recently, all FDA-approved drugs for HIV/AIDS treatment targeted either the viral reverse transcriptase (RT) or protease (PR) enzymes, and treatment guidelines specified that HAART multidrug cocktails should comprise one NNRTI or one PI in combination with two NRTIs.[5] Two of the most significant limitations on the effectiveness of these HAART combinations have been drug-related toxicities and the emergence of resistant viruses.[6] Drugs with novel modes of action—such as viral entry inhibitors and integrase (IN) inhibitors—have been sought to offer antiretroviral treatment-experienced HIV patients who harbor a drug-resistant virus or suffer toxicities with HAART an opportunity to achieve immunologic recovery and virologic suppression. IN inhibitors have been of particular interest to HIV/AIDS researchers because, unlike RT and PR, IN has no human homolog, and thus inhibitors of IN might be better tolerated at high doses.[7]

Raltegravir (1) is the first commercially available antiretroviral agent to target IN; at present, it is the only IN inhibitor approved for clinical use. Launched in 2007, raltegravir was originally indicated for combination therapy with other antiretroviral agents in treatment-experienced adults with evidence of viral replication and multidrug-resistant HIV-1 strains. In July 2009, the FDA approved an expanded indication for raltegravir to include treatment-naive adult patients, and in December 2009, the U.S. Department of Health and Human Services (DHHS) revised its HIV treatment guidelines to add a raltegravir combination to the preferred regimens for treatment-naive HIV patients.[5] In this chapter, the pharmacological profile and chemical syntheses of raltegravir will be described in detail.

1.2   Pharmacology

The HIV-1 replication cycle involves three key viral enzymes, all of which represent targets for antiretroviral drugs: RT, PR, and IN. RT and PR inhibitors are well represented in the HIV/AIDS treatment armamentarium, but until recently, inhibition of IN had not been successfully exploited in the clinic, despite nearly 20 years of intensive research.[8]

The enzymatic mechanisms of IN have been extensively reviewed.[9] IN catalyzes the insertion of viral genetic material into human DNA through a multistep process that includes 3'-processing, (removal of the terminal dinucleotide from each 3'-end of the viral DNA) and strand transfer (joining of the viral DNA to the host DNA). Both 3'-processing and strand transfer are catalyzed by a triad of acidic residues, D64, D116, and E152, that bind divalent metals such as Mg^{2+}. Divalent metals are required for 3'-processing and strand transfer and also for the assembly of the preintegration complex, which allows viral genetic material to cross the host nuclear membrane and access the host genome.

Raltegravir (1), like its precursors the 1,3-diketo acids (see Section 1.3), inhibits the strand-transfer activity of IN.[10] The IN-inhbitory activity of raltegravir is most likely the result of functional impairment of the active site arising from chelation of the critical divalent metal cofactors.[11] Raltegravir is both a potent (IC_{50} = 10 nM) and a highly selective inhibitor of integrase activity, found to be essentially inactive (IC_{50} > 50 μM) against related enzymes such as hepatitis C virus (HCV) polymerase; HIV RT; HIV RNase-H; and human α-, β-, and γ-polymerases. In a multiple-cycle cell-based assay (human T-lymphoid cells infected with a laboratory strain of HIV-1 in 50% human serum), raltegravir effectively inhibited (IC_{95} = 31 nM) the replication of HIV-1 infection.[12]

1.3   Structure–Activity Relationship (SAR)

The 4-aryl-2,4-diketobutanoic acid class of IN inhibitors (also known as 1,3-diketo acids, or DKAs) was discovered independently by researchers from Merck and Shionogi, with patents from both groups published in the same year.[13] From a random screen of > 250,000 compounds, the Merck group identified DKAs as the most active IN inhibitors. Compound 7 was the most potent compound found in this screen (Table 1), completely inhibiting HIV-1 infection in a cell-based assay at a concentration of 10 μM.[10]

The Shionogi group's first patent[13d] described compounds (e.g., 8) in which an isosteric tetrazole replaced the carboxylic acid group in the critical but biologically labile DKA pharmacophore. Compound 8, also known as 5CITEP, inhibited HIV-1 3'-processing as well as strand-transfer activity,[14] and was the first inhibitor co-crystallized with IN.[15] A subsequent patent from the Shionogi group described the systematic modification of the 5CITEP framework to include a variety of heterocyclic replacements for the indole and tetrazole moieties, culminating in the first clinically tested HIV-1 IN inhibitor, compound 9, also known as S-1360.[16,17] Clinical development of S-1360, undertaken by a Shionogi–GSK joint venture, was halted when the compound failed in efficacy studies in humans (due to formation of an inactive metabolite via reduction at the carbon adjacent to the triazole).[18]

The Merck group's efforts to find a more stable substitute for the DKA pharmacophore resulted in the design of 8-hydroxy-[1,6]naphthyridines such as compound 10,[19] wherein the keto-enol-acid triad was replaced with a 1,6-naphthyridine ketone bearing a phenolic hydroxyl group. Further refinement of compound 10—replacement of the naphthyridine phenyl ketone with a 4-fluorobenzyl carboxamide and addition of a six-membered sulfonamide at the 5-position of the naphthyridine core—resulted in compound 11, the second IN inhibitor to reach the clinic.[20] The discovery of liver toxicity in long-term safety studies of compound 11 in dogs led to the suspension of clinical development[21] of this compound.

Table 1. Activity of DKAs and Related Structures against H1V-1 Integrase

Entry	Compound	HIV-1 Strand Transfer IC$_{50}$ (μM)
1	**1**	0.01
2	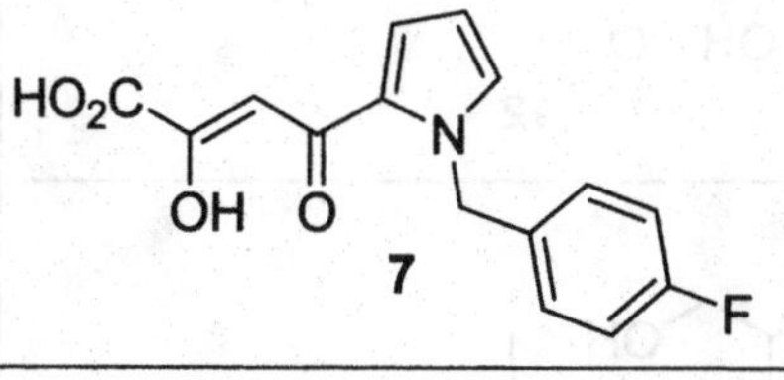 **7**	0.08
3	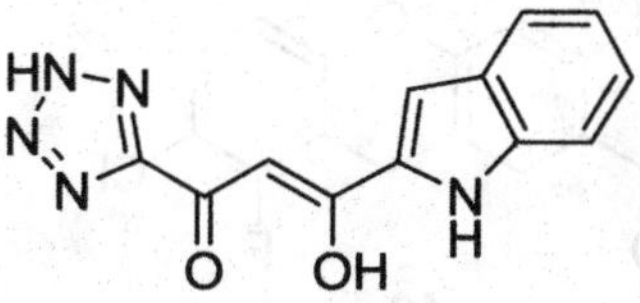 **8**	0.65
4	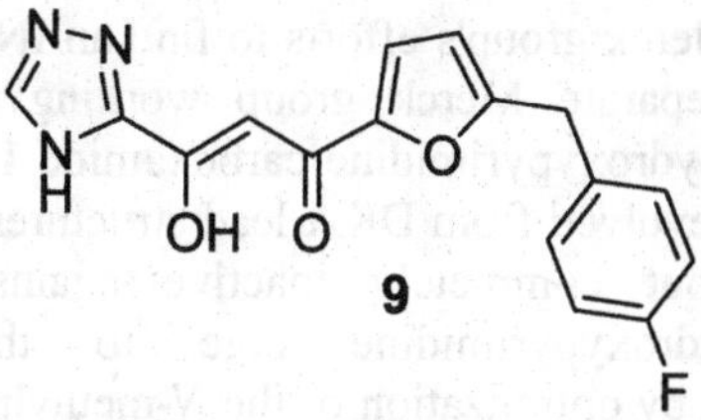 **9**	0.02
5	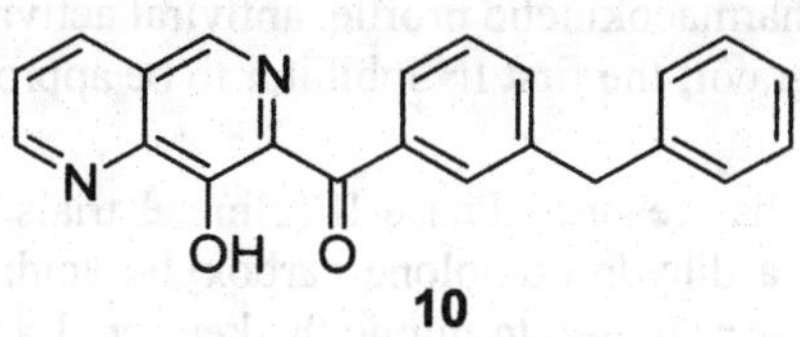 **10**	0.04
6	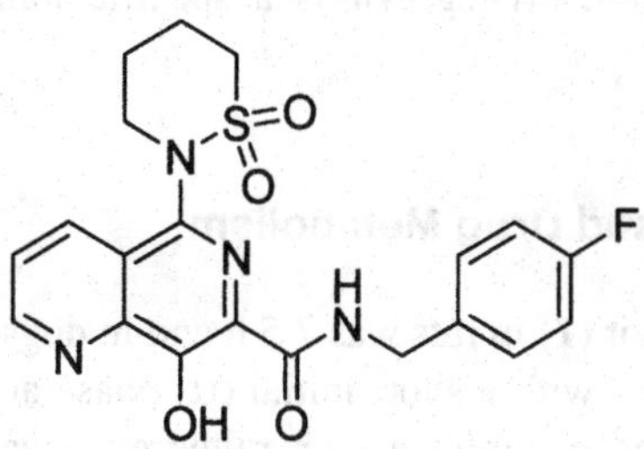 **11**	0.01

7	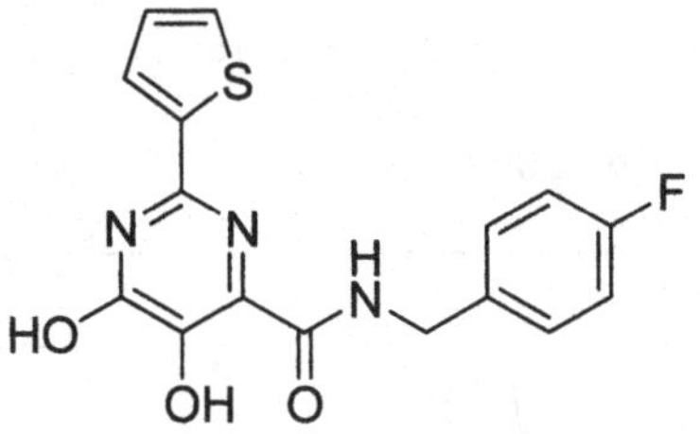 **12**	0.08
8	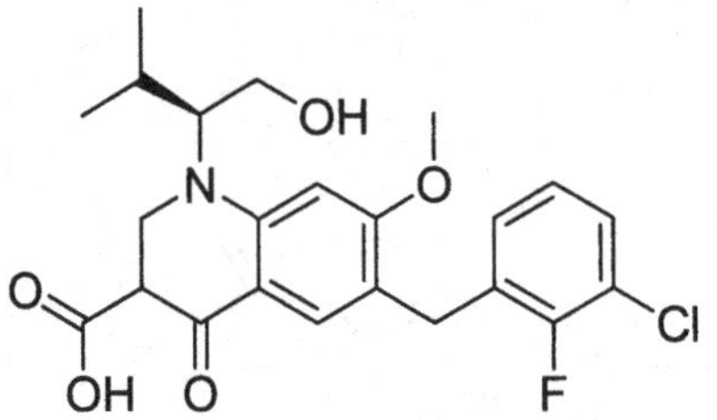 **13**	0.007

Concurrent with the Merck group's efforts to find an IN inhibitor that would be successful in the clinic, a separate Merck group working on inhibitors of HCV polymerase discovered that dihydroxypyrimidine carboxamide **12** (which, like the other compounds shown in Table 1, evolved from DKA lead structures) was a potent inhibitor of HIV-1 strand transfer, but completely inactive against HCV polymerase.[22] Modification of the dihydroxypyrimidine core to the corresponding *N*-methylpyrimidinone[23] followed by optimization of the *N*-methylpyrimidinone series with respect to metabolic stability, pharmacokinetic profile, antiviral activity, and genotoxicity led to the identification of raltegravir, the first IN inhibitor to be approved by the FDA.[12]

A second IN inhibitor has reached Phase III clinical trials since the launch of raltegravir. Elvitegravir (**13**) is a dihydroquinolone carboxylic acid; the monoketo acid motif of this series of inhibitors is proposed to mimic the keto-enol-acid triad of the DKA lead structures.[24] Like raltegravir, elvitegravir is a specific inhibitor of HIV-1 strand transfer.[25]

1.4   Pharmacokinetics and Drug Metabolism

The plasma half-life of raltegravir (**1**) in rats was 7.5 h and in dogs was 13 h. Plasma half-life was biphasic in both species, with a short initial (α) phase and a prolonged terminal (β) phase. The major route of metabolism for raltegravir was glucuronidation; the glucoronide was shown to be the conjugate of the hydroxyl group at C5 of the pyrimidinone ring.[12]

In humans, the pharmacokinetics of raltegravir was studied in both healthy HIV-negative subjects and in HIV-infected patients. In healthy HIV-negative subjects, raltegravir was rapidly absorbed; as was seen in preclinical species, concentrations declined from the mean maximum plasma concentration (C_{max}) in a biphasic manner, with an apparent half-life of approximately 1 h for the α phase and an apparent half-life of 7–12 h for the β phase. Raltegravir was generally well tolerated at doses of up to 1600 mg/day for 10 days. The mean plasma concentration of raltegravir at the end of a 12-h dose administration interval exceeded 33 nM (the in vitro IC_{95} in 50% human serum) after multiple doses $\geq$ 100 mg, supporting twice-daily dosing with doses $\geq$ 100 mg.[26]

In a double-blind, placebo-controlled, dose-ranging study in 35 antiretroviral-naive HIV-infected patients (Protocol 004), subjects were randomized to receive placebo or one of four raltegravir doses (100, 200, 400, or 600 mg) twice daily over 10 days. Although dose proportionality was not observed (possibly due to intersubject variability and the small number of patients), the pharmacokinetic data gathered in this study supported the selection of a 400-mg dose for raltegravir.[27]

While the NRTIs, NNRTIs, and PIs are primarily metabolized in humans via cytochrome P450 (CYP450), raltegravir is neither a substrate nor an inhibitor of CYP450 nor is it an inducer of CYP3A4, indicating that drug–drug interactions with drugs metabolized by CYP450 are unlikely. Instead, raltegravir is primarily metabolized via uridine diphosphate glucoronosyl transferase 1A1 (UGT1A1).[28]

1.5   Efficacy and Safety

Raltegravir (1) is a potent inhibitor of HIV integrase, originally approved for combination therapy with other antiretroviral agents in treatment-experienced adults with evidence of viral replication and multidrug-resistant HIV-1 strains, and recently (July 2009) approved for combination therapy with other antiretroviral agents in treatment-naive adult patients. Raltegravir is dosed orally twice daily (400 mg); no dose adjustment is required when it is co-administered with other antiretroviral agents.

Phase II clinical trials with raltegravir were conducted in both treatment-naive (Protocol 004) and treatment-experienced (Protocol 005) HIV patients. In Protocol 004, 201 treatment-naive HIV patients received either raltegravir or efavirenz (3, the current gold standard for treatment-naive patients) for 48 weeks on a background of tenofovir (2) and lamivudine combination therapy. Similar proportions of patients in the raltegravir and efavirenz groups achieved an HIV RNA level of < 50 copies/mL.[29] In Protocol 005, 179 treatment-experienced HIV patients with viral resistance received either raltegravir or placebo in combination with optimized background therapy (OBT) chosen by the investigator. In the raltegravir group, 57–67% of patients achieved an HIV RNA level of < 50 copies/mL, compared with 14% in the placebo group.[30]

In the two Phase III BENCHMRK clinical trials, 699 treatment-experienced patients with triple-class resistant HIV were randomized to receive either raltegravir (400 mg p.o., b.i.d.) plus OBT or OBT alone. In both trials, patients in the raltegravir group improved both virological (HIV RNA level < 50 copies/mL) and immunological (increased CD4+ cell counts) outcomes compared to patients in the placebo group.[31] In the Phase III STARTMRK study, 563 treatment-naive patients were randomized to receive either raltegravir (400 mg) or efavirenz (600 mg) in combination with tenofovir and emtricitabine.   Analysis of the 48-week results showed that raltegravir-based combination treatment was noninferior to efavirenz-based combination treatment; this study was the basis for the expansion of approved indications for raltegravir to include treatment-naive patients.[32] Recently, analysis of the 96-week results for STARTMRK confirmed the findings from the 48-week analysis.[33] Finally, 96-week results for Protocol 004 treatment-naive patients confirmed that the raltegravir group continued to show a suppression of viral replication comparable to that shown by the efavirenz group.[34]

The most commonly reported adverse events of moderate or severe intensity that occurred in > 2% of patients treated with raltegravir were headache, nausea, asthenia/weakness, and fatigue. It is interesting that the STARTMRK study showed that patients on raltegravir-based combination treatment showed significantly less impact on lipid levels than patients on efavirenz-based combination treatment.[32,33]

1.6   Syntheses

The discovery synthesis of raltegravir,[12] started from nitrile **14**, which was converted to the corresponding amidoxime **15**. Reaction of **15** with dimethylacetylenedicarboxylate provided the dihydroxypyrimidine core (**16**)[35] of raltegravir. Benzoylation followed by methylation provided the *N*-methylpyrimidone **17**, with the *O*-methylated pyrimidine analog (not shown) as a minor product. The discovery route to many analogs of raltegravir proceeded via subsequent installation of the fluorobenzylamide at C4 of the pyrimidone ring, followed by deprotection and functionalization at the C2 position. However, in the case of raltegravir, the preferred route (due to scalability of the purification process) involved functionalization at the deprotected amine **18** to provide the oxadiazole amide at C2 followed by installation of the fluorobenzylamide at C4 to provide up to 10 g of **19**, the free hydroxyl form of raltegravir. Conversion to the potassium salt with potassium hydroxide provided raltegravir potassium **1**.

The process synthesis of raltegravir[36] followed a convergent approach. Synthesis of the pyrimidone amine **22** started from 2-hydroxy-2-methylpropanenitrile, which was converted to the corresponding aminocyanohydrin with ammonia (30 psi, 10 °C, 97% yield), and then protected with benzylchlorofomate to provide Cbz-amine **14** in 88% yield. The process synthesis paralleled the discovery route from **14** through **15** to provide **16**, with a somewhat improved yield (52%) for the dihydroxypyrimidine. Deprotonation of **16** with magnesium methoxide followed by treatment with methyl iodide provided the *N*-methylpyrimidone **20** with < 0.5% of the *O*-methylpyrimidine side product. Installation of the fluorobenzylamide was accomplished by heating in ethanol followed by crystallization to provide **21**, and hydrogenolysis of the Cbz-protected amine provided amine **22**. Synthesis of the oxadiazole **23** began with the reaction of methyltetrazole with ethyl oxalyl chloride to provide the ethyl oxalyltetrazole intermediate, which rearranged with loss of nitrogen on heating in toluene. The crude ethyl ester was saponified with potassium hydroxide to give the oxadiazole carboxylate salt **23**. Finally, reaction of the acid chloride of **23** (formed using oxalyl chloride) with amine **22** provided the free hydroxyl form of raltegravir, which was converted to raltegravir potassium **1** with potassium hydroxide.

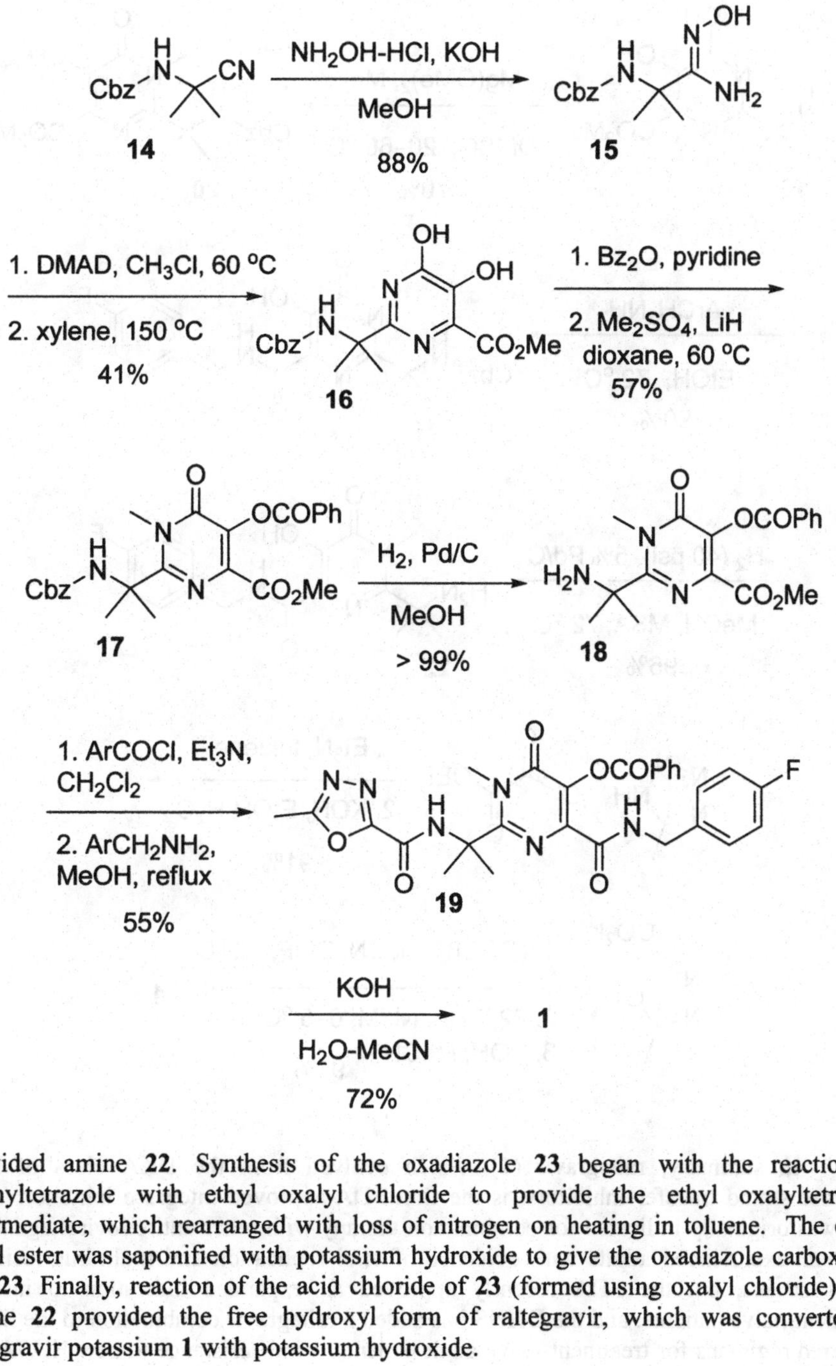

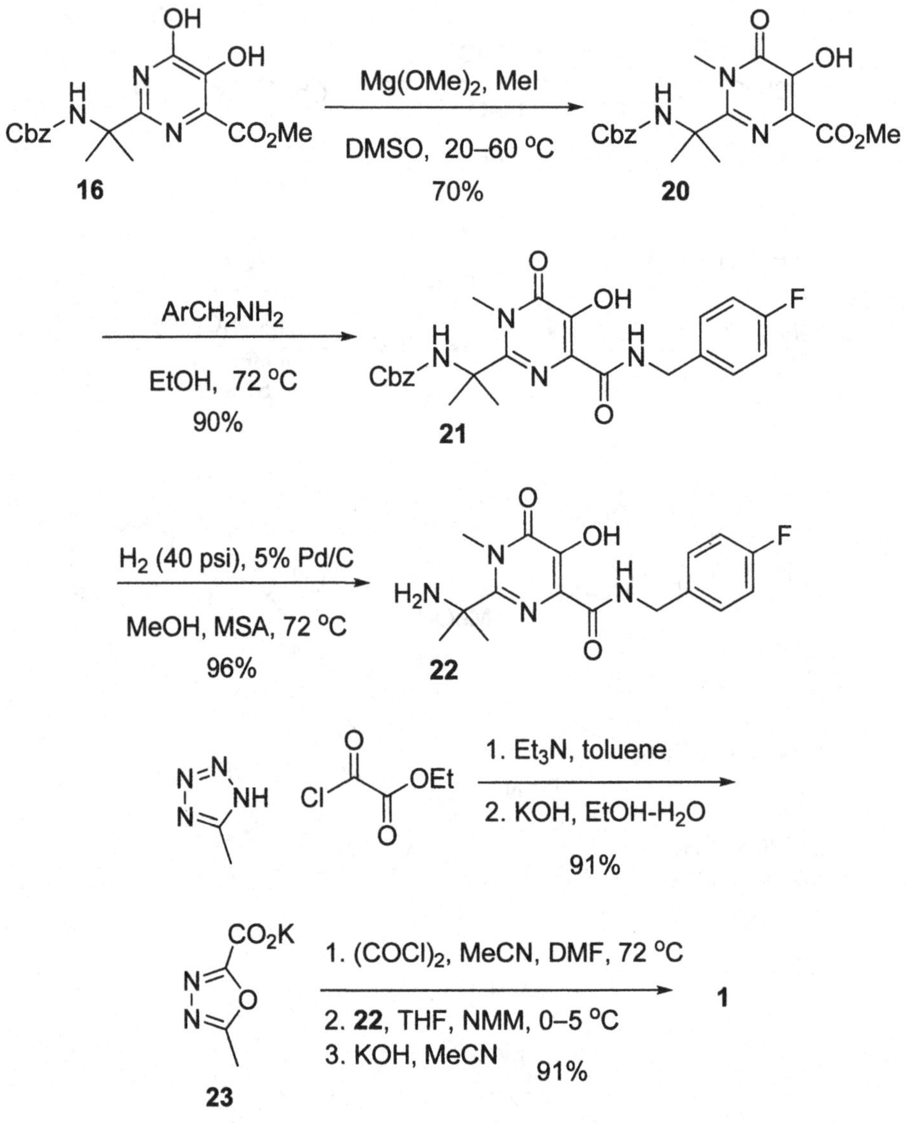

In summary, raltegravir (1), which evolved from the DKA class of HIV integrase strand transfer inhibitors, is the first FDA-approved integrase inhibitor. The drug was originally indicated for combination therapy with other antiretroviral agents in treatment-experienced adults with evidence of viral replication and multidrug-resistant HIV-1 strains, but the FDA recently approved an expanded indication to include treatment-naive patients, and the DHHS has added a raltegravir combination to the list of preferred regimens for treatment-naive adult patients. Viral resistance to new HIV/AIDS drugs is inevitable, and, indeed, multiple viral amino acid mutations have been identified that confer resistance to raltegravir,[37] necessitating a continuing search for improved

HIV/AIDS therapies. However, the development of raltegravir, the first member of an entirely new class of drugs that target a viral mechanism not previously exploited in the clinic, has provided physicians and patients a welcome addition to the HIV/AIDS treatment armamentarium.

1.7 References

1. Mocroft, A.; Vella, S.; Benfiedl, T. L.; Chiesi, A.; Miller, V.; Gargalianos, P.; d'Arminio Monforte, A.; Yust, I.; Bruun, J. N.; Phillips, A. N.; Lundgren, J. D. *Lancet*, **1998**, *353*, 1725–1730.

2. Kallings, L. O. *J. Intern. Med.* **2008**, *263*, 218–243.

3. UNAIDS "2007 AIDS Epidemic Update," http://data.unaids.org/pub/ EPISlides/2007/2007_epiupdate_en.pdf; accessed 2009-11-15.

4. CDC, "New estimates of U.S. HIV Prevalence, 2006" http://www.cdc.gov/hiv/ topics/surveillance/resources/factsheets/pdf/prevalence.pdf; accessed 2009-11-15.

5. Panel on Antiretroviral Guidelines for Adults and Adolescents. "Guidelines for the Use of Antiretroviral Agents in HIV-1-Infected Adults and Adolescents." DHHS. December 1, 2009; 1-161.
http://www.aidsinfo.nih.gov/ContentFiles/AdultandAdolescentGL.pdf; accessed 2009-12-02.

6. Tozzi, V.; Zaccarelli, M.; Bonfigli, S.; Lorenzini, P.; Liuzzi, G.; Trotta, M. P.; Forbici, F.; Gori, C.; Bertoli, A.; Bellagamba, R.; Narciso, P.; Perno, C. F.; Antinori, A. *Antivir. Ther.* **2006**, *11*, 553–560.

7. Havlir, D. V. *N. Engl. J. Med.* **2008**, *359*, 416–441.

8. Neamati, N. *Exp. Opin. Ther. Pat.* **2002**, 12, 709–724.

9. (a) Asante-Appiah, E.; Skalka, A. M. *Adv. Virus Res.* **1999**, *12*, 2331–2338. (b) Esposito, D.; Craigie, R. *Adv. Virus Res.* **1999**, *52*, 319–333. (c) Chiu, T. K.; Davies, D. R. *Curr. Top. Med. Chem.* **2004**, *4*, 965–977. (d) Lewinski, M. K.; Bushman, F. D. *Adv. Genet.* **2005**, *55*, 147–181.

10. Hazuda, D. J.; Felock P.; Witmer. M.; Wolfe, A.; Stillmock, K.; Grobler, J. A.; Espeseth, A.; Gabryelski, L.; Schleif, W.; Blau, C.; Miller, M. D. *Science* **2000**, *287*, 6466–6650.

11. (a) Espeseth, A. S.; Felock P.; Wolfe, A.; Witmer, M.; Grobler, J.; Anthony, N.; Egbertson, M.; Melamed, J. Y.; Young, S.; Hamill, T.; Cole, J. L.; Hazuda, D. J. *Proc. Natl. Acad. Sci. USA.* **2000**, *97*, 11244–11249. (b) Grobler, J. A.; Stillmock, K.; Hu, B.; Witmer, M.; Felock, P.; Espeseth, A. S.; Wolfe, A.; Egbertson, M.; Bourgeois, M.; Melamed, J.; Wai, J. S.; Young, S.; Vacca, J.; Hazuda, D. J.; *Proc. Natl. Acad. Sci. USA.* **2002**, *99*, 6661–6666.

12. Summa, V.; Petrocchi, A.; Bonelli, F; Crescenzi, B.; Donghi, M.; Ferrara, M.; Fiore, F.; Gardelli, C.; Gonzalez Paz, O.; Hazuda, D. J.; Jones, P.; Kinzel, O.; Laufer, R.; Monteagudo, E; Muraglia, E.; Nizi, E.; Orvieto, F.; Pace, P.; Pescatore, G.; Scarpelli, R.; Stillmock, K.; Witmer, M. V.; Rowley, M. *J. Med. Chem.* **2008**, *51*, 5843–5855.

13.   (a) Merck & Co., Inc. WO 9962513 (1999). (b) Merck & Co., Inc. WO 9962520
       (1999); (c) Merck & Co., Inc.: WO 9962897 (1999). (d) Shionogi & Co., Ltd.
       WO 9950245 (1999).
14.   Marchand. C.; Zhang, X.; Pais, G. C.; Cowansage, K.; Neamati, N.; Burke, T. R.
       Jr.; Pommier, Y. *J. Biol. Chem.* **2002**, *277,* 12596–12603.
15.   Goldgur, Y.; Craigie, R.; Cohen, G. H.; Fujiwara, T.; Yoshinaga, T.; Fujishita,
       T.; Sugimoto, H.; Endo, T.; Murai, H.; Davies, D. R. *Proc. Natl. Acad. Sci. U. S.
       A.* **1999**, *96,* 13040–13043.
16.   Shionogi & Co. Ltd. WO 0039086 (2000).
17.   Billich, A. *Curr. Opin. Investig. Drugs* **2003**, *4*, 206–209.
18.   Rosemond, M. J.; St John-Williams, L.; Yamaguchi, T.; Fujishita, T.; Walsh, J.
       S. *Chem. Biol. Interact.* **2004**, *147*, 129–139.
19.   Zhuang, L.; Wai, J. S.; Embrey, M. W.; Fisher, T. E.; Egbertson, M. S.; Payne,
       L. S.; Guare, J. P. Jr.; Vacca, J. P.; Hazuda, D. J.; Felock, P. J.; Wolfe, A. L.;
       Stillmock, K. A.; Witmer, M. V.; Moyer, G.; Schleif, W. A.; Gabryelski, L. J.;
       Leonard, Y. M.; Lynch, J. J. Jr.; Michelson, S. R.; Young, S. D. *J. Med. Chem.*
       2003, *46*, 453–456.
20.   Hazuda, D. J.; Anthony, N. J.; Gomez, R. P.; Jolly, S. M.; Wai, J. S.; Zhuang,
       L.; Fisher, T. E.; Embrey, M.; Guare, J. P. Jr.; Egbertson, M. S.; Vacca, J. P.;
       Huff, J. R.; Felock, P. J.; Witmer, M. V.; Stillmock, K. A.; Danovich, R.;
       Grobler, J.; Miller, M. D.; Espeseth, A. S.; Jin, L.; Chen, I. W.; Lin, J. H.;
       Kassahun, K.; Ellis, J. D.; Wong, B. K.; Xu, W.; Pearson, P. G.; Schleif, W. A.;
       Cortese, R.; Emini, E.; Summa, V.; Holloway, M. K.; Young, S. D. *Proc. Natl.
       Acad. Sci. USA* **2004**, *101*, 11233–11238.
21.   Little, S.; Drusano, G.; Schooley, R. Paper presented at 12th Conference on
       Retroviruses and Opportunistic Infections, Feb. 22–25, 2005, Boston, U.S.A.
22.   Summa, V.; Petrocchi, A.; Matassa, V. G.; Gardelli, C.; Muraglia, E.; Rowley,
       M.; Paz, O. G.; Laufer, R.; Monteagudo, E.; Pace, P. *J. Med. Chem.* **2006**, *49*,
       6646–6649.
23.   Gardelli, C.; Nizi, E.; Muraglia, E.; Crescenzi, B.; Ferrara, M.; Orvieto, F.;
       Pace, P.; Pescatore, G.; Poma, M.; Rico Ferreira, M. R.; Scarpelli, R.; Hommick,
       C. F.; Ikemoto, N.; Alfieri, A.; Verdirame, M.; Bonelli, F.; Gonzalez Paz, O.;
       Monteagudo, E.; Taliani, M.; Pesci, S.; Laufer, R.; Felock, P.; Stillmock, K. A.;
       Hazuda, D.; Rowley, M.; Summa, V. *J. Med. Chem.* **2007**, *50*, 4953–4975.
24.   (a) Sato, M.; Motomura, T.; Aramaki, H.; Matsuda, T.; Yamashita, M.; Ito, Y.;
       Kawakami, H.; Matsuzaki, Y.; Watanabe, W.; Yamataka, K.; Ikeda, S.; Kodama,
       E.; Matsuoka, M.; Shinkai, H. *J. Med. Chem.* **2006**, *49*, 1506–1506. (b) Sato,
       M.; Kawakami, H.; Motomura, T.; Aramaki, H.; Matsuda, T.; Yamashita, M.;
       Ito, Y.; Matsuzaki, Y.; Yamataka, K.; Ikeda, S.; Shinkai, H. *J. Med. Chem.* **2009**,
       52, 4869–4882.
25.   DeJesus, E.; Berger, D.; Markowitz, M.; Cohen, C.; Hawkins, T.; Ruane, P.;
       Elion, R.; Farthring, C.; Zhong, L.; Chen, A. K.; McColl, D.; Kearney, B. P. *J.
       Acquir. Immune Defic. Syndr.* **2006**, *43*, 1–5.
26.   Iwamoto, M.; Wenning, L. A.; Petry, A. S.; Laethem, M.; De Smet, M.; Kost, J.
       T.; Merschman, S. A.; Strohmaier, K. M.; Ramael, S.; Lasseter, K. C.; Stone, J.

A.; Gottesdiener, K. M.; Wagner, J. A. *Clin. Pharmacol. Ther.* **2008**, *83*, 293–299.

27.   Markowitz, M.; Morales-Ramirez, J. O.; Nguyen, B.-Y.; Kovacs, C. M.; Steigbigel, R. T.; Cooper, D. A.; Liporace, R.; Schwartz, R.; Isaacs, R.; Gilde, L. R.; Wenning, L.; Zhao, J.; Teppler, H. *J. Acquir. Immune Defic. Syndr.* **2006**, *43*, 509–515. [Erratum *J. Acquir. Immune Defic. Syndr.* **2007**, *44*, 492.]

28.   Kassahun, K.; McIntosh, I.; Cui, D.; Hreniuk, D.; Merschman, S.; Lasseter, K.; Azrolan, N.; Iwamoto, M.; Wagner, J. A.; Wenning, L. A. *Drug Metab. Dispos.* **2007**, *35*, 1657–1663.

29.   Markowitz, M.; Nguyen, B. Y.; Gotuzzo, E.; Mendo, F.; Ratanasuwan, W.; Kovacs, C.; Prada, G.; Morales-Ramirez, J. O.; Crumpacker, C. S.; Isaacs, R. D.; Gilde, L. R.; Wan, H.; Miller, M. D.; Wenning, L. A.; Teppler, H. *J. Acquir. Immune Defic. Syndr.* **2007**, *46*, 125–133.

30.   Grinsztejn, B.; Nguyen, B.-Y.; Katlama, C.; Gatell, J. M.; Lazzarin, A.; Vittecoq, D.; Gonzalez, C. J.; Chen, J.; Harvey, C. M.; Isaacs, R. D. *Lancet* **2007**, *369*, 1261–1269.

31.   Steigbigel, R. T.; Cooper, D. A.; Kumar, P. N. Eron, J. E.; Schechter, M.; Markowitz, M.; Loutfy, M. R.; Lennox, J. L.; Gatell, J. M.; Rockstroh, J. K.; Katlama, C.; Yeni, P.; Lazzarin, A.; Clotet, B.; Zhao, J.; Chen, J.; Ryan, D. M.; Rhodes, R. R.; Killar, J. A.; Gilde, L. R.; Strohmaier, K. M.; Meibohm, A. R.; Miller, M. D.; Hazuda, D. J.; Nessly, M. L.; DiNubile, M. J.; Isaacs, R. D.; Nguyen, B.-Y.; Teppler, H. *N. Engl. J. Med.* **2008**, *359*, 339–354.

32.   Lennox, J. L.; DeJesus, E.; Lazzarin, A.; Pollard, R. B.; Valdez, J.; Madruga, R.; Berger, D. S.; Zhao, J.; Xu, X.; Williams-Diaz, A.; Rodgers, A. J.; O Barnard, R. J. O.; Miller, M. D.; DiNubile, M. J.; Nguyen, B.-Y.; Leavitt, R.; Sklar, P. *Lancet* **2009**, *374*, 796–806.

33.   Lennox, J.; DeJesus, E.; Lazzarin, A.; Berger, D.; Pollard, R.; Madruga, J. ; Zhao, J.; Gilbert, C.; Rodgers, A.; Teppler, H.; Nguyen, B.-Y.; Leavitt, R.; Sklar, P. Paper presented at 49th ICCAC, Sept. 12–15, 2009.

34.   Markowitz, M.; Nguyen, B.-Y.; Gotuzzo, E.; Mendo, F.; Ratanasuwan, W.; Kovacs, C.; Prada, G.; Morales-Ramirez, J. O.; Crumpacker, C. S.; Isaacs, R. D.; Campbell, H.; Strohmaier, K. M.; Wan, H.; Danovich, R. M.; Teppler, H. *J. Acquir. Immune Defic. Syndr.* **2009**, *52*, 350–356.

35.   Pye, P. J.; Zhong, Y.-L.; Jones, G. O.; Reamer, R. A.; Houk, K. N.; Askin, D.A. *Angew. Chem., Int. Ed.* **2008**, *47*, 4134–4136.

36.   (a) Merck & Co., Inc.: WO 060712A2 (2006); (b) Liu, K. K.-C.; Sakya, S. M.; O'Donnell, C. J.; Li, J. *Mini-Reviews Med. Chem.* **2008**, *8*, 1526–1548.

37.   (a) Malet, I.; Delelis, O.; Valantin, M. A.; Montes, B.; Soulie, C.; Wirden, M.; Tchertanov, L.; Peytavin, G.; Reynes, J.; Mouscadet, J. F.; Katlama, C.; Calvez, V.; Marcelin, A. G. *Antimicrob. Agents Chemother.* **2008**, *52*, 1351–1358. (b) Goethals, O.; Clayton, R.; Van ginderen, M; Vereycken, I.; Wagemans, E.; Geluykens, P.; Dockx, K.; Strijbos, R.; Smits, V.; Vos, A.; Meersseman, G.; Jochmand, D.; Vermeire, K.; Schols, D.; Hallenberger, S.; Hertogs, K. *J. Virol.* **2008**, *82*, 10366–10374.

2

Maraviroc (Selzentry): The First-in-Class CCR5 Antagonist for the Treatment of HIV

David Price

2.1 Background

Infection with HIV leads, in the vast majority of cases, to progressive disease and ultimately death. By 2004, just 23 years after AIDS was first recognized, the Joint United Nations Programme on HIV/AIDS estimated that 42 million people worldwide were infected with HIV, with more than 20 million dead since the beginning of the epidemic. Furthermore, rates of infection are once again on the increase in the developed world.[1] Despite the undoubted achievements of highly active antiretroviral therapy (HAART) using cocktails of reverse transcriptase and protease inhibitors,[2] there is still a high unmet medical need for better tolerated, conveniently administered agents to treat HIV and AIDS.[3,4]

HIV enters the host cell by fusing the lipid membrane of the virus with the host cell membrane. This fusion is triggered by the interaction of proteins on the surface of the HIV envelope with specific cell surface receptors. One of these is CD4, the main receptor for HIV-1 that binds to gp120, a surface protein on the virus particle.[5] CD4 alone, however, is not sufficient to permit HIV fusion and cell entry–an additional co-receptor from the chemokine family of G-protein coupled receptors (GPCRs) is required. The chemokine receptor CCR5 has been demonstrated to be the major co-receptor for the fusion and entry of macrophage tropic (R5-tropic) HIV-1 into cells. R5-tropic strains are prevalent in the early asymptomatic stages of infection. Shifts in tropism do occur during progression, mainly to X4 viruses that use CXCR4 as co-receptor, however, approximately 50% of individuals are infected with strains that maintain their requirement for CCR5. There is evidence that homozygotes possessing a 32-base pair deletion in the CCR5 coding region are resistant to infection with R5-tropic HIV-1. These homozygotes do not express functional CCR5 receptors on the cell surface. Individuals who are heterozygous for the 32-base pair deletion display significantly longer progression times to the symptomatic stages of infection and evidence is emerging that they respond better to HAART. Moreover, CCR5-deficient individuals are apparently fully immunocompetent, indicating that absence of CCR5 function may not be detrimental and that a CCR5 antagonist should be well tolerated. New mechanisms of action are particularly attractive to avoid issues of viral resistance and CCR5 antagonists have been an intense area of research within the HIV arena over the last decade with a number of compounds progressing to clinical trials from differing research institutions.[6–8]

Ancriviroc (2) was the first CCR5 antagonist to advance to clinical studies, where it was safe and demonstrated a clear antiviral effect, although cardiac side effects (QT prolongation) were noted at the highest dose tested (400 mg twice daily). It is of interest that the discovery and development timelines of CCR5 antagonists within various research groups mirrors the increased interest and understanding in hERG pharmacology, and often the two areas are closely intertwined.[9] Indeed concerns around the cardiac liabilities that may be attributed to off-target pharmacology (whether mediated through hERG or other pharmacologies) has complicated and stopped the development of a number of CCR5 antagonists in the clinic. A high therapeutic index is particularly important in the HIV arena as drugs to treat HIV are rarely given in isolation, rather in combination with other agents to prevent emergence of viral resistance. Many of the agents that a CCR5 antagonist could be co-administered with have known interactions with cytochrome P450 enzymes that may affect circulating levels of other compounds in the treatment regime. This information, combined with the need to maintain a free plasma concentration of the CCR5 antagonist above the antiviral IC_{90} drives the need for a large safety window. Thus achieving high selectivity with respect to hERG affinity to avoid cardiac liabilities was a key objective for the project from the outset. Further safety concerns surfaced in the development of aplaviroc (4) with the occurrence of severe hepatotoxicity in several patients leading to the stopping of any further studies in October 2005. This finding has lead to concerns about possible class-specific long-term adverse effects of CCR5 antagonists, particularly regarding hepatotoxicity or malignancy. Maraviroc (1) and aplaviroc (4) were proceeding through development at approximately the same time and with the halting of aplaviroc, there was high scrutiny of the safety profile of maraviroc from the clinical data. Post-analysis, it was decided that there was sufficient confidence in the profile of maraviroc to warrant the approval of the agent, and maraviroc was launched in 2007. The clinical history of vicriviroc (3) has been complicated, and it would appear the compound is currently in Phase 3, with an estimated completion date of the studies in 2011.

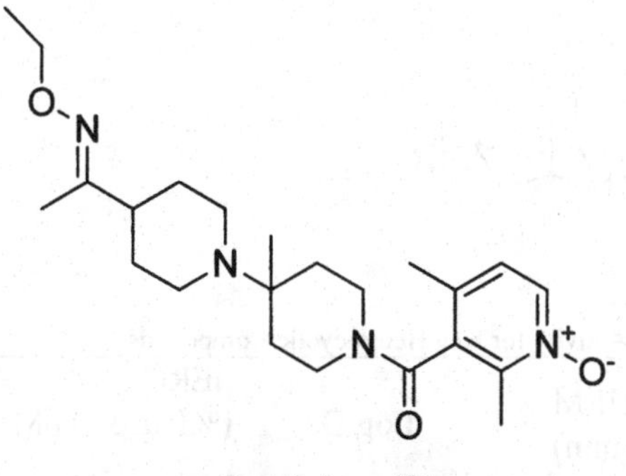

Ancriviroc (**2**)
Schering-Plough
Discontinued

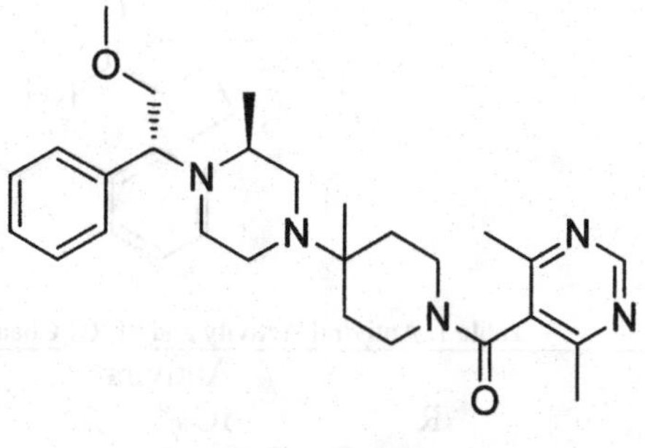

Vicriviroc (**3**)
Schering-Plough
Discontinued 2005

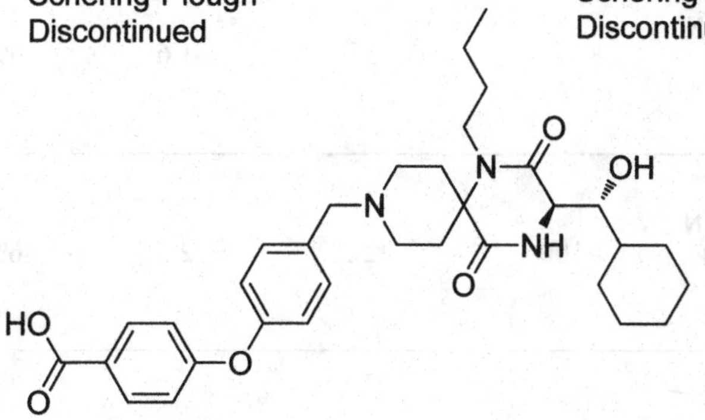

Aplaviroc (**4**)
ONO Pharmaceuticals/GlaxoSmithKline
Discontinued 2005

2.2   Structure–Activity Relationship (SAR)

The SAR around maraviroc (**1**) has been extensively reported, in particular the tactics used to successfully deliver a compound with a maximum therapeutic window over QT prolongation (Tables 1 and 2).[6,9] The validation and utility of a high-throughput binding assay for the hERG ion channel was an essential component of the drug discovery process. The design of synthetic routes to enable the late-stage variation of differing structural elements is a key skill in the drug discovery process, in particular ensuring that there are large monomer pools available for the final reaction to allow a large coverage of chemical space. The design criteria for selection of acid monomers was strict in terms of physicochemical properties of the final products (log D range 1.5–2.3), so that compounds prepared would have the best chance of possessing good pharmacokinetic and biopharmaceutical properties. This log D range was decided on from analysis of previous compounds prepared within the project as being most appropriate. In terms of the structural features of the monomers used as diverse a range as possible was selected to identify possible deleterious interactions with the hERG channel (Table 2).

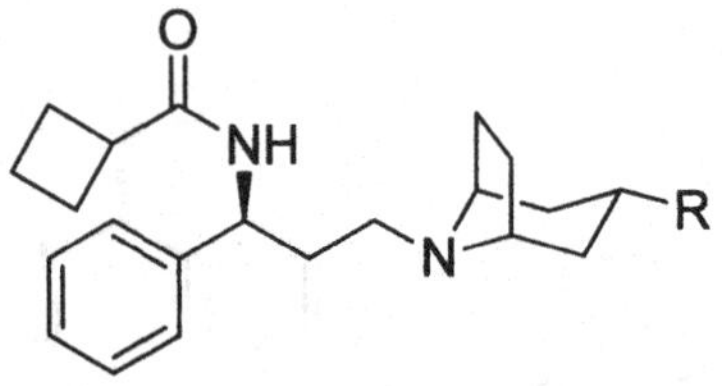

Table 1. Antiviral Activity and hERG Channel Activity for Key Heterocyclic Compounds

	R	Antiviral IC_{90}^{a} (nM)	HLM (min)	Log D	hERG[b] (%I @ 300 nM)
5	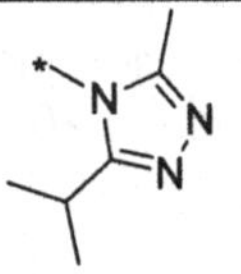	8	55	1.6	30%
6	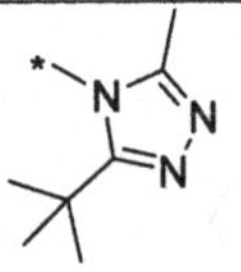	>100	22	2.3	46%
7	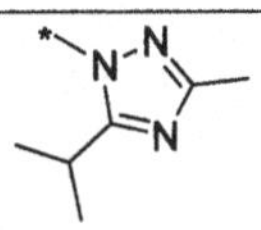	29	56	2.1	65%
8	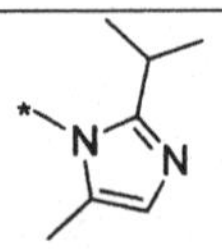	1	13	2.3	85%

[a]The concentration required to inhibit replication of HIV_{BaL} in PM-1 cells by 90%.
[b]Percent inhibition of [^{3}H]-dofetilide binding to hERG stably expressed on HEK-293 cells.

Ring homologation of **5** (log D 1.6) to compound **9** (log D 2.1) immediately gave an indication that an increase in antiviral activity was possible with a small decrease in affinity for the hERG channel (Table 2). This reduction is presumably due to some deleterious steric clash with residues in the ion channel with the increased size of the cyclopentyl substituent. This result also confirmed that key structural elements could reduce hERG affinity regardless of an increase in overall lipophilicity. Compound **10** (log D 1.8) also showed a small decrease in affinity for the hERG channel compared to the lead **5**, and this suggested that there could be a further reduction in affinity for hERG by fluorination of the amide substituent. Combining the data from **9** and **10** lead to the design and synthesis of the 4,4-difluorocyclohexyl group of maraviroc **1** (log D 2.1), which possessed no binding to the hERG channel when screened at 300 nM. The level of antiviral potency displayed was outstanding with a nanomolar IC_{90}. Within the triazole series the 4,4-difluorocyclohexyl group is unique in its antiviral profile and lack of affinity for the hERG channel. The 4,4-difluorocyclohexyl group is clearly not tolerated within the ion channel due to the steric demands of the cyclohexyl group and also the dipole generated by the difluoro moiety. Maraviroc also possessed high aqueous solubility (> 1 mg/mL across the pH range) and could be crystallized as the free base or the besylate salt. The early identification and optimization of stable crystalline polymorphs of maraviroc was a key contribution to the rapid development timelines and the importance of strong biopharmaceutical properties should not be underestimated.

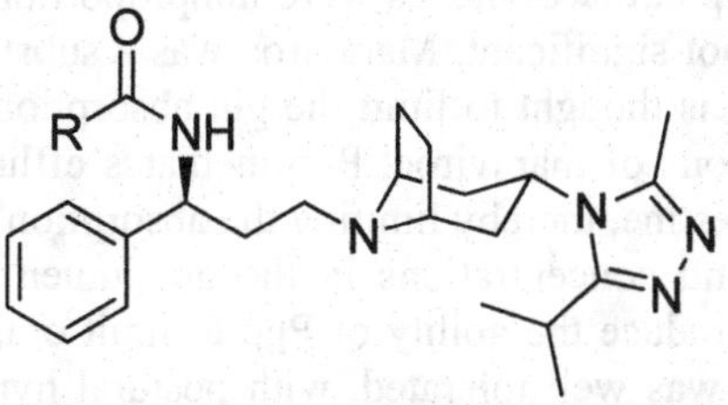

Table 2. Antiviral Activity and hERG Channel Activity for Key Triazole Compounds

	R	Antiviral IC$_{90}$[a] (nM)	hERG[b] (%I @ 300 nM)
5	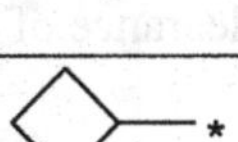 ▱—*	8	30%
9	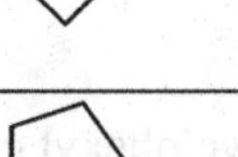 ⬠—*	2	18%
10	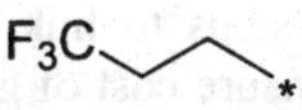 F$_3$C⌒⌒*	14	14%
1	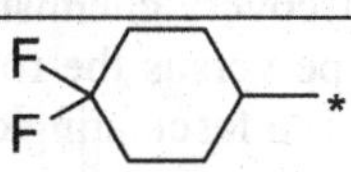 F,F—⬡—*	2	0%

[a]The concentration required to inhibit replication of HIV$_{BaL}$ in PM-1 cells by 90%.
[b]Percent inhibition of [^{3}H]-dofetilide binding to hERG stably expressed on HEK-293 cells.

2.3   Pharmacokinetics and Safety

In preclinical studies maraviroc (1) showed no significant activity against a range of pharmacologically relevant enzymes, ion channels, and receptors up to 10 μM, as measured in binding competition and functional assays. It was also well tolerated in mouse and rat studies, with no significant effects on the central, peripheral, renal, or respiratory systems. Further profiling for cardiac effects and in particular those mediated via hERG showed only very weak affinity using whole-cell voltage clamp techniques with 19% inhibition at 10 μM. In canine isolated Purkinje fiber assays, maraviroc was devoid of any effects at 1 μM on action potential morphology. Upon progression into in vivo cardiovascular assessment, there were no biologically relevant or statistically significant haemodynamic or ECG changes in conscious freely moving dogs at free plasma levels > 100-fold the antiviral IC_{90} of maraviroc.

Maraviroc is rapidly absorbed after oral administration and plasma T_{max} was achieved within 0.5–4.0 h post dose; steady-state plasma concentrations were achieved after 7 days of consecutive dosing. Maraviroc was metabolized by CYP3A4, although there is also a component of renal clearance of unchanged drug. As a CYP3A4 substrate, drug levels of maraviroc decrease when co-administered with strong CYP3A4 inducers and increase when administered with CYP3A4 inhibitors. Over the lower dose range studied (3–1200 mg) the pharmacokinetics were nonproportional; however, at clinically relevant doses, this was not significant. Maraviroc was a substrate for the transporter P-glycoprotein (Pgp) which is thought to limit the gut absorption of many compounds. At low intestinal concentrations of maraviroc, Pgp mediates efflux of maraviroc back into the lumen of the small intestine, thereby limiting the absorption of maraviroc. As the dose of maraviroc increases and concentrations in the gut lumen increase Pgp efflux may become saturated and so reduce the ability of Pgp to limit oral bioavailability. At doses up to 900 mg maraviroc was well tolerated, with postural hypotension being the dose-limiting effect. No clinically significant increases in QT prolongation were noted at relevant doses. Maraviroc does not modulate the activity of CYP2D6 or CYP3A4 at clinically relevant doses. This is very important as maraviroc is co-administered with other antiviral agents and so does not effect the CYP-mediated clearance of other agents.

2.4   Syntheses

Maraviroc (1) can be simply disconnected into the 4,4-difluorocyclohexyl carboxylic acid (11), the phenpropyl linker 12 and the tropane triazole motif 13. Once these individual fragments have been prepared, it is relatively straightforward to link them together to complete the synthesis of maraviroc. The ordering of the steps to link the fragments together can be altered to minimize purification issues and ensure cost of goods (CoG) is not a hindrance. A comparison of the routes used by the discovery chemists that allowed them to investigate SAR relationships within the chemotype versus the route developed by the process chemists to deliver multikilogram quantities is a fascinating case history.

The initial synthesis of maraviroc that delivered material for preclinical studies has been published both in journals[10,11] and in patent applications.[12] It is to the credit of the discovery and development chemists that the large-scale routes are similar with the necessary improvements in yields and ease of operation on a large scale. The communication between the chemists in discovery and development was high, ensuring that the route used to deliver relatively small amounts of preclinical material could be adopted as the backbone of the development routes. Throughout the whole pharmaceutical industry, this close relationship between discovery and development is desired if delivery of bulk active pharmaceutical ingredient (API) is not going to be a rate limiting factor in development timelines.

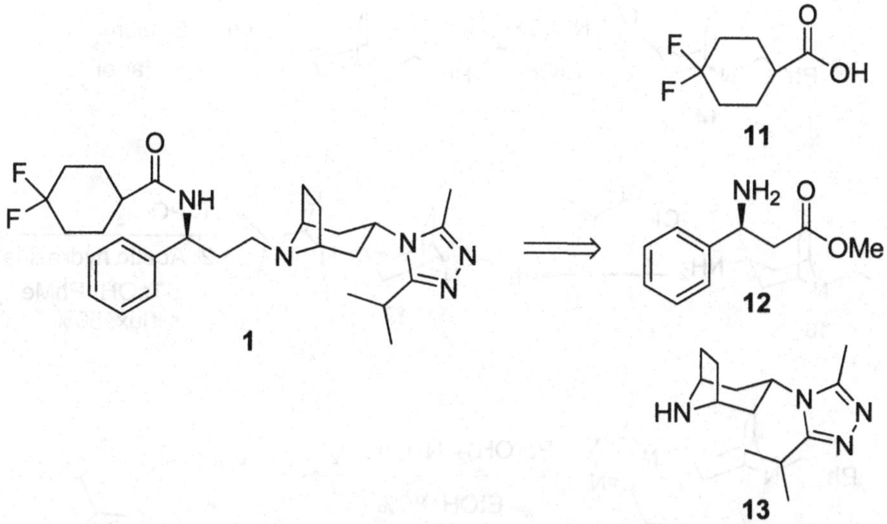

Before engagement from the development chemistry community, the challenges facing the discovery synthesis group were optimization of robust conditions for the preparation of the triazole motif and the purification of 4,4-difluorocyclohexyl carboxylic acid before amide coupling in the final step. The triazole methodology at the time of initial synthesis required significant effort, and the starting material was commercially available *N*-benzyl tropinone (**14**). Oxime formation and single electron reduction furnished **16** with the required equatorially disposed primary amine. The amide **17** was prepared using the necessary acid chloride and Schotten–Baumann conditions. The iminoyl chloride was prepared by treatment of the amide **17** with phosphorous oxychloride, which could then be reacted with acetic hydrazide. Cyclization to the required triazole **18** was completed by heating under reflux in toluene with a catalytic quantity of *p*-toluene sulfonic acid. The benzyl protecting group could be removed by transfer hydrogenation to furnish the required intermediate **13**. The triazole formation required significant optimization within the development phase. Even though the fundamental disconnection remained the same, modification of reagents and conditions were essential for the successful preparation of maraviroc. In particular, it was found that replacing phosphorous oxychloride with phosphorous pentachloride and minimizing the thermal instability of the intermediate iminoyl chloride before reaction with acetic hydrazide were critical to success.[13,14]

The use of fluorine in molecules of pharmacological interest is well known and there has been widespread research into methodologies for the introduction of this atom. Within the discovery synthesis of maraviroc, the initial preparation required the use of diethyl amino sulfur trifluoride (DAST), which is commercially available but does require careful handling due its well known thermal instability. Initial use of DAST for the fluorination of the ketone **19** gave an inseparable 1:1 mixture of the required difluoro compound **20** and the vinyl fluoride **21**. The formation of vinyl fluoride co-products from the treatment of ketones with DAST is known in the literature and is difficult to control.

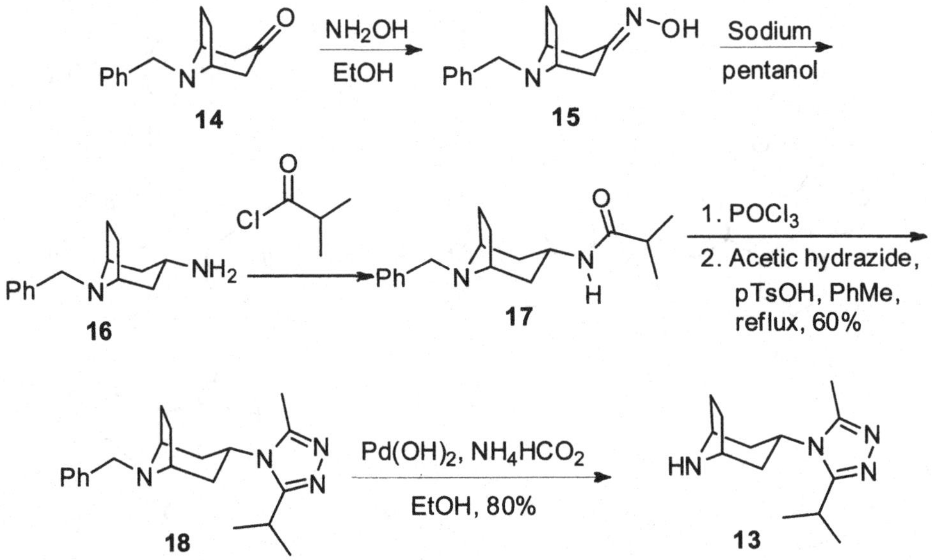

Optimization studies undertaken to influence the ratio of products, including temperature, reagent stoichiometry, and solvents were unsuccessful. It was decided that for early evaluation of maraviroc the use of DAST was acceptable. Within the discovery setting, the inseparable mixture of the difluoro and vinyl fluoro cyclohexyl esters **20** and **21** was subjected to Upjohn conditions for dihydroxylation of vinyl fluoro side product **21**. After an overnight reaction, the required difluorocyclohexyl ester **20** could be isolated in high purity by flash column chromatography. Saponification gave the acid **11**, which could be recrystallized from cyclohexane to furnish analytically pure material.

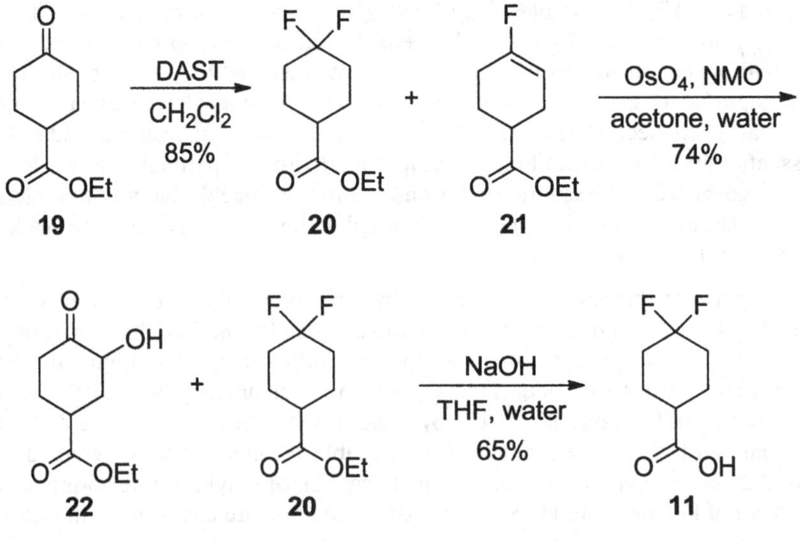

The aldehyde **25** was a key intermediate and could be prepared on scale from the commercially available ester **24** by partial reduction using DIBAL-H, this was a reaction where careful control of temperature was essential to prevent overreduction to the alcohol. With the aldehyde in hand, a simple reduction amination linked the two halves of maraviroc together for completion of the synthesis. Throughout the discovery phase of the project high-throughput chemistry was applied whenever possible to rapidly generate knowledge for the next design cycle. With this desire for the use of parallel chemistry, the reagent of choice for the final amide coupling was a polymer-bound carbodiimide reagent using dichloromethane as solvent. Once the amide coupling reaction was complete simple filtering through a pad of silica furnished maraviroc **1**, which was recrystallized to analytical purity from toluene/hexane. It is interesting to note that alternative amide coupling reagents were investigated in the discovery laboratories; however, the polymer-bound reagent was the reagent of choice even on relatively large-scale reactions (up to 10 g) due to the ease of workup and the high purity of the product, which could be easily recrystallized to high purity. Alternative amide coupling reagents were investigated and rejected as the samples of maraviroc contained unacceptable impurity profiles for the studies required.

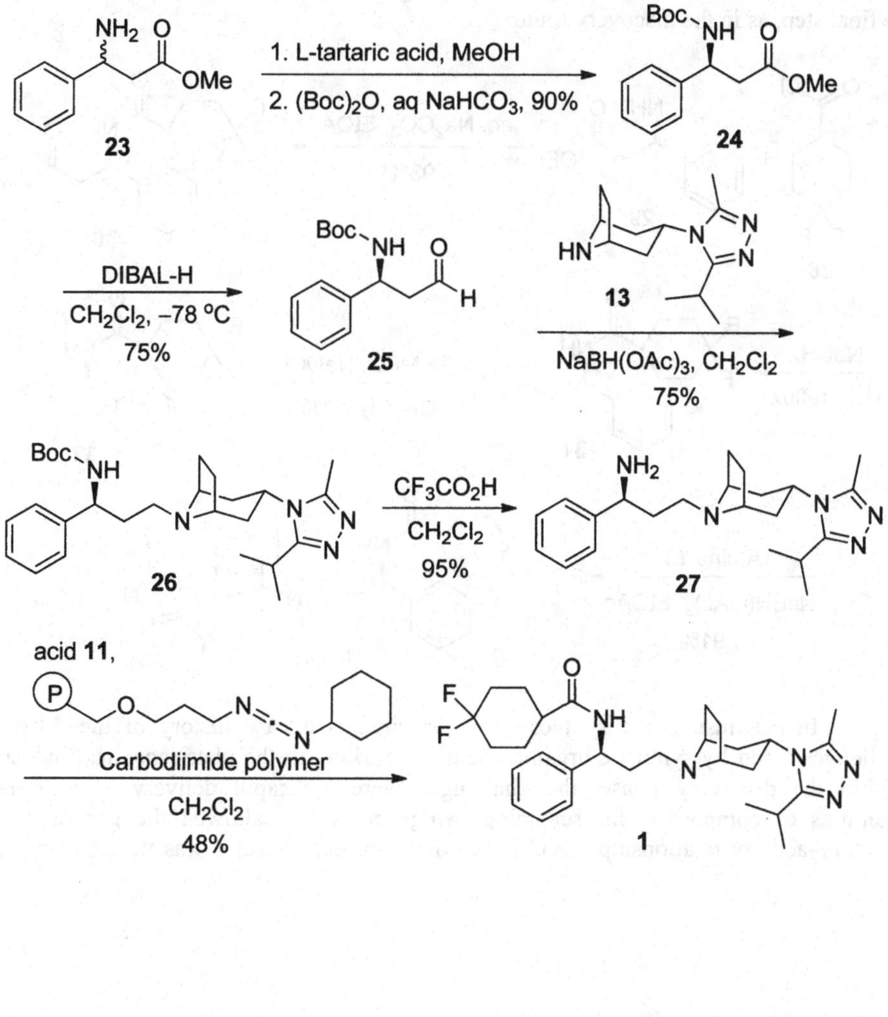

For large-scale development of maraviroc, the fluorination chemistry required modification due to concerns around safety, reproducibility of the DAST reaction, and the need for column chromatography to purify intermediates. The fluorination chemistry was eventually transferred to a specialist company who used the same starting material but was capable of using HF on large scale to produce intermediate **28**. The starting material for the phenyl propyl linker was the commercially available β-amino ester **29**, which was amidated in high yield. As the project progressed and large quantities of the ester **29** were needed, the material was delivered via an L-tartaric acid resolution of the racemic β-amino ester. A single-step partial reduction of the ester **30** to yield the required aldehyde **32** using DIBAL-H proved to be unsuccessful on larger scales, with overreduction to the alcohol **31** occurring under all conditions investigated. It was decided to follow this strategy, and by using sodium borohydride, it was possible to completely reduce the ester to the alcohol, which was then reoxidized to give the required aldehyde. Within the discovery and development routes, the same key reductive amination strategy was used to prepare maraviroc; however, the order of steps was modified, and the 4,4-difluorocyclohexyl group was introduced early in the synthesis. This was enabled by the now ready availability of the 4,4-difluorocyclohexyl carboxylic acid, so that this motif was no longer of such high value that it could be installed only in the final step, as in the discovery route.

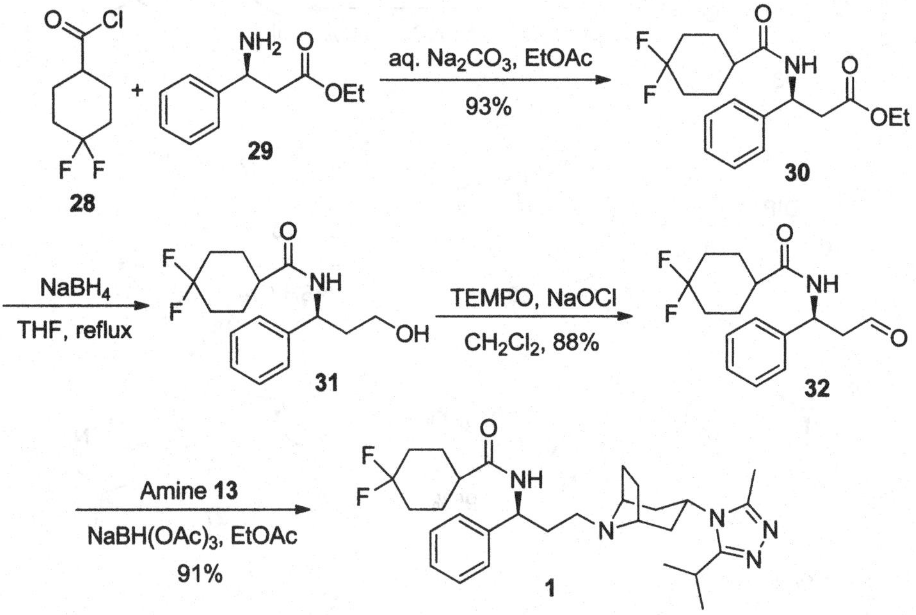

In conclusion, the synthesis of maraviroc is a case history of the differing challenges faced by synthetic organic chemists working in the pharmaceutical industry. Within the discovery phase, the challenges were the rapid delivery of milligram quantities of compounds for screening using routes that allowed the generation of structure–activity relationships. Within the development phase, it was the delivery of a route capable of delivering kilogram quantities safely and in a cost-effective manner. The initial route used in the discovery of maraviroc was not trivial to transfer into the pilot plant and eventually into manufacturing facilities. Reagents and solvents needed to be replaced, impurities eliminated, and crystalline intermediates identified. It is to the credit of the creativity and imagination of the synthetic community that all these goals were achieved in a remarkably short period of time. Maraviroc was first prepared and screened in 2000 and was ultimately approved in 2007.[16]

2.5   References

1.   WHO. "AIDS Epidemic Update", December 2005, UNAIDS/03.39E, www.who.int/hiv/epi-update2005_en.pdf, accessed June, 2009.
2.   Fauci, A. S. *Nat. Med.* **2003**, *9*, 839–843.
3.   Maddon, P. J.; Dalgleish, A. G.; McDougal, J. S.; Clapham, P. R.; Weiss, R. A.; Axel, R. *Cell* **1986**, *47*, 33–48.
4.   Berger, E. A.; Murphy, P. M.; Farber, J. M. *Annu. Rev. Immunol.* **1999**, *17*, 657–700.
5.   Liu, R.; Paxton, W. A.; Choe, S.; Ceradini, D.; Martin, S. R.; Horuk, R.; Macdonald, M. E.; Stuhlmann, H.; Koup, R. A.; Landau, N. R. *Cell* **1996**, *86*, 367–377.
6.   Armour, D.; de Groot, M. J.; Edwards, M.; Perros, M.; Price, D. A.; Stammen, B. L.; Wood, A. *ChemMedChem* **2006**, *1*, 706–709.
7.   Palani, A.; Tagat, J. R. *J. Med. Chem.* **2006**, *49*, 2851–2855.
8.   Armour, D.; de Groot, M. J.; Perros, M.; Price, D. A.; Stammen, B. L.; Wood, A.; Burt, C. *Chem. Biol. Drug. Des.* **2006**, *67*, 305–308.
9.   Price, D. A.; Armour, D.; de Groot, M. J.; Leishman, D.; Napier, C.; Perros, M.; Stammen, B. L.; Wood, A. *Bioorg. Med. Chem. Lett.* **2006**, *16*, 4633–4637.
10.   Price, D. A.; Gayton, S.; Selby, M. D.; Ahman, J.; Haycock-Lewandowski, S. *Synlett* **2005**, *7*, 1133–1134.
11.   Price, D. A.; Gayton, S.; Selby, M. D.; Ahman, J.; Haycock-Lewandowski, S.; Stammen, B. L.; Warren, A. *Tetrahedron Lett.* **2005**, *46*, 5005–5007.
12.   Perros, M.; Price, D. A.; Stammen, B. L. C.; Wood, A. PCT Int. Appl. WO 01/90106, **2001**.
13.   Haycock-Lewandowski, S. J.; Wilder, A.; Ahman, J. *Org. Process Res. Dev.* **2008**, *12*, 1094–1103.
14.   Ahman, J.; Birch, M.; Haycock-Lewandowski, S. J.; Long, J.; Wilder, A. *Org. Process Res. Dev.* **2008**, *12*, 1104–1113.
15.   Dorr, P.; Westby, M.; Dobbs, S.; Griffin, P.; Irvine, B.; Macartney, M.; Mori, J.; Rickett, G.; Smoth-Burchnell, C.; Napier, C.; Webster, R.; Armour, D.; Price, D.; Stammen, B.; Wood, A; Perros, M. *Antimicrob. Agents Chemoth.* **2005**, *49*, 4721–4732.
16.   Abel, S.; Van der Ryst, E.; Rosario, M. C.; Ridgway, C. E.; Medhurst, C. G.; Taylor-Worth, R. J.; Muirhead, G. J. *Br. J. Pharmacol.* **2008**, *Suppl. I*, 5–18.

3

Darunavir (Prezista): A HIV-1 Protease Inhibitor for Treatment of Multidrug-Resistant HIV

Arun K. Ghosh and Cuthbert D. Martyr

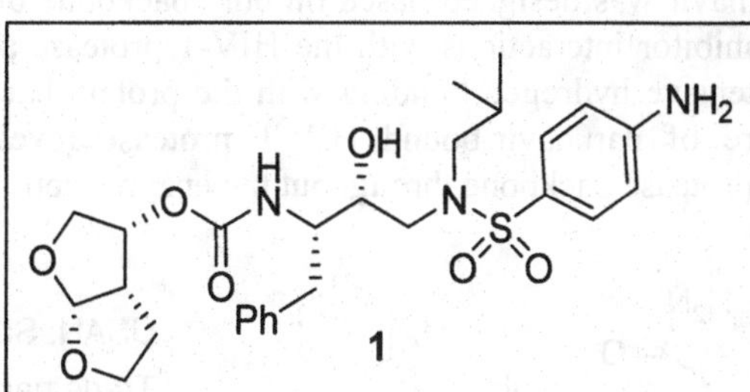

USAN: Darunavir
Trade name: Prezista®
Tibotec Pharmaceuticals
Launched: 2006

3.1 Background

Human immunodeficiency virus type 1 (HIV-1) infections continue to be a major global challenge in medicine in the 21st century. The World Health Organization (WHO) estimates that about 35 million people are living with HIV/AIDS (acquired immunodeficiency syndrome) with nearly 2.5 million new cases of HIV infection in 2008.[1] These statistics indicate the magnitude of the global epidemic of HIV/AIDS today.

Early on, biochemical events critical to HIV replication revealed a number of important targets for therapeutic intervention of HIV/AIDS.[2] Subsequently, drug development efforts in academic and industrial laboratories targeting retroviral enzymes, reverse transcriptase (RT), integrase (IN), and HIV protease (HIV-PR), led to the discovery of a variety of antiretroviral treatment regimens.[3] Of particular interest, HIV-1 protease inhibitors (PIs) have proven to be highly effective in combination with reverse transcriptase inhibitors for the treatment of HIV/AIDS. Various approved PIs are shown in Figure 1. This combination therapy or highly active antiretroviral therapy (HAART) greatly improved the quality of life of HIV/AIDS patients, decreased mortality, and improved the course of HIV management in the United States and other industrialized nations.[4]

Despite the marked progress since the advent of HAART, the long-term treatment of HIV/AIDS remains a critical issue. A variety of impediments, including debilitating side effects, drug toxicities, low oral bioavailability, higher therapeutic doses and the emergence of drug resistance, seriously hampers the long-term effectiveness of HAART treatment regimens. Current statistics show that 40–50% of patients who initially achieved favorable viral suppression due to HAART, experienced treatment failure due to drug resistance.[5] Also, 20–40% of antiretroviral therapy-naive individuals infected with HIV-1 have persistent viral replication under HAART, most likely due to transmission of multidrug-resistant HIV-1 variants.[6] As mentioned previously, in addition to the issues of drug resistance, the first-generation PIs are faced with severe limitations with respect to tolerance, drug side effects, and toxicities. Therefore, the development of a new generation of PIs with broad-spectrum activity and minimal drug side effects is crucial for effective management of current and future HIV/AIDS.

In this context, our research efforts addressing these issues led to the design and development of darunavir (1), a new generation of nonpeptide PI. Darunavir is exceedingly potent and has shown impressive broad-spectrum activity against highly cross-resistant HIV mutants. It received FDA approval in 2006 for the treatment of HIV/AIDS patients who are harboring drug-resistant HIV that does not respond to other therapies.[7] More recently, darunavir received full approval for all HIV/AIDS patients, including pediatrics.[8] Darunavir was designed based on our "backbone binding" design concept to maximize the inhibitor interactions with the HIV-1 protease active site and, particularly, to promote extensive hydrogen bonding with the protein backbone atoms.[9] Indeed, the X-ray structure of darunavir-bound HIV-1 protease revealed extensive hydrogen bonding with the protease backbone throughout the enzyme active site.[10]

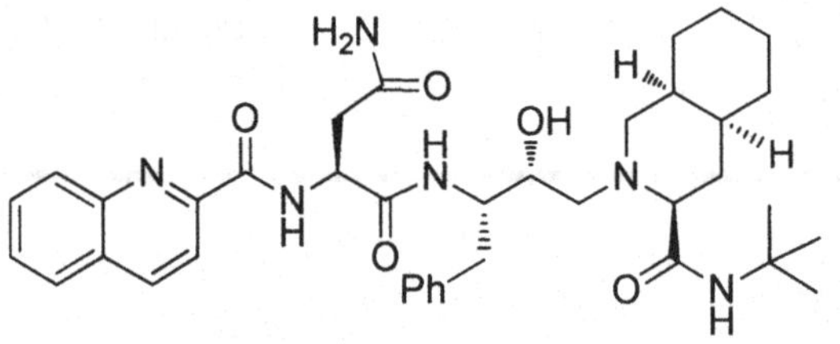

USAN: Saquinavir
Trade name: Fortovase®
Hoffmann-La Roche
Launched: 1995

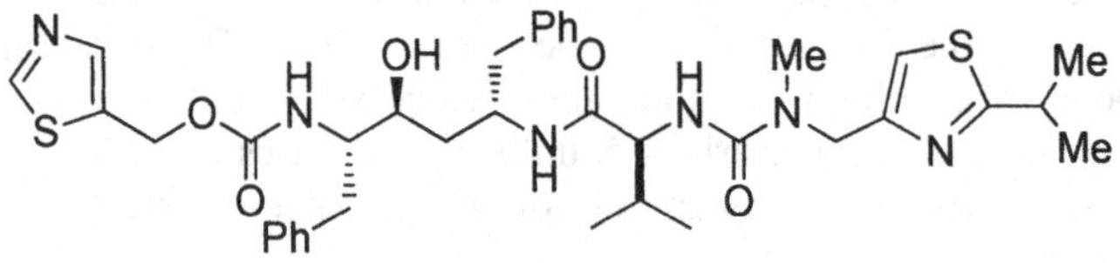

USAN: Ritonavir
Trade name: Norvir®
Abbott Laboratories
Launched: 1996

Figure 1. HIV protease inhibitors in order of FDA approval time up to 2005.

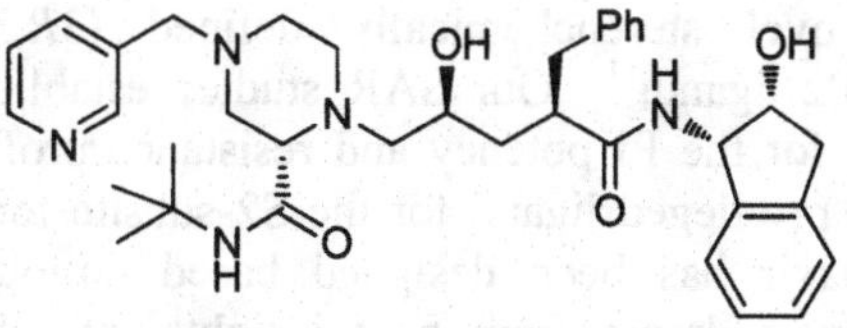

USAN: Indinavir
Trade name: Crixivan®
Merck & Co.
Launched: 1996

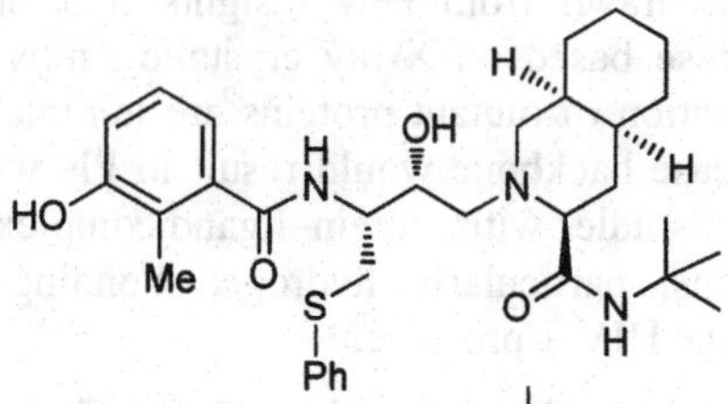

USAN: Nelfinavir
Trade name: Viracept®
Japan Tobacco/Agouron/Pfizer
Launched: 1997

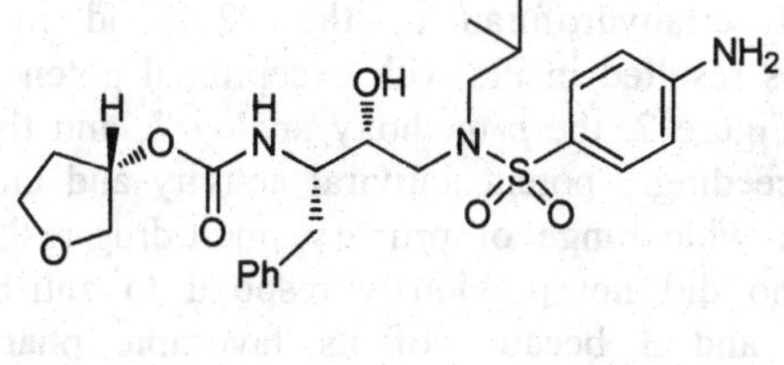

USAN: Amprenavir
Trade name: Agenerase®
GlaxoSmithKline
Launched: 1999

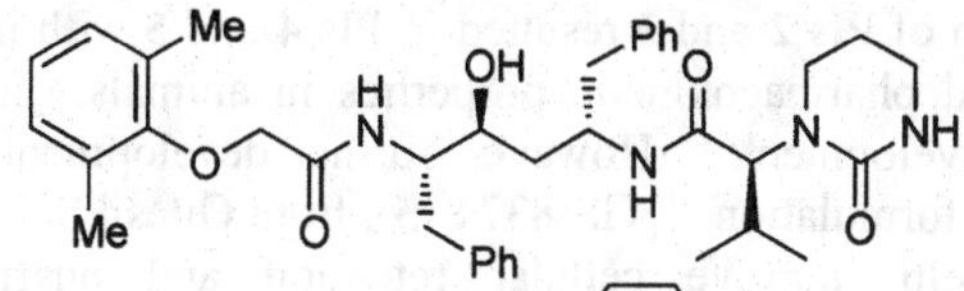

USAN: Lopinavir
Trade name: Kaletra®
Abbott Laboratories
Launched: 2000

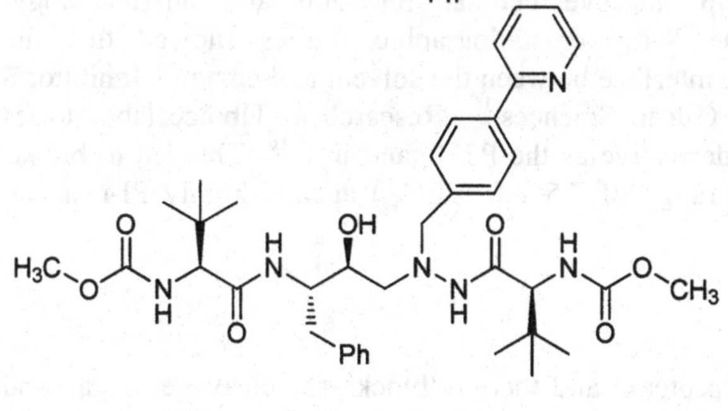

USAN: Atazanavir
Trade name: Reyataz®
Bristol-Myers-Squibb
Launched: 2003

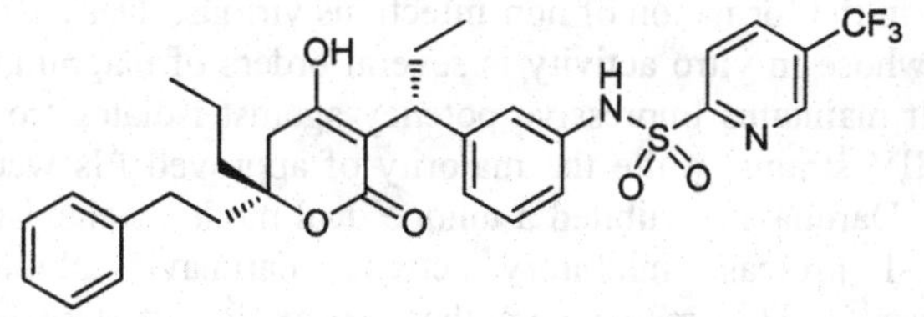

USAN: Tipranavir
Trade name: Aptivus®
Boehringer-Ingelheim
Launched: 2005

Figure 1 (cont.).

3.2   Structure–Activity Relationship and Darunavir Derivatives

Darunavir (1) incorporated a novel stereochemically defined (3R,3aS,6aR)-bis-tetrahydrofuran (bis-THF) as the P2-ligand.[11] Our SAR studies established that the (3R,3aS,6aR)-enantiomer is critical for the PI potency and resistance profile. We have demonstrated that the bis-THF is a privileged ligand for the S2-subsite for a variety of transition-state isosteres.[11] Darunavir has been designed based on our "backbone binding" concept. This concept was drawn from new insights into the molecular interaction of PIs with the HIV protease based on X-ray crystallography studies. In particular, since the backbone conformation of mutant proteins are minimally distorted, maximizing interactions with the protease backbone would result in PIs with improved resistance profiles. The X-ray structural studies with protein–ligand complexes show that darunavir makes extensive interactions, particularly hydrogen bonding interactions throughout the active site backbone of the HIV-1 protease.

Incorporation of the bis-tetrahydrofuran as the P2-ligand into modified hydroxyethyl sulfonamide isosteres resulted in PIs with exceptional potency and drug-resistance profiles. As shown in Figure 2, the p-methoxy analog 2, and the dioxolane derivative 3 have also shown exceedingly potent antiviral activity and unprecedented broad-spectrum activity against a wide range of primary, multidrug-resistant HIV-1 strains isolated from patients who did not previously respond to anti-HIV agents. Darunavir was selected over 2 and 3 because of its favorable pharmacokinetic properties.[11] Further modification of PIs 2 and 3 resulted in PIs 4 and 5 with impressive drug-resistance profiles and good pharmacokinetic properties in animals. Inhibitor 4 advanced to phase III clinical development.[12] However, further development has been halted due to lack of suitable oral formulation.[13] GS-8374 (5), from Gilead Sciences, was designed as a pro-drug to help improve cellular retention and pharmacology/pharmacokinetic properties.[14] The X-ray crystallographic studies showed that the phosphonate moiety remains at the interface between the solvent and enzyme. Inhibitor 5 is under clinical development by Gilead Sciences.[15] Research at Tibotec laboratories have incorporated a benzoxazole derivative as the P2'-ligand in 6.[16] This led to broad-spectrum antiviral activity in the range of 7.5–8.0 (pEC$_{50}$) against highly PI-resistant mutants.

3.3   Pharmacology

Darunavir (1) inhibits the HIV-1 protease and thereby blocks the cleavage of gag and gag-pol polyproteins, which results in the formation of non-infectious virions. Darunavir is an exceedingly potent inhibitor whose in vitro activity is several orders of magnitude higher than first generation PIs. It maintains impressive potency against isolates from patients with multidrug-resistant HIV strains, while the majority of approved PIs were less active against those isolates.[17] Darunavir exhibited a unique dual mode mechanism of action. In addition to HIV-1 protease inhibitory activity, darunavir inhibits dimerization of the HIV-1 protease.[18] This may be another reason for its superior antiviral activity and durability against the emergence of drug-resistance. Darunavir shows plasma protein binding properties of approximately 95%. It primarily binds to plasma alpha-1-acid glycoprotein (AAG or AGP).[19]

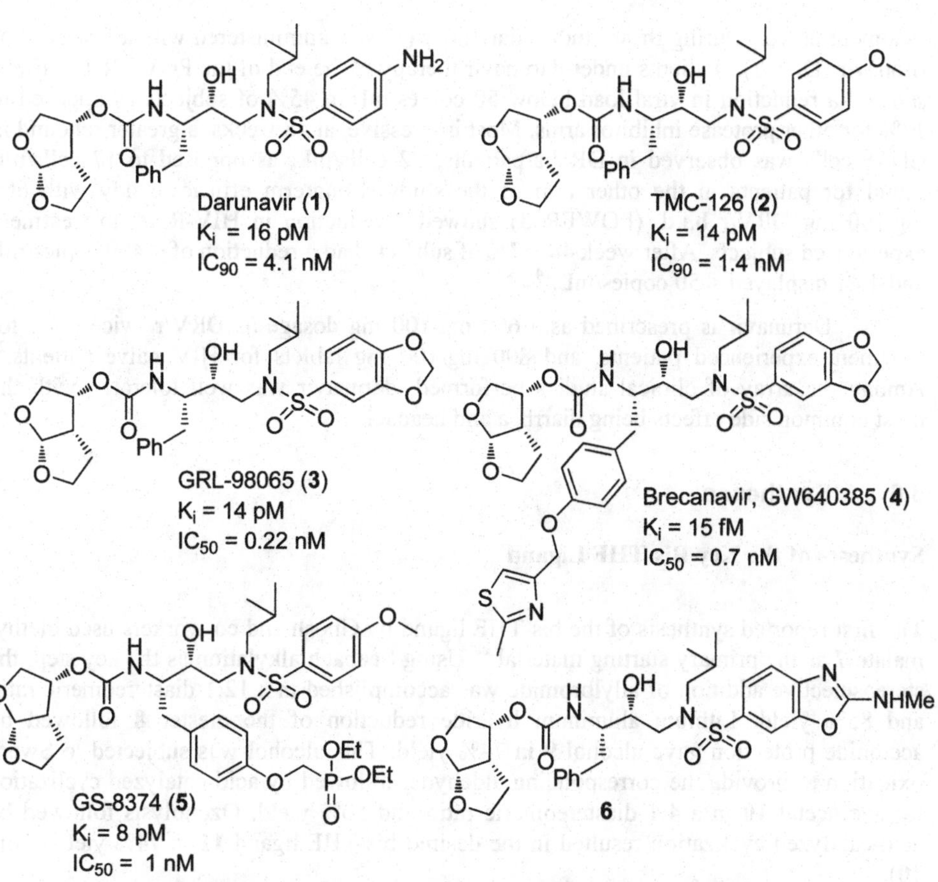

Figure 2. Potent PIs with bis-tetrahydrofuranylurethanes as the P2-ligand.

3.4 Pharmacokinetics and Drug Metabolism

When taken with meals, a 30% increase in absorption of the drug was observed, with peak plasma concentrations occurring within 2.5–4 h after oral administration. When co-administered with ritonavir, the overall half-life is improved to 15 h.[20] Darunavir (1) has been shown to be metabolized by liver enzyme CYP450 (3A4).[21,22] Thus when administered with low-dose ritonavir—a CYP450 and protease inhibitor—bioavailability increases from 37% to 84%. Absorption of darunavir occurs primarily in the intestine through passive intracellular diffusion. Darunavir and its metabolites are excreted primarily in the feces and urine.[23] Metabolism occurs via, carbamate hydrolysis, aliphatic hydroxylation, aromatic hydroxylation, and other metabolites.

3.5 Efficacy and Safety

Safety and efficacy studies were carried out in three different clinical studies—POWER 1–3, TITAN, and ARTEMIS—with patients who were highly treatment experienced and treatment naive. During these studies darunavir (1) was administered with a low dose of ritonavir (DRV/r). Patients under darunavir therapy at the end of the POWER 1–2 trials, showed a reduction in viral load below 50 copies/mL in 45% of subjects as opposed to 12% for other protease inhibitor arms. Most impressive, at 48 weeks, a greater rebound in CD4+ cells was observed in DRV/r patients, 92 cells/mL, as opposed to 17 cells/mL found for patients in the other arm of the study. Longterm efficacy study with 600 mg/100 mg DRV/r b.i.d. (POWER 3) showed a reduction in HIV-RNA in treatment experienced subjects. After week 48, 57% of subjects had a reduction of < 400 copies/mL and 40% displayed < 50 copies/mL.[24–26]

Darunavir is prescribed as a 600 mg/100 mg dosage as DRV/r twice daily for treatment-experienced patients, and 800 mg/100 mg tablets for HIV naive patients.[26] Among the array of clinical studies performed, darunavir was well tolerated with the most common side effects being diarrhea and headache.[27]

3.6   Syntheses

Syntheses of the Key Bis-THF Ligand

The first reported synthesis of the bis-THF ligand by Ghosh and co-workers used diethyl malate 7 as the primary starting material.[28] Using Seebach alkylation as the key step, the stereoselective addition of allylbromide was accomplished in a 12:1 diastereomeric ratio and 85% yield. Lithium aluminum hydride reduction of the diester 8 followed by acetonide protection gave alcohol 9 in 74% yield. This alcohol was subjected to Swern oxidation to provide the corresponding aldehyde, followed by acid-catalyzed cyclization to give acetal 10 in a 4:1 diastereomeric ratio and 50% yield. Ozonolysis followed by acid-catalyzed cyclization resulted in the desired bis-THF ligand 11 in 74% yield (from 10).

The malate-based synthesis was not convenient for scaleup. Subsequently, a three-step racemic synthesis of the bis-THF ligand was developed by Ghosh and Chen.[29] The racemic ligand was resolved by an enzymatic resolution to obtain optically active bis-THF. As shown in Scheme 2, commercially available 2,3-dihydrofuran 12 was reacted with N-iodosuccinimide and propargyl alcohol to result in the corresponding racemic alkyne 13 (91–95% yield). Radical cyclization using cobaloxime (10 mol%) and NaBH$_4$ in ethanol at 65 °C gave the desired bis-THF precursor 14 in 70–75% yield. Radical cyclization has also been achieved using tributyltin hydride and AIBN in refluxing toluene in 75–80% yield. Ozonolysis with reductive workup in the presence of NaBH$_4$ gave the racemic bis-THF ligand (±)-11 in 74–78% yield. Lipase resolution resulted in the desired chiral ligand (–)-11 in 42% yield and acetate (+)-15 in 45% yield. Optical purity of (–)-11 was determined to be 95% ee by Mosher ester analysis.

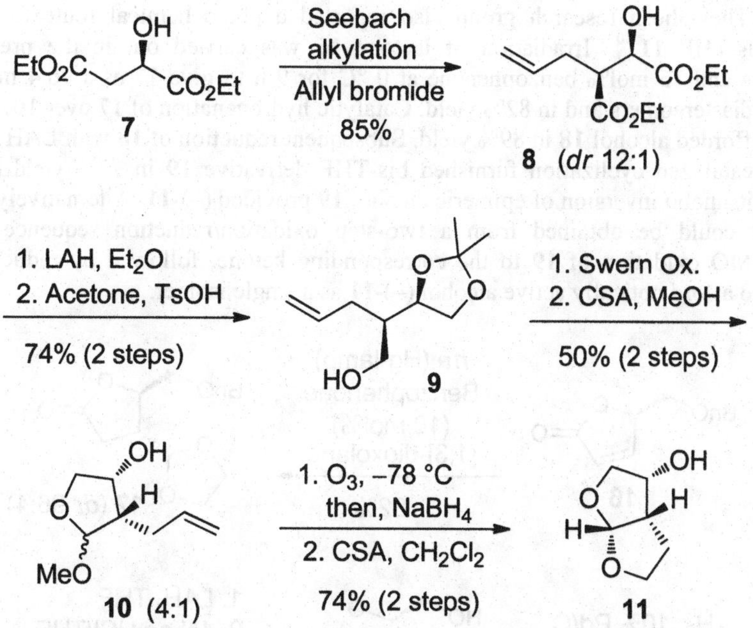

Scheme 1. Malic ester-derived synthesis of bis-tetrahydrofuranyl alcohol 11

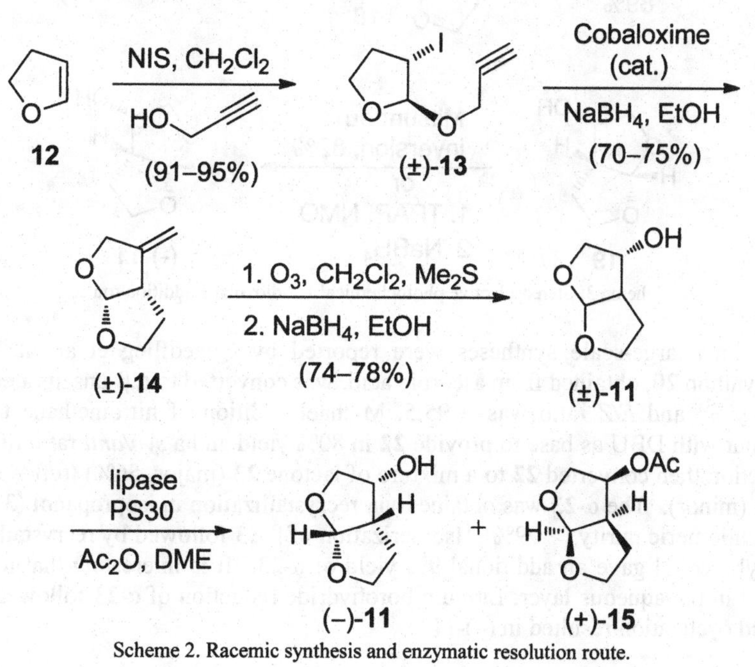

Scheme 2. Racemic synthesis and enzymatic resolution route.

The Ghosh research group also reported a photochemical route to optically active bis-THF **11**.[30] Irradiation of lactone **16** was carried out in the presence of dioxolane and 15 mol% benzophenone at 0 °C for 9 h to give **17** as a 96:4 mixture of *anti/syn* diastereomers and in 82% yield. Catalytic hydrogenation of **17** over 10% Pd/C in MeOH afforded alcohol **18** in 89% yield. Subsequent reduction of **18** with LAH followed by acid-catalyzed cyclization furnished bis-THF derivative **19** in 77% yield over two steps. Mitsunobu inversion of epimeric alcohol **19** provided (–)-**11**. Alternatively, the bis-THF **11** could be obtained from a two-step oxidation/reduction sequence using a TPAP/NMO oxidation of **19** to the corresponding ketone, followed by reduction with NaBH$_4$ to afford optically active alcohol (–)-**11** as a single isomer.

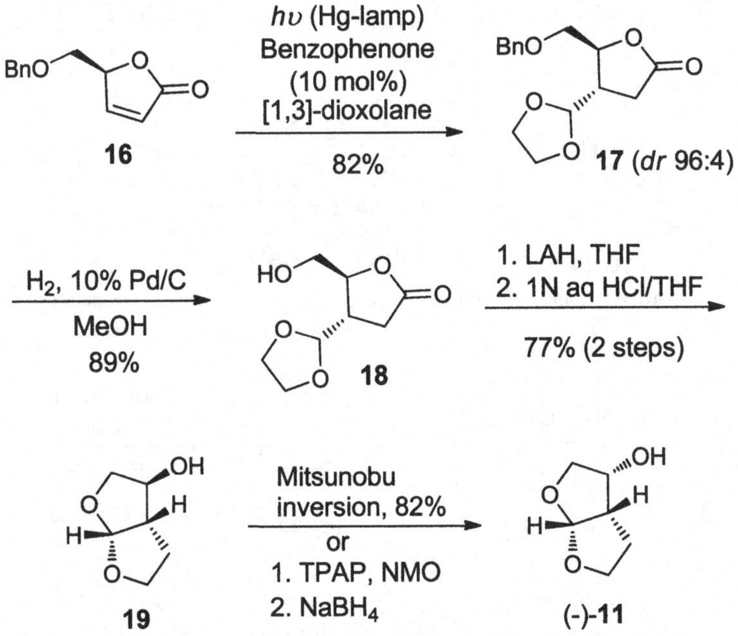

Scheme 3. Stereoselective photochemical 1,3-dioxolane addition route.

Two large-scale syntheses were reported by Quaedflieg et al. at Tibotec.[31] Chiral synthon **20**, obtained from ascorbic acid, was converted to α,β-unsaturated ester **21** in 92% yield and *E/Z* ratio was > 95:5. Michael addition of nitromethane to **21** was carried out with DBU as base to provide **22** in 80% yield and a *syn/anti* ratio of 5.7:1. A Nef reaction then converted **22** to a mixture of lactone **23** (major, 56%) (α/β = 3.8:1) and ester **24** (minor). The α-**23** was obtained via recrystallization in isopropanol (37%), with high enantiomeric purity (> 99%). Isomerization of β-**23** followed by recrystallization in isopropyl alcohol gave an additional 9% yield of α-**23**. It is interesting that most of **24** remained in the aqueous layer. Lithium borohydride reduction of α-**23** followed by acid-catalyzed cyclization resulted in (–)-**11**.

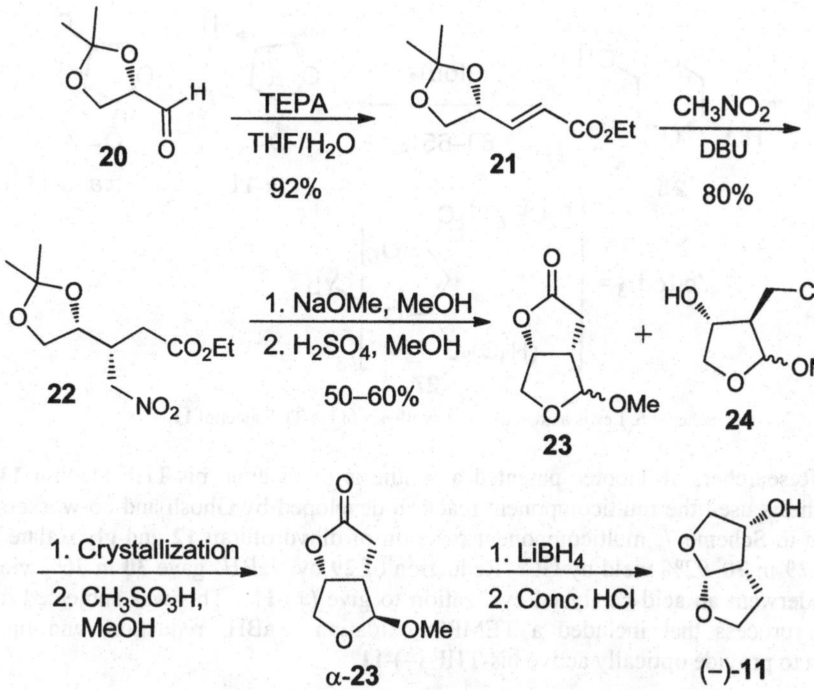

Scheme 4. Large-scale synthesis of bis-THF alcohol 11.

Yu and co-workers[32] have reported bis-THF syntheses using Lewis acid–catalyzed reaction of commercially available 2,3-dihydrofuran (12) and glycolaldehyde dimer 25, as shown in Scheme 5. The use of Sc(OTf)₃ and (S)-26 as chiral ligand in CH_2Cl_2 at 0 °C provided bis-THF alcohol 11 in a 85:15 mixture of enantiomers by GC analysis. The use of Cu[Pybox] gave similar results.

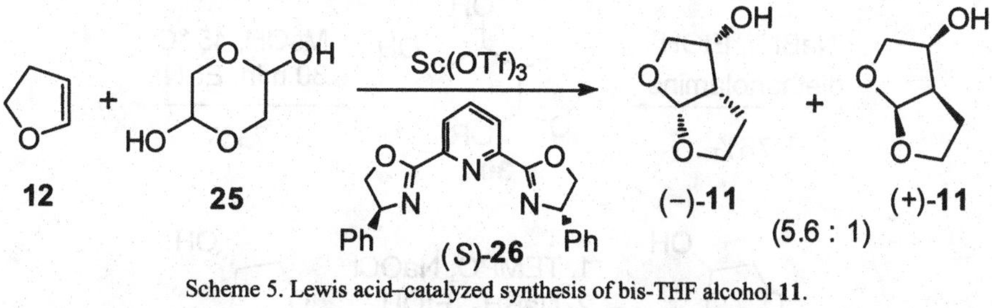

Scheme 5. Lewis acid–catalyzed synthesis of bis-THF alcohol 11.

The use of 0.75 mol% Yb(fod)₃ (27) at 50 °C offered more reproducible results (Scheme 6). A diastereomeric mixture (65:35) of cis-11 to trans-11 was isolated in 60–65% yields as racemic materials. This process, however, required an oxidation (TEMPO, NaOCl) followed by NaBH₄ reduction, protection as acetate derivative and lipase resolution (PS-C, amino-I) to afford the bis-THF alcohol in 97–98% ee and 28–35% yields.

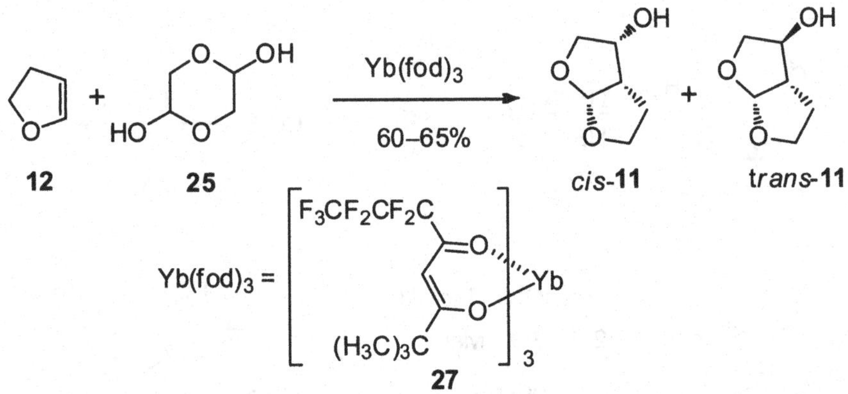

Scheme 6. Lewis acid–catalyzed synthesis of bis-THF alcohol 11.

Researchers at Tibotec patented a synthesis of racemic bis-THF alcohol 11.[33] This synthesis used the multicomponent reaction developed by Ghosh and co-workers.[34] As shown in Scheme 7, multicomponent reaction of dihydrofuran 12 and glyoxalate 28 provided 29 in 70–92% yield by GC. Reduction of 29 by NaBH$_4$ gave 30 in 76% yield, which underwent an acid-catalyzed cyclization to give (±)-11. This was subjected to a three-step process that included a TEMPO oxidation, NaBH$_4$ reduction, and lipase resolution to provide optically active bis-THF (–)-11.

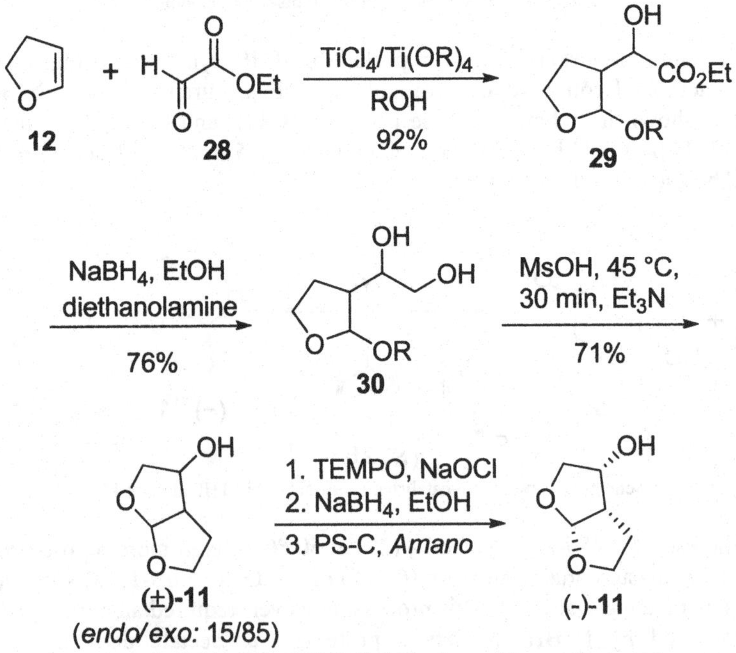

Scheme 7. Synthesis of bis-THF alcohol 11 using multicomponent reaction.

A patented synthesis by Doan et al. at GlaxoSmithKline used a Paterno-Buchi photochemical reaction to obtain acetal **33** in 96% yield, as shown in Scheme 8.[35] Hydrogenation followed by acid-catalyzed cyclization resulted in a diastereomeric mixture of bis-THF alcohol (±)-**11**. Formation of acetate **15**, followed by lipase resolution resulted in the formation of the acetyl bis-THF, which was subsequently deprotected to afford the desired product (−)-**11** in 98% *ee*.[35]

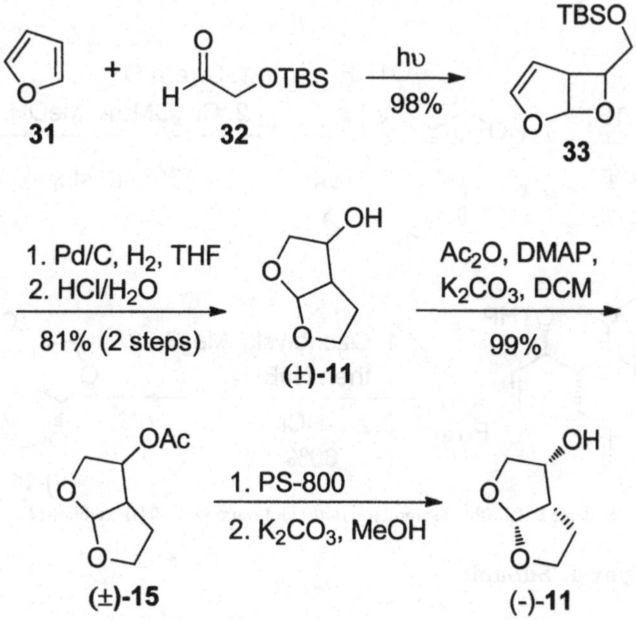

Scheme 8. A Paterno-Buchi photochemical route to bis-THF alcohol **11**.

Another approach used the *anti*-aldol reaction developed by Ghosh et al.[36] Treatment of **34** with TiCl₄ and *N,N*-diisopropylethylamine in CH₂Cl₂ at 0 °C created the corresponding titanium enolate. To this mixture was added a precomplexed cinnamaldehyde/titanium to give the *anti*-aldol product in a 4:1 diastereomeric ratio and 60% yield. The alcohol was protected as the THP ether **35**. Reduction of **35** by LAH effectively removed the chiral auxiliary and provided alcohol **36** in near quantitative yield. Subsequently, Swern oxidation and dimethoxyacetal protection resulted in **37**. Ozonolysis of **37** followed by reduction with dimethyl sulfide and NaBH₄ provided the corresponding diol. Exposure of the resulting diol to aqueous HCl effected the deprotection of the THP group and acid-catalyzed cyclization to the desired bis-THF alcohol **11** in 60% yield and 99% *ee*.

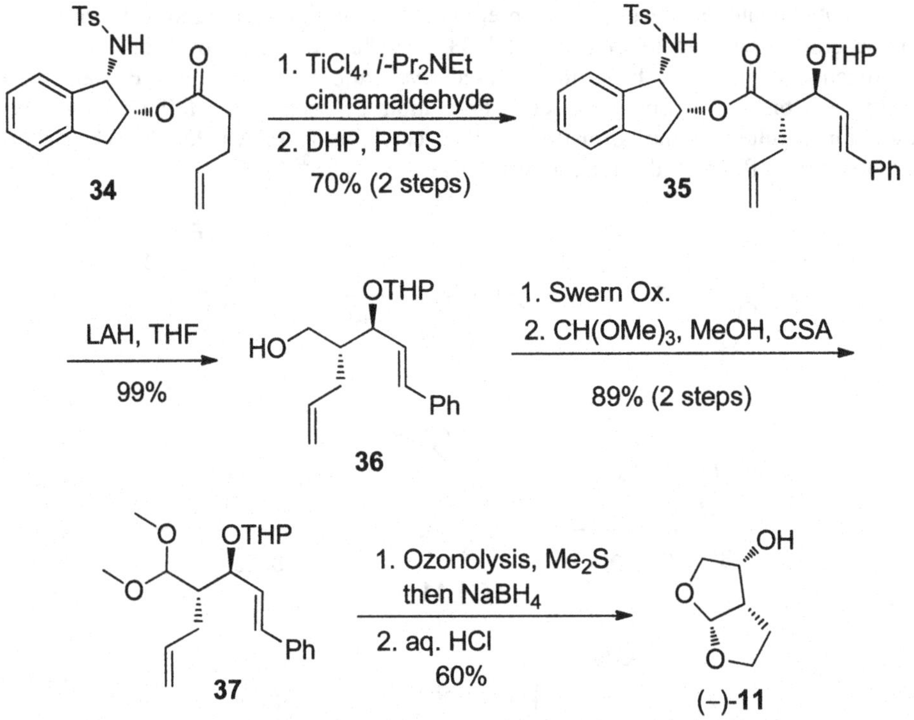

Scheme 9. Ghosh's asymmetric *anti*-aldol route to bis-THF alcohol **11**.

Synthesis of Epoxide Subunit

Darunavir (**1**) was synthesized in a convergent manner using optically active epoxide **38** and the bis-THF alcohol (−)-**11**. Scheme 10 outlines an asymmetric synthesis of **38** using a Sharpless epoxidation as the key step.[37] As shown, phenyl magnesium bromide was reacted with commercially available 1-butadiene monoxide (**39**) at −78 °C in the presence of catalytic CuCN to furnish the allylic alcohol **40**. Sharpless epoxidation of **40** gave the corresponding (*R*)-epoxide **41**. It was then subjected to an epoxide ring opening reaction, as described by Sharpless, to provide azidoalcohol **42**. This was treated with 2-acetoxybutyryl chloride to give azidoepoxide **43**. Catalytic hydrogenation of **43** in the presence of Boc$_2$O afforded epoxide **38**.

A convenient synthesis of epoxide **38** was developed using commercially available Boc-protected phenylalanine.[38] As shown in Scheme 11, protected phenylalanine **44** was converted to alcohol **45** in a three-step sequence, involving (1) reaction of **44** with isobutyl chloroformate followed by diazomethane to give the corresponding diazoketone, (2) treatment of the diazoketone with HCl to provide the corresponding chloroketone, and (3) reduction of the resulting ketone with NaBH$_4$ to provide **45** in 52% overall yield. Treatment of **45** with alcoholic KOH furnished epoxide **38** in 99% yield.

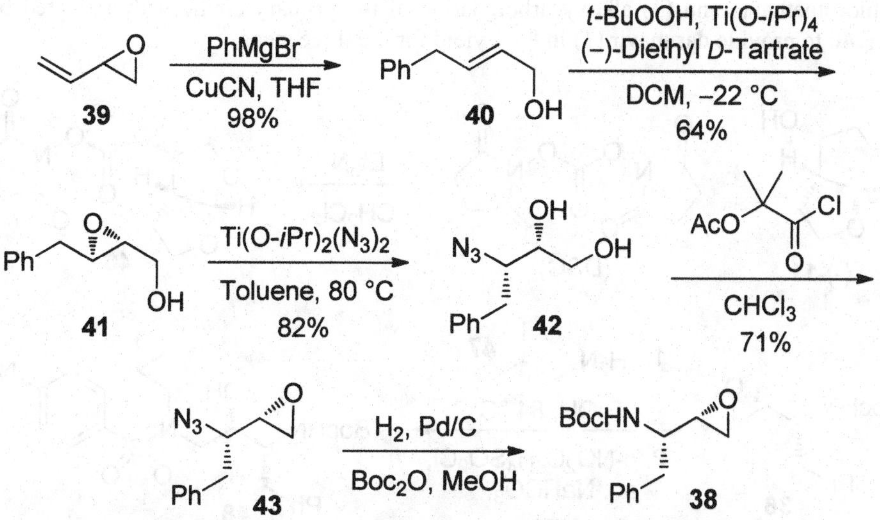

Scheme 10. Asymmetric synthesis of azido epoxide **43** and epoxide **38**.

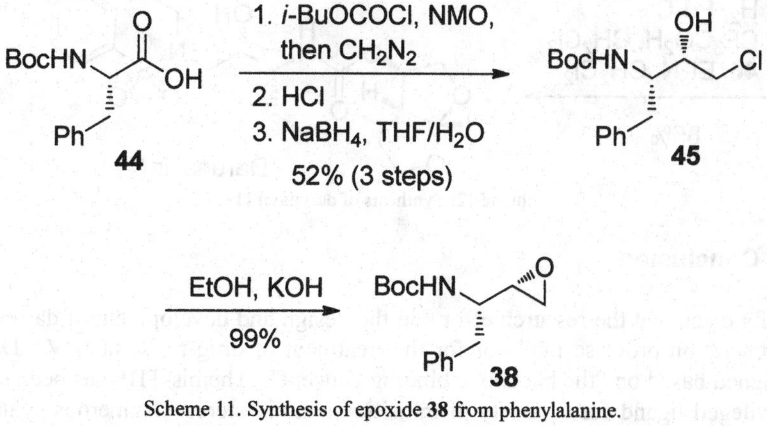

Scheme 11. Synthesis of epoxide **38** from phenylalanine.

Assembly of Darunavir

The synthesis of darunavir (**1**) is shown in Scheme 12. Optically active bis-THF alcohol (–)-**11** was converted to activated mixed carbonate **46** by treatment with *N,N*-disuccinimidyl carbonate (DSC) in the presence of triethylamine.[30] For the synthesis of the hydroxyethylsulfonamide isostere, epoxide **38** was treated with isobutyl amine (**47**) in 2-propanol at reflux to provide the corresponding amino alcohol. Reaction of the resulting amino alcohol with *p*-nitrophenylsulfonyl chloride in the presence of aqueous NaHCO₃ afforded the sulfonamide derivative **48** in 95% yield for the two steps. This was converted to darunavir in a three-step process, involving (1) catalytic hydrogenation of nitro to an amine, (2) removal of the Boc group by exposure to trifluoroacetic acid in dichloromethane, and (3) alkoxycarbonylation of the primary amine with activated bis-THF **46** to provide darunavir (**1**) in 85% yield for the three steps.[30]

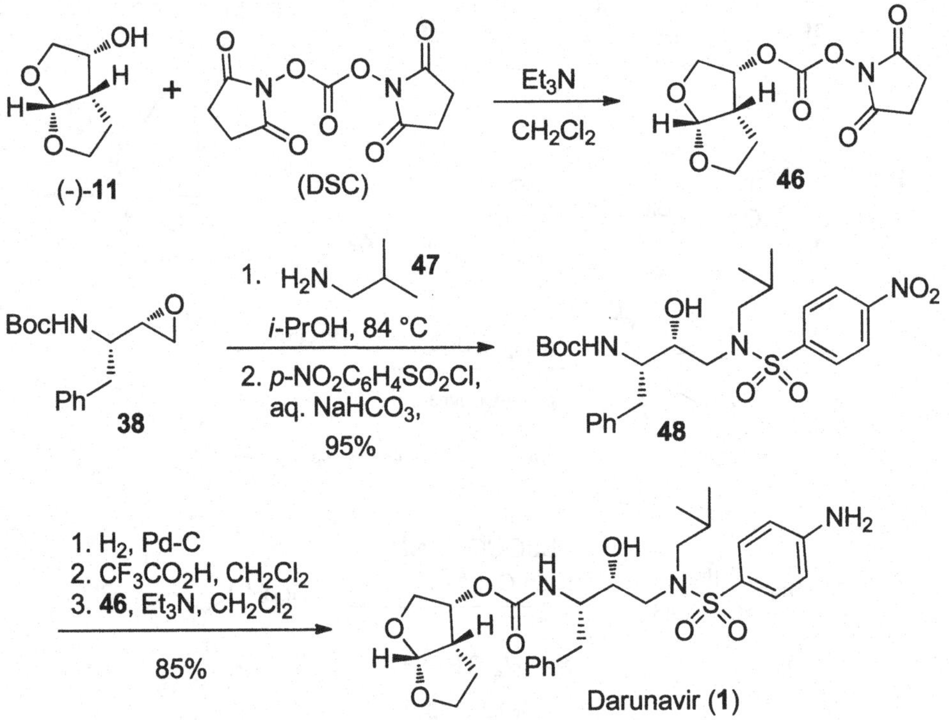

Scheme 12. Synthesis of darunavir (**1**).

3.7   Conclusion

We briefly discussed the research efforts in the design and development of darunavir (**1**), a new generation protease inhibitor for the treatment of drug-resistant HIV. Darunavir was designed based on "the backbone binding concept". The bis-THF has been shown to be a privileged ligand for a variety of transition-state isosteres. Numerous syntheses of the bis-THF ligand have been carried out in academic and industrial laboratories to meet the challenges of cost-effective preparation of darunavir. We have reviewed various syntheses of this important ligand. Furthermore, we have outlined the syntheses of the hydroxyethylamine isostere and assembly of darunavir. Darunavir continues to make a significant impact on the current treatment of HIV/AIDS.

Acknowledgment. This work was supported by a grant from the National Institutes of Health (GM53386).

3.8 References

1. UNAIDS/WHO. "Report on Annual AIDS Epidemic Update, December 2008." www.unaids.org/en/KnowledgeCentre/HIVData/GlobalReport/2008/2008_ Global_report.asp, accessed August 2009.
2. Coffin, J.; Hughes, S.; Varmus, H. *Retroviruses*, New York: Cold Spring Harbor Laboratory Press, **1997**.
3. Clercq, E. D. *J. Med. Chem.* **2005**, *48*, 1297–1313.
4. *Antiviral Agents Bull.* **1995**, *8*, 353–55.
5. Staszewski, S.; Morales-Ramirez, J.; Tashima, K. T.; Rachlis, A.; Skiest, D.; Stanford, J.; Stryker, R.; Johnson, P.; Labriola, D. F.; Farina, D.; Manion, D. J.; Ruiz, N. M. (for the study 006 Team), *N. Engl. J. Med.* **1999**, *341*, 1865–1873.
6. Wainberg, M. A.; Friedland, G. *JAMA* **1998**, *279*, 1977–1983.
7. FDA. "FDA Approves New HIV Treatment for Patients Who Do not Respond to Existing Drugs, June 2006." www.fda.gov/NewsEvents/Newsroom/ PressAnnouncements/2006/ucm108676.htm, accessed April 2010.
8. On October 21, 2008, FDA granted traditional approval to Prezista (darunavir), co-administered with ritonavir and with other antiretroviral agents, for the treatment of HIV-1 infection in treatment-experienced adult patients. In addition to the traditional approval, a new dosing regimen for treatment-naive patients was approved.
9. Ghosh, A. K.; Chapsal, B. D.; Weber, I. T.; Mitsuya, H. *Acc. Chem. Res.* **2008**, *41*, 78–86.
10. Kovalevsky, A. Y.; Tie, Y.; Liu, F.; Boross, P. I.; Wang, Y. F.; Leshchenko, S.; Ghosh, A. K.; Harrison, R. W.; Weber, I. T. *J. Med. Chem.* **2006**, *49*, 1379–1387.
11. Ghosh, A. K.; Sridhar, P. R.; Kumaragurubaran, N.; Koh, Y.; Weber, I. T.; Mitsuya, H. *ChemMedChem* **2006**, *1*, 939–950.
12. (a) Miller, J. F.; Andrews, C. W.; Brieger, M.; Furfine, E. S.; Hale, M. R.; Hanlon, M. H.; Hazen, R. J.; Kaldor, I.; et al. *Bioorg. Med. Chem. Lett.* **2006**, *16*, 1788–1794. (b) Ford, S. L.; Reddy, S. S.; Anderson, M. T.; Murray, S. C.; Fernandez, P.; Stein, D. S.; Johnson, M. A. *Antimicrob. Agents Chemother.* **2006**, *50*, 2201–2206.
13. GSK. "Corporate Responsibility Report 2006." www.gsk.com/responsibility/ downloads/CR-Report-2006.pdf, accessed April 2010.
14. Cihlar, T.; He, G.-X.; Liu, X.; Chen, J. M.; Hatada, M.; Swaminathan, S.; McDermott, M. J.; Yang, Z.-Y.; et al. *J. Mol. Biol.* **2006**, *363*, 635–647.
15. Callebaut, C.; Stray, K.; Tsai, L.; Xu, L.; Lee, W.; Cihlar, T. Paper presented at 16th International HIV Drug Resistance Workshop, June 12–16, 2007, Barbados.
16. Surleraux, D. L. N. G.; de Kock, B. A.; Verschueren, W. G.; Pille, G. M. E.; Maes, L. J. R.; Peeters, A.; Vendeville, S.; De Meyer, S.; Azijn, H.; Pauwels, R.; de Bethune, M-P.; King, N. M.; Jeyabalan, M. P.; Schiffer, C. A.; Wigernick, P. B. T. P. *J. Med. Chem.* **2005**, *48*, 1965–1973.
17. (a) Koh, Y.; Nakata, H.; Maeda, K.; Ogata, H.; Bilcer, G.; Devasamudram, T.; Kincaid, J. F.; Boross, P.; et al. *Antimicrob. Agents Chemother.* **2003**, *47*, 3123–3129. (b) De Meyer, S.; Azijn, H.; Surleraux, D.; Jochmans, D.; Tahri, A.;

Pauwels, R.; Wigerinck, P.; de Bethune, M.-P. *Antimicrob. Agents Chemother.* **2005**, *49*, 2314–2321.

18. Koh, Y.; Matsumi, S.; Das, D.; Amano, M.; Davis, D. A.; Li, J. F.; Leschenko, S.; Baldridge, A.; et al. *J. Biol. Chem.* **2007**, *282*, 28709–28720.

19. Fujimoto, H.; Higuchi, M.; Watanabe, H.; Koh, Y.; Ghosh, A. K.; et al. *Biol. Pharm. Bull.* **2009**, *32*, 1588–1593.

20. Rittweger, M.; Arasteh, K.; et al. *Clinical Pharmacokin.* **2007**, 46, 739–56.

21. Benet, L. Z.; Izumi, T.; Zhang, Y.; Silverman, J. A.; Wacher, V. J. *J. Controlled Release* **1999**, *62*, 25–31.

22. Cooper, C. L.; van Heeswijk, R. P.; Gallicano, K.; Cameron, D. W. *Clin. Infect. Dis.* **2003**, *36*, 1585–1592.

23. Vermeir, M.; Lachau-Durand, S; Mannens, G.; Cuyckens, F.; van Hoof, B.; Raoof, A. *Drug Metab. Dispos.* **2009**, *37*, 809–820.

24. Molina, J.-M.; Cohen, C.; Katlama, C.; Grinsztejn, B.; Timerman, A.; Pedro, R.; et al. Abstract P4, Paper presented at the 12th Annual BHIVA, March 29–April 1, 2006, Brighton, UK.

25. Hazen, R.; Harvey, R.; Ferris, R.; Craig, C.; Yates, P.; Griffin, P.; Miller, J.; Laldor, I.; et al. *Antimicrob. Agents Chemother.* **2007**, *51*, 3147–3154.

26. Label information for PREZISTA (Darunavir Ethanolate). www.accessdata.fda.gov/scripts/cder/drugsatfda/index.cfm?fuseaction=Search.Label_ApprovalHistory, accessed January 2010.

27. McKeage, K.; Perry, C. M.; Keam, S. J. *Drugs* **2009**, *69*, 477–503.

28. Ghosh, A. K.; Kincaid, J. F.; Walters, D. E.; Chen, Y.; Chaudhuri, N. C.; Thompson, W. J.; Culberson, C.; Fitzgerald, P. M. D.; et al. *J. Med. Chem.* **1996**, *39*, 3278–3290.

29. Ghosh, A. K.; Chen, Y. *Tetrahedron Lett.* **1995**, *36*, 505–508.

30. Ghosh, A. K.; Leshchenko, S.; Noetzel, M. *J. Org. Chem.* **2004**, *69*, 7822–7829.

31. Quaedflieg, P. J. L. M.; Kesteleyn, B. R. R.; Wigerinck, P. B. T. P.; Goyvaerts, N. M. F.; Vijn, R. J.; Liebregts, C. S. M.; Kooistra, J. H. M. H.; Cusan, C. *Org. Lett.* **2005**, *7*, 5917–5920.

32. (a) Yu, R. H.; Polniaszek, R. P.; Becker, M. W.; Cook, C. M.; Yu, L. H. L. *Org. Proc. Res. Dev.* **2007**, *11*, 972–980. (b) Black, D. M.; Davis, R.; Doan, B. D.; Lovelace, T. C.; Millar, A.; Toczko, J. F.; Xie, S. P. *Tetrahedron Asym.* **2008**, *19*, 2015–2019.

33. Lemaire, S. F. E.; Horvath, A.; Aelterman, W. A. A.; Rammeloo, T. J. L. PCT/EP2007/ 062119, Nov. 9, **2007**.

34. Ghosh, A. K; Kawahama, R. *Tetrahedron Lett.* **1999**, *40*, 1083–1086.

35. Doan, B. D.; Davis, R. D.; Lovelace, T. C. PCT/US02/29315 and WO 03/024974 A2, March 27, **2003**.

36. Ghosh, A. K.; Li, J.; Perali, R. S. *Synthesis* **2006**, 3015–3018.

37. Ghosh, A. K.; Thompson, W. J.; Holloway, M. K.; McKee, S. P.; Duong, T. T.; Lee, H. Y.; Munson, P. M.; Smith, A. M.; Wai, J. M.; Darke, P. L.; et al. *J. Med. Chem.* **1993**, *36*, 2300–2310.

38. Ghosh, A. K.; Bilcer, G.; Schlitz, G. *Synthesis* **2001**, 2203–2229.

II

CANCER

4

Decitabine (Dacogen): A DNA Methyltransferase Inhibitor for Cancer

Jennifer A. Van Camp

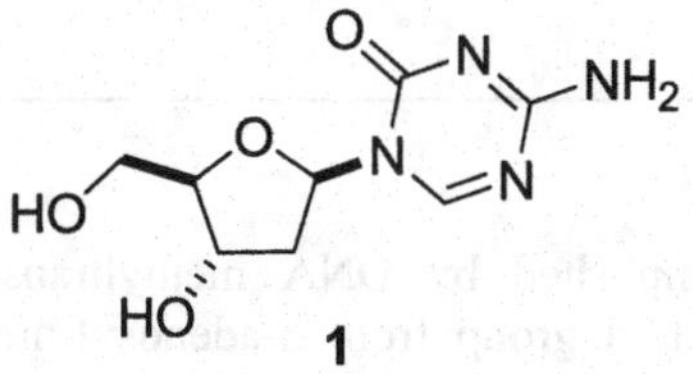

USAN: Decitabine
Trade name: Dacogen®
Pharmachemie / SuperGen
Launched: 2006

4.1   Background

Myelodysplastic syndrome (MDS) is the name given to a group of closely related diseases that arise in the bone marrow from hematopoietic stem cells. In patients with MDS, the bone marrow stops making healthy blood cells and instead produces abnormal, poorly functioning blood cells. MDS can affect red blood cells, white blood cells, or platelets, or any combination of the three. Each year between 15,000 and 20,000 new cases of MDS are diagnosed in the United States. Patients tend to be between the ages of 60 and 75 at the time of diagnosis. In the United States, the number of new cases of MDS is rising, both because older people make up a growing segment of the population and because people now live longer after treatment of their first cancer. Disease mechanisms can be divided into two main groups: those underlying the increased apoptosis of bone marrow progenitors, and those associated with progressive blast proliferation and transformation to acute myeloid leukemia.[1–3]

With the exception of intensive chemotherapy, including allogeneic stemcell transplantation, there are no curative therapeutic options for patients with MDS. The majority of MDS patients is elderly and cannot tolerate intensive chemotherapy. Thus there has been no effective therapy available. Recently, epigenetic therapy using DNA-methylation inhibitors has been extensively studied. Two drugs, decitabine (**1**) and azacitidine (**2**), were recently approved by the Food and Drug Administration (FDA) in the United States for the treatment of MDS. Zebularine (**3**) is another nucleoside analog that has weak hypomethylating activity. No clinical data are available on **3** at this time.

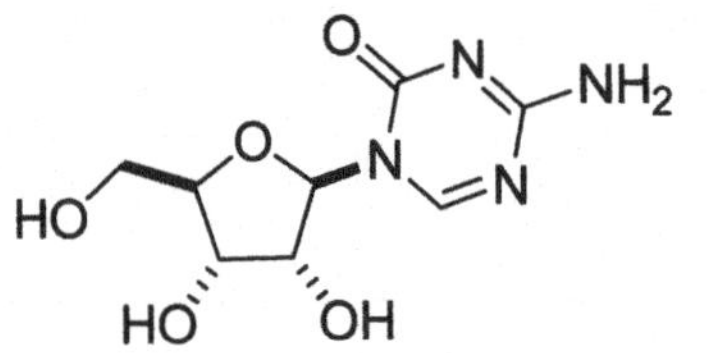

USAN: Azacitidine
Trade name: Vidaza®
Celgene
Launched: 2004

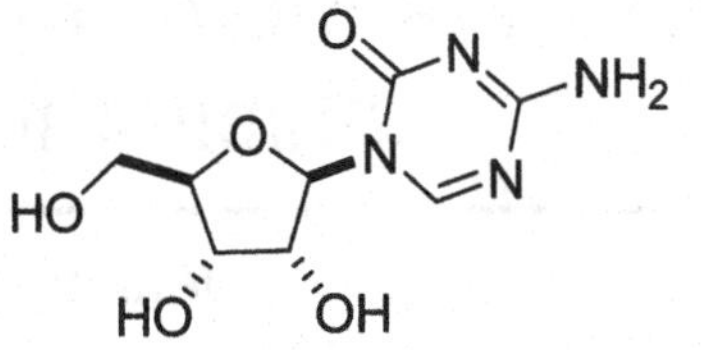

2

USAN: Zebularine
NSC-309132
University of Norh Carolina
Preclinical

3

DNA methylation is accomplished by DNA methyltransferases (DNMTs), which catalyze the transfer of a methyl group from S-adenosyl-methionine to the 5-position of cytosine in the CpG dinucleotide. DNA hypermethylation within the promoter region of tumor suppressor genes involved in cell proliferation and differentiation is a common phenomenon in numerous solid tumors.[4–5]

An important difference between the two compounds is that azacytidine (**2**) incorporates into RNA, while decitabine (**1**) acts on DNA. Within cells, decitabine (**1**) is phosphorylated by deoxycytidine kinase, and after conversion to decitabine triphosphate, it is incorporated into DNA in place of deoxycytidine triphosphate. Azacitidine (**2**) is phosphorylated by uridine-cytidine kinase and eventually incorporated into RNA, inhibiting the processing of ribosomal RNA and ultimately protein synthesis. Azacitidine (**2**) can also inhibit DNMTs when azacytidine diphophate is reduced to decitabine diphophate, which is further phosporylated by kinases to dcitabine triphosphate and incorporated into DNA. Because of the inefficiency of these extra steps, the hypomethylating potency of **2** is believed to be one fifth to one tenth that of **1**.[6–8]

DNA methyltransferase inhibitors represent promising new drugs for cancer therapies. Approval of the hypomethylating compounds **1** and **2** marked a recent, significant advance in the treatment of MDS. In this chapter, the pharmacological profile and syntheses of decitabine (1) are profiled.

4.2   Pharmacology

Decitabine (1) is a pyrimidine nucleoside analog. It differs from the natural nucleoside deoxycytidine by the presence of a nitrogen at position 5 of the cytosine ring. Decitabine is phosphorylated by deoxycytidine kinase to form an active intermediate, 5-aza-deoxycytidine triphosphate. This analog is then incorporated into DNA. The substitution at the 5'-position prevents methylation at this site, consequently affecting transcriptional activity of genes sensitive to silencing through methylation.

To assess the antineoplastic potency of decitabine, the effect of different concentrations of 1 on DNA replication in the human MDA-MB-231 breast and Calu-6 lung carcinoma cell lines was observed. The concentration of 1 that produced 50% of inhibition (IC_{50}) was 5 ng/mL for MDA-MB-231 cells and 50 ng/mL for Calu-6 cells as shown in Table 1.[9]

Table 1. Effect of 1 on the Inhibition of DNA Synthesis

Cell Line	IC_{50} (ng/mL)
MDA-MB-231	5
Calu-6	50

To assess the cytotoxic profile of 1, the effect of different concentrations of decitabine on colony formation in the human cell lines MDA-MGB-231, Calu-6, and DU-145 were observed (Table 2). The CC_{50} (50% cytotoxic concentration) values of 1 for MDA-MB-231, Calu-6, and DU-145 were 50 ng/mL.

Table 2. Effect of 1 on the loss of clonogenicity of cell lines

Cell Line	CC_{50} (ng/ml)
MDA-MB-231	50
Calu-6	50
DU-145	50

4.3   Structure–Activity Relationship (SAR)

Table 3. IC_{20} Concentrations for Various Cancer Cell Lines (µmol/L)

	TK6	Jurkat	KG-1	HCT116	Mean IC_{20}
Decitabine (1)	0.1	0.05	0.1	0.3	0.1
Azacytidine (2)	0.4	0.1	0.4	0.5	0.4
Zebularine (3)	20	10	20	5	10

In general, DNA methyltransferase in inhibitors can be divided into two subgroups: nucleoside and nonnucleoside inhibitors. The first group consists of decitabine (1), azacitidine (2), and zebularine (3). These three inhibitors were directly compared in four human cancer cell lines representing lymphoid, myeloid, or colorectal targets in a proliferation assay to determine their cellular IC_{20} concentrations. The results in Table 3 indicate the variability in IC_{20} concentrations among these compounds.[10]

While the potencies of 1 and 2 are quite similar, azacitidine (2) requires incorporation into DNA, which requires extensive modification of the compound through metabolic pathways. On the other hand, decitabine (1) becomes more directly incorporated into DNA and causes more efficient inhibition of DNA methyltransferases.[11]

4.4   Pharmacokinetics and Drug Metabolism

Decitabine (1) exhibits poor oral bioavailability and is consequently administered intravenously. It is rapidly eliminated in animals, with a half-life of 30 minutes in mice, 42 min in rabbits, and 75 min in dogs. For both rabbits and dogs, the apparent volume of distribution at steady state is 800 mL/kg. This is within the range of total body water and consistent with a weak protein-binding drug.[12]

In humans, the mean maximum plasma concentration (C_{max}) is 79 ng/mL and the time to achieve C_{max} was 2.67 h after a single 3-h infusion of decitabine (1) at 15 mg/m^2. At this dose, the volume of distribution at steady state is 148 mL/kg, and the total plasma clearance is 122 L/kg/m^2. The terminal half-life is approximately 35 min as decitabine (1) is primarily metabolized in the liver by cytidine deaminase to yield noncytotoxic 5-aza-2'-deoxyuridine. Urinary excretion of unchanged decitabine (1) is low (0.01–0.9% of total dose).[13]

4.5   Efficacy and Safety

Decitabine (1) is a hypomethylating agent that exhibits potent antileukemic activity by inducing epigenetic changes in chromatin or reactivating tumor suppressor genes. It is administered intravenously at 15 mg/m^2 over 3 h every 8 h for 3 consecutive days. The frequency of this administration is every 6 weeks.

In Phase II studies of agent 1 in patients with myelodysplastic syndrome (MDS), overall response rates of 42% to 54% were achieved, including complete responses in 20% to 28% of patients.[14,15]

In a randomized, nonblind, multicenter Phase III clinical trial, a total of 170 patients with MDS received either agent 1 or supportive care (RBC and/or platelet infusions and hematopoetic colony-stimulating factors). MDS patients treated with 1 achieved a significantly higher response rate (17%) compared with supportive care (0%). The median times to response and duration of response were 3.3 and 10.3 months. In addition, patients treated with decitabine (1) had a trend toward a longer median time to acute myelogenous leukemia (AML) progression or death compared to those receiving supportive care alone. Overall, decitabine (1) is well tolerated with a manageable toxicity profile. In the Phase III study, serious adverse events were experienced by 69% of decitabine (1) patients compared with 56% of patients receiving supportive care. The most common adverse event was myelosuppresion (including neutropenia, thorombocytopenia, and anemia). Gastrointestinal toxicities were generally mild and occurred infrequently (> 5% of patients).[16]

4.6 Syntheses

Decitabine (1) was first synthesized in 1964 by a multistep procedure described by Pliml and Sorm.[17] In this preliminary communication, 5-aza-2′-deoxycytidine was prepared via cyclization of the peracylglycosyl isocyanate 5, which was prepared from silver cyanate and chloride 4.[18,19] Addition of 2-methylisourea to 5 afforded the crystalline intermediate 6. Condensation of isobiuret derivative 6 with ethyl orthoformate resulted in the formation of crystalline triazine 7. Prolonged treatment with methanolic ammonia at room temperature afforded 1-(2-deoxy-D-ribofuranosyl)-5-azacytosine (8) as a mixture of α and β anomers. No yield was reported as the authors did not separate or distinguish between decitabine (1) and its α-anomer.

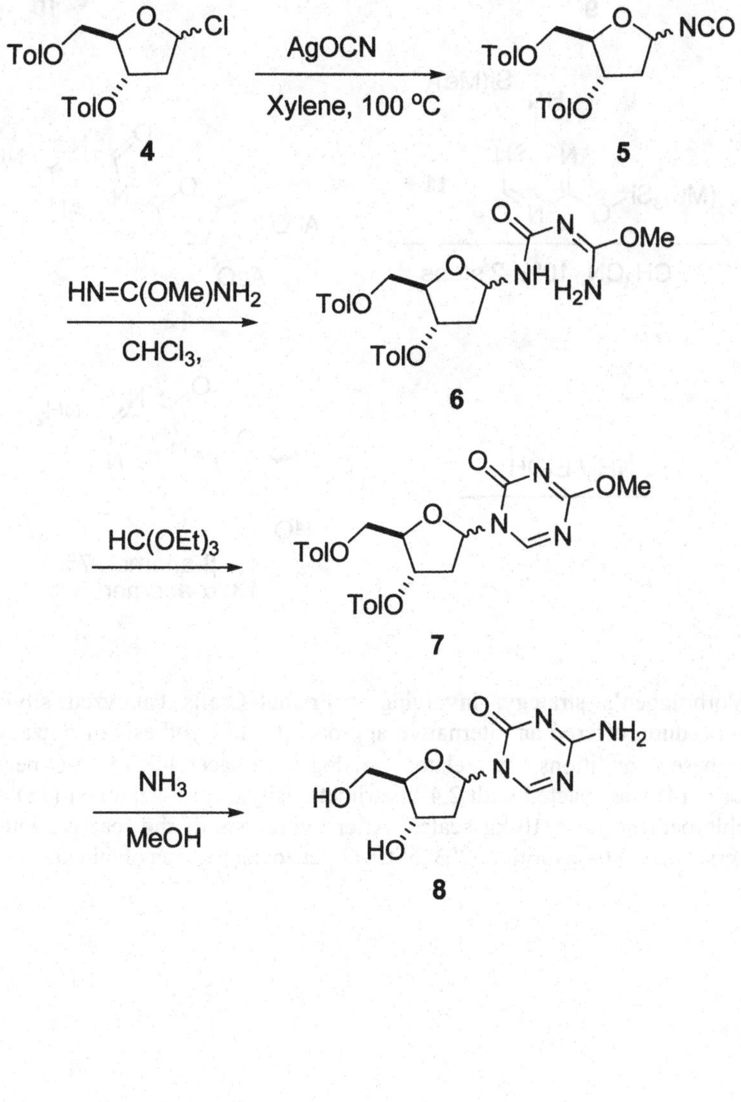

In 1970, **1** was prepared by direct glycosylation of the silylated 5-azacytosine with acylated 1-halo sugars, but the yields were very low.[20] In this procedure, 1,3,5-tri-*O*-acetyl-2-deoxy-D-ribofuranose (**9**) was converted into 3,5-di-*O*-acetyl-2-deoxy-D-ribofuranosyl chloride (**10**).[21] The trimethylsilyl derivative of 5-azacytosine (**11**),[22] prepared from 4-amino-6-pyrimidine by treatment with hexamethyldisilazane, was then allowed to react with intermediate **10** in acetonitrile over 7 days to give a mixture of anomers of 1-(3,5-di-*O*-acetyl-2-deoxy-D-ribofuranosyl)-5-azacytosine (**12**) in 10% overall yield. The anomeric mixture was treated with ethanolic ammonia to remove the acetyl groups. The resulting α and β anomers were separated by a combination of fractional crystallization and preparative layer chromatography on silica gel to give pure decitabine (**1**) and its α-anomer **13** in 7% and 52% yield, respectively.

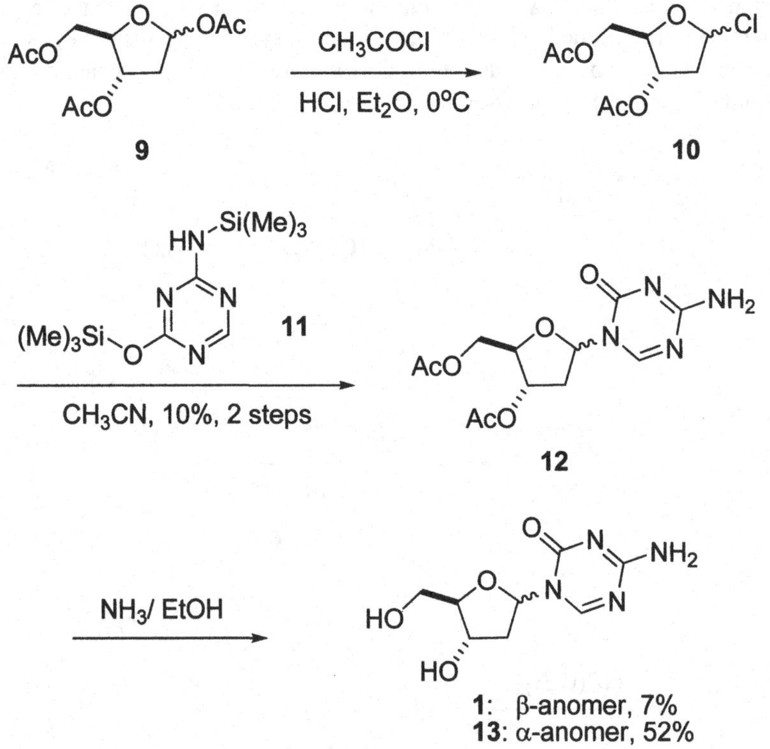

Vorbruggen's strategy, involving a Friedel–Crafts catalyzed silyl Hilbert–Johnson procedure, offered an alternative approach to the synthesis of 5-azacytidines.[23] Utilizing these conditions, a related analog, 1-*O*-acetyl-2,3,5-tri-*O*-benzoyl-β-D-ribofuranose (**14**) was reacted with 2,4-bis(trimethylsilyloxy)-6-azauracil (**15**) and SnCl₄ in 1,2-dichloroethane on a 10-kg scale. After hydrolysis of the reactive intermediate, 93% of recrystallized 6-azauridine-2′,3′,5′-tri-*O*-benzoate(**16**) was obtained.

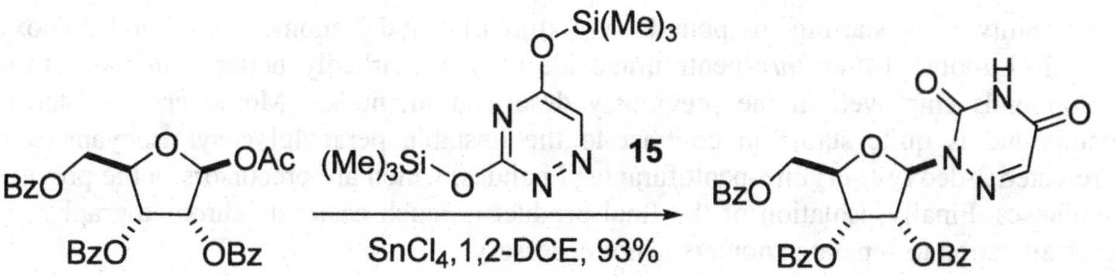

In the above example, only the β-anomer of the N_1-nucleoside could be detected and isolated in the presence of a 2α-acyloxy substituent in the sugar moiety. However, in the case of acylated 2-deoxy-D-ribose, as well as benzylated D-arabinose derivatives without a 2-participating group, both anomeric nucleosides were formed. The contributing factors in this regulation seem to be the steric and the electronic effects exerted by the sugar protecting groups on the C-1 position.[24] Aroyl groups such as benzoyl, nitrobenzoyl, chlorobenzoyl, and especially toluoyl have been successfully used to date. The latter was found to be the protecting group of choice as it allows for easy separation of the β-anomer during crystallization.[23] This effect is evident in the synthesis of 1-(2-deoxy-3,5-di-O-p-tolyl-β-D-ribofuanosyl)-4-amino-1,2-dihydro-1,3,5-triazin-2-one (18), the protected analog of 1. Reaction of the silylated 5-azacytosine 11 with the crystalline 1α-chloro-2-deoxy-3,5-bis(p-toluoyl-α-D-ribofuanosyl chloride (17)[25] afforded 77% yield of the anomeric mixture of nucleosides from which 42% crystalline β anomer 18 could be readily separated by crystallization from the α anomer 19. These conditions consistently provided a ratio of anomers α/β ≈ 1.

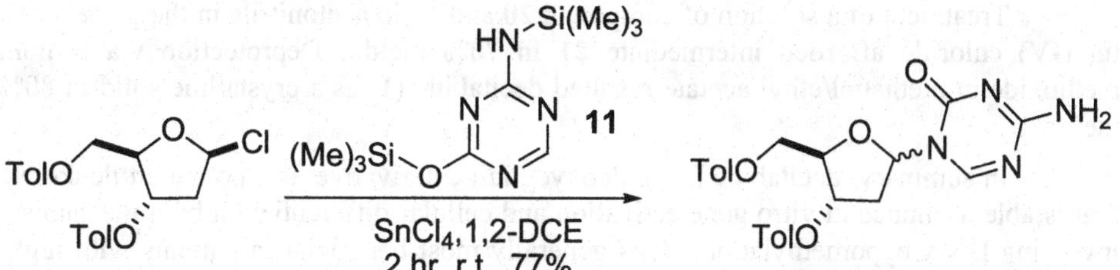

The synthesis of 1, as reported by Piskala et al., optimized this direct glycosylation procedure to afford predominantly the desired β anomer.[26] It was found that if the glycosylation procedure was carried out in acetonitrile and in the presence of mercuric oxide and bromide, the total yield increased a little and the β anomer predominated the mixture.

The manufacturing method of decitabine (1) was patented by Piskala et al.[27] The process synthesis affords the required product selectively in high yield and purity. The availability of the starting compound—a mixture of α and β anomers of methyl-2-deoxy-3,5-di-*O*-*p*-toluoyl-D-*erythro*-pentofuranoside (**20**) is markedly better than the starting compounds employed in the previously described methods. Moreover, this starting compound is quite stable in contrast to the unstable peracylglycosyl isocyanates or protected 2-deoxy-D-erythro-pentofuranosyl halides, which are precursors in the previous syntheses. Finally, isolation of the final product is much easier as chromatography and crystallization to separate anomers is not necessary.

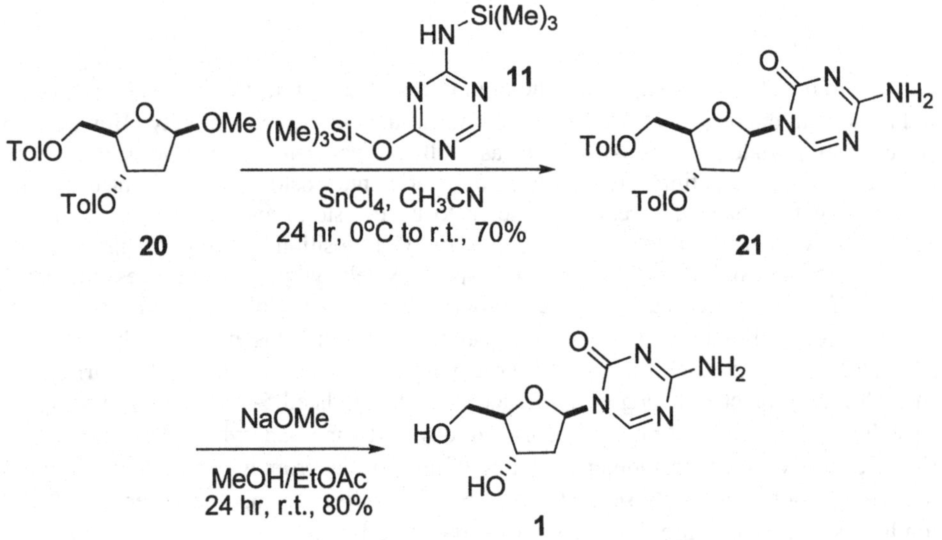

Treatment of a solution of compound **20** and **11** in acetonitrile in the presence of tin (IV) chloride afforded intermediate **21** in 70% yield. Deprotection via sodium methoxide in methanol/ethyl acetate resulted decitabine (**1**) as a crystalline solid in 80% yield.

In summary, decitabine (**1**), a deoxycytidine derivative, is a potent antileukemic that is able to induce in vitro gene activation and cellular differentiation by a mechanism involving DNA hypomethylation. It is generally most beneficial in patients with high-risk MDS or in MDS patients with intermediate-2 or high-risk features. Its process synthesis and academic synthetic approaches have been summarized in this chapter.

4.7   References

1.   Steensma, D. P.; Bennett, J. M. *Mayo Clin Proc.* **2006**, *81*, 104–130.
2.   Lindberg, E. H. *Curr Drug Targets* **2005**, 6, 713–725.
3.   Aul, C.; Giagounidis, A.; Germing, U. *Int. J.Hematol.* **2001**, *73*, 405–410.
4.   Herman, J. G.; Baylin, S. B. *N. Engl. J. Med.* **2003**, *349*, 2042–2054.
5.   Herman, J. G.; Jen, J.; Merlo, A.; Baylin, S. B. *Cancer Res.* **1996**, *56*, 722–727.
6.   Christman, J. K. *Oncogene* **2002**, *21*, 5483–5495.

7.   Li, L. H.; Lin, E. J.; Buskirk, H. H.; Reineke, L. M. *Cancer Res.* **1970**, *30*, 2760–2769.

8.   Jones, P. A.; Taylor, S. M. *Cell* **1980**, *20*, 85–93.

9.   Hurtubise, A.; Momparler, R. L. *Anti-Cancer Drugs* **2004**, *15*, 161–167.

10.   Stresemann, C.; Brueckner, B.; Musch, T.; Stopper, H. Lyko, F. *Cancer Res.* **2006**, *66*, 2794–2800.

11.   Momparler, R. L.; Momparler, L. F.; Samson, J. *Leuk. Res.* **1984**, *8*, 1043–1049.

12.   Chabot, G. G.; Rivard, G. E.; Momparler, R. L. *Cancer Res.* **1983**, *43*, 592–597.

13.   Cashen, A. F.; Shah, A.; Helget, A. *Blood* **2005**, *106*, 527–528.

14.   Wijermans, P. W.; Krulder, J. W. M.; Huijgens, P.C.; Neve, P. *Leukemia* **1997**, *11*, 1–5.

15.   Wijermans, P. W.; Lubbert, M.; Verhoef, G.; Bosly, A.; Ravoet, C.; Andre, M.; Ferrant, A. *J. Clin. Oncol.* **2000**, *18*, 956–962.

16.   Kantarjian, H.; Issa, J.-P. J.; Rosenfeld, C. S.; Bennett, J. M.; Albitar, M.; DiPersio, J.; Klimek, V.; Slack, J.; deCastro, C.; Ravandi, F.; Helmer III, R.; Shen, L.; Nimer, S. D.; Leavitt, R.; Raza, A.; Saba, H. *Cancer* **2006**, *106*, 1794–1803.

17.   Pliml, J.; Sorm, F. *Collect. Czeh. Chem. Commun.* **1964**, *29*, 2576–2577.

18.   Piskala, A. ; Sorm, F. *Collect. Czeh. Chem. Commun.* **1964**, *29*, 2060–2076.

19.   Fischer, E. *Chem. Ber.* **1914**, *47*, 1377–1393.

20.   Winkley, M. W.; Robins, R. K. *J. Org. Chem.* **1970**, *35*, 491–495.

21.   Robins, R. K.; Robins, M. J. *J. Amer. Chem. Soc.* **1965**, *87*, 4934–4940.

22.   Winkley, M. W.; Robins, R. K. *J. Org. Chem.* **1969**, *34*, 431–434.

23.   Vorbruggen, H.; Niedballa, U. *J. Org. Chem.* **1974**, *39*, 3654–3660.

24.   Wierenga, W.; Skulnick, H. I. *Carbohydr. Res.* **1981**, *90*, 41–52.

25.   Hoffer, M. *Chem. Ber.* **1960**, *93*, 2777–2781.

26.   Piskala, A.; Synackova, M.; Tomankova, H.; Fiedler, P.; Zizkovsky, V. *Nucleic Acids Res.* **1978**, *54*, 109–114.

27.   Piskala, A.; Holy, A.; Otmar, M. WO 08101445 (2008).

5

Capecitabine (Xeloda): An Oral Chemotherapy Agent

R. Jason Herr

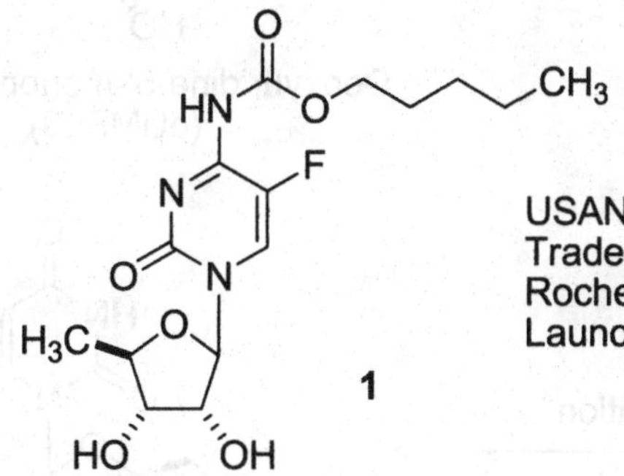

USAN: Capecitabine
Trade name: Xeloda
Roche Laboratories, Inc.
Launched: 1998

1

5.1   Background

Most anticancer agents in clinical use today are molecules that disturb regulation of cell growth and replication, directly interfere with DNA function or block DNA synthesis indirectly through inhibition of nucleic acid precursor biosynthesis. Like all human cells, cancer cells have the capability to reclaim cytoplasmic purines and pyrimidines for the synthesis of deoxyribonucleotides, which in turn are used to synthesize DNA as part of the cell replication process. If the selective delivery of altered nucleotide precursor analogues into the cancer cell matrix can be achieved, the new amine base molecules may be incorporated into the growing DNA chain. When the resulting mutant is not compatible with the further cellular machinery, the ensuing inhibition of replication is recognized by DNA repair enzymes, setting up a nonproductive cycle that ultimately results in cell death. This strategy of "antimetabolite" molecules as chemotherapeutic agents has proven to be a successful means for the development of cancer drugs approved for clinical use.[1] In total, 14 purine and pyrimidine antimetabolite compounds have been approved by the FDA for the treatment of cancers, which account for nearly 20% of all drugs that are used for oncology therapy.[2]

5-Fluorouracil (5-FU, **5**) is one of the first antimetabolite oncology drugs that was rationally designed founded on available biochemical data. It was known that the biosynthesis of thymidine from uracil (**2**) is critical to DNA replication and repair and relies on the function of thymidylate synthase (TS). This enzyme mediates the replacement of the C-5 proton of deoxyuridine monophosphate (dUMP, **3**) with a methyl group obtained from methylene tetrahydrofolate to make deoxythymidine monophosphate (dTMP, **4**), a precursor of thymidylate. Thus Heidelberger and co-workers proposed that 5-fluorouracil (**5**), in which the C-5 carbon of uracil was replaced with fluorine, should inhibit the function of thymidylate synthase due to the inability of the enzyme to remove the 5-fluorine atom.[3] As a result, 5-FU should selectively disrupt DNA replication in tumor cells due to selective metabolism to F-dUMP (**6**), which would in turn unbalance the cell growth and ultimately lead to programmed cell death processes that kill the tumors.

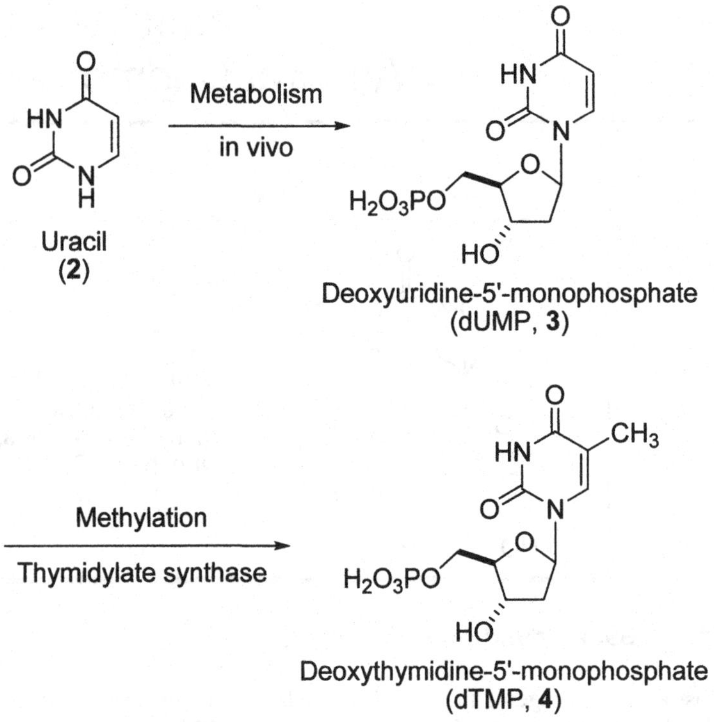

Uracil
(**2**)

Metabolism
in vivo

Deoxyuridine-5'-monophosphate
(dUMP, **3**)

Methylation

Thymidylate synthase

Deoxythymidine-5'-monophosphate
(dTMP, **4**)

5-Fluorouracil
(5-FU, **5**)

5-Fluoro-2'-deoxyuridine monophosphate
(FdUMP, **6**)

Further studies have validated this hypothesis, in part,[4] and ultimately this inventive premise was borne out in clinical practice. As a result, 5-FU (**5**) was eventually approved for treatment of solid tumors, such as breast, colorectal, and gastric cancers. Marketed as Adrucil when administered intravenously, 5-FU can be used either as monotherapy or combination therapy with various cytotoxic drugs and biochemical modulators, such as leucovorin and methotrexate.[5] Because 5-fluorouracil is not orally bioavailable, it must be administered by continuous infusion to optimize its efficacy due to its short half-life in plasma. In addition, 5-FU has poor selectivity toward tumors in vivo, and its distribution into tissues such as bone marrow, the gastrointestinal tract, the liver and skin causes high incidences of toxicity. In addition, in spite of its limited lipid solubility, 5-fluorouracil diffuses readily across the blood–brain barrier into cerebrospinal fluid and brain tissue.[1,5]

To address the efficacy and side effect profile of 5-fluorouracil as well as to make inroads into oral dosing regimens, several research programs introduced prodrug strategies seeking to selectively deliver 5-fluorouracil into tumor tissues. In this approach, the cytotoxic agents are physicochemically masked through the attachment of chemical functionality that is removed by endogenous enzymes preferentially contained in neoplasm. In this way, the active drug substance is released within the cancer, lowering the toxicity and side effect liabilities of interactions with healthy cells. Several 5-FU prodrugs, including carmofur (**7**), floxuridine (**8**), and doxifuridine (**9**), have been developed to improve such issues, have been approved by the FDA for clinical use, and are even approved for oral dosing in some countries. Floxuridine (**8**), for example, exhibits good tumor selective action due to its conversion to 5-FU by pyrimidine nucleoside phosphorylases (PyNPase), enzymes preferentially expressed in many tumor cell types. Many of these unmasking enzymes, however, are also present in the intestinal tract, generating 5-FU from prodrugs in healthy tissues and, as a result, can cause dose-limiting toxicity when given orally at high doses for long durations. While early prodrug approaches have had limited success for delivering cytotoxic agents preferentially to neoplasm, a great need still existed for the improvement of flurouracil prodrugs for greater tumor selectivity, efficacy and safety profiles.[1,2]

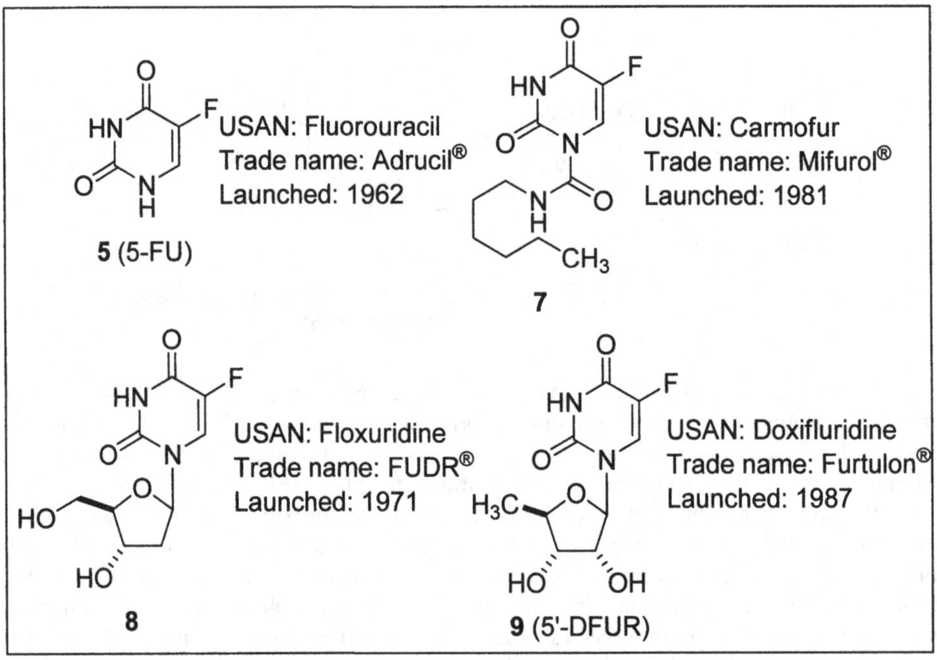

USAN: Fluorouracil
Trade name: Adrucil®
Launched: 1962

5 (5-FU)

USAN: Carmofur
Trade name: Mifurol®
Launched: 1981

7

USAN: Floxuridine
Trade name: FUDR®
Launched: 1971

8

USAN: Doxifluridine
Trade name: Furtulon®
Launched: 1987

9 (5'-DFUR)

Capecitabine (**1**) represents an N^4-carbamate prodrug of a pyrimidine nucleoside structure that was created to improve the selectivity and bioavailability of its parent compound, 5-fluorouracil. This compound uses a multilayer of prodrug strategies that not only avoids the relatively quick systemic clearance of the parent compound, thus enhancing the bioavailability (and therefore increased therapeutic effect), but has the advantage over intravenous 5-FU in that it can be administered orally. It is currently approved by the FDA under the trade name of Xeloda for use in the treatment of cancer of the colon after surgery, metastatic colorectal cancer, metastatic breast cancer (along with docetaxel), and metastatic breast cancer that is resistant to standard chemotherapy with paclitaxel or anthracycline-containing cytotoxic regimens.[6–8] In this chapter, the pharmacological profile and syntheses of capecitabine are profiled in detail.

5.2 Pharmacology

Doxifluridine (5'-deoxy-5-fluorouridine, 5'-DFUR, **9**) was designed and developed by researchers at Roche Laboratories as a prodrug of antimetabolite 5-fluorouracil (**5**) with the specific purpose of delivering the active component selectively to tumor cells in vivo.[9,10] In this strategy, 5-FU was masked as a pyrimidine nucleoside that allowed the inactive drug substance **9** to be administered orally to human subjects, after which the anabolizing enzyme thymidine phosphorylase (dThdPase) selectively converted the inactive parent form into the active cytotoxic drug substance **5**. Key to the tissue selectivity of this tactic is that thymidine phosphorylase is highly overexpressed in solid tumors versus the surrounding healthy cells, thus generating 5-FU largely within neoplasm. Although ultimately approved for the treatment of certain cancers, doxifluridine is constrained in its use owing to its dose-limiting intestinal toxicity (diarrhea) when orally given at high doses for long periods. This is a result of the existence of thymidine phosphorylase in the human intestinal tract, where the enzymes generate significant amounts of 5-FU in the healthy tissue. To improve the efficacy and side effect profile of 5′-DFUR (**9**), researchers at Roche Laboratories took the antimetabolite prodrug approach several steps further, relying on a cascade of biochemical transformations that ultimately generates 5-fluorouracil from the 5′-deoxycytidine-type prodrug precursor capecitabine (**1**).[9–14]

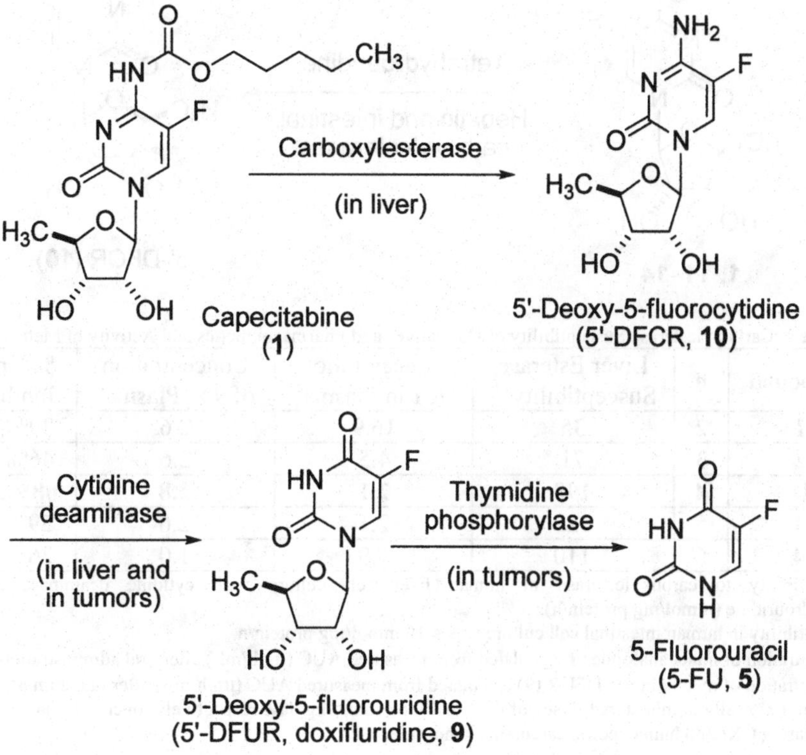

The in vivo metabolism of capecitabine (**1**) to the active tumor cytotoxic substance 5-fluorouracil (**5**) is now fairly well understood. When capecitabine is administered orally it is delivered to the small intestine, where it is not a substrate for thymidine phosphorylase in intestinal tissue, and so passes through the intestinal mucosa as an intact molecule and into the bloodstream. When **1** reaches the liver, the carbamate moiety is hydrolyzed through the action of carboxylesterase enzymes, liberating 5′-deoxy-5-fluorocytidine (5′-DFCR, **10**). DFUR is partially stable in systemic circulation, but eventually diffuses into tumor cell tissue where it is transformed into 5′-deoxy-5-fluorouridine (5′-DFUR, **9**) by cytidine deaminase, an enzyme present in high concentrations in various types of human cancers compared to adjacent healthy cells (although it is present in significantly lower levels in the liver). Within the tumor, 5-

DFUR is further converted into 5-fluorouracil (5-FU, **5**) by thymidine phosphorylase, thus providing selective delivery of this cytotoxic agent.[10,12–14]

5.3   Structure–Activity Relationship (SAR)

1, 11–14   5'-DFCR (**10**)

Table 1. Carboxylesterase Susceptibility of Derivatives and Pharmacokinetics and Activity of Metabolites.

Compound	n	Liver Esterase Susceptibility[a,b]	Concentration of **1** in Plasma[c]	Concentration of **9** in Plasma[d]	% Growth Inhibition[e]
11	2	35	16.9	2.6	39%, 75%
12	3	71	4.5	2.6	46%, 84%
1	4	190	2.1	2.8	68%, 86%
13	5	220	2.1	2.6	59%, 79%
14	6	110	3.9	1.0	26%, 84%

[a]Susceptibility to carboxylesterase in human liver cell culture with cytidine deaminase inhibitor tetrahydrouridine (mmol/mg protein/h).
[b]Susceptibility in human intestinal cell culture was < 10 mmol/mg protein/h.
[c]Concentration of intact analogues **1** calculated from measured AUC (μg h/mL) after oral administration.
[d]Concentration of metabolite 5'-DFUR (**9**) calculated from measured AUC (μg h/mL) after oral administration.
[e]Measured at orally administered doses of 0.13 and 0.67 mmol/kg/day, respectively, over a 21-day interval in mice bearing CXF280 human colon carcinoma xenografts.

To develop an effective approach for the oral delivery of a fluorocytidine prodrug that would pass unchanged through intestinal tissue, it was important to identify N^4-acyl functionality that would be selectively hydrolyzed by enzymes specific to the liver, but not susceptible to action by enzymes in the intestine or in other organs. Screening of various N^4-substituted 5'-DFCR derivatives eventually led to the identification of carbamate derivatives **1** and **11–14**, which showed higher specificity for human liver enzyme ester hydrolysis, but were not changed by intestinal enzymes (Table 1).[9] This extremely high specificity for human liver intestinal enzyme suggested that these compounds would undergo efficient biotransformation to 5'-DFCR only after passing intact through the intestinal mucosa when given orally, thus limiting toxicity associated with 5-fluorouracil generated in the intestine. The susceptibility of 5'-DFCR N^4-carbamate derivatives **1** and **11–14** to enzymes converting them to 5'-DFCR was measured by incubating the compounds with extracts from human liver and intestinal tissue samples in the presence of the cytidine deaminase inhibitor tetrahydrouridine. Mixtures were cultured at 37°C for 60 min, after which the concentrations of the parent compound and 5′-DFCR (**10**) were measured by HPLC analysis. The optimal carbamate chain lengths for the human liver enzyme were found to be *n*-pentyl (**1**, capecitabine) and *n*-hexyl (**13**), and in all cases the analogues were resistant to action by intestinal carboxylesterases.

At the same time, carbamate compounds **1** and **11–14** were orally administered to cynomolgous monkeys to determine plasma concentrations of the intact molecules and the active metabolite 5′-DFUR (**9**) generated in vivo (Table 1).[9] It was found that the measured AUC concentrations of **9** generated from carbamate precursors and their susceptibility to liver carboxylesterase both correlated to the carbamate alkyl chain lengths. Capecitabine (**1**), which exhibited one of the highest propensities to hydrolysis by hepatic carboxylesterase (but not to intestinal enzymes), resulted in one of the best pharmacokinetic profiles, as described by the lowest amount of parent drug in systemic circulation, but with the highest concentration of 5′-DFUR.

In parallel to these studies, carbamate compounds **1** and **11–14** were assessed for their antitumor efficacy in mouse cancer xenograft models.[15] When implanted human colon cancer CXF280 xenografts were grown within mice for fourteen days, doses of test compounds were administered orally. After a three-week regimen, excised tumor volumes were measured and the percent inhibition of tumor growth was calculated. From this investigation, capecitabine (**1**) was found to be the most effective treatment, and was furthermore found not to cause intestinal toxicity.[16] All of these preclinical observations contributed to the selection of capecitabine as a candidate for further development.

5.4   Pharmacokinetics and Efficacy

Pharmacokinetic parameters of the 5-fluorouracil prodrug capecitabine (**1**) were measured in preclinical experiments and were compared to results with the direct administration of 5-FU (**5**) itself. In mice bearing the HCT116 human colon cancer xenograft, intraperitoneal administration of 5-fluorouracil at the maximum tolerated dose produced approximately uniform concentrations of 5-FU in tumors as well as in plasma and muscle tissues. However, when capecitabine was administered orally at equitoxic doses, 5-fluorouracil concentrations were measured to be considerably higher within tumor xenografts relative to other tissues (AUCs for released 5-FU in neoplasm were 114- and 209-fold higher than in plasma and muscle, respectively).[17,18] The pharmacokinetics of capecitabine was evaluated in over 200 human cancer patients over a daily dosage range between 500 and 3,500 mg/m^2. Over this range, the pharmacokinetic behaviors of capecitabine and its first metabolite 5′-DFCR (**10**) were dose proportional and did not change over time. On the other hand, the increases in the concentrations of 5′-DFUR (**9**) and the cytotoxic end product 5-FU (**5**) were greater than proportional to the dose increase, and the AUC of 5-FU was 34% higher after two weeks than on the first day of administration. The interpatient variability in the measured C_{max} and AUC of 5-FU was greater than 85%.

In human patients, capecitabine has almost 100% oral bioavailability and reaches peak blood levels in 1.5 h, with levels of 5-fluorouracil reaching a peak at abut 2 h. Capecitabine is less than 60% plasma protein bound (primarily to human serum albumin), and its elimination half-life in plasma is about 45 min. Capecitabine and its metabolites are 95% excreted in urine, with 3% of the administered dose representing the unchanged drug. In vitro enzymatic studies with human liver microsomes have indicated that capecitabine and its metabolites have no inhibitory effects on cytochrome P450 isozymes.[14,16]

The recommended dose following human Phase I/II studies was determined as 1,250 mg/m^2 administered orally twice daily (morning and evening) for 14 days, followed by a seven-day rest period, given as three-week cycles. For example, a patient with a median body surface area of 1.75 m^2 (patient ranges calculated from 1.25 to 2.18) would require a total daily dose of 4,300 mg, translating to taking one 150-mg tablet and four 500-mg tablets both in the morning and in the evening. The most common adverse events reported for capecitabine therapy were diarrhea and nausea as well as hand and foot syndrome (numbness and/or painful swelling on the palms of the hands or soles of the feet).[13,14]

Two Phase III clinical studies of orally administered capecitabine in over 1,200 patients with untreated metastatic colorectal cancer demonstrated at least equal efficacy and improved tolerability versus the Mayo Clinic regimen of intravenous 5-fluorouracil/leucovorin administration. The overall response rate for patients taking capecitabine orally was 21%, versus 14% for the intravenous 5-FU/leucovorin regimen. A median 53-month follow-up revealed a three-year disease-free survival rate of 66% for capecitabine versus 63% for 5-FU/leucovorin patients. International Phase II trials also demonstrated therapeutic benefits of capecitabine monotherapy for women with metastatic breast cancer that was either resistant to both paclitaxel and anthracycline therapy. Orally administered at the twice-daily 1,250 mg/m^2 regimen (cycles of two weeks of therapy followed by a week of rest), the tumor response rate was in the range of 20–25%. In addition, combination of capecitabine with a taxane yielded a unique survival benefit compared to the previous standard of taxane monotherapy for anthracycline-resistant breast cancer.[13,14]

5.5   Syntheses

The de novo discovery synthesis of capecitabine (1) was reported by the Nippon Roche Research Center scientists[9,19] and was followed up with a preparation invented by a team at the Hoffmann-La Roche laboratories in New Jersey for the conversion to 1 from 5'-DFCR (10).[20] In the first route, 5-fluorocytosine (15) was mono-silated using one equivalent of hexamethyldisilazane in toluene at 100 °C followed by stannic chloride-catalyzed glycosidation with known 5-deoxy-1,2,3-tri-O-acetyl-β-D-ribofuranoside (17) in ice-cooled methylene chloride. While this procedure provided the 2',3'-di-O-acetyl 5-fluorocytidine 18 in 76% yield on a 25-g scale, an alternative method was also devised using in situ-generated trimethylsilyl iodide in acetonitrile at 0°C to provide a 49% yield of 18 on smaller scale. Acylation of the N^4-amino group of the bis-protected 5'-DFCR derivative was accomplished by the slow addition of two equivalents of n-pentyl chloroformate to a solution of 18 in a mixture of pyridine and methylene chloride at –20 °C, followed by a quench with methanol at room temperature to provide the penultimate intermediate 19 on 800-g scale. The yield of intermediate 19 was assumed to be quantitative and was subjected to the final deprotection step, with only a trituration to remove pyridine hydrochloride. Selective hydrolysis of the two ester moieties of **19** in methanol at –20°C with aqueous sodium hydroxide, followed by acidification to pH 5 with concentrated HCl smoothly provided crude capecitabine (**1**). Precipitation of crude **1** from ethyl acetate with *n*-hexane (2:5 ratio) provided 170 g pure 5'-deoxy-5-fluoro-*N*^{4}-(pentyloxycarbonyl)cytidine (**1**) as colorless crystals in 75% yield on a 300-g scale.

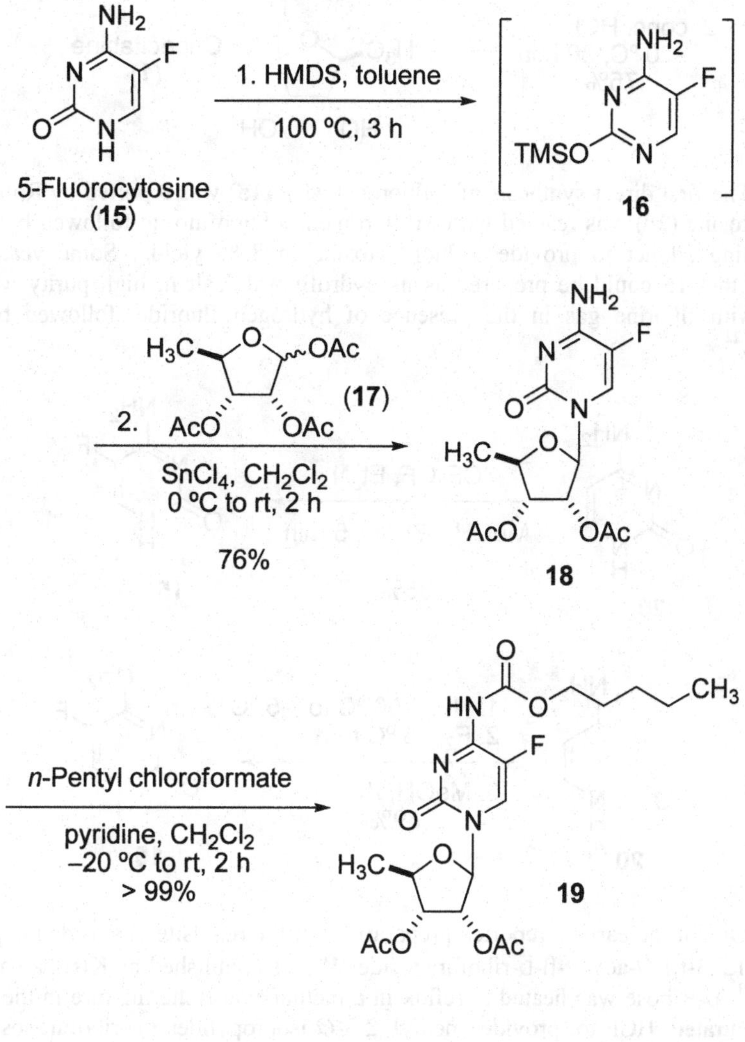

5-Fluorocytosine
(15)

1. HMDS, toluene
100 °C, 3 h

16

2.

(17)

SnCl$_4$, CH$_2$Cl$_2$
0 °C to rt, 2 h

76%

18

n-Pentyl chloroformate

pyridine, CH$_2$Cl$_2$
–20 °C to rt, 2 h
> 99%

19

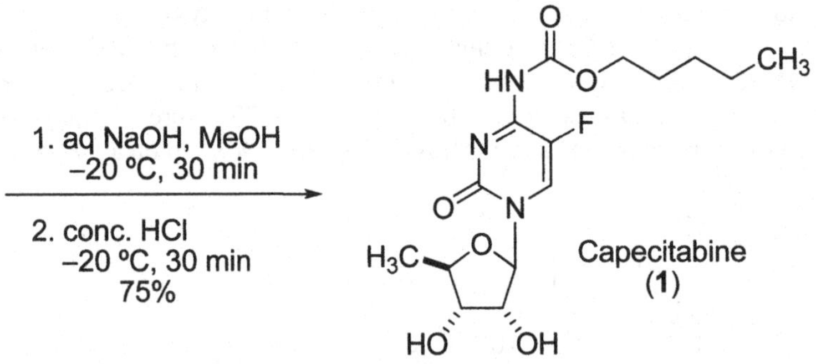
1. aq NaOH, MeOH
−20 °C, 30 min
2. conc. HCl
−20 °C, 30 min
75%
O
HN
O
CH3
F
N
O
N
H3C
O
HO
OH

Capecitabine
(1)

The first direct synthesis of 5-fluorocytosine (**15**) was reported by Robins and Naik,[21] in which cytosine (**20**) was reacted with trifluoromethyl hypofluorite followed by decomposition of the resulting adduct to provide 5-fluorocytosine in 85% yield. Some years later, Takahara disclosed that **15** could be prepared as its hydrofluoride salt in high purity when cytosine was reacted with fluorine gas in the presence of hydrogen fluoride, followed by treatment with methanol.[22]

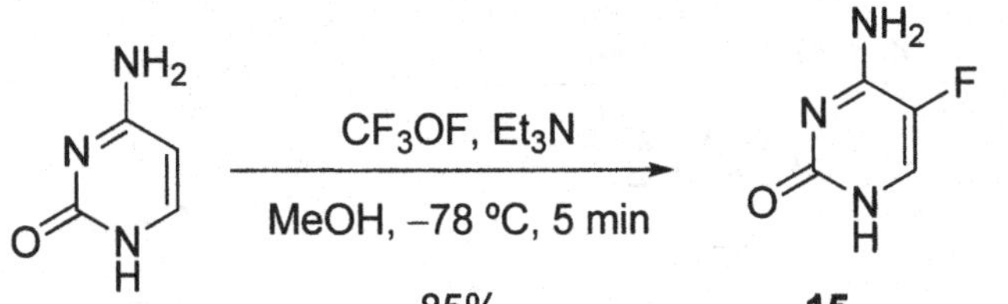
NH2
N
O
N
H
CF3OF, Et3N
MeOH, −78 °C, 5 min
85%
NH2
F
N
O
N
H

20 **15**

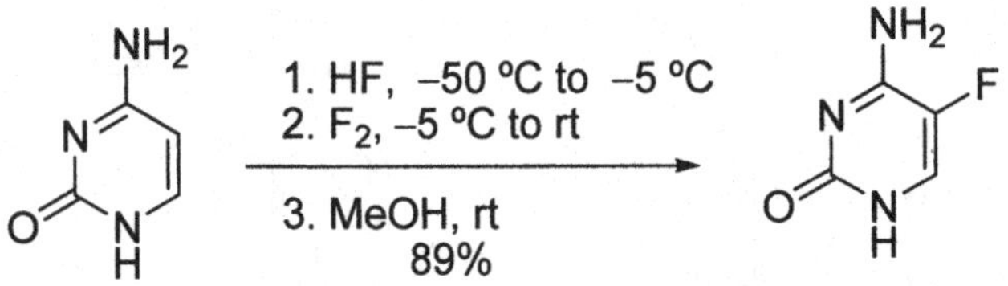
NH2
N
O
N
H
1. HF, −50 °C to −5 °C
2. F2, −5 °C to rt
3. MeOH, rt
89%
NH2
F
N
O
N
H

20 **15**

One of the earliest reported preparations of the requisite glycosidation precursor 5-deoxy-1,2,3-tri-*O*-acetyl-β-D-ribofuranoside (**17**) was published by Kissman and Baker in 1957.[23] D-Ribose was heated at reflux in a methanol/acetone mixture in the presence of concentrated HCl to provide methyl 2,3-*O*-isopropylidene-D-ribofuranoside (**21**), which was in turn converted to the corresponding 5-*O*-mesyl ribofuranoside **22** with methanesulfonyl chloride in pyridine in 63% yield. The sulfonate moiety of **22** was then displaced with sodium iodide in refluxing DMF to provide 5-deoxy-5-iodo derivative **23** in 76% yield on a multigram scale. Reductive dehalogenation of **23** was accomplished under heterogeneous catalytic hydrogenation conditions to provide the reduced 2,3-*O*-protected intermediate **24** in 56% yield, which was subjected to hydrolysis conditions in refluxing aqueous HCl to provide 5-deoxy-D-ribose (**25**) in almost quantitative yield. The crude sugar **25** was not purified, but directly subjected to acetic anhydride in pyridine at room temperature for 3 days to provide the *tris*-protected D-ribose **17** in 64% yield.

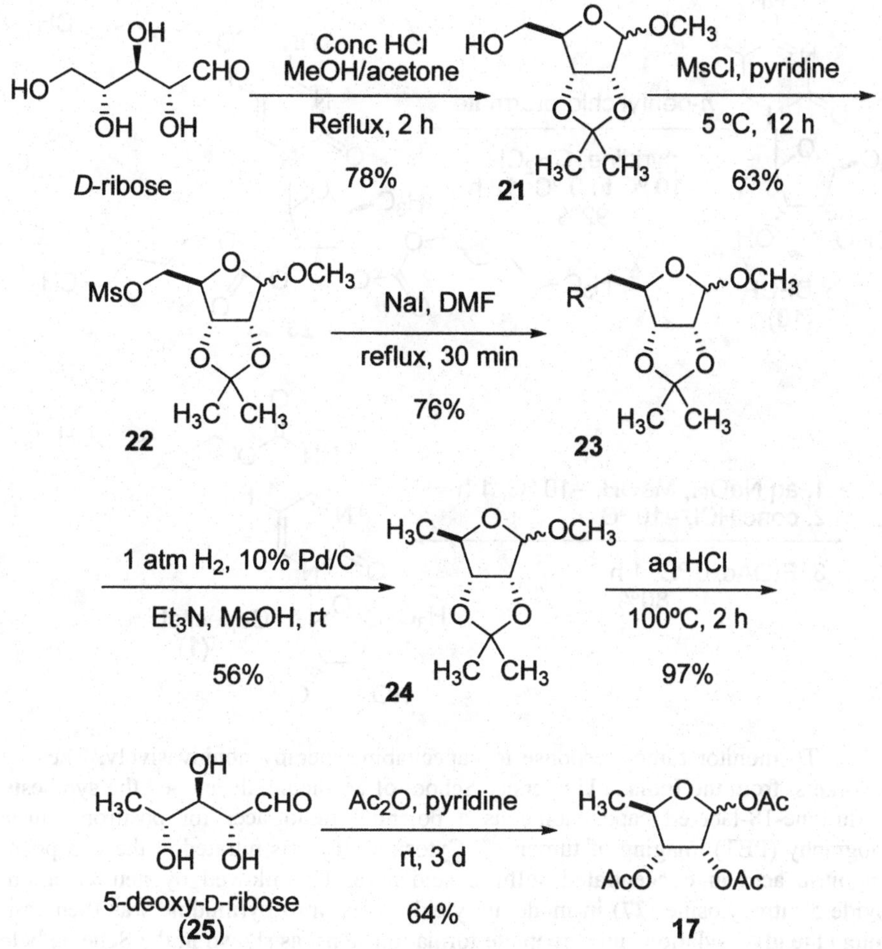

The route described by Hoffmann-La Roche scientists in 1995 was designed to make use of the drug substance 5'-DFCR (**10**) to prepare capecitabine in a few steps without multiple protection/deprotection transformations.[19] In this process, 5'-deoxy-5-fluorocytodine (**10**)[24-27] was added to three equivalents of *n*-pentyl chloroformate and pyridine in methylene chloride while maintaining an internal temperature below –5 °C, ultimately providing the tris-acylated cytodine adduct **26** in a 92% isolated yield. Removal of the two ester functional groups by selective hydrolysis with aqueous sodium hydroxide in methanol at –10 °C for a short time followed by adjustment to pH5 with concentrated HCl provided the crude carbamate **1** in quantitative yield. Purification of crude **1** was accomplished by precipitation from ethyl acetate at 0 °C to provide pure capecitabine (**1**) as a white solid in 80% isolated yield from **20**. This process constituted a 62% overall yield of **1** from 5′-DFCR when exemplified on a 25-g scale.

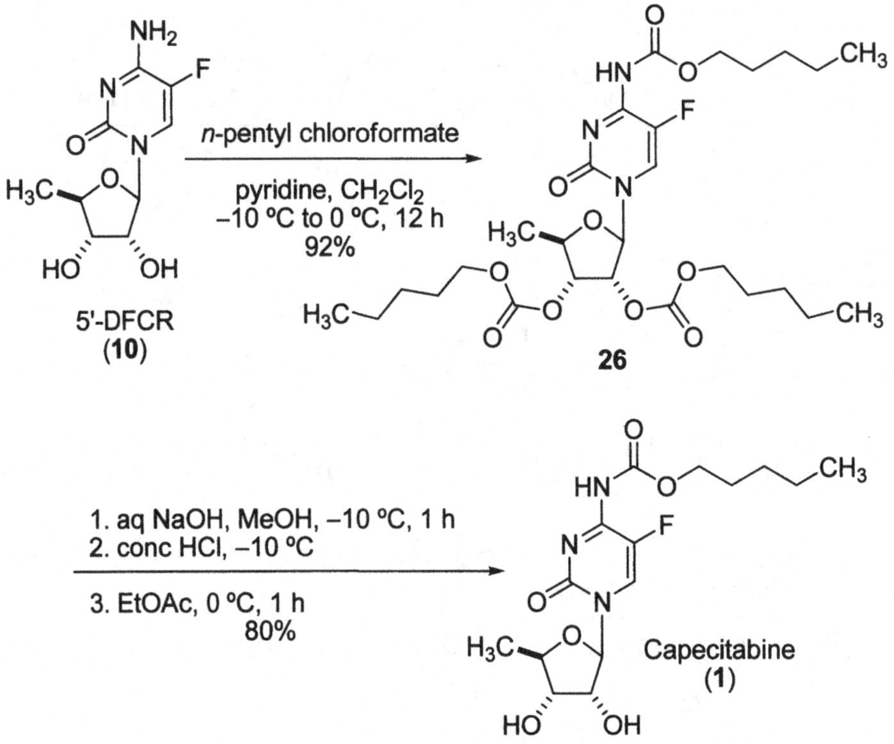

To monitor tumor response to capecitabine therapy noninvasively, Zheng and co-workers, from the Indiana University School of Medicine, developed the synthesis of the fluorine-18-labeled capecitabine as a potential radiotracer for positron emission tomography (PET) imaging of tumors.[28] Cytosine (**20**) was nitrated at the C-5 position with nitric acid in concentrated sulfuric acid at 85°C, followed by neutralization to provide 5-nitrocytosine (**27**) in moderate yield. This nitro pyrimidine was then carried through the glycosylation and carbamate formation steps, as shown in the Scheme below, to provide the *bis*-protected 5-nitro cytidine **28** in 47% for the three-step process. Precursor **28** was then labeled by nucleophilic substitution with a complex of ^{18}F-labeled potassium fluoride with cryptand Kryptofix 222 in DMSO at 150 °C to provide the fluorine-18-labeled adduct. This intermediate was not isolated, but semi-purified and deprotected with aqueous NaOH in methanol to provide [^{18}F]-capecitabine in 20–30% radiochemical yield for the 3-mg-scale process. The synthesis time for fluorine-18 labeled capecitabine (including HPLC purification) from end of bombardment to produce K[18]F to the final formulation of [^{18}F]-**1** for in vivo studies was 60–70 min.

20 27

28

[^{18}F]capecitabine ([^{18}F]-1)

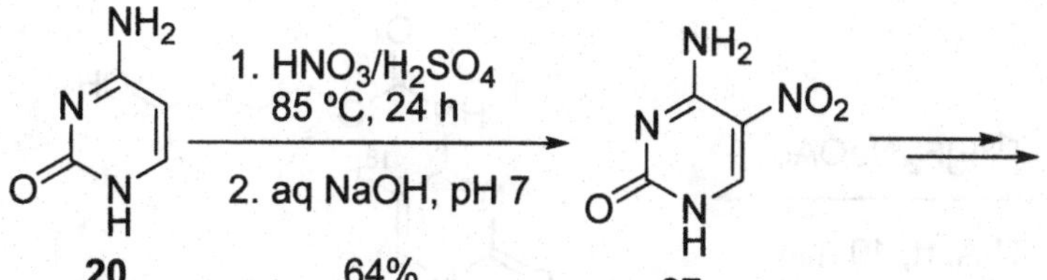

A team of researchers led by Chun, at the Korea Institute of Radiological and Medical Sciences, also devised a method to prepare [^{18}F]-capecitabine using electrophilic fluorination with ^{18}F-labeled fluorine gas.[29] In this route, cytosine (**20**) was carried through the glycosylation, carbamate formation and ester deprotection procedures outlined in the Scheme to provide 5′-deoxy-N^4-(pentyloxycarbonyl)cytidine (**29**) in an overall 17% isolated yield for the four chemical steps. Cytidine **29** was then fluorinated at the C-4 position on a 10-mg-scale by dissolution in trifluoroacetic acid at room temperature followed by bubbling radioisotope [^{18}F]-F_2 gas through a sodium acetate column and into the mixture for 40 min. Following removal of the TFA solvent, the reaction residue was purified by HPLC to provide [^{18}F]-capecitabine in 5–15% radiochemical yield. The total elapsed labeling time (including HPLC purification) from end of bombardment to produce $^{18}F_2$ to the final formulation of [^{18}F]-**1** was 90–110 min.

20 29

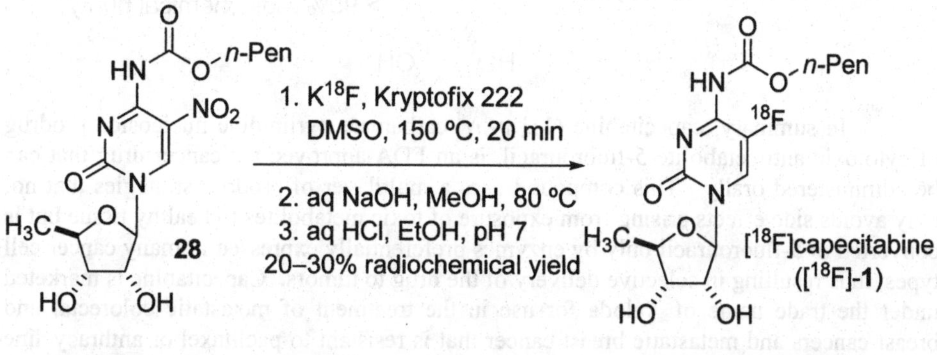

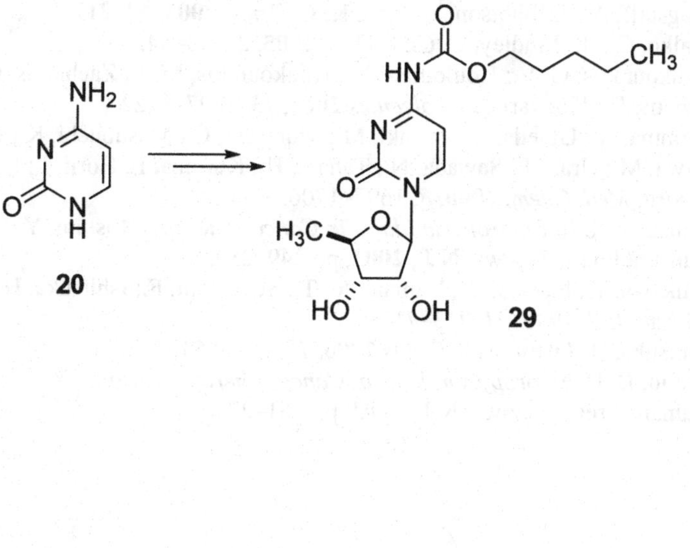

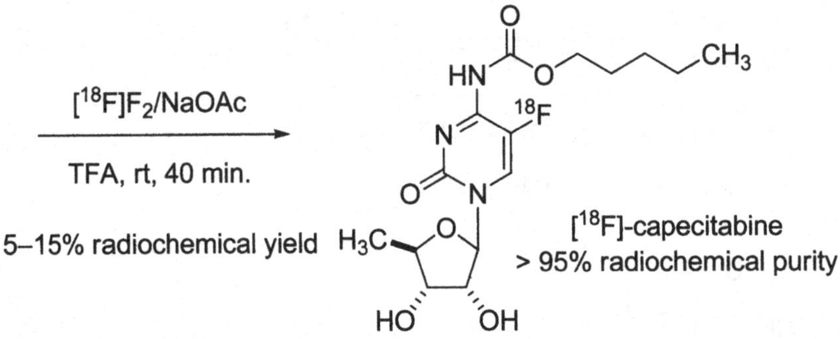

In summary, capecitabine (1), an N^4-carbamate pyrimidine nucleoside prodrug of cytotoxic antimetabolite 5-fluorouracil, is an FDA-approved anticancer drug that can be administered orally. This compound uses a multilayer of prodrug strategies that not only avoids side effects arising from exposure of toxic metabolites to healthy tissue but is converted to 5-fluorouracil only by enzymes preferentially expressed in many cancer cell types, thus resulting in selective delivery of the drug to tumors. Capecitabine is marketed under the trade name of Xeloda for use in the treatment of metastatic colorectal and breast cancers and metastatic breast cancer that is resistant to paclitaxel or anthracycline therapies.

5.6   References

1.   Henry, J. R.; Mader, M. M. In *Annual Reports in Medicinal Chemistry* Doherty, A. M., ed.; Elsevier; San Diego, CA, **2004**; pp 161–172.
2.   Parker, W. B. *Chem. Rev.* **2009**, *109*, 2880–2893.
3.   Heidelberger, C.; Chaudhuri, N. K.; Danenberg, P.; Mooren, D.; Griesbach, L.; Duschinsky, R.; Schnitzer, R. J.; Pleven, E.; Scheiner, J. *Nature* **1957**, *179*, 663–666.
4.   Parker, W. B.; Cheng, Y. C. *Pharmacol. Ther.* **1990**, *48*, 381–395.
5.   Chabner, B. A.; Longo, D. L. In *Cancer Chemotherapy and Biology: Principles and Practice* 2nd ed; Lippincott-Raven: Philadelphia, **1996**, pp 149–211.
6.   Wagstaff, A. J.; Ibbotson, T.; Goa, K. L. *Drugs* **2003**, *63*, 217–236.
7.   Walko, C. M.; Lindley, C. *Clin. Ther.* **2005**, *27*, 23–44.
8.   Koukourakis, G. V.; Kouloulias, V.; Koukourakis, M. J.; Zacharias, G. A.; Zabatis, H.; Kouvaris, J. *Molecules* **2008**, *13*, 1897–1922.
9.   Shimma, N.; Umeda, I.; Arasaki, M.; Murasaki, C.; Masubuchi, K.; Kohchi, Y.; Miwa, M.; Ura, M.; Sawada, N.; Tahara, H.; Kuruma, I.; Horii, I.; Ishitsuka, H. *Bioorg. Med. Chem.* **2000**, *8*, 1697–1706.
10.   Ishitsuka, H. In *Fluoropyrimidines in Cancer Therapy*, Rustum, Y. M., ed. Humana Press: Totowa, N.J., **2003**, pp 249–259.
11.   Ishikawa, T.; Fukase, Y.; Yamamoto, T.; Sekiguchi, F.; Ishitsuka, H. *Biol. Pharm. Bull.* **1998**, *21*, 713–717.
12.   Ishitsuka, H. *Invest. New Drugs* **2000**, *18*, 343–354.
13.   Samid, D. In *Fluoropyrimidines in Cancer Therapy*, Rustum, Y. M., ed; Humana Press: Totowa, N.J., **2003**, pp 261–273.

14. Budman, D. R. In *Fluoropyrimidines in Cancer Therapy* Rustum, Y. M., ed. Humana Press: Totowa, N.J., **2003**, pp 275–284.

15. Arasaki, M.; Ishitsuka, H.; Karuma, I.; Miwa, M.; Murasaki, C.; Shimma, N.; Umeda, I. U.S. Pat 5,472,949, Dec.5, **1995**.

16. Xeloda® (capecitabine) prescribing information. Roche Laboratories, Inc. www.gene.com/gene/products/information/xeloda, revised: Nov. 2009, accessed Dec. 1, 2009.

17. Ishitsuka, H.; Shimma, N. In *Modified Nucleosides: in Biochemistry, Biotechnology and Medicine* Herdewijn, P., ed., Wiley-Verlag: Weinheim, **2008**, pp 587–600.

18. Hoshi, A.; Castaner, J. *Drugs Fut.* **1996**, *21*, 358–360.

19. Kamiya, T.; Ishiduka, M.; Nakajima, H. U.S. Pat 5,453,497, Sept. 26, **1995**.

20. Brinkman, H. R.; Kalaritis, P.; Morrissey, J. F. U.S. Pat 5,476,932, Dec. 19, **1995**.

21. Robins, M. J.; Naik, S. R. *J. Chem. Soc., Chem. Commun.* **1972**, *1*, 18–19.

22. Takahara, T. U.S. Pat 4,473,691, Sept. 25, **1984**.

23. Kissman, H. M.; Baker, B. R. *J. Am. Chem. Soc.* **1957**, *79*, 5534–5540.

24. Cook, A. F.; Holman, M. J. *J. Med. Chem.* **1979**, *22*, 1330–1335.

25. Cook, A. F. *J. Med. Chem.* **1977**, *20*, 344–348.

26. Robins, M. J.; MacCoss, M.; Naik, S. R.; Ramani, G. *J. Am. Chem. Soc.* **1976**, *98*, 7381–7389.

27. Wempen, I.; Duschinsky, R.; Kaplan, L.; Fox, J. J. *J. Am. Chem. Soc.* **1961**, *83*, 4755–4766.

28. Fei, X.; Wang, J.-Q.; Miller, K. D.; Sledge, G. W.; Hutchins, G. D.; Zheng, Q.-H. *Nuclear Med. Biol.* **2004**, *31*, 1033–1041.

29. Moon, B. S.; Shim, A. Y.; Lee, K. C.; Lee, H. J.; Lee, B. S.; An, G. I.; Yang, S. D.; Chi, D. Y.; Choi, C. W.; Lim, S. M.; Chun, K. S. *Bull. Korean Chem. Soc.* **2005**, *26*, 1865–1868.

6

Sorafenib (Nexavar): A Multikinase Inhibitor for Advanced Renal Cell Carcinoma and Unresectable Hepatocellular Carcinoma

Shuanghua Hu and Yazhong Huang

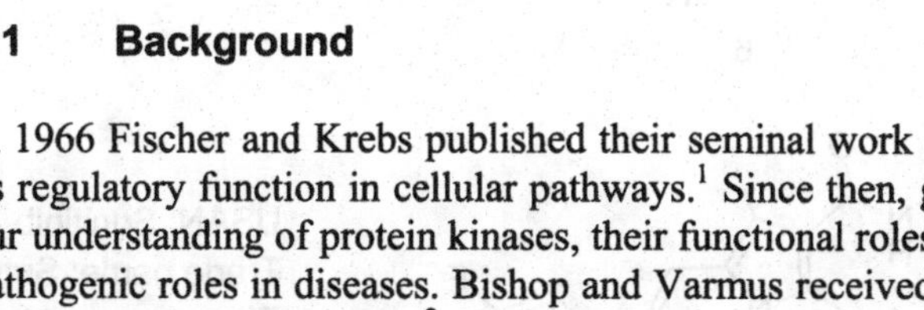

USAN: Sorafenib
Trade name: Nexavar®
Bayer Pharmaceuticals
& Onxy Pharmaceuticals
Launched: 2005

6.1   Background

In 1966 Fischer and Krebs published their seminal work on protein phosphorylation and its regulatory function in cellular pathways.[1] Since then, great strides have been made in our understanding of protein kinases, their functional roles in cellular processes, and their pathogenic roles in diseases. Bishop and Varmus received the 1989 Nobel Prize for their work on oncogenic kinases.[2] In 2001, Nurse and Hunt were also awarded the Nobel Prize for their work in elucidating the role of cyclins and cyclin-dependent kinases in regulating cell cycles.[3]

Protein kinases represent a large, diverse family of ATP-regulated cellular or membrane-bound proteins. They regulate critical cellular processes including gene transcription and cell growth, proliferation, and differentiation, often through complex and interactive processes. Kinases have been implicated strongly in cancer. Oncogenic kinases are known to cause aberrant and often uncontrollable, cellular phosphorylation and lead to tumor formation.   Therefore kinase inhibitors hold great potential as anticancer therapeutics. So far, more than 100 small molecule kinase inhibitors have entered clinical studies. Most of these target the highly conserved ATP binding site,

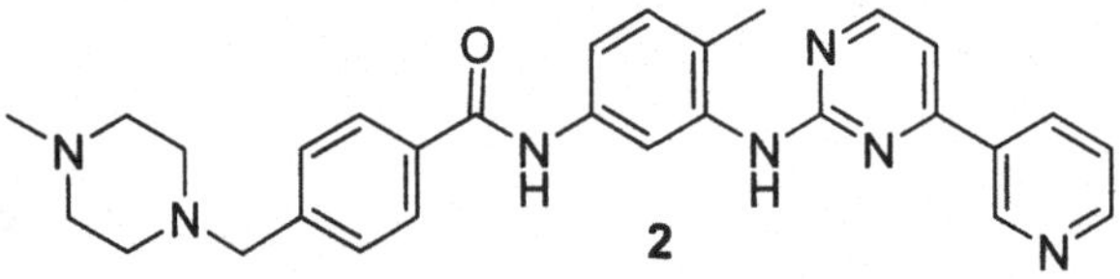

USAN: imatinib
Trade name: Gleevec®
Novartis
Approved: 2001

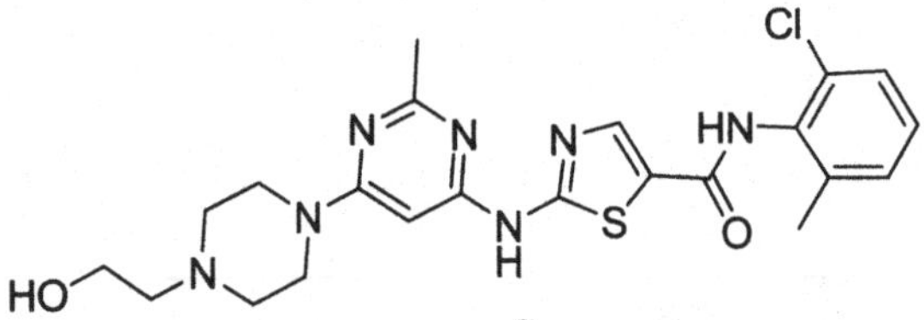

2

USAN: Dasanitib
Trade name: Sprycel®
Bristol-Myers Squibb
Approved: 2006

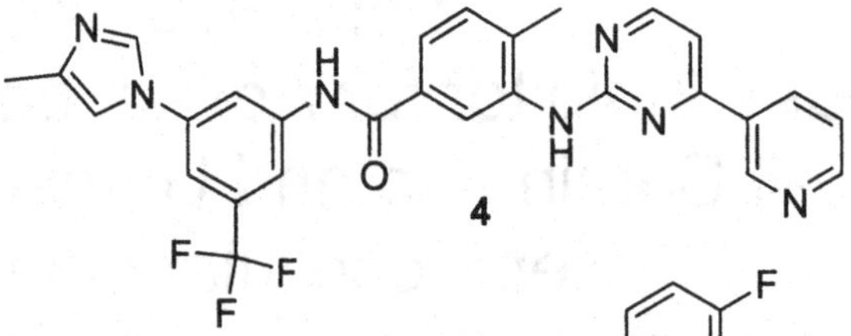

3

USAN: Nilotinib
Trade name: Tasignal®
Novartis
Approved: 2007

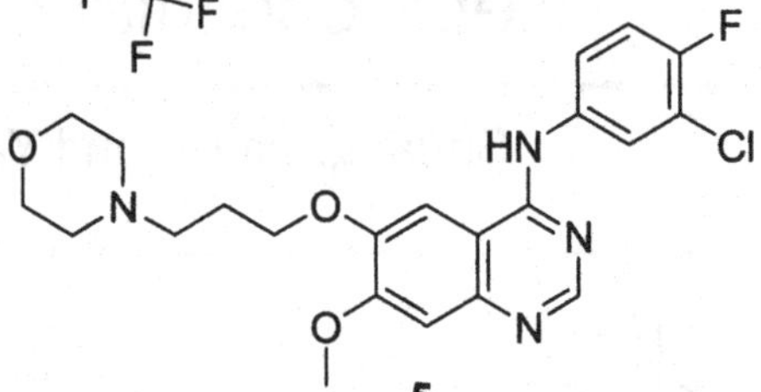

4

USAN: Gefitinib
Trade name: Iressa®
AstraZeneca
Approved: 2002 (Japan)

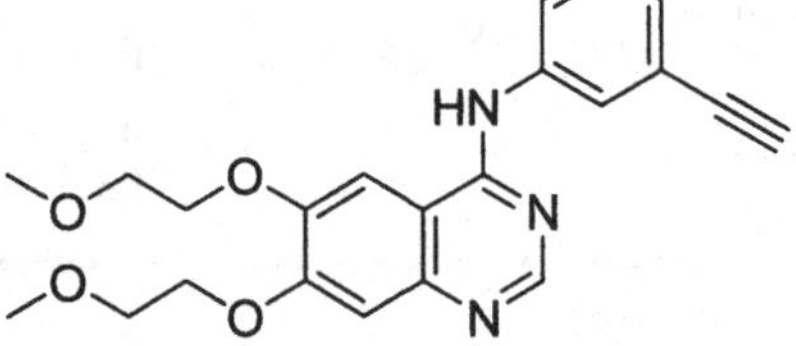

5

USAN: Erlotinib
Trade name: Tarceva®
OSI/Genentech/Roche
Approved: 2004

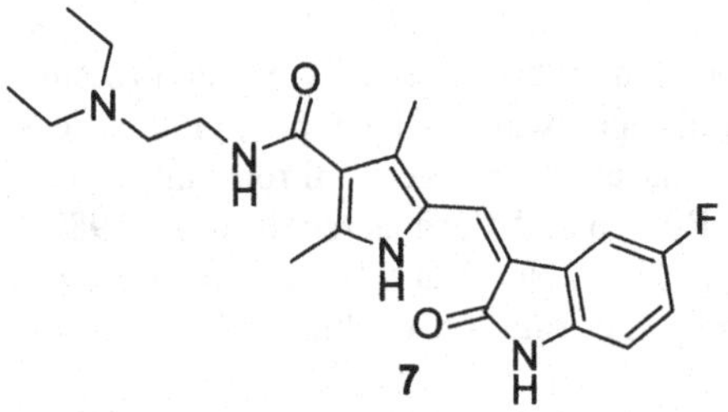

6

USAN: Sunitinib
Trade name: Sutent®
Pfizer
Approved: 2006

7

leading to concerns about kinase selectivity. However, several have overcome this early skepticism to achieve regulators' approval as targeted antitumor therapeutics.

Imatinib (**2**),[5] marketed by Novartis since 2001, is the first tyrosine kinase inhibitor approved as a treatment for chronic myelogenous leukemia (CML). It inhibits the BCR-ABL kinase that is highly specific to leukemic cells. Dasatinib (**3**)[6] and nilotinib (**4**)[7] are two other BCR-ABL kinase inhibitors marketed by Bristol-Myers Squibbs and Novartis, respectively. They are effective for the treatment of imatinib-resistant CML patients.

Gefitinib (**5**)[8] and erlotinib (**6**)[9] are human epidermal growth factor receptor (EGFR) kinase inhibitors marketed for the treatment of non-small-cell lung cancer (NSCLC). Unlike BCR-ABL kinase, EGFR kinases are trans-membrane proteins with a ligand-binding extracellular domain and a catalytic intracellular domain, which triggers the downstream cellular signaling cascades when activated by extracellular ligands such as human epidermal growth factor (EPF). Gefitinib and erlotinib function to block this process, which plays a key role in tumorigenesis and disease progression.

Sunitinib (**7**),[10] developed by Sugene/Pfizer, was approved in 2006 for the treatment of renal cell carcinoma (RCC). It is a VEGFR kinase inhibitor that slows down the angiogenesis of tumor endothelial cells, which provide nutrients and help tumor growth and metastasis. Sunitinib also blocks KIT, an oncogenic kinase that causes gastrointestinal stromal cell tumors. It was approved as a second-line anti-GSIT therapy for imatinib-resistant patients.

Sorafenib (**1**)[11] is a multikinase inhibitor marketed by Bayer and Onyx. Sorafenib blocks tyrosine kinases as well as serine/threonine kinases. Its story began in 1994 when Bayer and Onyx entered a collaboration to discover novel Raf/MEK/ERK inhibitors. They first discovered a very mildly active compound **8** (IC_{50}: 17 μM) against Raf1 kinase in 1995 from screening a collection of 200,000 compounds. The optimization of its potency and its ADMET profile using medicinal chemistry and combinatorial chemistry methods led to the identification of sorafenib (**1**) in 1999 as a preclinical development candidate. Multiple phase I studies started in 2000, when sorafenib tosylate (**19**) was evaluated in patents with advanced solid tumors of different types. In December 2005, Sorafenib tosylate (**19**) received U.S. FDA approval for the treatment of advanced renal cell carcinoma (RCC). Two years later, it was approved for the treatment of unresectable hepatocellular carcinoma (HCC).

As demonstrated by these marketed anticancer drugs, ATP competitive inhibitors are generally not as selective as people had thought; however, inhibition of multiple kinases appears to be manageable and potentially an advantage when applied to the treatment of cancer.

6.2   Pharmacology

Sorafenib (**1**) was initially discovered as a serine/threonine Raf kinase inhibitor against isoforms wild-type Raf1 (IC_{50}: 6 nM), B-Raf (IC_{50}: 25 nM) and oncogenic B-raf V600E (IC_{50}: 38 nM).[12] It was later found to block multiple important kinases, including VEGFR-2 and VEGFR-3, and platelet-derived growth factor receptor B (PDGFR-B), c-

KIT, Fit-3, and RET.[13] It exhibited selectivity against other kinases, including MEK-1, ERK1, epidermal growth factor receptor and human insulin receptor (Table 1).

In cell mechanistic studies, sorafenib dose dependently inhibited basal MEK1/2 and ERK 1/2 phosphorylation in MDA-MB-231 breast cancer cells and selectively blocked the MAPK pathway over the PKB pathway. Sorafenib also showed potent inhibitory effects on ERK1/2 phosphorylation in human pancreatic tumor cell lines (Mia PaCa 2), colon tumor cell lines (HCT116 and HT29), and LOX melanoma and pancreatic BxPC3 cell lines. Beside its antiproliferative effect by blocking the MAPK pathway, sorafenib potently inhibited the autophosphorylation of several other tyrosine kinases that play key role in angiogenesis, including VEGFR2 (NIH3T3 fibroblasts), PDGFR (SMC) in human cells, and VEGFR3 in mouse HEK-293 cells. Furthermore, sorafenib was shown to induce apoptosis in human cancer lines, including ACHN renal carcinoma, MDA-MB-231 breast carcinoma, HT29 colon carcinoma, and A549 NSCLC in addition to KMCH cholangiocarcinoma cells, Jurkat (acute T-cells), K562 (chronic myelogenous), and MEC2 (chronic lymphocytic) leukemia cells.

Table 1. In Vitro Inhibitory Profile of Sorafenib[13]

Kinase Targets	IC_{50}(nM)
Biochemical kinase assay	
Raf1	6
Wild-type B-Raf	25
Oncogenic B-raf V600E	38
VEGFR-1	26
VEGFR-2	90
Murine VEGFR3	20
Murine PDGFRβ	57
Flt3	33
p38	38
c-KIT	68
MEK1, ERK1, EGFR, HER2/neu, c-met, IGFR1, PKA, PKB, CDK1/cyclin B, pim1, PKC-α, PKC-λ	> 10,000
Cellular kinase assay	
MEK phosphorylation in MDA-MB-231 cells	40
ERK1/2 phosphorylation in MDA-MB-231 cells	90
VEGFR2 phosphorylation in human NIH3T3 fibroblasts	30
VEGF–ERK1/2 phosphorylation in human HUVEC cells	60
PDGFRβ phosphorylation in HAoSMC	80
VEGFR3 phosphorylation in mouse HEK-293 cells	100
Flt3 phosphorylation in mouse HEK-293 cells with human ITDs	20

In in vivo studies, sorafinib (1) demonstrated efficacy in multiple cancer models. It inhibited the tumor growth of human melanoma; renal, colon, pancreatic, hepatocellular, thyroid, and ovarian carcinomas; and NSCLC, presumably through its anti-angiogenic effect on tumor endothelial cells that provide nutrients and help tumor growth. In addition, sorafenib (1) was shown to induce substantial tumor regression in a breast cancer model harboring B-Raf and K-Ras oncogenic mutations and to a lesser degree in a mice hepatocellular carcinoma model bearing PLC/PRF/5HCC.

It is likely that sorafenib (**1**) exhibits such broad antitumor effects in preclinical setting through its antiproliferative, anti-angiogenic and pro-apopototic action.[13]

6.3   Structure–Activity Relationship (SAR)

Table 2. The Optimization of Inhibitory Potency of the Initial Hit against Wild-Type Raf1 Kinase[14]

Entry	Compound	IC$_{50}$ (µM), Raf1
1	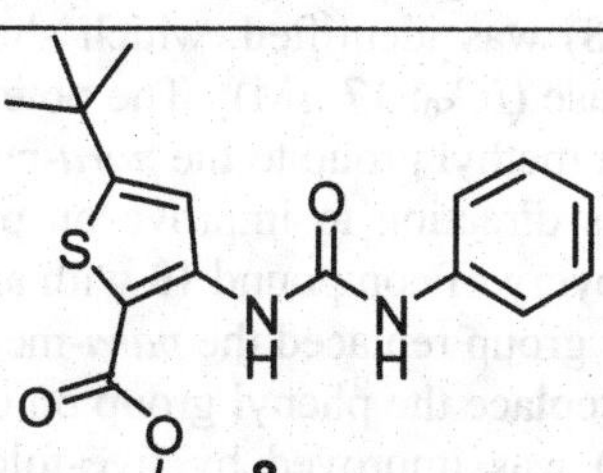 **8**	17
2	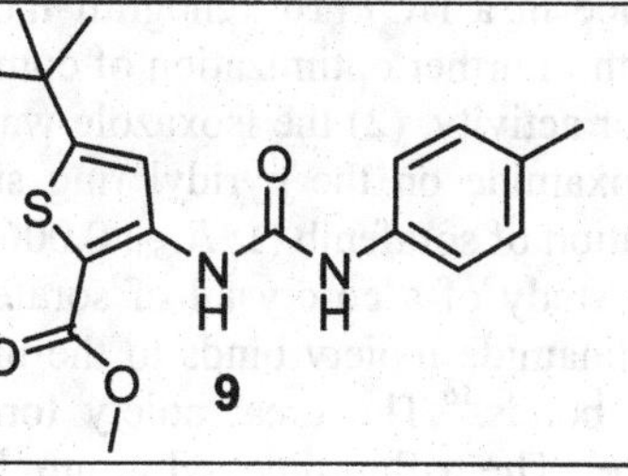 **9**	1.7
3	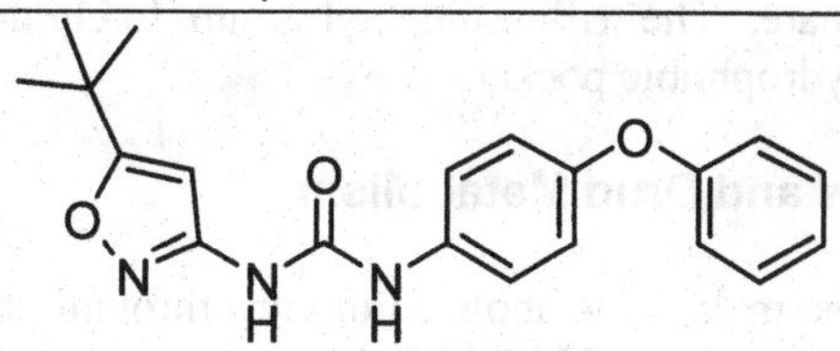 **10**	1.1
4	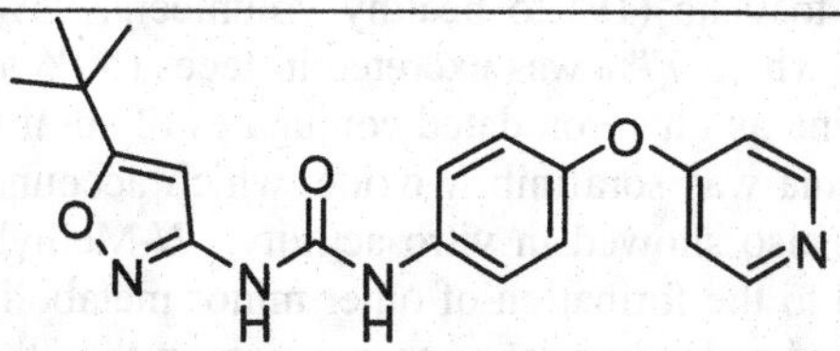 **11**	0.23

| 5 | | 0.006 |

Bayer and Onyx started their collaboration in 1994 to discover selective Raf/MEK/ERK inhibitors. Using the high throughput screening method developed by McDonald et al.,[15] an initial hit (8) was identified, which showed only low inhibitory potency against wildtype Raf1 kinase (IC_{50}: 17 μM). The potency was first improved to 1.7 μM in compound 9 by adding a methyl group to the *para*-position of the distal phenyl, indicating a possible vector in this direction to improve on potency of the lead. This rational was supported by the discovery of compound 10 with an enhanced potency (IC_{50}: 1.1 μM), wherein a bulky phenoxy group replaced the *para*-methyl group. Furthermore, when a pyridyl was introduced to replace the phenyl group on compound 10, the potency of compound 11 (IC_{50}: 0.23 μM) was improved by five-fold. As a tool compound, compound 11 was dosed in mice in a HCT116 xenograft model and demonstrated an inhibitory effect on tumor growth. Further optimization of compound 11 revealed that (1) the urea moiety was essential for activity, (2) the isoxazole was replaceable by a regular phenyl ring, and (3) a 2'-carboxamide on the pyridyl ring substantially increased the potency that led to the identification of sorafenib (1; IC_{50}: 0.006 μM).

X-ray crystallographic study of a co-crystal of sorafenib (1) with Raf1 kinase revealed that the *N*-methylpicolinamide moiety binds to the hinge region in a bidentate fashion through two hydrogen bonds.[16] The urea moiety forms two crucial hydrogen bonds to aspartate and glutamate. The trifluoromethyl group, likely the *tert*-butyl and in the initial hit, 8, occupies a hydrophobic pocket.

6.4   Pharmacokinetics and Drug Metabolism

Sorafenib (1) was shown to be mainly metabolized in vitro through phase I oxidation by CYP3A4 and phase II conjugation by UGT1A9. Following oral administration of a 100-mg dose of [^{14}C]-sorafenib tosylate (19) to healthy volunteers, 96% of the dose was recovered within 14 days, of which 77% was excreted in feces (50% as unchanged drug) and 19% was excreted in urine as glucuronidated conjugates of sorafinib or metabolites. The main metabolite in plasma was sorafinib *N*-oxide, which accounted for 17% of the circulating radioactivity and also showed in vitro activity. *N*-Methylhydroxylation and *N*-demethylation contributed to the formation of other minor metabolites. In continuous daily dosing, sorafenib reached a steady-state concentration on the 7th day. For example, at the FDA-approved dosing (400 mg b.i.d.), C_{max} and AUC values on the 7th day were 6.2 mg/L and 56.6 mg•h/L compared to 2.3 mg/L and 18.0 mg•h/L on day 1, respectively. Sorafenib had a relatively long elimination half-life of 20.0–27.4 h at the 400 mg b.i.d. dosing.

6.5   Efficacy and Safety

Because sorafenib inhibited multiple kinase targets and showed a broad antitumor effect in numerous preclinical models,[13] it was first evaluated in multiple phase I trials in patients with different solid tumor types. Efficacy was observed in patients with advanced RCC[18] and the efficacy was further confirmed in a RCC patients-enriched phase II trial.[19] In a larger phase III trial of 903 RCC patients, sorafenib significantly prolonged progression-free survival (PFS) in patients receiving drug compared to those receiving placebo (24 versus 12 weeks). The overall survival (OS) of patients on the drug was improved by 38% relative to those on the placebo. In this trial, sorafenib was well tolerated by patients with advanced RCC. Common adverse events were skin rash, hand–foot syndrome, fatigue, diarrhea, and hypertension. All symptoms are mild (grade 1–2) and easily manageable.[20]

In another phase III trial in patients with advanced HCC, sorafenib increased OS from 7.9 months on placebo to 10.7 months on sorafinib, and prolonged progression-free survival (PFS) from 2.8 months on the placebo to 5.5 months on the drug. These results indicate that sorafenib provides clinical benefits by primarily delaying disease progression.[21,22]

6.6   Syntheses

Banstin et al.[23] reported a four-step synthesis of sorafinib based on the original route of the discovery chemistry program. They significantly improved the synthesis from an overall yield of 10% to 63% and avoided chromatographic purifications all together. In this synthesis, picolinic acid 12 was first dichlorinated by the Vilsmeier reagent at 72 °C (formed by adding small amount of anhydrous DMF to neat thionyl chloride) to afford 4-chloropicolinoyl chloride hydrochloric acid salt (13), which precipitated out of toluene as an off-white solid. Excessive heating should be avoided in this step to prevent multichlorination, which led to a much lowered yield and the need for chromatographic purification under the original conditions. 4-Chloropicolinoyl chloride salt (13) was then converted to the methyl amide 14 in two steps. The methyl ester formed first when 13 was added into methanol at a temperature < 55 °C and precipitated out of the solution as a hydrochloric salt when ethyl ether was added. The methyl ester salt, as a suspension in methanol, was added to an excess amount of methylamine in THF at 5 °C to neatly form 4-chloro-N-methylpicolinamide (14). Alternatively, 4-chloro-N-methylpicolin-amide (14) was obtained in one step when 4-chloropicolinoyl chloride (13) was treated directly with methylamine in THF. Next, 4-chloro-N-methylpicolinamide (14) was substituted with 4-aminophenol (15) at 80°C in the presence of one equivalent of *tert*-butoxide and half an equivalent of potassium carbonate. Potassium carbonate was found to shorten the reaction time from 16 h to 6 h. Simple aqueous workup with ethyl acetate as the extraction solvent afforded the advanced intermediate 16 as a light brown solid at the yield of 87%. Compound 16 was then treated with isocyanate 17 in methylene chloride at room temperature. Sorafenib (1) crystallized out of the reaction mixture gradually as an off-white crystal at the yield of 92%.

The team also used N,N-carbonyldiimidazole (CDI) as the coupling reagent to generate sorafinib (1) directly from amines—namely, 4-chloro-3-trifluoromethylaniline

(**18**) and the advanced intermediate **16**. The CDI approach is particularly useful when desirable isocynates are not commercially available or difficult to prepare at an acceptable purity. In this method, 4-chloro-3-trifluoromethylaniline (**18**) reacted with a small excess amount of CDI (1.06 equiv) in methylene chloride at room temperature to form an imidazole–carboxamide intermediate, then smaller than one equivalent of intermediate **16** (0.95 equiv was added into the reaction mixture. Sorafenib (**1**) crystallized with time to afford the pure product. However, caution should be exercised to maintain appropriate concentrations of the reactants in order to eliminate the formation of symmetrical ureas as impurities. This improved synthesis of sorafinib (**1**) by Banstin et al.[23] was used to prepare up to hundreds of grams of sorafenib for preclinical studies.

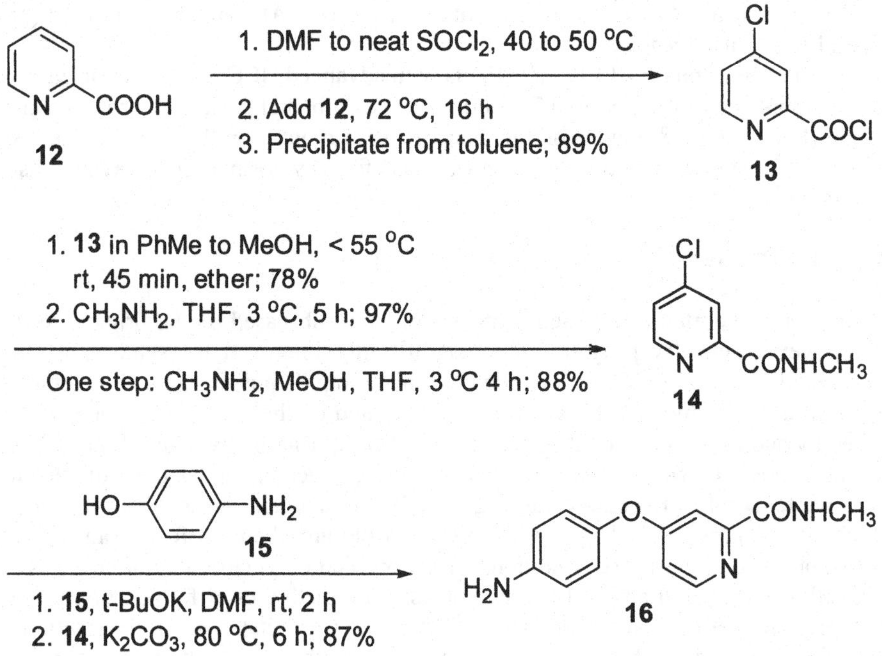

1. DMF to neat SOCl$_2$, 40 to 50 °C

2. Add **12**, 72 °C, 16 h
3. Precipitate from toluene; 89%

12   **13**

1. **13** in PhMe to MeOH, < 55 °C
   rt, 45 min, ether; 78%
2. CH$_3$NH$_2$, THF, 3 °C, 5 h; 97%

One step: CH$_3$NH$_2$, MeOH, THF, 3 °C 4 h; 88%

14

15

1. **15**, t-BuOK, DMF, rt, 2 h
2. **14**, K$_2$CO$_3$, 80 °C, 6 h; 87%

16

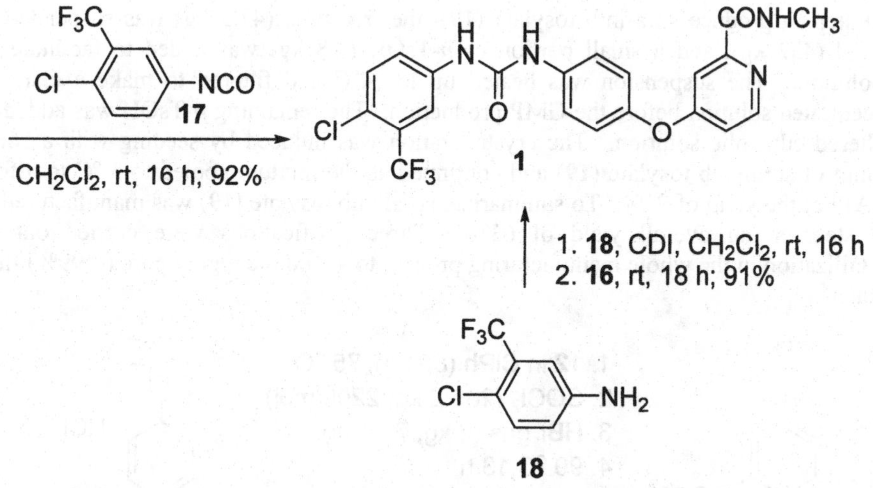

Furthermore, Logers et al.[24] were able to optimize this rather linear route to manufacture highly pure sorafinib tosylate (**19**), the API of Naxavar, on an industry scale. Optimizations were made to improve on the industry employability, environmental compatibility, safety and volume yields, as detailed below.

For the chlorination of picolinic acid (**12**, 65 kg) on an industry scale, bromide compounds such as hydrogen bromide was used to replace dimethylformamide (DMF) in the Vilsmeier reagent so as to better control the reaction course. The crude 4-chloropicolinoyl (**13**) was obtained after removing the excess thionyl chloride by azeotropic distillation with toluene (150 kg × 2) and was used directly to avoid isolation of this corrosive and unstable material. Thus crude **13** in toluene (~ 335 kg) was added into a concentrated methylamine solution in water (21.8%, 214.5 kg) at below 30°C to form the methyl amide product **14**. The crude oily product was harvested (95 kg) after the separation of the toluene layer from water, wash with aqueous NaCl solution, and concentration. It was surprising that water did not complicate this reaction although **13** is a highly water-reactive acyl chloride. Furthermore, the use of concentrated aqueous solution increased the volume yield substantially, which is otherwise hard to achieve using organic solvents because of the low solubility of methylamine in them. Substitution of chloride **14** with 4-aminophemol (**15**) took place effectively at 60 °C in a two-phase reaction system using *tetra-n-*butylammonium hydrogensulfate as the phase-transfer catalyst and sodium hydride as the base. Compound **16** first crystallized as a dichloride salt when concentrated aqueous hydrochloric solution was added to neutralize the two-phase reaction mixture. The dichloride salt was redissolved in water and recrystallized as a free base by adding aqueous NaOH solution. To this end, a highly pure product **16** (88 kg, purity > 95%) was manufactured at a total yield of 74% in three steps on large scale. Next, ethyl acetate, inert to isocynates, was used in replacement of methylene chloride for the coupling between isocynate **17** and **16**. The coupling was carried out at an elevated temperature (~ 60 °C) to increase the volume yield and accelerate the reaction rate. Sorafenib (**1**, 93 kg) crystallized when the reaction mixture cooled to 20 °C to afford a colorless to slightly brownish crystal at a yield of 93%. In the final step to produce sorafinib tosylate (**19**), the free base (47.5 kg) was suspended in ethanol (432 kg), and a small portion of p-TsOH (6.8 kg) was added to facilitate its dissolution. The suspension was heated up to 74°C and filtered to make a clarified concentrated solution before the GMP production. The remaining p-TsOH was added as a filtered ethanolic solution. The crystallization was induced by seeding with a small amount of sorafenib tosylate (**19**) and continued as the mixture cooled to 3 °C to afford the API at the yield of 91%. To summarize, sorafenib tosylate (**19**) was manufactured in five steps at an overall yield of 63%. Three purifications were carried out by crystallization in the whole manufacturing process to afford the highly pure (> 99%) final product.

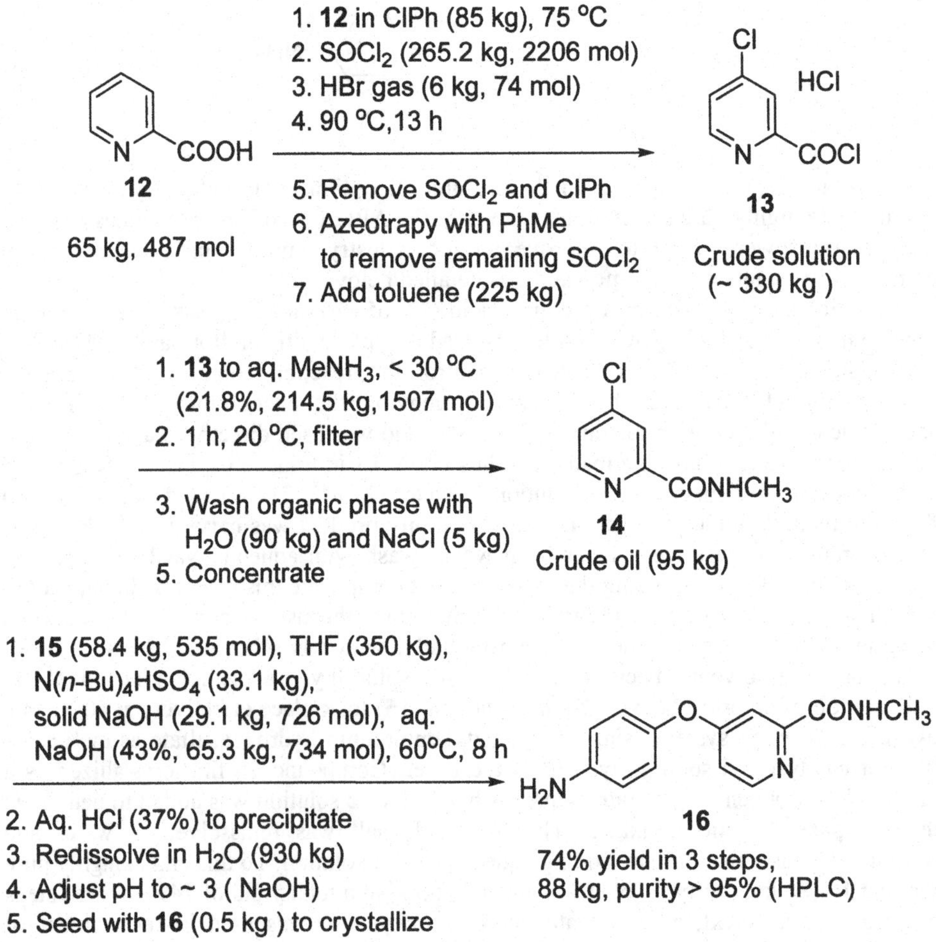

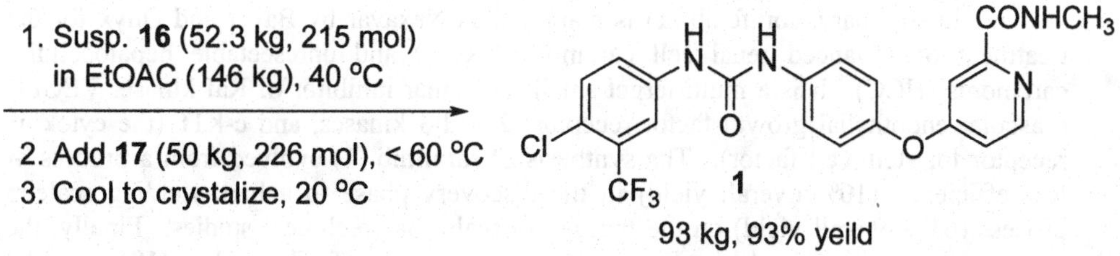

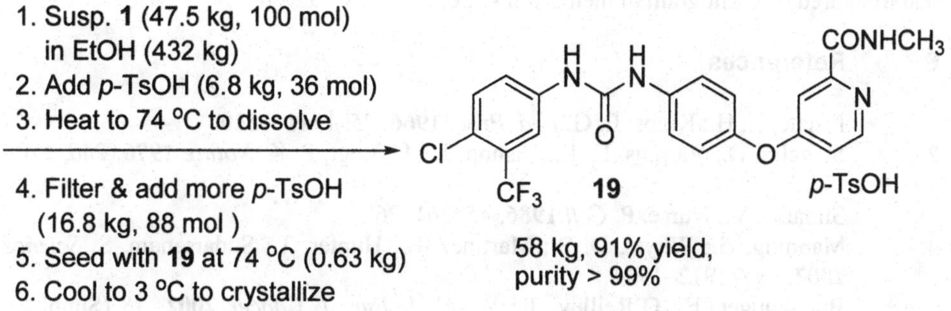

1. Susp. 16 (52.3 kg, 215 mol) in EtOAC (146 kg), 40 °C
2. Add 17 (50 kg, 226 mol), < 60 °C
3. Cool to crystalize, 20 °C
CONHCH3
H
N
O
Cl
CF3
1
93 kg, 93% yeild

1. Susp. 1 (47.5 kg, 100 mol) in EtOH (432 kg)
2. Add p-TsOH (6.8 kg, 36 mol)
3. Heat to 74 °C to dissolve
4. Filter & add more p-TsOH (16.8 kg, 88 mol)
5. Seed with 19 at 74 °C (0.63 kg)
6. Cool to 3 °C to crystallize
CONHCH3
H
N
O
Cl
CF3
19
p-TsOH
58 kg, 91% yield,
purity > 99%

More recently, Rossetto et al.[25] improved the CDI method so that the approach can synthesis very pure sorafinib without contamination for asymmetric ureas. In their method, CDI first reacted with one equivalent of 4-chloro-3-trifluoromehtylaniline (**18**) to form an isolable imidazole complex **20** as a well-characterized crystal. Complex **20** then treated with advanced intermediate **16** to afford sorafenib (**1**) as an ultra pure colorless crystal.

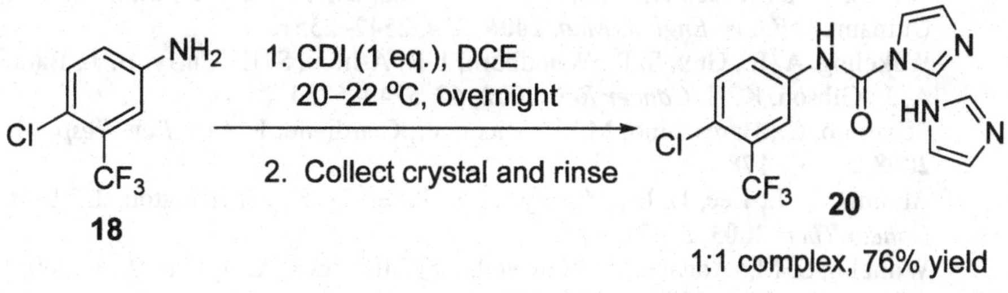

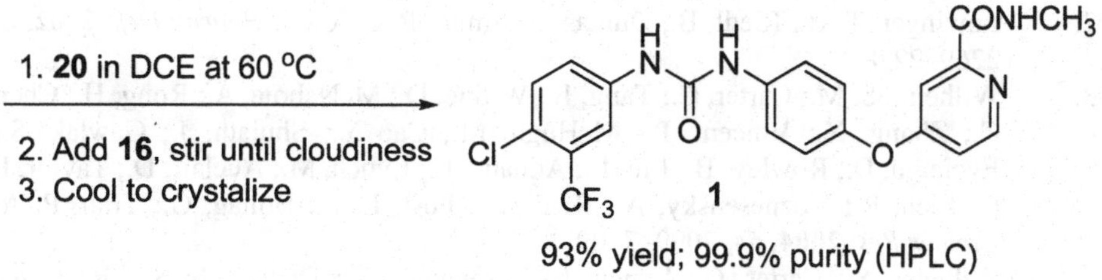

NH2
Cl
CF3
18
1. CDI (1 eq.), DCE
20–22 °C, overnight
2. Collect crystal and rinse
H
N
O
HN
Cl
CF3
20
1:1 complex, 76% yield

1. 20 in DCE at 60 °C
2. Add 16, stir until cloudiness
3. Cool to crystalize
CONHCH3
H
N
O
Cl
CF3
1
93% yield; 99.9% purity (HPLC)

In summary, sorafenib (**1**) is marketed as Nexavar by Bayer and Onyx for the treatment of advanced renal cell carcinoma (RCC) and unresectable hepatocellular carcinoma (HCC). It is a multi-target small molecular inhibitor of Raf kinase, VEGFR (vascular endothelial growth factor receptor) 2 and 3 kinases, and c-KIT (the cytokine receptor for stem cell factor). The synthesis of sarafinib (**1**) evolved from a process of low efficiency (10% overall yield) in the discovery phase to a high-yielding scaleup process (63% overall yield) on the multigram scale for preclinical studies. Finally, the same route was optimized by process chemists so that sorafenib tosylate (**19**) could be manufactured on a kilogram to metric ton scale.

6.7   References

1.   Fischer E. H.; Krebs, E. G. *Fed. Proc.* **1966**, *25*, 1511–1520.
2.   Stehelin, D.; Varmus, H. E.; Bishop, J. M.; Vogt, P. K. *Nature* **1976**, *260*, 170–173.
3.   Simanis, V.; Nurse, P. *Cell* **1986**, *45*, 261–268.
4.   Manning, G.; Whyte, D. B.; Martinez, R.; Hunter, T.; Sudarsanam, S. *Science* **2002**, *298*, 1912–1934
5.   Buchdunger, E.; O'Reilley, T.; Wood, J. *Eur. J. Cancer* **2002**, *38* (Suppl 5), S28–S36.
6.   Talpaz, M.; Shah, N. P.; Kantarjian, H.; Donato, N.; Nicoll, J.; Paquette, R.; Cortes, J.; O'Brien, S.; Nicaise, C.; Bleickardt, E.; Blackwood-Chirchir, M. A.; Iyer, V.; Chen, T.; Huang, F.; Decillis, A. P.; Sawyers C. L. *N. Engl. J. Med.* **2006**, *354*, 2531–2541.
7.   Kantarjian, H.; Giles, F.; Wunderle, L.; Bhalla, K.; O'Brien, S.; Wassmann, B.; Tanaka, C.; Manley, P.; Rae, P.; Mietlowski, W.; Bochinski, K.; Hochhaus, A.; Griffin, J. D.; Hoelzer, D.; Albitar, M.; Dugan, M.; Cortes, J.; Alland, L.; Ottmann, O. G. *N. Engl. J. Med.* **2006**, *354*, 2542–2551.
8.   Wakeling, A. E., Guy, S. P., Woodburn, J. R. Ashton, S. E.; Curry, B. J.; Barker, A. J.; Gibson, K. H. *Cancer Res.* **2002**, *62*, 5749–5754.
9.   Schettino, C.; Bareschino, M. A.; Ricci, V.; Ciardiello, F. *Exp. Rev. Resp. Med.* **2008**, *2*, 167–178.
10.   Abrams, T. J.; Lee, L. B.; Murray, L. J.; Pryer, N. K.; Cherrington, J. M. *Mol. Cancer Ther.* **2003**, *2*, 471–478.
11.   Wilhelm, S. M.; Adnane, L.; Newwell, P.; Villanueva, A.; Llovet, J.; Lynch, M. *Mol. Cancer Ther.* **2008**, *7*, 3129–3140.
12.   Lowinger, T. B., Riedl, B., Dumas, J.; Smith, R. A. *Curr. Pharm. Des.* **2002**, *8*, 2269–2278.
13.   Wilhelm, S. M.; Carter, C.; Tang, L.; Wilkie, D.; McNabola, A.; Rong, H.; Chen, C.; Zhang, X.; Vincent, P.; McHugh, M.; Cao Y.; Shujath, J.; Gawlak, S.; Eveleigh, D.; Rowley, B.; Liu, L.; Adnane, L.; Lynch, M.; Auclair, D.; Taylor, I.; Gedrich, R.; Voznesensky, A.; Riedl, B.; Post, L. E.; Bollag, G.; Trail, P. A. *Cancer Res.* **2004**, *64*, 7099–7109.
14.   Wilhelm, S.; Carter, C.; Lynch, M.; Lowinger, T.; Dumas, J.; Smith, R. A.; Schwartz, B.; Simantov, R.; Kelley, S. *Nat. Rev. Drug Disc.* **2007**, *6*, 126.

15.   McDonald, O. B.; Chen, W.; Ellis, B.; Hoffman, C.; Overton, L.; Rink, M.; Smith, A.; Marshalland, C. J.; Wood, E. R. *Anal. Biochem.* **1999**, *268*, 318–329.

16.   Wan, P. T. C.; Garnett, M. J.; Roe, S. M.; Lee, S.; Niculescu-Duvaz, D.; Good, V. M.; Jones, C. M.; Marshall, C. J.; Springer, C. J.; Barford, D.; Marais, R. *Cell* **2004**, *116*, 855–867.

17.   Strumberg, D.; Clark, J. W.; Awada, A.; Moore, M. J.; Richly, H.; Hendlisz, A.; Hirte, H.W.; Eder, J. P.; Lenz, H.; Schwartz, B. *Oncologist* **2007**, *12*, 426–437.

18.   Clark, J. W.; Eder, J. P.; Ryan, D.; Lee, R.; Lenz, H.-J. *Clin. Cancer Res.* **2005**, *11*, 5472–5480.

19.   Ratain, M. J.; Eisen, T.; Stadler, W. M.; Flaherty, K. T.; Kaye, S. B.; Rosner, G. L.; Gore, M.; Desai, A. A.; Patnaik, A.; Xiong, H. Q.; Rowinsky, E.; Abbruzzese, J. L.; Xia, C.; Simantov, R.; Schwartz, B.; O'Dwyer, P. J. *J. Clin. Oncol.* **2006**, *24*, 2505–2512.

20.   Escudier, B.; Eisen, T.; Stadler, W. M.; Szczylik, C.; Oudard, S.; Staehler, M.; Negrier, S.; Chevreau, C.; Desai, A. A.; Rolland, F.; Demkow, T.; Hutson, T. E.; Gore, M.; Anderson, S.; Hofilena, G.; Shan, M.; Pena, C.; Lathia, C.; Bukowski, R. M. *J. Clin. Oncol.* **2009**, *27*, 3312–3318.

21.   Llovet, J. M.; Ricci, S.; Mazzaferro, V.; Hilgard, P.; Gane, E.; Blanc, J.-F.; Oliveira, A. C.; Santoro, A.; Raoul, J.-L.; Forner, A.; Schwartz, M.; Porta, C.; Zeuzem, S.; Bolondi, L.; Greten, T. F.; Galle, P. R.; Seitz, J.-F.; Borbath, I.; Häussinger, D.; Giannaris, T.; Shan, M.; Moscovici, M.; Voliotis, D.; Bruix, J. *N. Engl. J. Med.* **2008**, *359*, 378–390.

22.   Cheng, A.-L.; Kang, Y.-K.; Chen, Z.; Tsao, C.-J.; Qin, S.; Kim, J. S.; Luo, R.; Feng, J.; Ye, S.; Yang, T.-S.; Xu, J.; Sun, Y.; Liang, H.; Liu, J.; Wang, J.; Tak, W. Y.; Pan, H.; Burock, K.; Zou, J.; Voliotis, D.; Guan, Z. *Lancet Oncol.* **2009**, *10*, 25–34.

23.   Bankston, D.; Dumas, J.; Natero, R.; Riedl, B.; Monahan, M.-K.; Sibley, R. *Org. Pro. Res. Deve.* **2002**, *6*, 777–781.

24.   Loegers, M.; Gehring, R.; Kuhn, O.; Matthaeus, M.; Mohrs, K.; Mueller-Gliemann, M.; Stiehl, J.; Berwe, M.; Lenz, J.; Heilmann, W. PCT Int. Appl. **2006**, WO2006034796.

25.   Rossetto, P.; Macdonald, P. L.; Canavesi, A. PCT Int. Appl. **2009**, WO2009111061.

7

Sunitinib (Sutent): An Angiogenesis Inhibitor

Martin Pettersson

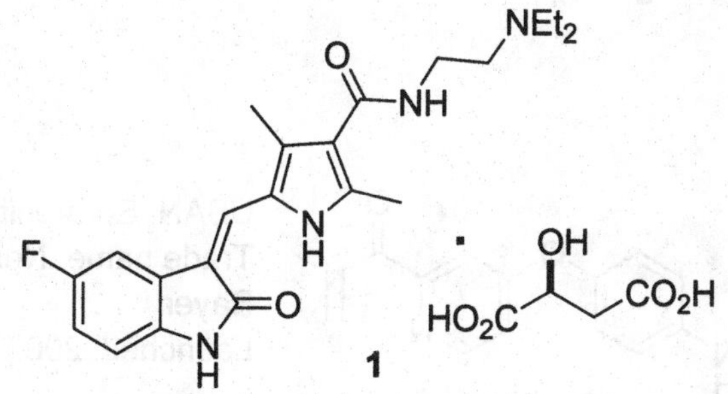

USAN: Sunitinib Maleate
Trade name: Sutent®
Pfizer Inc.
Launched: 2006

7.1   Background

In a seminal paper in 1971, Folkman proposed that inhibition of angiogenesis may be an effective therapy for cancer.[1] Angiogenesis is a process by which new capillary blood vessels are formed from endothelial cells in pre-existing blood vessels. This process is critical to supply oxygen and nutrients and allow tumors to grow larger than 0.2–2 mm in diameter.[2] In healthy individuals, there is an intricate balance between pro-angiogenic and anti-angiogenic factors, whereas in many cancers, these substances and their cell-surface receptors (receptor tyrosine kinases, RTKs) are mutated or over-expressed. Vascular endothelial growth factors (VEGFs) and platelet-derived growth factors (PDGFs) constitute key pro-angiogenic ligands, which upon binding to their corresponding receptors (VEGFRs and PDGFRs) initiate a signaling cascade that promotes angiogenesis.[3] Drugs that target these signaling pathways generally have lower toxicity compared to standard cytotoxic cancer therapy that block cell division in cancer cells and normal cells alike.[4]

Sunitinib maleate (1, Sutent, SU11248) was discovered by Sugen, which later became part of Pfizer. It is an orally active drug that exhibits potent anti-angiogenic activity through the inhibition of multiple RTKs. Specifically, 1 inhibits vascular endothelial growth factor receptors VEGFR1, VEGFR2, and VEGFR3 and platelet-derived growth factor receptors PDGFR-α and PDGFR-β. In addition, sunitinib also targets receptors implicated in tumerogenesis including fetal liver tyrosine kinase receptor 3 (Flt3) and stem-cell factor receptor (c-KIT).[5,6] A drug with a multitarget profile is less likely to develop drug resistance than a compound that displays selectivity for a single kinase. Furthermore, drugs that act on multiple pathways simultaneously may be more efficacious than single-targeted agents.[4] Additional multitargeted kinase inhibitors approved for the treatment of solid tumors include imatinib[7] (2, Gleevec; Novartis) and sorafenib[8] (3, Nexavar; Bayer). Sunitinib was approved by the FDA in 2006 for the treatment of gastrointestinal stromal tumors (GIST) and cases of renal cell carcinoma (RCC) for which standard treatments have failed. In 2007 the use of sunitinib was also approved as a first-line treatment for advanced RCC.[9]

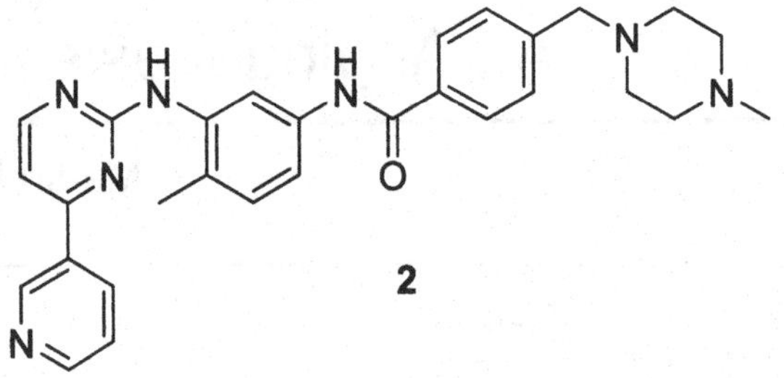

USAN: Imatinib
Trade name: Gleevec®
Novartis
Launched: 2001

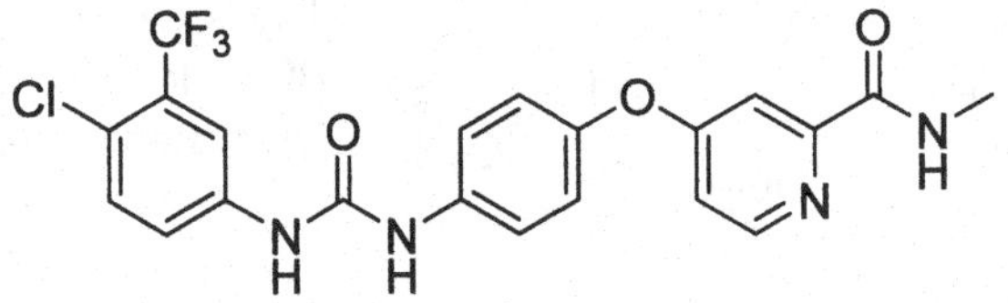

2

USAN: Sorafenib
Trade name: Nexavar®
Bayer
Launched: 2005

3

Cancer of the kidney and the renal pelvis was estimated to account for 3.9% of all new malignancies in the United States, with more than 57,000 new cases and almost 13,000 deaths predicted to occur in 2009.[10] RCC is the most common form of kidney cancer among adults and accounts for 85% of all cases.[11] Surgery remains an important method of treatment, but in cases where the cancer has metastasized, additional therapies are required. RCC is resistant to radiation therapy and exhibits only a 4–6% response rate to various chemotherapeutic agents. Standard methods of treatment have involved immunomodulatory therapies such as interferon-α (INFα) and interleukin-2 (IL2), however; their effectiveness is limited.[11] In addition, IL2 is poorly tolerated among patients with metastatic RCC.[11] Thus, given the high lethality of metastatic RCC, this disease represents a highly unmet medical need.

Recent advances in the treatment of RCC stem from major breakthroughs in the underlying molecular biology of this disease. It has been shown that many patients with clear-cell RCC, which accounts for the majority of cases, have a mutation in the von Hippel–Lindau gene. The corresponding gene product, the VHL protein, functions as a tumor-suppressor. Loss of the VHL protein results in an increase in the hypoxia-inducible factor (HIF), and this in turn leads to overexpression of pro-angiogenic growth factors such as VEGF.[11]

Proof of concept for using an anti-angiogenic agent in cancer therapy was first established in patients with metastatic colorectal cancer using bevacizumab (Avastin; Genentech/Roche).[12,13] Launched in 2004, this recombinant humanized monoclonal antibody selectively targets VEGF and prevents it from binding to the VEGF receptors. It was shown to be highly efficacious in combination with 5-fluorouracil chemotherapy in patients with metastatic colorectal cancer.[12] After receiving FDA approval for non-small cell lung cancer, metastatic breast cancer, and glioblastoma, bevacizumab was approved as a first-line therapy for renal cell carcinoma in 2009.[9]

. The first small-molecule multi-kinase inhibitor for the treatment of advanced RCC, sorafenib (3), reached the market in 2005 only 5 years after entering phase I clinical trials.[8] This orally active drug targets various Raf isoforms as well as RTKs implicated in angiogenesis (VEGFR1, VEGFR2 and VEGFR3, and PDGFR-β) and tumerogenesis (Flt3 and c-KIT). Sorafenib (3) was evaluated in combination with several commonly used chemotherapeutic agents, such as carboplatin/pacilitaxel, and was found not to increase the toxicity of these drugs. The favorable safety profile of 3 coupled with its broad spectrum of activity in preclinical cancer models prompted additional clinical studies, which led to the FDA approval in 2007 for the treatment of hepatocellular carcinoma.[9]

7.2   Discovery and Development of Sunitinib

The discovery path to sunitinib (1) commenced with random screening of a compound collection, which resulted in the identification of several leads that contain an indoline-2-one core. Using structure-guided design, Sun and co-workers prepared a number of 3-substituted oxindole derivatives that were potent and selective VEGFR and PDGFR inhibitors. SU-5416 (4)[14] was found to be selective for VEGFR2, whereas SU-6668 (5),[15] in which the 4'-position bears a carboxylic acid, was selective for PDGFR-β. Upon entering clinical trials, however, 4 failed because of toxicity and solubility issues, while 5 was found to have an inadequate pharmacokinetic profile.[6] Further structural optimization was required to improve solubility, reduce protein binding, and increase VEGFR2 and PDGFR-β activity while also mitigating toxicity.[16] Examination of a co-crystal structure of 5 and FGFR1-kinase, which has high amino acid sequence homology to VEGFR2, suggested that substituents in the 4'-position may be partially solvent exposed. This was seen as an opportunity to modulate physicochemical and pharmacokinetic properties, and efforts focused on this region of the molecule eventually resulted in the discovery of SU-11248 (sunitinib free base, 6).[16] Compared to 5, indolinone 6, which incorporates a C4' amide moiety bearing a secondary amine, had dramatically improved solubility: At pH 2, the solubilities of 5 and 6 were < 5 μg/mL and 2582 μg/mL respectively, while at pH 6 the solubilities were 18 μg/mL and 511

$\mu g/mL.^{16}$ Moreover, **6** exhibited favorable kinase activity in cell-based assays for VEGFR2 (IC_{50} = 10 nM) and PDGFRβ (IC_{50} = 10 nM).[17]

4, R = H; SU-5416

5, R = $CH_2CH_2CO_2H$; SU-6668

6, R' = Et; SU-11248, sunitinib free base

7, R' = H; SU-12662

In vivo studies using athymic nude mice demonstrated that **6** inhibits the growth of several human tumor xenografts.[17] The minimum effective plasma concentration required to inhibit the VEGF and PDGF receptors was found to be at or above 125–250 nM. At the efficacious oral dose of 40 mg/kg/day, receptor phosphorylation was blocked for 12 h but recovered before administration of the next dose. Based on this observation, Mendel and co-workers concluded that continuous inhibition of VEGFR2 and PDGFR-β was not always necessary to achieve maximum efficacy. Measurements of plasma protein binding indicated that **6** was 95% bound in both mouse and human plasma, which translates into a free drug concentration of 6–12 nM at the 40 mg/kg/day dose. These drug levels are similar to the in vitro IC_{50} values that were established in the cell-based assays.

Preclinical toxicology studies revealed some adverse effects including bone marrow depletion and pancreatic toxicity in rats and monkeys.[6] Because of these findings, a dosing regimen involving 4 weeks of 50 mg/day oral administration followed by 2 weeks of rest was established for the clinical trials. Metabolism of **6** was found to be mediated primarily by cytochrome P450 3A4 (CYP 3A4) resulting in *N*-dealkylation to afford SU-12662 (**7**).[6] It is interesting that this primary metabolite had in vitro VEGFR, PDGFR, and c-KIT activity that was similar to the activity of the parent compound **1**.

In the clinic, sunitinib was not only efficacious in treating renal cell carcinoma[18] but also showed significant antitumor activity toward gastrointestinal stromal tumors (GIST).[19] GIST is relatively rare and only accounts for 1–3% of all gastrointestinal tumors, with approximately 5000 new cases per year in the United States.[5] Nevertheless, GIST represents a high unmet medical need due to limited treatment options. Imatinib (**2**) has been the drug of choice for the treatment GIST, but many patients develop resistance or intolerance to **2**. Sunitinib (**1**) was evaluated in a large randomized placebo-controlled phase III study in patients with resistance or intolerance to **2**, and tumor progression was found to be more than four times longer in patients taking **1** compared to placebo. The effectiveness of **1** and **2** in the treatment of GIST is a result of their potent inhibition of stem cell factor receptor c-KIT, which is activated through mutation in a majority of patients with GIST. Due to its multitarget kinase activity, sunitinib (**1**) is currently in clinical trials for a number of cancers including breast, lung, colorectal, and intestinal cancer.

7.3 Syntheses

7.3.1 Discovery Route

The discovery route to the sunitinib free base **6** as described by Sun and co-workers commenced with a two-step synthesis of 5-fluorooxindole (**10**) starting from the corresponding 5-fluoroisatin (**8**) (Scheme 7.1).[16,20] Heating a neat mixture of **8** and hydrazine hydrate to 110 °C effected a Wolff–Kishner reduction and ring opening to give acyl hydrazide **9**. The crude material was subjected to an intramolecular acylation reaction by exposure to aqueous HCl at room temperature to afford the oxindole **10** in 73% yield over two steps. The requisite pyrrole subunit **14** was prepared using the Knorr pyrrole synthesis (Scheme 7.2).[21,22] *tert*-Butyl acetoacetate (**11**) was reacted with sodium nitrite in acetic acid to furnish oxime **12**. Exposure of this intermediate to ethyl acetoacetate (**13**) under reductive cyclization conditions using zinc and acetic acid afforded the tetra-substituted pyrrole **14** in 65% yield over two steps. Selective hydrolysis of the *tert*-butyl ester, followed by decarboxylation, was accomplished by stirring **14** in HCl and ethanol to provide intermediate **15** in 87% yield. The unsubstituted position of pyrrole **15** was then formylated in quantitative yield by treatment with Vilsmeier reagent generated from DMF and POCl$_3$ in dichloromethane. Hydrolysis of the ethyl ester using aqueous potassium hydroxide in methanol provided the key intermediate **17** in 94% yield, thus setting the stage for the introduction of the amide side chain and the final coupling reaction with oxindole **10**. The EDCI-mediated amidation of carboxylic acid **17** with diamine **18** afforded the desired amide **19** in 43% yield.[16] With sufficient quantities of **19** in hand, the synthesis could be completed via condensation of **19** with 5-fluorooxindole (**10**) to furnish the sunitinib free base **6** in 88% yield.

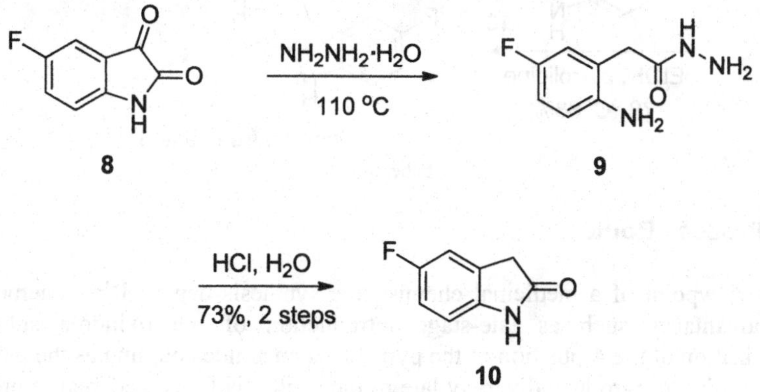

Scheme 7.1

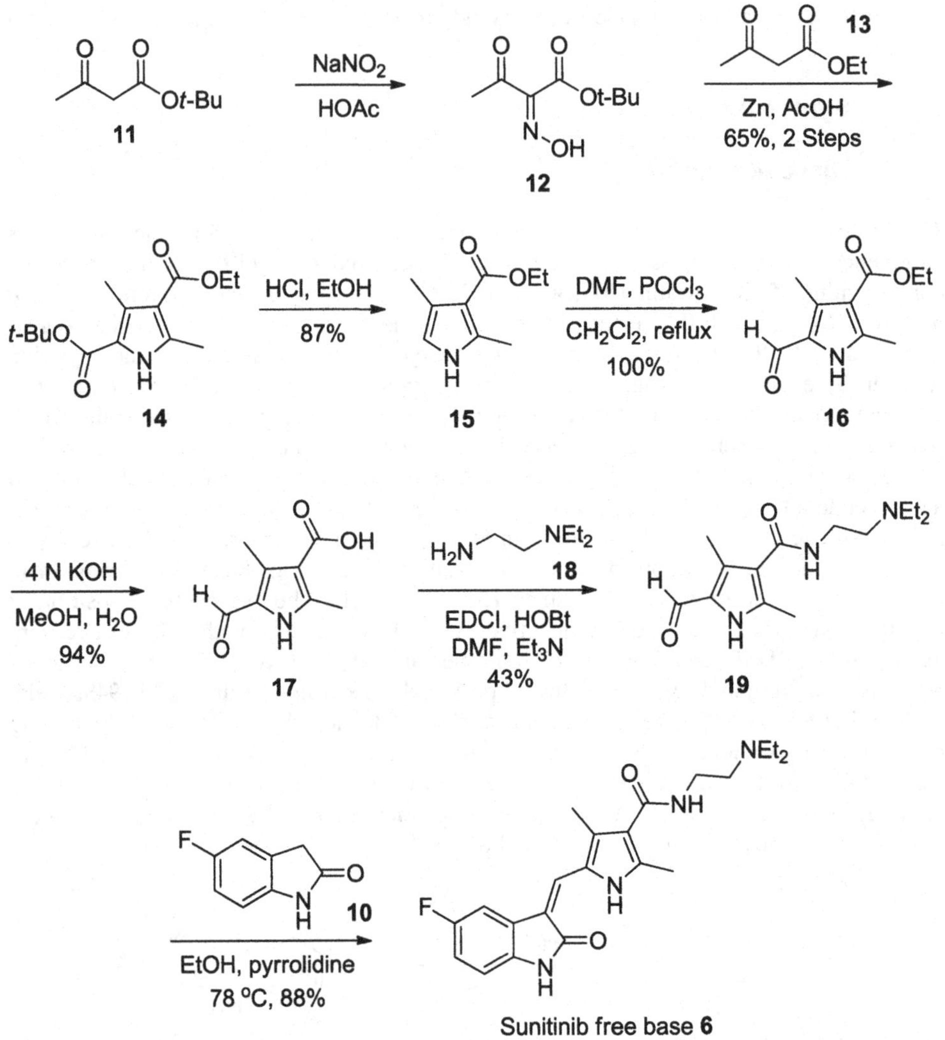

Scheme 7.2

7.3.2 Process Route

From the viewpoint of a medicinal chemist, the synthesis depicted in Scheme 7.2 had several advantages such as late-stage introduction of the oxindole subunit and functionalization of the 4-position of the pyrrole via an amide coupling as the penultimate step. The ability to introduce diversity late in the synthesis facilitated analog preparation and enabled rapid exploration of chemical space. However, for developing a large-scale process, there were several aspects of the late-stage amidation approach that needed to be addressed.[23,24] For example, formation of **19** was complicated by competing imine formation resulting from condensation of diamine **18** with aldehyde **17**. Moreover, the amide coupling reaction required activation of the carboxylic acid using reagents such as

EDCI. For a commercial synthesis it is preferable to avoid the use of carbodiimide coupling reagents because they are strong sensitizers.[24] In addition, these reagents generate stoichiometric amounts of urea by-products that can complicate isolation and purification. Early process development efforts were focused on improving the coupling reaction through activation of **17** as either an acid chloride or a mixed anhydride, but these efforts were not fruitful. Treatment of carboxylic acid **17** with isobutyl chloroformate (**20**) afforded the mixed anhydride **21** (Scheme 7.3), but its subsequent exposure to diamine **18** did not provide the desired coupling product **19** but instead returned the starting acid **17** along with carbamate **22**.[24]

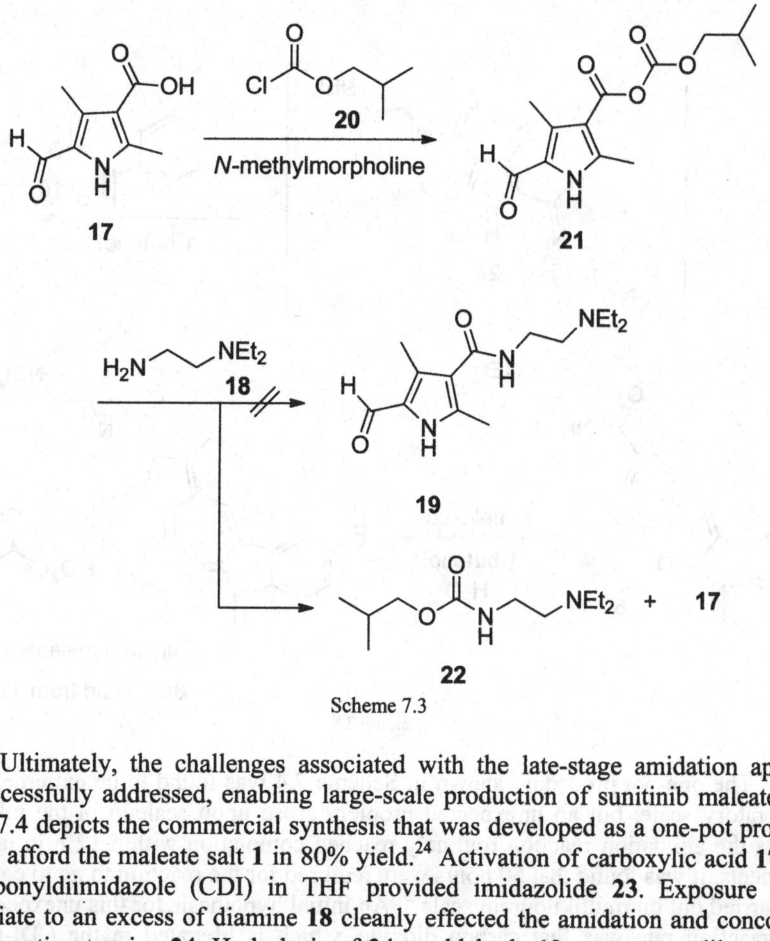

Scheme 7.3

Ultimately, the challenges associated with the late-stage amidation approach were successfully addressed, enabling large-scale production of sunitinib maleate **1**.[24,25] Scheme 7.4 depicts the commercial synthesis that was developed as a one-pot process to routinely afford the maleate salt **1** in 80% yield.[24] Activation of carboxylic acid **17** using *N,N*'-carbonyldiimidazole (CDI) in THF provided imidazolide **23**. Exposure of this intermediate to an excess of diamine **18** cleanly effected the amidation and concomitant imine formation to give **24**. Hydrolysis of **24** to aldehyde **19** was not readily achieved, but this was found to be unnecessary since the imine function of **24** was able to undergo condensation with oxindole **10** to afford the sunitinib free base (**6**) directly. Imidazole and small amounts of impurities were then removed by aqueous workup. The final isolation of the active pharmaceutical ingredient (API) was accomplished by forming the maleate salt **1**, which exhibited superior filtration properties as compared to the free base **6**.

Following addition of L-malic acid, the reaction mixture was concentrated through vacuum distillation, which resulted in precipitation of sunitinib maleate (1).[24]

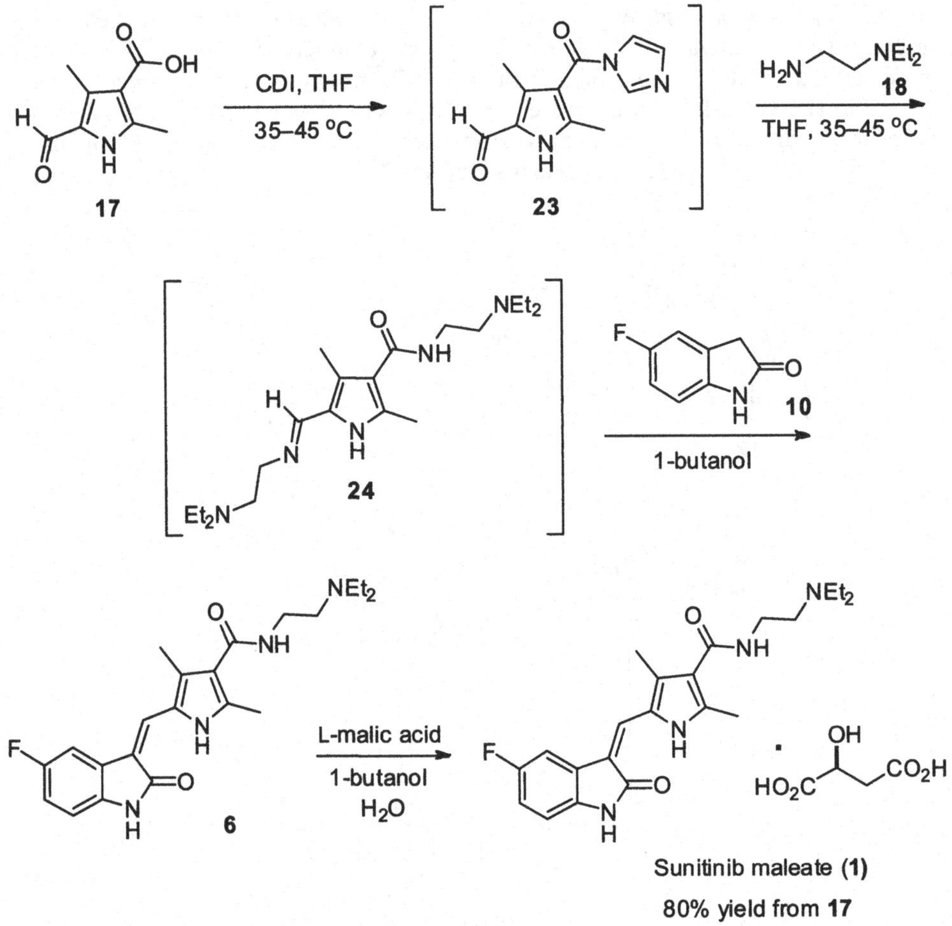

Scheme 7.4

The one-pot procedure shown in Scheme 7.4 was found to be extremely robust on laboratory scale, but an unexpected problem arose upon scaleup in the pilot plant. Whereas the amidation reaction routinely reached completion within 12 h in laboratory experiments, it was found that 50 hours were required for the reaction to go to completion when carried out on multikilogram scale.[24] An initial hypothesis for this unexpected drop in the reaction rate was that carbon dioxide, which is liberated in the CDI-mediated coupling, may play an important role in the reaction. It was postulated that CO_2 may be removed more slowly from the large-scale reaction and would be available to react with diamine **18**, thereby reducing the rate of the amide coupling. In an attempt to test this hypothesis, imidazolide **23** was prepared and used as the starting material for two parallel amidation reactions. In the first experiment, the reaction mixture was sparged with CO_2

for 15 min before adding diamine **18**, while the second amidation experiment was carried out in the absence of CO_2. Unexpectedly, the reaction rate was found to be considerably slower under CO_2-free conditions as compared to the control experiment in which the reaction mixture had been saturated with CO_2. After 4 h, the two reactions reached 37% and 88% completion, respectively. Having made the critical observation that CO_2 actually enhances the rate of the reaction, the chemistry team modified the process so as to prevent or slow the release of CO_2 from the reactor by stirring the reaction more slowly and lowering the nitrogen flow-rate. Gratifyingly, these modifications solved the problem and allowed the large-scale amidation to reach completion within 12 h. Vaidyanathan and co-workers later explored the generality of this CO_2-mediated amidation of imidazolides and proposed a possible mechanism for the observed rate-enhancement.[26]

During early process development, an alternative route was explored that circumvented the problems initially encountered in the late-stage amidation approach. This strategy involved introducing the amide side chain before pyrrole formation and the subsequent formylation step. Toward this end, the process chemistry team at Pfizer designed an elegant approach starting from diketene (**25**).[23,24,27] As depicted in Scheme 7.5, nucleophilic addition of diamine **18** to diketene (**25**) afforded the β-ketoamide **26**. Because of the instability of **26**, this material was either used directly in the ensuing Knorr pyrrole synthesis or stored at –20 °C. Heating a mixture of **26** and oxime **12**, in the presence of zinc and acetic acid furnished pyrrole **27** in ~60% yield. Again, whereas this reaction worked well on small scale, limitations to this procedure were encountered when the reaction was carried out on larger scale (> 10 g).

Due to the presence of the secondary amine in pyrrole **27**, the reaction mixture had to be made basic to enable product isolation. However, upon reaching pH 9, gelatinous zinc salts precipitated out of the reaction, and all attempts to remove the salts through filtration were unsuccessful. Adjusting the pH to 14 dissolved the zinc salts, whereupon pyrrole **27** could be isolated by extraction with dichloromethane. Although these efforts allowed for isolation of the desired product in reasonable yield, it was anticipated that the precipitation of zinc salts during the workup could be problematic on large scale by binding to the mechanical stirrer in the reactor. In addition, minimization of the waste stream is not only important in process development from an environmental perspective, but directly impacts the cost-of-goods (COG) by reducing the costs associated with disposal of chemical waste. Therefore, the generation of stoichiometric amounts of zinc salts was undesirable, and an alternative method to forming pyrrole **27** was preferred.

Toward this end, the process team envisioned that the oxime reduction and subsequent cyclization to form pyrrole **27** could be effected using hydrogenative conditions, which would avoid the problems associated with stoichimetric zinc salts and therefore greatly simplify the workup and isolation of the product. This strategy was successfully realized by stirring a solution of oxime **12** and β-ketoamide **26** in acetic acid under 45 psi of H_2 in the presence of 10% palladium on carbon. The desired product **27** could then be isolated in 77% yield after filtration to remove the catalyst, pH adjustment to 11–13, and subsequent extraction into CH_2Cl_2.

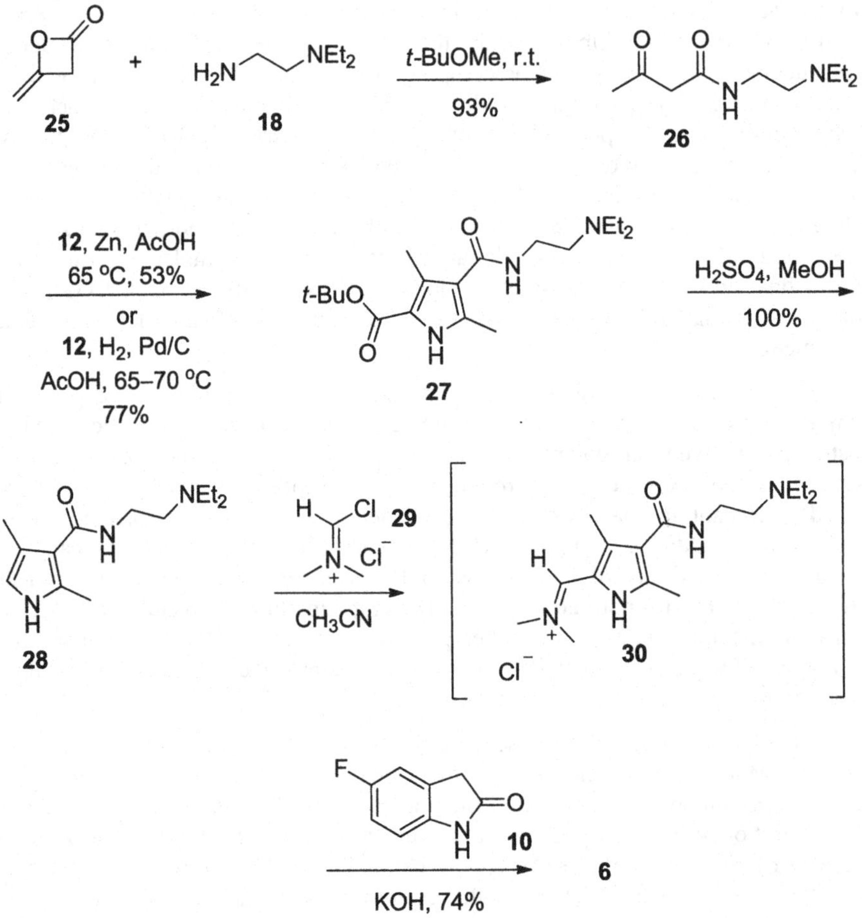

Scheme 7.5

With the key pyrrole **27** in hand, the team was in a position to complete the synthesis. Hydrolysis and decarboxylation of the *tert*-butyl ester was initially attempted using the conditions employed in the discovery route for the conversion of **14** to **15**. While these reaction conditions effected the hydrolysis and decarboxylation of **27** to **28** in good yield, the formation of impurities resulting from dimerization of the pyrrole was also observed. After screening various acids, the team eventually found that side product formation could be completely suppressed using 1 M H_2SO_4 in 3:1 MeOH/H_2O at 65 °C to afford the pyrrole **28** in quantitative yield.

All that remained at this point was the formylation of pyrrole **28** followed by condensation with 5-fluorooxindole (**10**). While the initial discovery route had accomplished this in two separate steps, the process team developed a one-pot procedure. Thus pyrrole **28** was added to a solution of the Vilsmeier reagent **29** in acetonitrile at room temperature (Scheme 7.5). Following the completion of the formylation reaction, the resulting intermediate **30** was treated with 5-fluorooxindole (**10**) and pulverized KOH, which led to precipitation of the desired product **6**. Filtration of the reaction mixture afforded the sunitinib free base (**6**) in 74% yield.

In summary, sunitinib maleate (**1**) is a multitargeted receptor tyrosine kinase inhibitor with potent anti-angiogenic and antitumor activity. It is approved for the treatment of advanced renal cell carcinoma and gastrointestinal stromal tumors, and is currently undergoing clinical trials for a number of additional malignancies. The discovery synthesis of **1** along with its process development approaches were described in this chapter.

7.4   References

1.   Folkman, J. *N. Engl. J. Med.* **1971**, *285*, 1182–1186.
2.   Roskoski, R. Jr. *Crit. Rev. Oncol. Hematol.* **2007**, *62*, 179–213.
3.   Cherrington, J. M.; Strawn, L. M.; Shawver, L. K. *Adv. Cancer Res.* **2000**, *79*, 1–38.
4.   Faivre, S.; Djelloul, S.; Raymond, E. *Semin. Oncol.* **2006**, *33*, 407–420.
5.   Atkins, M.; Jones, C. A.; Kirkpatrick, P. *Nat. Rev. Drug. Discov.* **2006**, 5, 279–280.
6.   Faivre, S.; Demetri, G.; Sargent, W.; Raymond, E. *Nat. Rev. Drug Discov.* **2007**, *6*, 734–745.
7.   Capdeville, R.; Buchdunger, E.; Zimmerman, J.; Matter, A. *Nat. Rev. Drug Discov.* **2002**, *1*, 493–502.
8.   Wilhelm, S.; Carter, C.; Lynch, M.; Lowinger, T.; Dumas, J.; Smith, R.; Schwartz, B.; Simantov, R.; Kelley, S. *Nat. Rev. Drug Discov.* **2006**, *5*, 835–844.
9.   FDA labeling information, www.fda.gov, accessed 11/04/2009.
10.   Jemal, A.; Siegel, R.; Ward, E.; Hao, Y.; Xu, J.; Thun, M. *J. CA Cancer J. Clin.* **2009**, *59*, 225–249.
11.   Cohen, H. T.; McGovern, F. J. *N. Engl. J. Med.* **2005**, *353*, 2477–2490.
12.   Hurwitz, H.; Fehrenbacker, L.; Novotny, W.; Cartwright, T.; Hainsworth, J.; Heim, W.; Berlin, J.; Baron, A.; Griffin, S.; Holmgren, E.; Ferrara, N.; Fyfe, G.; Rogers, B.; Ross, R.; Kabbinavar, F. *N. Engl. J. Med.* **2004**, *350*, 2335–2342.
13.   Ferrara, N.; Hillan, K. J.; Gerber, H.-P.; Novotny, W. *Nat. Rev. Drug Discov.* **2004**, *3*, 391–400.
14.   Sun, L.; Tran, N.; Tang, F.; App, H.; Hirth, P.; McMahon, G.; Tang, C. *J. Med. Chem.* **1998**, *41*, 2588–2603.
15.   Sun, L.; Tran, N.; Liang, C.; Tang, F.; Rice, A.; Schreck, R.; Waltz, K.; Shawver, L. K.; McMahon, G.; Tang, C. *J. Med. Chem.* **1999**, *42*, 5120–5130.
16.   Sun, L.; Liang, C.; Shirazian, S.; Zhou, Y.; Miller, T.; Cui, J.; Fukuda, J. Y.; Chu, J.-Y.; Nematalla, A.; Wang, X.; Chen, H.; Sistla, A.; Luu, T. C.; Tang, F.; Wei, J.; Tang C. *J. Med. Chem.* **2003**, *46*, 1116–1119.
17.   Mendel, D. B.; Laird, A. D.; Xin, X.; Louie, S. G.; Christensen, J. G.; Li, G.; Schreck, R. E.; Abrams, T. J.; Ngai, T. J.; Lee, L. B.; Murray, L. J.; Carver, J.; Chan, E.; Moss, K. G.; Haznedar, J. O.; Sukbuntherng, J.; Blake, R. A.; Sun, L;

Tang, C.; Miller, T.; Shirazian, S.; McMahon, G.; Cherrington, J. M. *Clin. Cancer Res.* **2003**, *9*, 327–337.

18. Thompson Coon, J. S.; Liu, Z.; Hoyle, M.; Rogers, G.; Green, C.; Moxham, T.; Welch, K.; Stein, K. *Br. J. Cancer*, **2009**, *101*, 238–243.

19. Demetri, G. D.; van Oosterom, A. T.; Garrett, C. R.; Blackstein, M. E.; Shah, M. H.; Verweij, J.; McArthur, G.; Judson, I. R.; Heinrich, M. C.; Morgan, J. A.; Desai, J.; Fletcher, C. D.; George, S.; Bello, C. L.; Huang, X.; Baum, C. M.; Casali, P. G. *Lancet* **2006**, *368*, 1329–1338.

20. Tang, P. C.; Miller, T.; Li, X.; Sun, L.; Wei, C. C.; Shirazian, S.; Liang, C.; Vojkovsky, T.; Nematalla, A. S.; WO 01/060814, **2001**.

21. Fisher, H. *Organic Synthesis*, Wiley: New York, **1943**, 2: 202.

22. Treibs, A.; Hintermeier, K. *Chem. Ber* **1954**, *86*, 1167.

23. Manley, J. M.; Kalman, M. J.; Conway, B. G.; Ball, C. C.; Havens, J. L.; Vaidyanathan, R. *J. Org. Chem.* **2003**, *68*, 6447–6450.

24. Vaidyanathan, R. In *Process Chemistry in the Pharmaceutical Industry*, Ed. Braish, T. and Gadamasetti, K. CRC Press, Boca Raton, Fl, **2008**, 49–63.

25. Jin, Q.; Mauragis, M. A.; May, P. D. WO 03/070725, **2003**.

26. Vaidyanathan, R.; Kalthod, V. G.; Ngo, D. P.; Manley, J. M.; Lapekas, S. P. *J. Org. Chem.* **2004**, *69*, 2565–2568.

27. Havens, J. L.; Vaidyanathan, R. U.S. Pat. 06/0009510, **2006**.

8

Bortezomib (Velcade):
A First-in-Class Proteasome Inhibitor

Benjamin S. Greener and David S. Millan

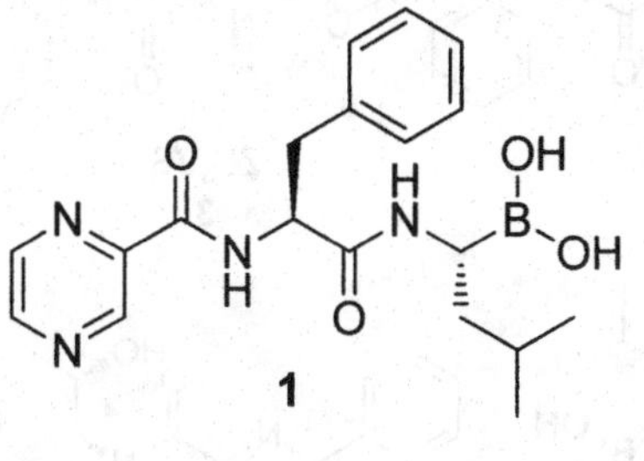

USAN: Bortezomib
Trade name: Velcade®
Millennium Pharmaceuticals
Launched: 2003

8.1   Background

Bortezomib (**1**, Velcade, formerly PS-341) is an intravenously administered first-in-class proteasome inhibitor that was approved by the FDA in 2003 for the treatment of patients with multiple myeloma (MM) who had received at least two prior therapies. In 2007 its maker, Millennium Pharmaceuticals, was granted FDA approval for bortezomib (**1**) as a front-line therapy for MM. MM is the second most common hematological disease, which is characterized by neoplastic proliferation of plasma cells in the bone marrow. It affects 2–3 people per 100,000, with patients having a median survival time of 3 years. Treatment regimens, before the introduction of bortezomib (**1**), relied on alkylating agents (melphalan) and glucocorticoids (prednisone), which prolonged survival times but did not offer a cure. By contrast, in the bortezomib Phase II Study of Uncontrolled Multiple Myeloma Managed with Proteasome Inhibtion Therapy (SUMMIT) trial, remarkably 35% of the 192 patients demonstrated a response to treatment, and therefore bortezomib offered a new hope in the treatment of this disease. Bortezomib (**1**) was also later approved for the treatment of patients with the rarer blood disease, mantle cell lymphoma, who have received at least one prior therapy. The drug is currently being evaluated in several clinical trials for solid tumors, other hematological malignancies, and in particular, as part of combination therapies.[1–5]

During the last two decades there has been extensive research into proteasome inhibitors, with several synthetic and natural product based inhibitors being identified. Most synthetic proteasome inhibitors are dipeptides or tripeptides ending in a "warhead" designed to bind to a specific residue in the active site of the proteasome.[6–9] One of the first reported inhibitors was MG132 (2), with its aldehyde warhead reversibly binding to a threonine in the active site of the proteasome.[10,11] However, compounds such as 2 suffered from a lack of selectivity over other proteases, such as cathepsin B and calpains, and also displayed poor oral pharmacokinetics.[11] Advances in specificity were made with ZL₃VS (3), which is armed with a vinyl sulfone warhead that irreversibly binds to the proteasome.[12] Probably the biggest leap forward came with the boronic acid–containing warhead MG262 (4), with a 100-fold potency improvement over its aldehydic analogue (MG132), and with selectivity over cysteine proteases.[11] Improvements in ligand efficiency and selectivity resulted in bortezomib (1), and together with its ease of synthesis made it a suitable candidate for clinial evaluation.[11] More recently another boronic acid proteasome inhbitor, CEP-18770 (5), has been described. The compound is reported to be potent, selective for proteasomal activity and is also orally bioavailable in rodents, and may represent a significant step forward.[13]

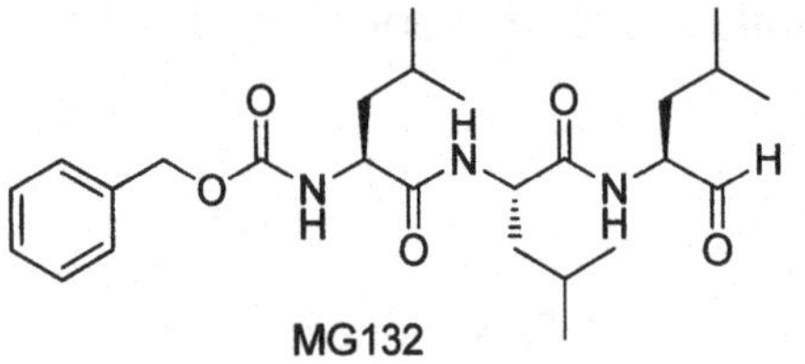

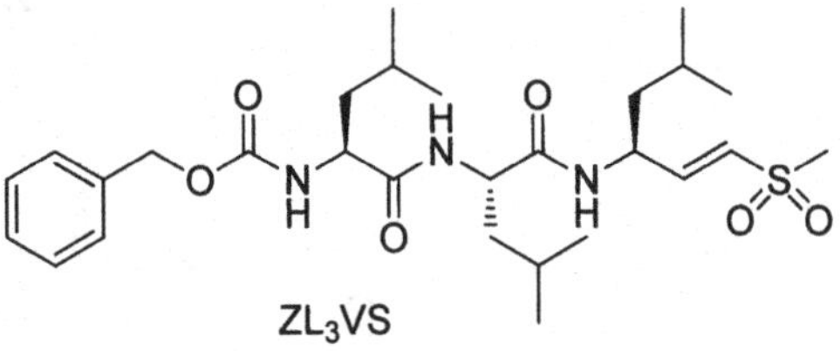

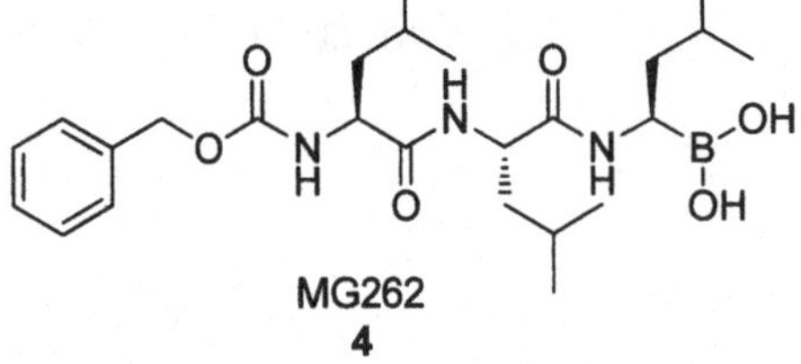

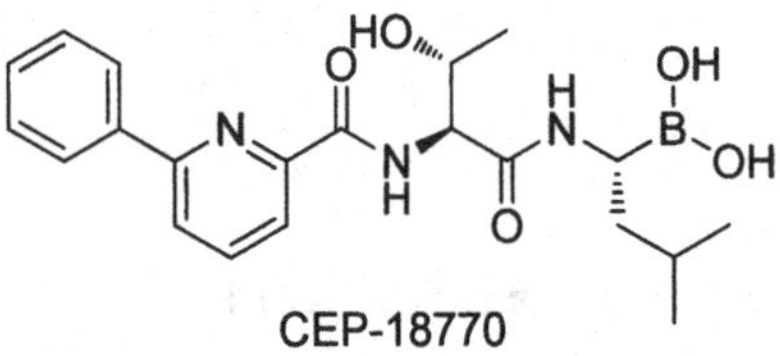

The literature also describes several proteasome inhibitors derived from natural products, with many of these irreversibly inhibiting the proteasome. Carfilzomib (formerly PR-171, 6) is an epoxyketone peptide, related to the natural product epoxomicin (7), which is currently in phase II clinical trials for patients with relapsed solid tumors, including nonsmall-cell lung cancer, small-cell lung cancer, ovarian and renal cancers. It is also in a single-agent phase II trial for patients with MM.[14] Lactacystin (8) is a natural product isolated from *Streptomyces lactacystinaeus* that exhibits proteasomal activity. The pharmacological activity is believed to orginate by nonenzymatic conversion of the inactive 8 into the active omuralide (9). It is 9 which is believed to react with a threonine residue in the catalytic site of the proteasome, through ring opening at the lactone carbonyl.[8] Salinosporamide (NPI-0052, **10**) is a related β-lactone that is currently in phase Ib clinical trials in combination therapy to treat patients with nonsmall-cell lung cancer, pancreatic cancer and melanoma.[9]

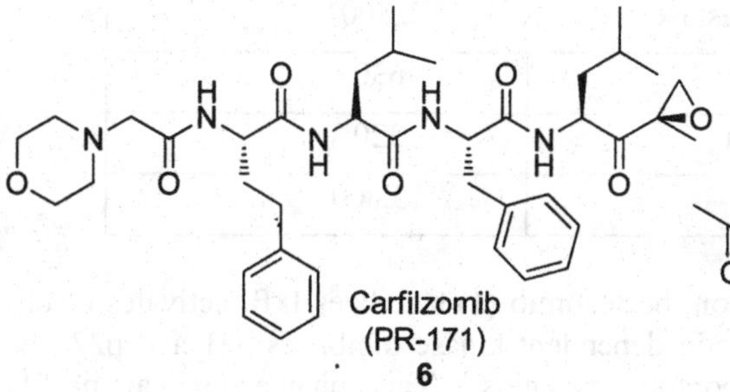

Carfilzomib
(PR-171)
6

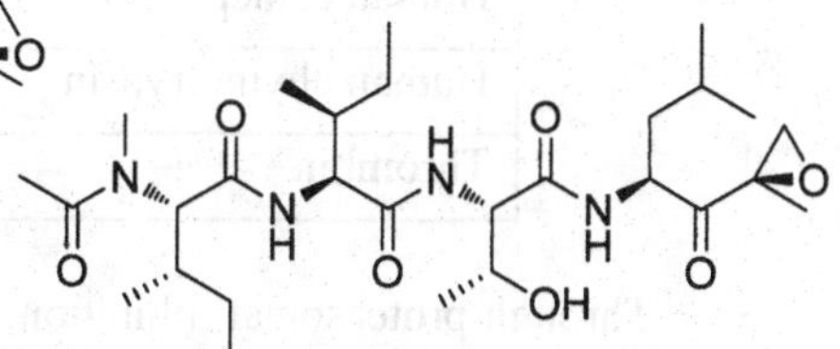

Epoxomicin
7

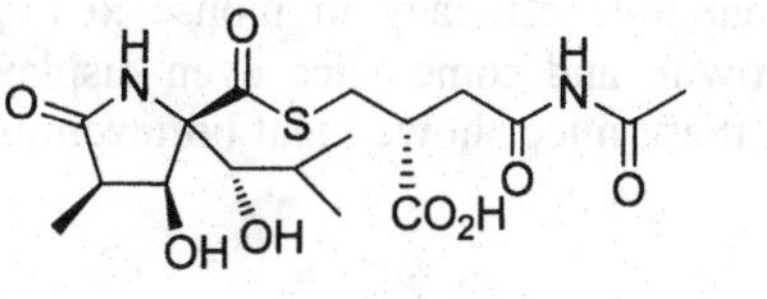

Lactacystin
8

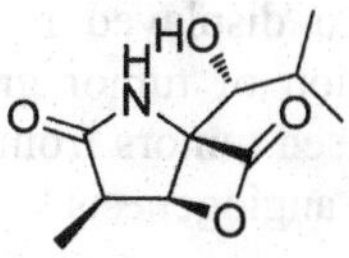

Omuralide
9

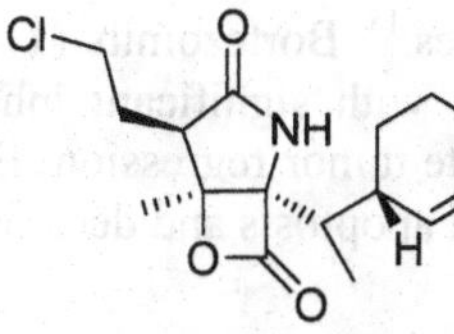

Salinosporamide
(NPI-0052)
10

8.2   Pharmacology

Protein synthesis and degradation are highly regulated cellular processes that are essential for normal cell division and cell survival. Many of the cellular processes underlying carcinogenesis and cancer progression are due to an imbalance in proteins that control cell division (e.g., cyclins), apoptosis (e.g., pro-apoptotic protein Bax), tumor suppression (e.g., p53), and stress response (e.g., NF-κB). In the normal healthy cell, the majority of intracellular proteins that require degradation, due to damage, misfolding, or transient signaling molecules, are labeled through polyubiquitination, targeting them for proteolysis within the multicatalytic 26S proteasome. The 26S proteasome is a cylindrical structure comprised of a 20S catalytic core, with caspase-like, trypsin-like, and chymotrypsin-like activities. The catalytic core is capped by two 19S regulatory subunits that are involved in directing the entry of polyubiquitin tagged proteins into the enzyme complex. Bortezomib (**1**) preferentially inhibits the chymotrypsin-like activity of the proteasome, resulting in accumulation of pro-apoptotic proteins in the cell, ultimately leading to apoptosis and cell death.[2,3] The compound binds strongly to the proteasomal complex, with a Ki of 620 pM. Bortezomib (**1**) also displays remarkable selectivity over other proteases (Table 1).[11]

Table 1. Potency and Selectivity Profile of Bortezomib (1)

Enzyme	Ki (nM)
20S Proteasome	0.62
Human Leukocyte elastase	2,300
Human cathepsin G	630
Human chymotrypsin	320
Thrombin	13,000

Through proteasomal inhibition, bortezomib (1) stabilizes IκB, activates c-Jun-terminal kinase, and stabilizes the cyclin dependent kinase inhibitors p21 and p27, the tumour suppressor p53 and pro-apoptotic proteins.[8] This phamacological profile translated into substantial activity across the National Cancer Institutes panel of 60 cancer cell lines.[15] Bortezomib (1) also displayed remarkable efficiacy in mouse xenograft models, with significant inhibition of tumor growth, and some mice even displaying complete tumor regression. Excised tumors from these mice showed that bortezomib (1) induced apoptosis and decreased angiogenesis.[16]

8.3   Structure–Activity Relationship (SAR)

Peptidyl aldehydes, such as MG132 (2), were first described as proteasomal inhibitors as early as 1993.[10] Researchers at ProScript (Millennium Pharmaceuticals legacy) took MG132 (2) and attempted to optimize the peptidic portion of the molecule by scanning each amino acid residue. They concluded that leucine is preferred at P1, but that larger, lipophilic nonnatural amino acids were more potent at P2 and P3. However, the selectivity for the proteasome over other endogenous proteases was not acceptable for further progression in the screening sequence. Moreover, the lack of configurational stability adjacent to the aldehyde functionality, prompted a search for a replacement warhead. Several reversible and irreversibly binding warheads (11–14) were designed and synthesized, but generally led to a reduction in proteasomal potency (Table 2). The introduction of the boronic acid functionality in MG262 (4) led to a 100-fold jump in potency, and crucially also produced a compound demonstrating cysteine protease selectivity. The potent binding is attributable to the availability of an empty p-orbital on boron, which accepts a lone pair of electrons from the donating threonine residue, thereby forming a stable and reversible tetrahedral complex. The selectivity over cysteine proteases was rationalized on the basis of a much weaker interaction between the thiol of cysteine and boron of 4. Improvements in selectivity over common serine proteases, as well as an acceptable oral pharmacokinetic profile were however still required. In an attempt to reduce the molecular weight of the compound, truncated versions of 4 were designed and synthesized, such as 15 and 1, which maintained the desired potency for the proteasome, and crucially were selective over serine proteases, such as chymotrypsin and elastase. This work ultimately led to the identification of the potent and selective boronic acid, bortezomib (1).[11]

Table 2. SAR Progression Leading to Bortezomib (1)

Compound	Structure	Ki (nM)
MG132 (2)	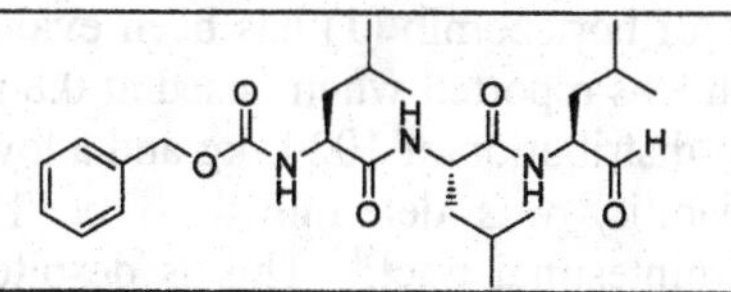	4
11	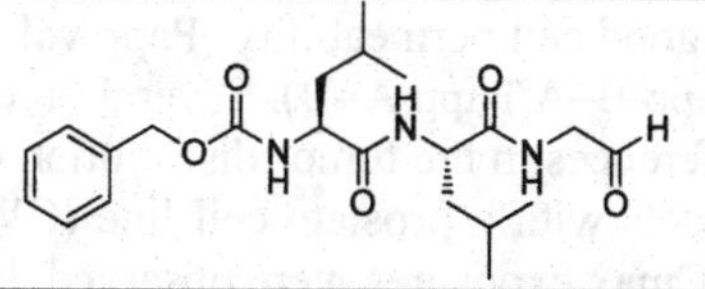	3800
12	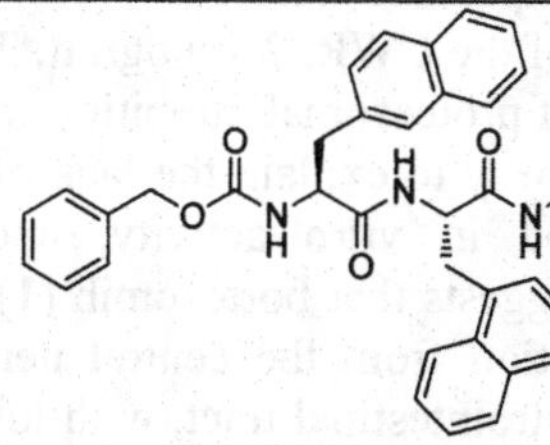	0.015
13	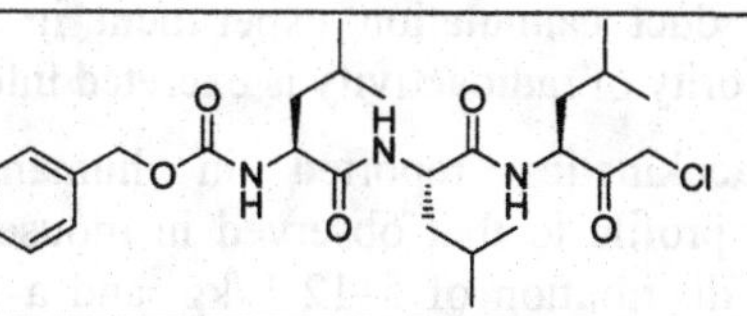	22000
14	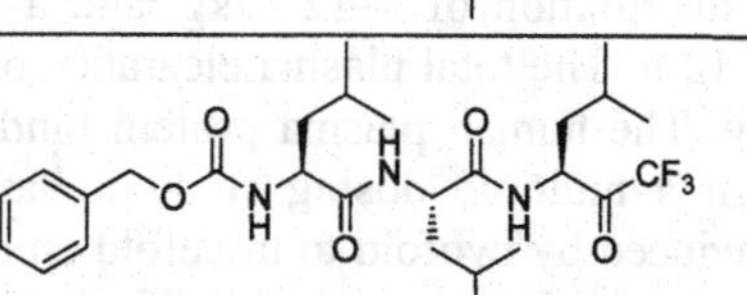	1400
MG262 (4)	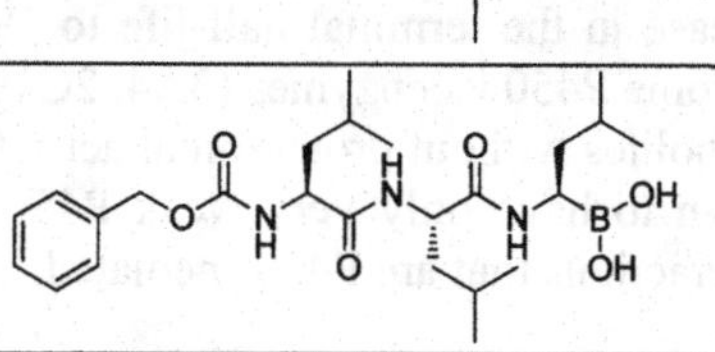	0.03
15	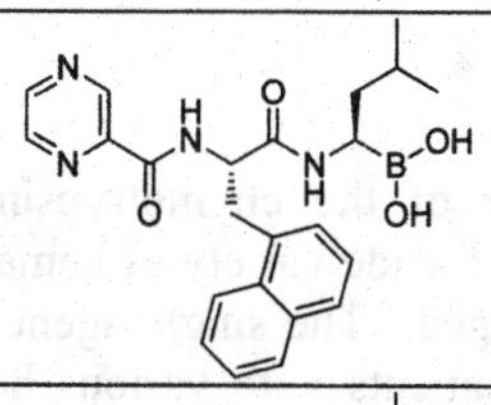	0.18
1	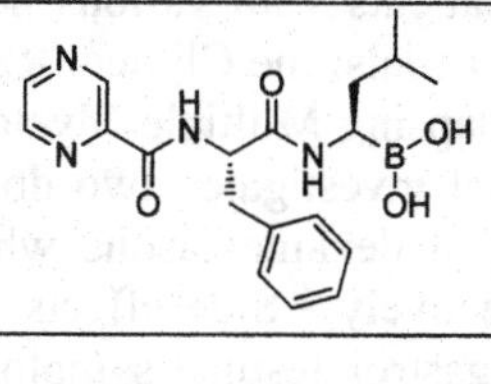	0.62

8.4   Pharmacokinetics and Drug Metabolism

The preclinical pharmacokinetics of bortezomib (1) has been evaluated intravenously in mouse. A plasma half-life of 98 h was reported when dosed at 0.8 mg/kg. The long half-life is driven by a large volume of distribution of 102 L/kg and a low total clearance of 14 mL/min/kg. The oral bioavailability was determined to be 11%, suggesting poor absorption from the mouse gastrointestinal tract.[13] This is despite Caco-2 experiments suggesting the compound to have good cell permeability (Papp value of ~10 × 10⁻⁶ cm/s), and with an efflux ratio of < 2 (Papp B–A/Papp A–B). Several preclinical studies in mice have also revealed interesting differences in the tumor distribution of bortezomib (1). For example, in mouse xenograft models with a prostate cell line (CWR22) and a lung cell line (H460), similar whole blood Cmax exposures were observed, but Cmax levels within the tumor differed by fivefold, in favor of the CWR22 xenograft. This disparity was also reflected in pharmacodynamic markers of proteasomal inhibition within the tumor. These differences in distribution may go some way to explain the lack of in vivo efficacy with certain xenograft models despite good in vitro activity, and may have clinical implications.[17] Another study in mice suggests that bortezomib (1) is rapidly and widely distributed peripherally but with restriction from the central nervous system. Highest levels were observed in the liver and gastrointestinal tract, with lower levels observed in the skin and muscles. A bile duct cannulation experiment in rat with radiolabeled bortezomib (1) suggests the majority of radioactivity is excreted into bile.[15]

The plasma pharmacokinetics reported in humans after intravenous administration shows a similar profile to that observed in mouse. A rapid distribution phase, leading to a volume of distribution of 5–12 L/kg, and a prolonged elimination phase affording a half-life of 6–12 h. The total plasma clearance on day 1 of dosing was determined to be 14 mL/min/kg. The human plasma protein binding of bortezomib (1) was shown to average 83%. In a multiple dosing study it has been noted that the clearance of bortezomib (1) is reduced by twofold to threefold on day 8 compared to day 1, which translated into an increase in the terminal half-life to 19 h. Bortezomib (1) is metabolized primarily by cytochrome P450 isoenzymes (3A4, 2C19, and 1A2), leading to oxidative loss of boron and metabolites without proteasomal activity. Bortezomib (1) and its major metabolites were shown to have only very weak P450 inhibition and so are unlikely to lead to drug–drug interactions that are P450-mediated.[1]

8.5   Efficacy and Safety

Bortezomib (1) is a potent inhibitor of the chymotrypsin-like activity of the 20S proteasome that shows activity against a wide variety of hematological malignancies and has side effects that can be well managed. The single-agent efficacy of bortezomib (1) was first noted in a Phase I trial of patients with various hematological malignancies. This was then followed by two phase II trials: the Clinical Response and Efficacy Study of Bortezomib in the Treatment of Relapsing Multiple Myeloma (CREST) trial and the larger SUMMIT trial. The CREST trial investigated two doses of bortezomib (1) (1.3 mg/m² and 1.0 mg/m²), which included dexamethasone when required, and led to a response rate of 50% and 37%, respectively. Side effects noted in the trial included myelosuppression, thrombocytopenia, gastrointestinal symptoms, and peripheral sensory neuropathy. The 5-year overall survival was 45% in the higher dose group, compared to 32% in the lower. In the larger SUMMIT trial, the 1.3 mg/m^2 dose (with dexamethasone if required) was investigated further in relapsing/refractory patients, leading to an astounding 35% response rate. Bortezomib (1) has also been tested in combination with the alkylating agent mephalan and the glucocorticoid agonist prednisone. The Phase III trial, Velcade as Initial Standard Therapy in Multiple Myeloma Assessment (VISTA) with mephalan and prednisone, demonstrated the utility of the bortezomib (1)/mephalan/prednisone combination in treatment naive MM patients. The overall reponse rate. was 71% in the triple combination arm, compared to 35% in the mephalan/prednisone arm of the trial. Crucially, the time to progression and overall survival was significantly improved in the triple combination group. However, incidences of side effects, such as neuropathy, were higher. Other common side effects observed were nausea, fatigue, and diarrhea, as well as the more serious thrombocytopenia, neutropenia, and lymphopenia.[1–5]

Bortezomib (1) is being investigated extensively for other hematological malignancies, both as a single agent and in combination with other drugs. The drugs approval for mantle cell lymphoma will hopefully mark the beginning of the expanding utility of this first-in-class drug. Bortezomib (1) has also been investigated for treating patients with solid tumors. Initial studies have resulted in disappointing clinical benefit, but many more trials are ongoing.[4,7]

8.6   Syntheses

Bortezomib (1) is an *N*-acylated dipeptide analogue of phenylalanyl-leucine in which a boronic acid replaces the *C*-terminal carboxylic acid. The Millennium discovery synthesis was patented by Adams et al.[18] in 1998, and Pickersgill et al.[19] reported the process synthesis in 2005. Despite both routes employing the same key fragments, the process route is superior in regard to ease of operation on scale. Isolating intermediates as salts and telescoping chemical steps together have allowed chromatographic purifications to be minimized.

The process synthesis commences with pinanediol boronic ester **16**. Organoboranes of this sort are routinely prepared by stirring equimolar quantities of alkyl boronic acid and glycol in a nonpolar solvent such as pentane or diethyl ether.[20,21] Chiral glycol (+)-pinanediol acts as a directing group for the subsequent homologation reaction to give α-chloro boronic ester **17**.[22] (Dichloromethyl)lithium, generated in situ from dichloromethane (DCM) and lithium diisopropylamide (LDA), alkylates **16**, generating intermediate boronate complex **18**.[23] A Lewis acid–catalyzed migration of the alkyl group affords **17**.[24]

The use of a Lewis acid such as zinc or ferric chloride offered an improvement to the original procedure reported by Matteson and Ray.[22] Rearrangements occur at room temperature with improved conversion. Perhaps more important, the use of 0.5–1.0 equivalent ZnCl$_2$ minimizes the rate of epimerization of the α-carbon center that occurs due to the presence of lithium chloride (LiCl).[25] Prolonged exposure of **17** to chloride ions produced in the reaction and during the aqueous workup can decrease the stereoselectivity of the process. When complexed to ZnCl$_2$, lithium chloride forms

LiZnCl$_3$ and Li$_2$ZnCl$_4$, both of which are less active catalysts toward epimerization.[25] Through the use of *tert*-butyl methyl ether (TBME) as an alternative to water miscible tetrahydrofuran (THF), the process route avoids the need to carry out a distillation to exchange solvents, a process that is detrimental to the diastereomeric ratio (*dr*) of the product. THF is instead added as a co-solvent to aid the reaction conversion but limit miscibility of the aqueous washes during workup. A methylcyclohexane solution of **17** is used directly in the next step.

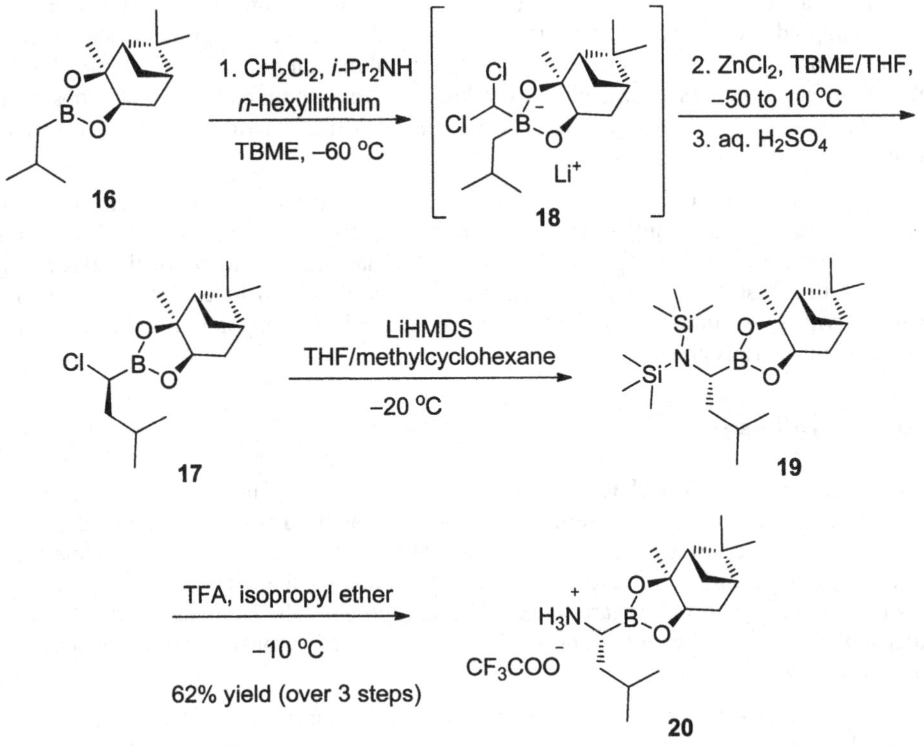

Nucleophilic substitution with lithium hexamethyldisilazane (LiHMDS) proceeds with inversion to give silylated amino boronic ester **19**.[26] A solution of **19** is passed through a short plug of silica before its use in the desilylation reaction. Due to the instability of underivatized α-amino boronic esters,[26] trifluoroacetic acid (TFA) is used to furnish the corresponding TFA salt **20**.[27] A second recrystallization from TFA and isopropylether further enhances the optical purity. A diastereomeric ratio (*dr*) > 97:3 is typically obtained from the process route.

Chiral amine salt **20** is coupled with *N*-Boc-*L*-phenylalanine (**21**) using 2-(*H*-benzotriazol-1-yl)-1,1,3,3-tetramethyluronium tetrafluoroborate (TBTU)[28] to afford Boc-protected dipeptide **22**. An ethyl acetate solution of **22** is dried azeotropically before treatment with gaseous HCl to cleave the Boc group and furnish **23**. Crystallization of the salt removes any tri-peptide impurity that may have formed during the peptide coupling.

A second peptide coupling employing 2-pyrazine carboxylic acid (**24**) and TBTU affords **25**. Finally the boronic ester moiety is removed under acidic conditions using isobutyl boronic acid. This regenerates pinanediol boronic ester **16**, which can be used in another batch run of the process. Crystallization from ethyl acetate gives bortezomib in its anhydride (boroxine) form **26**. The overall yield for the route is 35% with a typical purity of > 99% w/w.

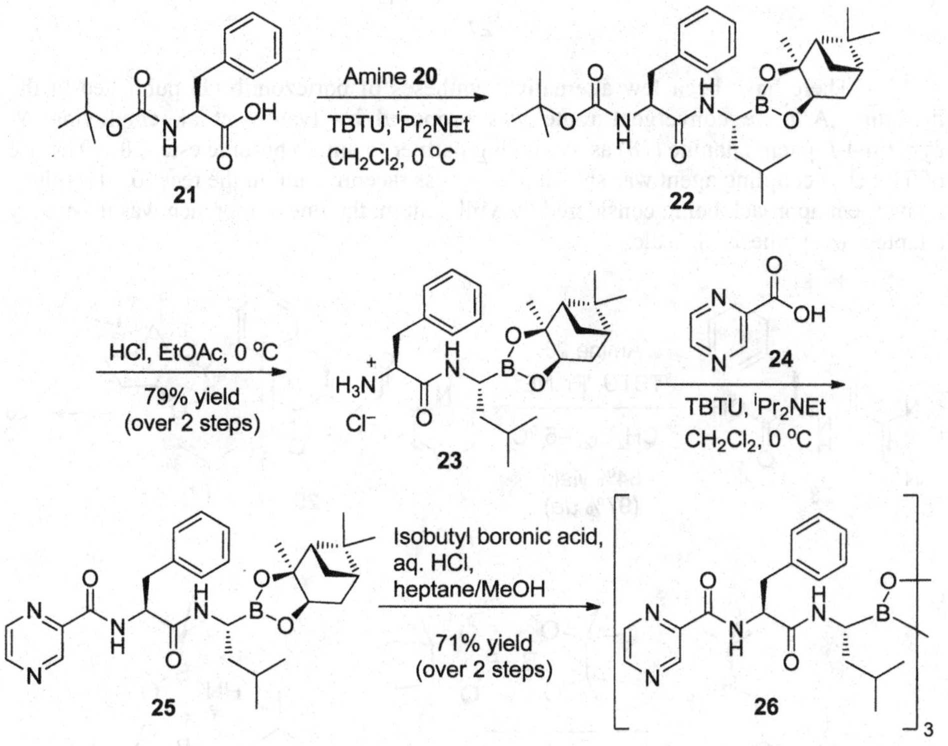

The relative instability of alkyl boronic acids and their tendency to form anhydrides under mild dehydrating conditions[29] poses a problem in their use as pharmaceutical agents. Obtaining and storing analytically pure forms of drug is of significant importance. To overcome these issues, bortezomib in its anhydride form **26** is lyophilized (freeze-dried) from an aqueous solution containing *D*-mannitol.[30] This affords **27**, a composition that is considerably more stable upon storage than anhydride form **26**. Bortezomib (**1**) is reconstituted by dissolution in saline before administration.

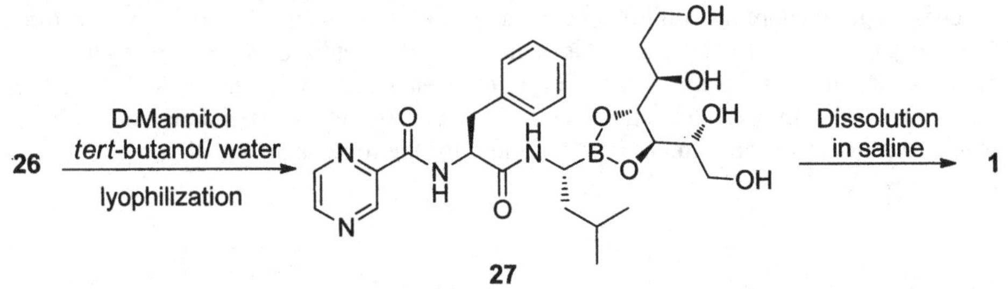

There have been few alternative syntheses of bortezomib (1) published in the literature. A more convergent route was proposed by Ivanov et al. employing *N*-pyrazinyl-*L*-phenylalanine (28) as a coupling partner to amino boronic ester 20.[31] The use of TBTU as coupling agent was shown to suppress racemization in the reaction. Despite a convergent approach being considered by Millennium, the linear approach was ultimately adopted for synthesis on scale.

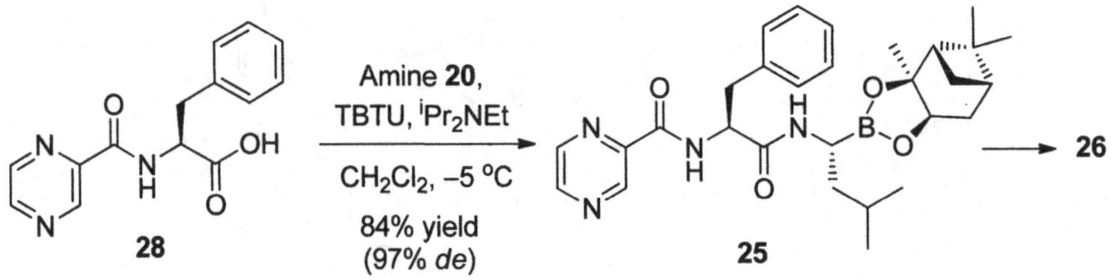

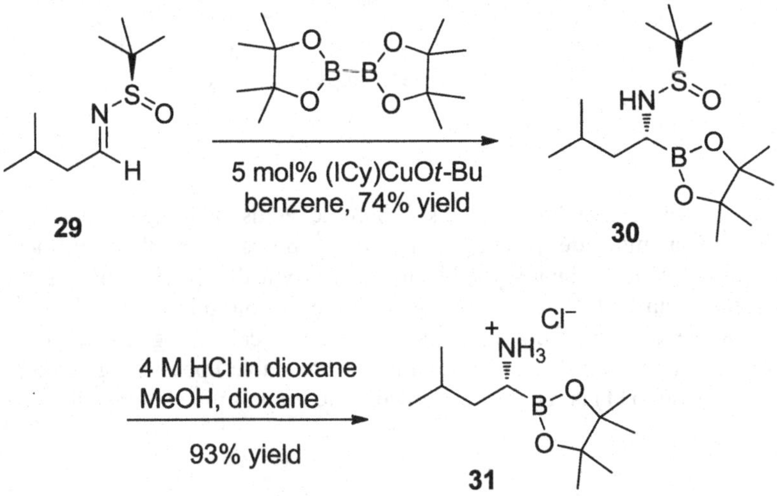

Until recently, the synthesis of chiral amino boronic acids has been limited to that reported by Matteson et al. Beenen and co-workers reported an asymmetric copper-catalyzed synthesis of α-amino boronic esters from *N-tert*-butanesulfinyl aldimines.[32] Addition of bis(pinacolato)diboron to **29** gives *N*-sulfinyl-α-amino boronate ester **30** in 74% yield with an excellent diastereomeric ratio of > 98:2. Cleavage of the *N*-sulfonyl group affords **31**, which was successfully elaborated to bortezomib (**1**).

8.7   References

1.   Popat, R.; Joel, S.; Oakervee, H.; Cavenagh, J. *Expert Opin. Pharmacother.* **2006**, *7*, 1337–1346.
2.   Albanell, J.; Adams, J. *Drugs Fut.* **2002**, *27*, 1079–1092.
3.   Adams, J. *Cancer Cell* **2003**, *5*, 417–421.
4.   Richardson, P. G. *Expert Opin. Pharmacother.* **2004**, *5*, 1321–1331.
5.   Paramore, A.; Frantz, S. *Nat. Rev. Drug Disc.* **2003**, *2*, 611–612.
6.   Voorhees, P. M.; Orlowski, R. Z. *Ann. Rev. Pharmcol. Toxicol.* **2006**, *46*, 189–213.
7.   Yang, H.; Zonder, J. A.; Dou, Q. P. *Expert Opin. Investig. Drugs* **2009**, *18*, 957–971.
8.   Huang, L; Chen, C. H. *Curr. Med. Chem.* **2009**, *16*, 931–939.
9.   Bennett, M. K.; Kirk, C. J. *Curr. Opin. Drug Disc. Dev.* **2008**, *11*, 616–625.
10. Tsubuki, S.; Kawasaki, H.; Saito, Y.; Miyashita, N.; Inomata, M.; Kawashima, S. *Biochem. Biophys. Res. Commun.* **1993**, *196*, 1195–1201.
11. Adams, J.; Behnke, M.; Chen, S.; Cruickshank, A. A.; Dick, L. R.; Grenier, L.; Klunder, J. M.; Ma, Y.-T.; Plamondon, L.; Stein, R. L. *Bioorg. Med. Chem. Lett.* **1998**, *8*, 333–338.
12. Bogyo, M.; McMaster, J. S.; Gaczynska, M.; Tortorella, D.; Goldberg, A.L.; Ploegh, H. *Proc. Natl. Acad. Sci. U.S.A.* **1997**, *94*, 6629–6634.
13. Dorsey, B. D.; Iqbal, M.; Chatterjee, S.; Menta, E.; Bernadini, R.; Bernareggi, A.; Cassara, P. G., D'Arasmo, G.; Ferretti, E.; De Munari, S.; Oliva, A.; Pezzoni, G.; Allievi, C.; Strepponi, I.; Ruggeri, B.; Ator, M. A.; Williams, M.; Mallamo, J. P. *J. Med. Chem.* **2008**, *51*, 1068–1072.
14. Kuhn, D. J.; Chen, Q.; Voorhees, P. M.; Strader, J. S.; Shenk, K. D.; Sun, C. M.; Demo, S. D.; Bennett, M. K.; van Leeuwen, F. W. B.; Chanan-Khan, A. A.; Orlowski, R. Z. *Blood* **2007**, *110*, 3281–3290.
15. Adams, J.; Palombella, V. J.; Sausville, E. A.; Johnson, J.; Destree, A.; Lazarus, D. D.; Maas, J.; Pien, C. S.; Prakash, S.; Elliott, P. J. *Cancer Res.* **1999**, *59*, 2615–2622.
16. LeBlanc, R.; Catley, L. P.; Hideshima, T.; Lentzsch, S.; Mitsiades, C. S.; Mitsiades, N.; Neuberg, D.; Goloubeva, O.; Pien, C. S.; Adams, J.; Gupta, D.; Richardson, P. G.; Munshi, N. C.; Anderson, K. C. *Cancer Res.* **2002**, *62*, 4996–5000.
17. Williamson, M. J.; Silva, M. D.; Terkelsen, J.; Robertson, R.; Yu, L.; Xia, C.; Hatsis, P.; Bannerman, B.; Babcock, T.; Cao, Y.; Kupperman, E. *Mol. Cancer Ther.* **2009**, *8*, 3234–3243.
18. Adams, J.; Ma, Y.; Stein, R.; Baevsky, M.; Grenier, L.; Plamondon, L. U.S. Pat. 5,780,454, **1998**.

19.   Pickersgill, I. F.; Bishop, J. E.; Koellner, C.; Gomez, J.; Geiser, A.; Hett, R.;
       Ammoscato, V.; Munk, S.; Lo, Y.; Chiu, F.; Kulkarni, V. R. US 2005/0240047,
       2005.

20.   Brown, H. C.; Bhat, N. G.; Somayaji, V. *Organometallics*, **1983**, *2*, 1311–1316.

21.   Janca, M.; Dobrovolny, P. WO 2009/004350, **2009**.

22.   Matteson, D. S.; Ray, R. *J. Am. Chem. Soc.* **1980**, *102*, 7590–7591.

23.   Matteson, D. S.; Majumdar, D. *J. Am. Chem. Soc.* **1980**, *102*, 7588–7590.

24.   Matteson, D. S.; Sandu, K. M. U.S. Pat. 4,525,309, **1985**.

25.   Matteson, D. S.; Erdik, E. *Organometallics* **1983**, *2*, 1083–1088.

26.   Matteson, D. S.; Sadhu, K. M.; Lienhard, G. E. *J. Am. Chem. Soc.* **1981**, *103*,
       5241–4242.

27.   Shenvi, A. B. US 4,537,773 (1985).

28.   Knorr, R.; Trzeciak, A.; Bannwarth, W.; Gillessen, D. *Tetrahedron Lett.* **1989**,
       30, 1927–1930.

29.   Snyder, H. R.; Konecky, M. S.; Lennarz, W. J. *J. Am. Chem. Soc.* **1958**, *80*,
       3611–3615.

30.   Plamondon, L.; Grenier, L.; Adams, J.; Gupta, S. L. WO 02/059131, **2002**.

31.   Ivanov, A. S.; Zhalnina, A. A.; Shishkov, S. V. *Tetrahedron* **2009**, *65*, 7105–
       7108.

32.   Beenen, M. A.; An, C.; Ellman, J. A. *J. Am. Chem. Soc.* **2008**, *130*, 6910–6911.

9

Pazopanib (Votrient): A VEGFR Tyrosine Kinase Inhibitor for Cancer

Ji Zhang and Jie Jack Li

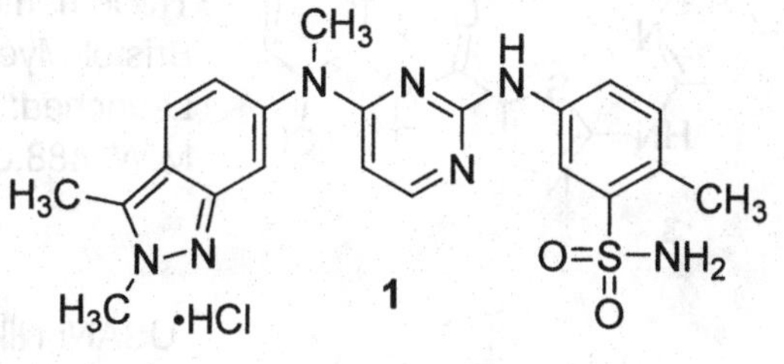

USAN: Pazopanib
Trade name: Votrient®
GlaxoSmithKline
Launched: 2010

9.1   Background

According to a report from the American Cancer Society,[1] beginning in 2005, cancer surpassed heart diseases as the number one killer in the United States. Furthermore, cancer will become the world's leading cause of death by 2010, based on a recently released document by the World Health Organization (WHO).[2] A key feature of the disease unlike other diseases, is that cancer is considered untreatable if it is diagnosed in later stages. In 2002, an estimated 208,000 new cases of kidney cancer were diagnosed globally. It is projected that about 57,700 people will be diagnosed with kidney cancer, and 13,000 people will die from the disease in the United States alone in 2009. Early prevention, diagnosis, and treatment remain as the best approach for treatment. Since solid tumors progression requires nutrients and oxygen received via angiogenesis, targeting this process by impeding angiogenesis, affords the ability to prevent subsequent tumor growth. Thus inhibition of the vascular endothelial growth factor (VEGF) signaling pathway has emerged as one of the most promising and attractive anticancer strategies.[3] In contract to the carpet-bombing approach of old chemotherapy, VEGFR tyrosine kinase inhibitors have provided a kinder, gentler, and more effective therapy for cancer treatment.[4]

After the first selective small-molecule tyrosine kinase inhibitor, imatinib **2** (Gleevec),[5] received FDA approval and launched in 2001, more than seven tyrosine kinase inhibitors have been granted approval by the FDA, including pyrimidine-based tyrosine kinase inhibitors disatinib **3** (Sprycel)[6] and nilotinib **4** (Tasigna)[7] as well as quinazoline-based tyrosine kinase inhibitors, such as gefitinib **5** (Iressa®),[8] erlotinib **6** (Tarceva)[9] and lapatinib **7** (Tykerb).[10] So far, there are approximately 100 kinase inhibitors currently at various stages of clinical investigation and more than 30 VEGFR inhibitors are under active development for the treatment of cancer.[11]

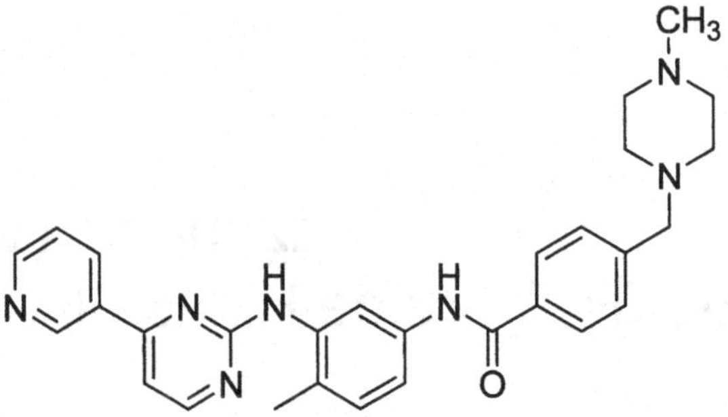

USAN: Imatinib
Trade name: Gleevec®
Novartis
Launched: 2001
M.W. 493.60

2

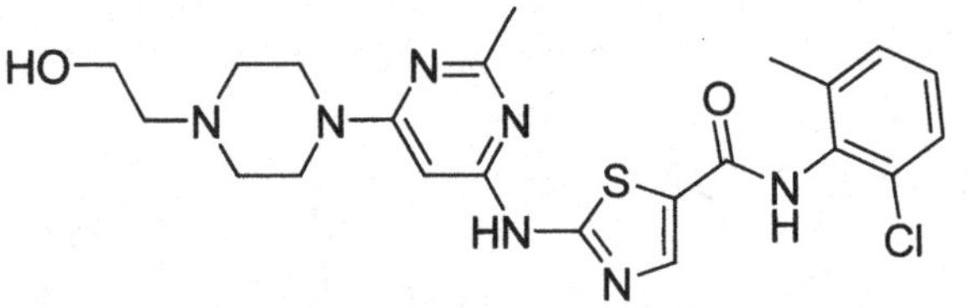

USAN: Dasatinib
Trade name: Sprycel®
Bristol-Myers Squibb
Launched: 2006
M.W. 488.01

3

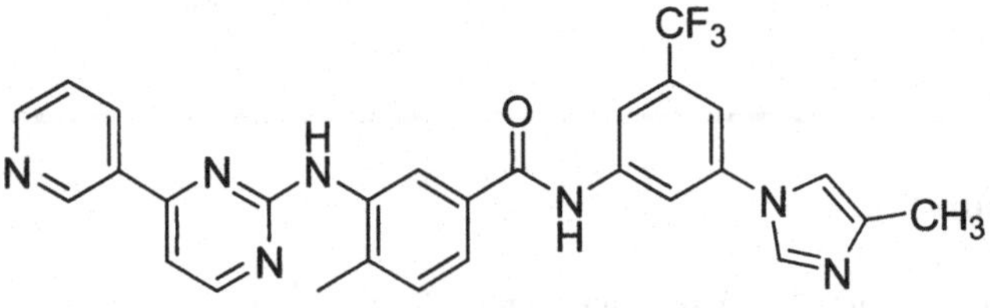

USAN: Nilotinib
Trade name: Tasigna®
Novartis
Launched: 2007
M.W. 529.52

4

Figure 1. Pyrimidine-based VEGFR Tyrosine Kinase Inhibitors.

Pazopanib hydrochloride (**1**, Votrient),[12] originally GW-786034, is a second-generation VEGFR tyrosine kinase inhibitor developed by GlaxoSmithKline (GSK). The compound was granted marketing approval by the U.S. FDA and was launched under the brand name Votrient in 2010 as a treatment for patients with advanced renal cell carcinoma (RCC), the most common type of kidney cancer. In this chapter, the pharmacological profile and syntheses of pazopanib (1) will be summarized.[13]

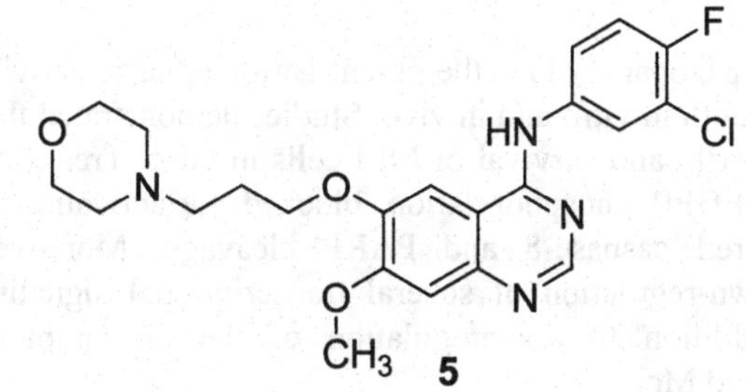

USAN: Gefitinib
Trade name: Iressa®
AstraZeneca
Launched: 2003
M.W. 446.91

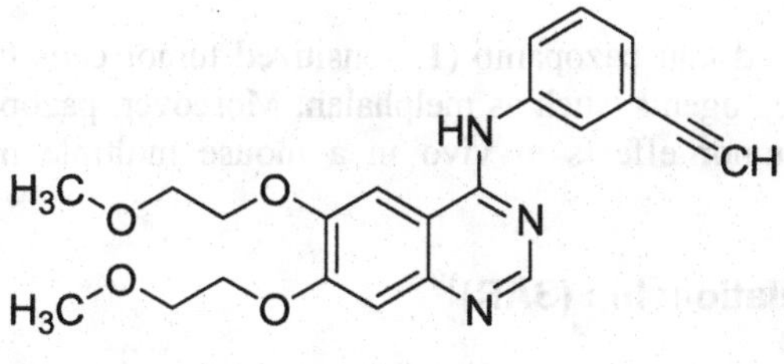

USAN: Erlotinib
Trade name: Tarceva®
OSI Pharmaceuticals
Launched: 2004
M.W. 429.90 (HCl salt)

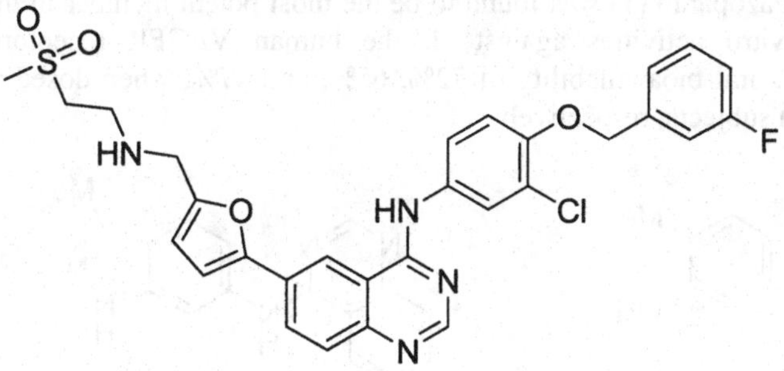

USAN: Lapatinib
Trade name: Tykerb®
GlaxoSmithKline
Launched: 2007
M.W. 581.06

Figure 2. Quinazoline-based Tyrosine Kinase Inhibitors.

9.2   Pharmacology[14,15]

Lead optimization of a series of pyrimidine derivatives led to the discovery of pazopanib (1). Basic screening tests for this compound demonstrated a highly potent and selective, direct inhibitory action on VEGFR1, VEGFR2, and VEGFR3 with IC_{50} values of 10, 30, and 47 nM, respectively. The agent inhibited VEGF-induced proliferation of human umbilical vein endothelial cells (HUVEC) more potently than bFGF-stimulated proliferation (IC_{50} = 21 nM vs. 721 nM) and concentration-dependently inhibited VEGF-induced VEGFR2 phosphorylation (IC_{50} = 7 nM) in these cells. Pazopanib (**1**) also showed over 1,400- and 48-fold selectivity for VEGF-induced HUVEC proliferation compared to several tumor cell lines and fibroblasts, respectively. In addition, it potently inhibited angiogenesis in Matrigel plug and corneal micropocket assays.

An addition vital feature that pazopanib (**1**) is the potent anti-neoplastic activity against multiple myeloma (MM) cells both in vitro and in vivo. Studies demonstrated that pazopanib inhibited the migration, growth, and survival of MM cells in vitro. Treatment of cells suppressed VEGF-induced VEGFR phosphorylation, blocked the activation of downstream Src kinase, and triggered caspase-8 and PARP cleavage. Moreover, pazopanib-treated cells exhibited down-regulation of several transcriptional signaling pathways, particularly c-Myc, in addition to downregulation of the pro-apoptotic molecules surviving, c-IAP1, c-IAP2 and Mcl1.

Additional results indicated that pazopanib (**1**) sensitized tumor cells bound to endothelial cells to DNA-damaging agents, such as melphalan. Moreover, pazopanib (**1**) exerted antiangiogenic and antitumor effects in vivo in a mouse multiple myeloma xenograft model.

9.3   Structure–Activity Relationship (SAR)[12]

The SAR around pazopanib **1** was extensively investigated by Harris and co-workers[15] at GlaxoSmithKline (GSK). Pazopanib (**1**) was found to be the most potent member in the series with excellent in vitro activities against all the human VEGFR receptors. Moreover it exhibited good oral bioavailability of 72%, 65%, and 47%, when dosed at 10, 5 and 1 mg/kg in human subjects, respectively.

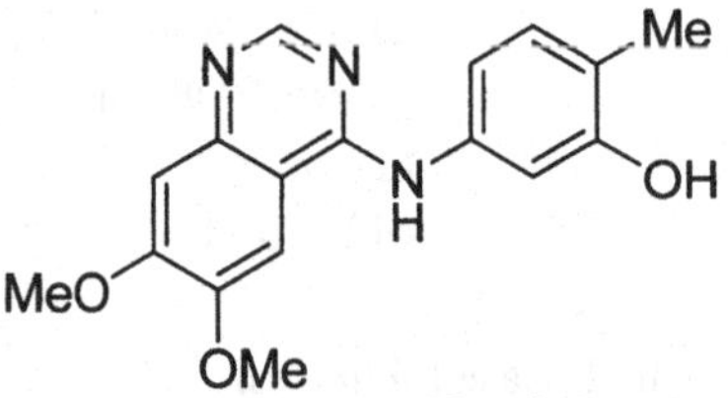

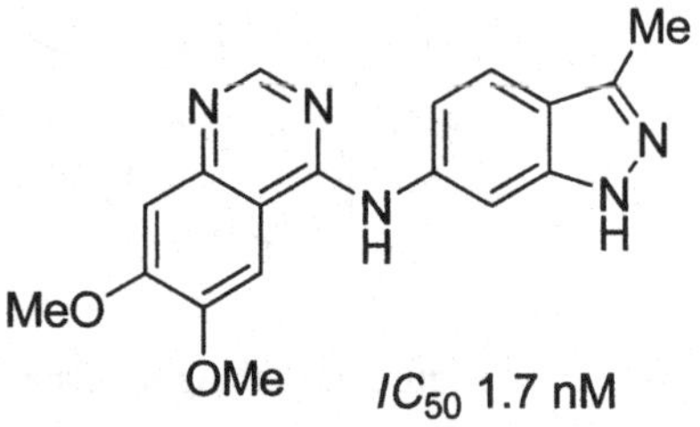

The lead compound N,N'-bis(3-bromoanilino)-5-fluoro-2,4-pyrimidinediamine **10** ($IC_{50} \sim 400$ nM) was derived from high throughput (HTS) screening of in-house kinase inhibitor libraries. In the initial optimization and based on the fibroblast growth factor receptor (FGFR) crystal structures as well as lead compounds from quinazoline-based kinase inhibitors **8** and **9**, a homology model was used to predict the binding outcome. The conformational studies demonstrated that the anilino group at the C-4 position of pyrimidine was necessary, which led to the preparation of pyrimidine **11** (IC_{50} = 6.3 nM). Introduction of a 4-methyl-3-hydroxyanilino group increased the binding by 100-fold, unfortunately, the pharmacokinetic profile was poor, presumably because of a rapid phase II glucuronidation or sulfation reaction of the phenol functionality. Incorporating 3-methyllindazole in the pyrimidine series produced **12** with good potency against VEGFR2 (IC_{50} = 6.3 nM) and HUVECs (IC_{50} = 0.18 μM), with a significantly improved pharmacokinetic profile (clearance of 16 mL/min/kg and an oral bioavailability of 85% at a dose of 10 mg/kg in rats).

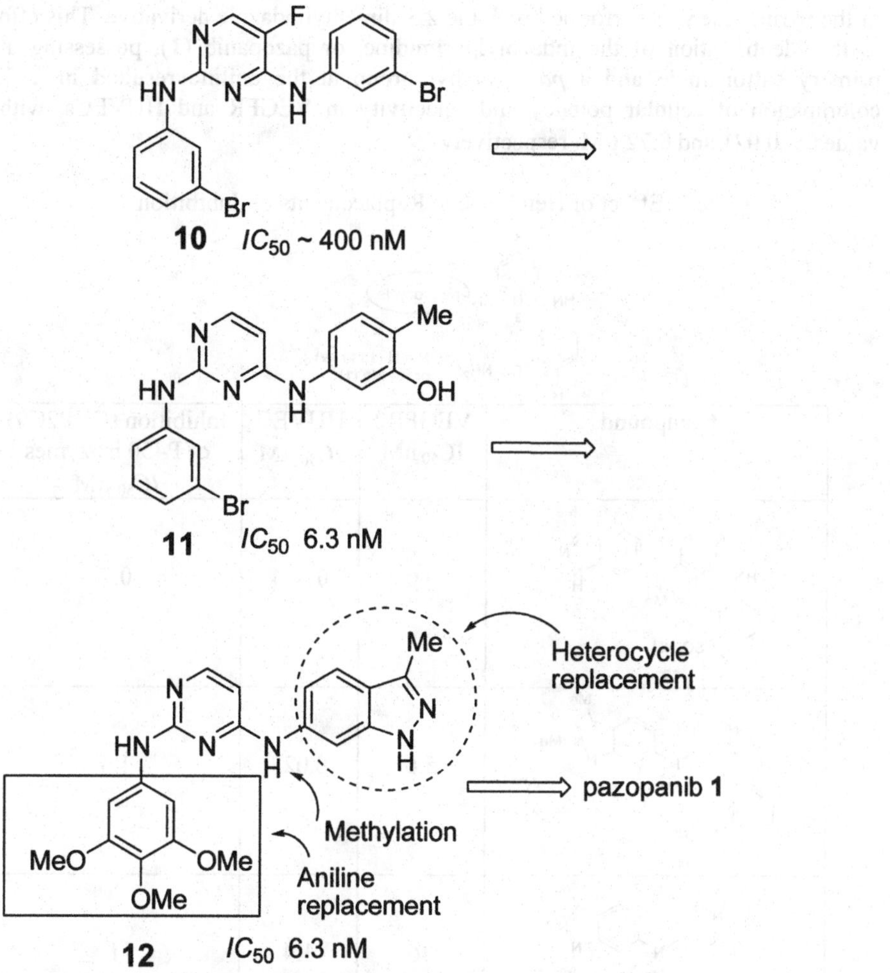

10 IC_{50} ~ 400 nM

11 IC_{50} 6.3 nM

12 IC_{50} 6.3 nM

Figure 3. Directions in the Lead Optimization and Discovery of Pazopanib (**1**).

After modification of lead compound **12**, it was found that pyrimidines containing the 3-methyllindazole heterocycle and methylated at the C-4 amino nitrogen possessed both good in vitro and cross-species pharmacokinetic profiles. The replacement of the 3,4,5-trimethoxy aniline with a 3-aminobenzenesulfonamide led to comparable potencies. However, significant inhibition (IC_{50} < 10 μM) was observed against a number of cytochrome (CYP) P450 isozymes, presumably resulting from binding of the nitrogens of the indazole heterocycle to the heme ion of the CYP450 enzyme. To reduce that binding, investigation directed toward adding steric hindrance to the heterocycle ring was performed. The identified 2,3-dimethylindazole derivatives **14** demonstrated improved cellular efficacy against HUVECs and a better profile against the CYP2C19, CYP2D6, and CYP3A4 isozymes, but not CYP2C9. With a focus on sulfonamide- and sulfone-substituted anilines, a second aniline survey of the C2 position of the pyrimidine was performed with the 2,3-dimethylindazole derivative. This effort led to the identification of the indazolylpyrimidine, or pazopanib (**1**), possessing a *meta* primary sulfonamide and a *para* methyl group at the aniline resulted in a desired combination of cellular potency and selectivity in VEGFR and HUVECs, with IC_{50} values of 0.021 and 0.72 µM, respectively

Table 1. Effect of Heterocyclic Replacements on Inhibition

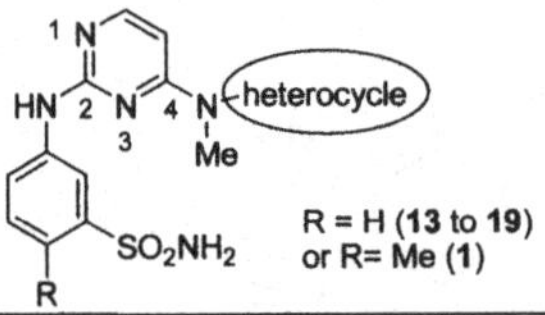

Compound	VEGFR2 IC_{50} nM	HUVEC IC_{50} µM	Inhibition (CYP2C9) of P450 isozymes IC_{50} µM
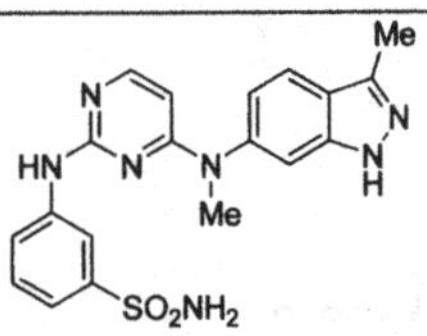 **13**	0.6	0.094	0.7
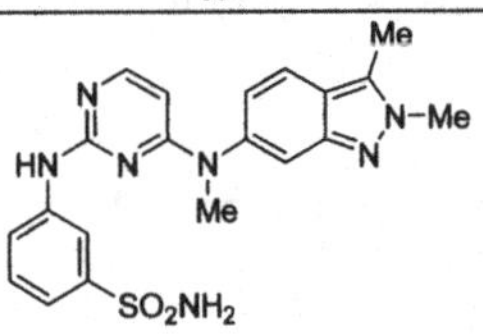 **14**	5.6	0.023	1.4
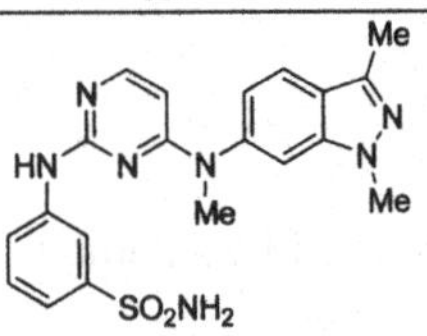 **15**	36	1.4	1
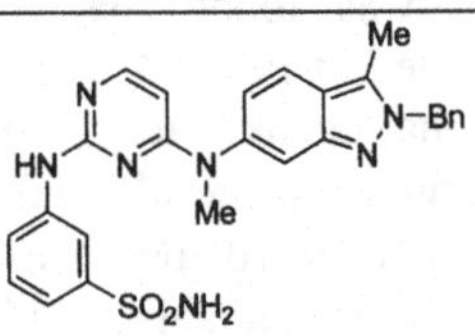 **16**	2.6	11	0.5

Structure			
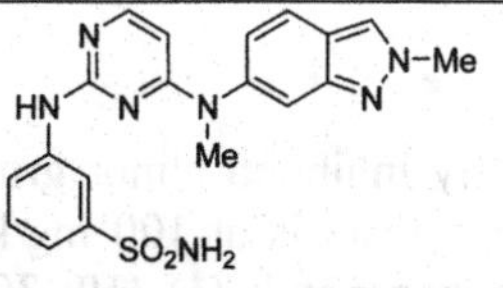 **17**	7.6	0.11	2.9
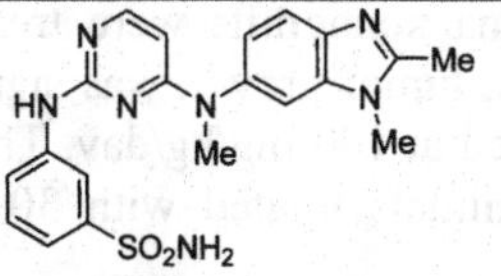 **18**	63	2.8	15
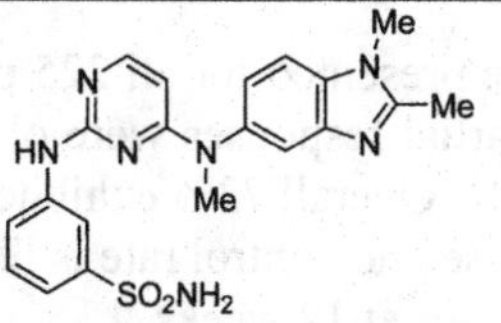 **19**	17	1.4	9
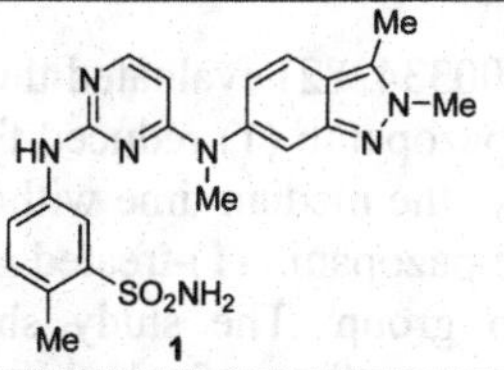 **1**	0.021	0.72	7.9

Source: Ref. 12.

9.4 Pharmacokinetics and Drug Metabolism[16,17]

Pazopanib (1) had oral bioavailability of 72%, 47%, and 65% in rats, dogs, and monkeys treated with 10, 1, and 5 mg/kg, respectively. Clearance was low in all species (1.4 to 1.7 mL/min/kg). In dogs treated with pazopanib (1 mg/kg i.v. and p.o.), the mean C_{max} value, AUC value, clearance, and volume of distribution were 0.8 μg/mL, 5.4 μg/mL, 1.4 mL/min/kg and 0 mL/min/kg, respectively.

It was found that pharmacokinetic-pharmacodynamic correlation in mice demonstrated that plasma concentrations of > 40 μmol/L after pazopanib (30 mg/kg p.o.) inhibited VEGF-induced VEGFR2 phosphorylation for more than 8 h. The mean plasma concentrations were 153.9, 47.5, 41.1, 17.4, and 4.3 μmol/L at times of 1, 4, 8, 16, and 24 h.

Table 2. Pharmacokinetic Properties of Pazopanib (1)

Study Type	in vivo	Model Used
Area Under the Curve (AUC)	1141 μg.h/mL	100 mg/kg p.o.
C_{max}	128 μg/mL	100 mg/kg p.o. (Mice)
$t_{1/2}$	~35 h	50–2000 mg/day p.o. (Patients)

Source: Ref. 13a.

9.5   Efficacy and Safety[18]

In vivo, it was found that pazopanib (**1**) dose-dependently inhibited tumor growth with 77% growth inhibition at 10 mg/kg/day, and complete cytostasis at 100 mg/kg/day in mice bearing Caki-2 RCC xenografts and treated with pazopanib (**1**) (10, 30, or 100 mg/kg/day p.o.). When mice bearing multiple myeloma xenografts were treated with pazopanib (**1**) (30 or 100 mg/kg/day p.o.) over 5 weeks, tumor growth was significantly delayed at 30 mg/kg/day and almost completely inhibited at 100 mg/kg/day. The overall survival was 20, 41, and 52 days in controls and animals treated with 30 and 100 mg/kg/day, respectively.

Efficacy and safety data for pazopanib (**1**) were presented for all 225 patients at the American Society of Clinical Oncology in 2007. Partial responses were observed in 27% at 12 weeks by independent review, with SD in 46%. Overall 73% exhibited disease control (CR + PR + SD) at 12 weeks. The overall response and control rate is likely to be higher with mature follow up, as these data reflect responses at 12 weeks.

Phase III clinical studies (435 patients in NCT00334282) evaluated the efficacy and safety of pazopanib in patients.[14] In these studies, pazopanib (**1**) reduced the risk of tumor progression or death by 54% compared to placebo, the median time without tumor growth or death (progression free survival or PFS) in the pazopanib (**1**)-treated group was 9.2 months compared to 4.2 months in the placebo group. The study shows that pazopanib (**1**) significantly improved PFS for patients regardless of whether they had prior therapy.

9.6   Synthesis[12]

The synthesis of pazopanib (**1**) involves sequential amination of 2,4-dichloropyrimidine **25** with 6-amino-2,3-dimethylindazole **24** and 5-amino-2-methyl-benzenesulfonamide **28**. The 6-amino-2,3-dimethylindazole **24**, on the other hand, was prepared from 2-ethylphenylamine **20** via 5-nitration with fuming nitric acid and concentrated sulfuric acid, followed by treatment with isoamyl nitrite and acetic acid to produce 6-nitro-3-methylindazole **22**. The 6-nitro group was reduced with stannous chloride and concentrated HCl in glyme and subsequently methylated at the C2 position of the indazole ring with trimethyloxonium tetrafluoroborate in acetone to produce 6-amino-2,3-dimethylindazole **24**. The resultant indazole **24** was condensed with 2,4-dichloropyrimidine **25** in the presence of sodium bicarbonate in ethanol/THF and subsequent *N*-methylation with iodomethane and cesium carbonate to produce **27**. The 2-chloro group of pyrimidine was then allowed to react with 5-amino-2-methyl-benzenesulfonamide **28** in catalytic HCl/isopropanol and heated to reflux to deliver pazopanib hydrochloride (**1**) in good yield.

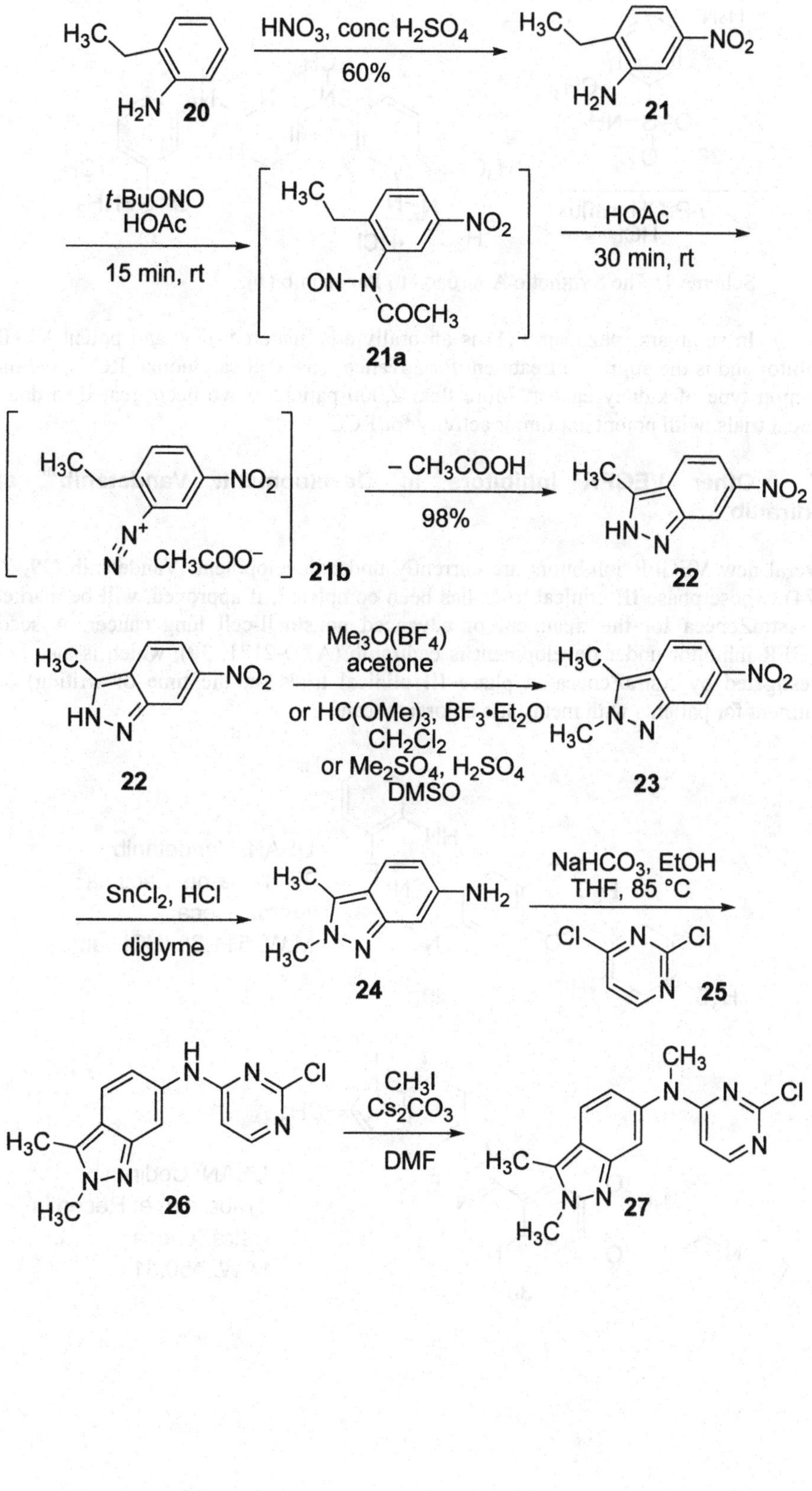
HNO3, conc H2SO4
60%
20
21
t-BuONO
HOAc
15 min, rt
21a
HOAc
30 min, rt
21b
– CH3COOH
98%
22
Me3O(BF4)
acetone
or HC(OMe)3, BF3•Et2O
CH2Cl2
or Me2SO4, H2SO4
DMSO
22
23
SnCl2, HCl
diglyme
24
NaHCO3, EtOH
THF, 85 °C
25
CH3I
Cs2CO3
DMF
26
27

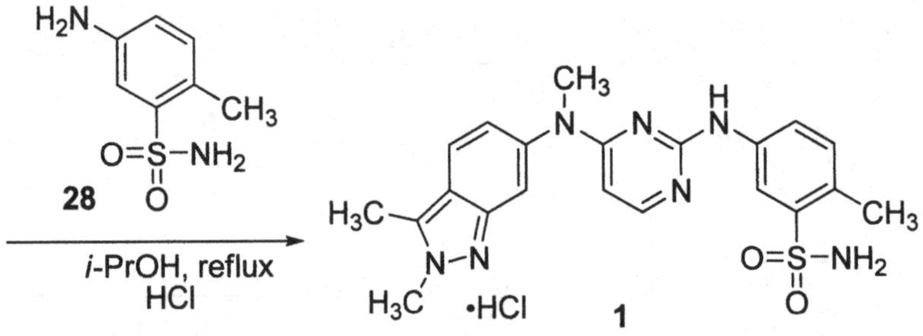

Scheme 1: The Synthetic Approach to Pazopanib (**1**).

In summary, pazopanib (**1**) is an orally administered, new and potent VEGFR inhibitor and is the approved treatment for advance renal cell carcinoma (RCC), the most common type of kidney cancer. More than 2,000 patients have been treated to date in clinical trials, with potent antitumor activity for RCC.

9.7   Other VEGFR Inhibitors in Development: Vandetanib[19] and Cediranib[20]

Several new VEGFR inhibitors are currently under development. Vandetanib (**29**, ZD-6474), whose phase III clinical trials has been completed, if approved, will be marketed by AstraZeneca for the treatment of advanced nonsmall-cell lung cancer. A second VEGFR inhibitor under development is cediranib (AZD-2171, **30**), which is also being investigated by AstraZeneca in phase III clinical trials (at the time of writing) as a treatment for patients with metastatic colorectal cancer.

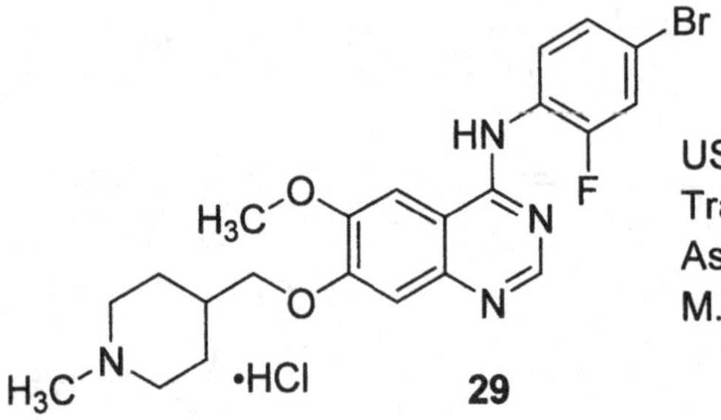

USAN: Vandetanib
Trade name: Zactima®
AstraZeneca
M.W. 511.81 (HCl salt)

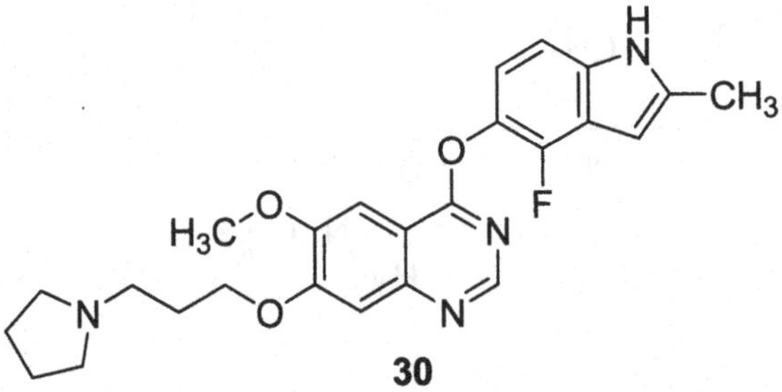

29

USAN: Cediranib
Trade name: Recentin®
AstraZeneca
M.W. 450.51

30

9.8 References

1. Li, J. J. *Cancer Drugs: From Nitrogen Mustards to Gleevec* in *Laughing Gas, Viagra, and Lipitor, The Human Stories behind the Drugs We Use,* Oxford University Press: New York, **2006**, pp 3–42.

2. Report of World Health Organization (WHO), 2009.

3. Folkman, J. *Nat. Rev. Drug Disc.* **2007**, *6*, 273–286. Sellis, L. M.; Hicklin, D. J. *Nat. Rev. Cancer* **2008**, *8*, 579. Li, R.; Pourpak, A.; Morris, S. W. *J. Med. Chem.* **2009**, *52*, 4981–5004. Pollard, J. R.; Mortimore, M. *J. Med. Chem.* **2009**, *52*, 2629–2651.

4. (a) Gilbert, M.; Mousa, S.; Mousa, S. A. *Drugs Fut.* **2008**, *33*, 515–525. (b) Revill, P.; Mealy, N.; Bayes, M.; Bozzo, J.; Serradell, N.; Rosa, E.; Bolos, J. *Drugs Fut.* **2007**, *32*, 389–398. (c) Kiselyov A.; Balakin K. V.; Tkachenko S. E. *Exp. Opin. Invest. Drugs* **2007**, *16*, 83–107

5. *Protein-Tyrosine Kinase Inhibitors: Imatinib (Gleevec) and Gefitinib (Iressa),* In *Contemporary Drug Synthesis,* Li, J. J.; Johnson, D. S.; Sliskovic, D. R.; Roth, B. D., John Wiley & Sons: Hoboken, **2004**, pp 29–38.

6. Lombardo, L. J.; Lee, F. Y.; Chen, P.; Norris, D.; Barrish, J. C.; Behnia, K.; Castaneda, S.; Cornelius, L. A. M.; Das, J.; Doweyko, A. M.; Fairchild, C.; Hunt, J. T.; Inigo, I.; Johnston, K.; Kamath, A.; Kan, D.; Klei, H.; Marathe, P.; Pang, S.; Peterson, R.; Pitt, S.; Schieven, G. L.; Schmidt, R. J.; Tokarski, J.; Wen, M.-Li.; Wityak, J.; Borzilleri, R. M. *J. Med. Chem.* **2004**, *47*, 6658–6661.

7. (a) Jabbour, E.; El Ahdab, S.; Cortes, J.; Kantarjian, H. *Exp. Opin. Invest. Drugs* **2008**, *17*, 1127–1136. (b) Davies, S. L.; Bolos, J.; Serradell, N.; Bayes, M. *Drugs Fut.* **2007**, *32*, 17–25. (c) Plosker, G. L.; Robinson, D. M. *Drugs* **2008**, *68*, 449–459.

8. Levin, M.; D'Souza, N.; Castaner, J. *Drugs Fut.* **2002**, *27*, 339–345.

9. Sorbera, L. A.; Castaner, J.; Silvestre, J. S.; Bayes, M. *Drugs Fut.* **2002**, *27*, 923–934.

10. Dhillon, S.; Wagstaff, A. J. *Drugs* **2007**, *67*, 2101–2108.

11. Li, R.; Stafford, J. A. "*Kinase Inhibitor Drugs,* in *Drug Discovery and Development"* Wang, B., Ed.; John Wiley & Sons: Hoboken, NJ, **2009**.

12. Harris, P. A.; Boloor, A.; Cheung, M.; Kumar, R.; Crosby, R. M.; Davis-Ward, R. G.; Epperly, A. H.; Hinkle, K. W.; Hunter III, R. N.; Johnson, J. H.; Knick, V. B.; Laudeman, C. P.; Luttrell, D. K.; Mook, R. A.; Nolte, R. T.; Rudolph, S. K.; Szewczyk, J. R.; Truesdale, A. T.; Veal, J. M.; Wang, L.; Stafford, J. A. *J. Med. Chem.* **2008**, *51*, 4632–4640.

13. a) Sloan, B.; Scheinfeld, N. S. *Exp. Opin. Invest. Drugs* **2008**, *9*, 1324–1335; b) Sorbera, L. A.; Bolos, J.; Serradell, N. *Drugs Fut.* **2006**, *31*, 585–589.

14. Cheung, M.; Baloor, A.; Chamberlain, S. D.; Hinkle, K. W.; Davis-Ward, R. G.; Harris, P. H.; Laudeman, C. P.; Mook, R. A.; Nailor, K. E.; Szewczyk, G. R.; Veal, J. M.; et al. *Clin. Cancer Res.* **2005**, *11(23, Suppl.),* Abstract C42.

15.   Kumar, R.; Knick, V. B.; Rudolph, S. K.; Johnson, J. H.; Crosby, R. M.;
       Crouthamel, M. C.; Hopper, T. M.; Miller, C. G.; Harrington, L. E.; Onori,
       J. A.; Mullin, R. J.; et al. *Mol. Cancer Ther.* **2008**, *6*, 2012–2021.

16.   Suttle, A. B.; Hurwitz, H.; Dowlati, A.; Fernando, N.; Savage, S.; Coviello,
       K.; Dar, M.; Ertel, P.; Whitehead, B.; Pandite, L. *Proc. Am. Soc. Clin.
       Oncol.* **2004**, *22(14 Suppl.)*, Abstract 3054.

17.   Hurwitz, H.; Dowlati, A.; Savage, S.; Fernando, N.; Lasalvia, S.;
       Whitehead, B.; Suttle, B.; Collins, D.; Ho, P.; Pandite, L *Proc. Am. Soc.
       Clin. Oncol.* **2005**, *23(16 Suppl.)*, Abstract 3012.

18.   Hutson, T. E.; Davis, I. D.; Machiels, J. P.; de Souza, P. L.; Hong, B. F.;
       Rottey, S.; Baker, K. L.; Crofts, T.; Pandite, L.; Figlin, R. *Am. Soc. Clin.
       Oncol.* **2007**, *25(18, Suppl.)*, Abstract 5031.

19.   (a) Sathornsumetee, S.; Rich, J. N. *Drugs Today* **2005**, *42*, 657–670. (b)
       Zareba, G.; Castaner, J.; Bozzo, J. *Drugs Fut.* **2005**, *30*, 138–145. (c)
       Herbst, R. S.; Heymach, J. V.; O'Reilly, M. S.; Onn, A.; Ryan, A. J. *Expert.
       Opin. Investig. Drugs* **2007**, *16*, 239–249.

20.   Sorbera, L. A.; Serradell, N.; Rose, E.; Bolos, J.; Bayes, M. *Drugs Fut.*
       2007, *32*, 577–589.

III

CARDIOVASCULAR AND METABOLIC DISEASES

10

Sitagliptin (Januvia): A Treatment for Type 2 Diabetes

Scott D. Edmondson, Feng Xu, and Joseph D. Armstrong III

USAN: Sitagliptin Phosphate
Trade Name: Januvia®
Merck & Co., Inc.
Launched: 2006

1

10.1 Background

Type 2 diabetes is a disease characterized by chronic hyperglycemia and is accompanied by insulin resistance and an inability of the pancreas to produce enough insulin to control blood sugar levels. Diabetes has been recognized as a worldwide epidemic, with a prevalence of over 180 million people worldwide that is expected to rise to greater than 360 million by 2030.[1,2] Long-term effects of chronic hyperglycemia include damage to the nervous system (e.g., diabetic neuropathy), the circulatory system (e.g., retinopathy, vascular disease), and the kidneys. In addition, co-morbidities such as hypertension, dyslipidemia, and obesity contribute to an increased overall risk of heart disease or stroke for diabetics.[3] Many therapies used to treat type 2 diabetes possess undesirable side effects such as weight gain, hypoglycemia, and inadequate durability. Consequently, there is an unmet medical need for improved therapies for the treatment of this disease.

The approval of sitagliptin (**1**) by the U.S. FDA in 2006 established dipeptidyl peptidase IV (DPP-4) inhibitors as an important new therapy for the treatment of type 2 diabetes.[4-7] DPP-4 inhibitors stimulate insulin secretion indirectly by enhancing the action of the incretin hormones glucagon-like peptide 1 (GLP-1) and glucose-dependent insulinotropic polypeptide (GIP). GLP-1 and GIP stimulate insulin secretion in a glucose-dependent manner, thus posing little or no risk for hypoglycemia. In addition, GLP-1 stimulates insulin biosynthesis, inhibits glucagon secretion, slows gastric emptying, reduces appetite, and stimulates the regeneration and differentiation of islet β-cells.[8,9] DPP-4 inhibitors increase circulating GLP-1 and GIP levels in humans, which leads to decreased blood glucose levels, hemoglobin A_{1c} (HbA$_{1c}$) levels, and glucagon levels. DPP-4 inhibitors such as sitagliptin possess advantages over alternative diabetes therapies, including a lowered risk of hypoglycemia, a potential for weight loss, and the potential for the regeneration and differentiation of pancreatic β-cells.[10,11] In addition to sitagliptin (**1**), DPP-4 inhibitors vildagliptin (**2**) and saxagliptin (**3**) are also marketed for the treatment of type 2 diabetes.[12] Moreover, alogliptin (**4**)[13] and linagliptin (**5**)[14] have also reached advanced stages of development.

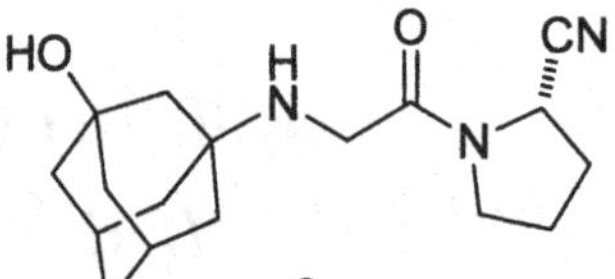

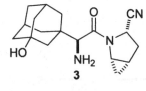

USAN: Vildagliptin
Trade Name: Galvus®
Novartis Pharmaceuticals
Launched: 2007 in the EU

USAN: Saxagliptin
Trade Name: Onglyza®
Bristol-Myers Squibb Co.
Launched: 2009

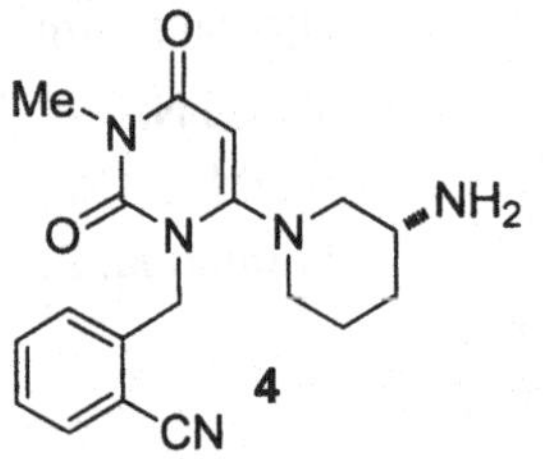

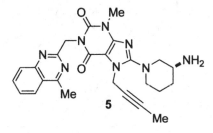

USAN: Alogliptin
Takeda Pharmaceutical Co.
NDA submitted in the USA

USAN: Linagliptin
Boeheringer Ingelheim
Ph III clinical trials

10.2   Pharmacology

Sitagliptin is a potent, competitive, and reversible DPP-4 inhibitor that exhibits > 2,600-fold in vitro selectivity over closely related enzymes such as QPP, DPP8, and DPP9 (Table 1).[15] Selectivity over DPP8 and DPP9 is important because inhibition of these enzymes has been associated with multiorgan toxicity in preclinical species.[16]

In lean mice, orally administered sitagliptin reduced blood glucose excursion dose dependently in an oral glucose tolerance test at doses ranging from 0.1 mg/kg to 3.0 mg/kg. At doses of 1 mg/kg and 3 mg/kg, plasma drug levels were 190 nM and 600 nM and inhibition of plasma DPP-4 activity was 69% and 84%, respectively. At maximally efficacious doses, an increase of postprandial, active GLP-1 concentrations of twofold to threefold was observed.[15] These studies demonstrate the relationship between inhibition of DPP-4, increases in active GLP-1 concentration, and an improvement in glucose tolerance after an oral glucose challenge.

Table 1. In Vitro Potency and Selectivity of Sitagliptin

Enzyme	IC_{50}	Selectivity
DPP-4	18 nM	—
QPP	> 100,000 nM	> 5,000
DPP8	48,000 nM	2,700
DPP9	> 100,000 nM	> 5,000

Inhibition of DPP-4 in rodent models for diabetes has also been shown to attenuate the decline of pancreatic β-cell function and mass.[17,18] DPP-4 inhibition with sitagliptin in a mouse model of type 2 diabetes resulted in improvements in glucose homeostasis and significant improvements in islet mass and β-cell function. In the same rodent model, sitagliptin demonstrated improved β-cell preserving effects and superior glucose-lowering efficacy compared to glipizide, a commonly used insulin secretagogue.[19]

10.3   Structure–Activity Relationship (SAR)

The structural origin of sitagliptin can be traced to a proline-derived HTS lead (6) and a piperazine-derived HTS lead (7). Dramatic improvements in potency of each of these compounds were achieved by introducing an *ortho*-fluoro homophenylalanine amide moiety to ultimately afford proline amide 8[20] and piperazine amide 9.[21] Unfortunately, both of these compounds possess low oral bioavailability in rats ($F_{rat} \leq 1\%$). DPP-4 inhibitors with moderate potencies are achievable with simplified *ortho* fluoro homophenylalanine amides such as 10[22] and 11;[21] however, these compounds also possess low oral bioavailabilities in rats due to metabolic instability of the thiazolidine and piperazine rings, respectively. Consequently, a variety of fused heterocycles were screened in an effort to improve metabolic stability and DPP-4 potency. This effort led to the discovery of the fused triazole 12, with improved metabolic stability and similar DPP-4 potency to 10 and 11.[15] Although 12 is stable to metabolism in rat hepatocytes, its oral bioavailability in rats is still low. Replacement of the ethyl substituent with a trifluoromethyl substituent and installation of the 2,4,5-trifluorophenyl homophenyl-alanine moiety improved oral bioavailability in rats as well as potency against DPP-4. Overall, sitagliptin (1) was superior to other fused heterocycle derivatives in terms of potency, off-target selectivity, and preclinical pharmacokinetic profile.

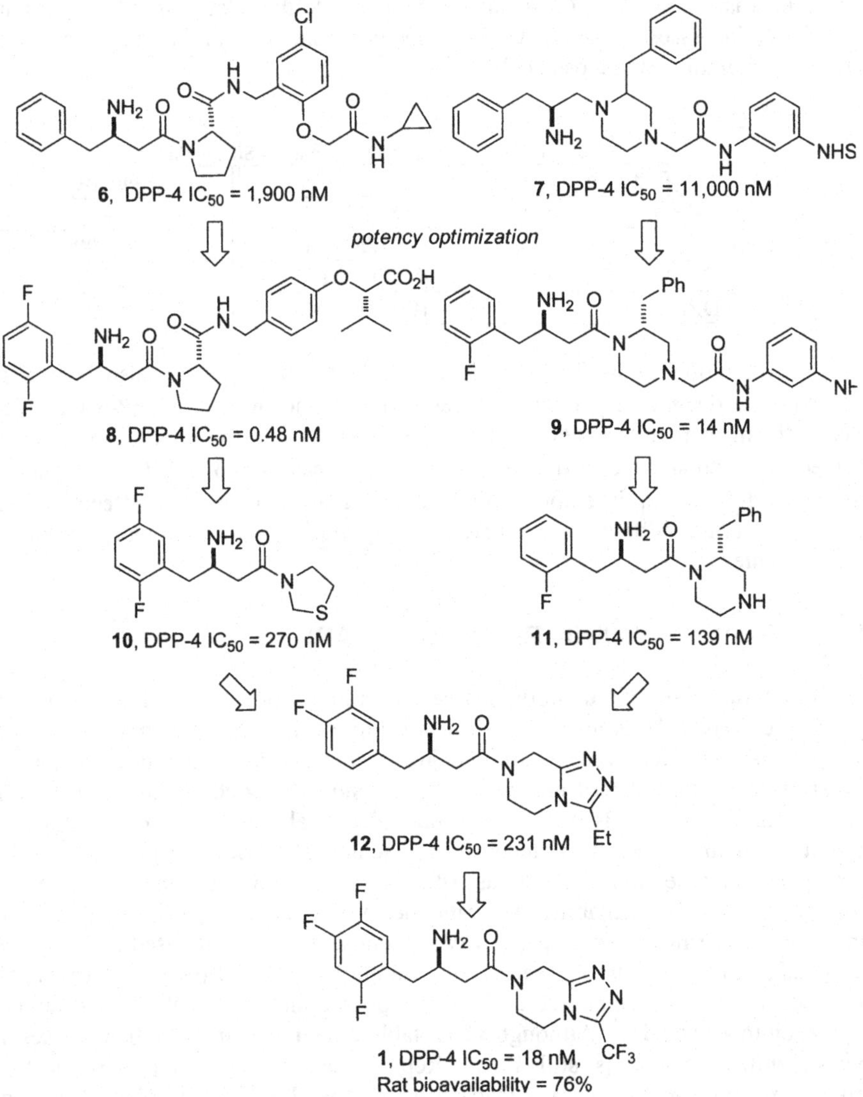

6, DPP-4 IC$_{50}$ = 1,900 nM

7, DPP-4 IC$_{50}$ = 11,000 nM

potency optimization

8, DPP-4 IC$_{50}$ = 0.48 nM

9, DPP-4 IC$_{50}$ = 14 nM

10, DPP-4 IC$_{50}$ = 270 nM

11, DPP-4 IC$_{50}$ = 139 nM

12, DPP-4 IC$_{50}$ = 231 nM

1, DPP-4 IC$_{50}$ = 18 nM,
Rat bioavailability = 76%

10.4   Pharmacokinetics and Drug Metabolism

Sitagliptin (**1**) demonstrated plasma half-lives in rats, dogs, and rhesus monkeys of 1.7–4.9 h, and oral bioavailabilities of 68–100%. In rats and monkeys, clearance of **1** was relatively high (60 and 28 mL/min/kg, respectively) and was lower in dogs (6 mL/mg/kg).[15] In rats and dogs, **1** was systemically cleared primarily by renal elimination of intact parent drug, with additional contributions from biliary excretion (rats) and metabolism (minor in both species).[23] In rats, data suggest that active transport mechanisms are involved in the renal elimination of **1**.

In humans, the bioavailability of sitagliptin is 87% and peak plasma concentrations occur between 1 and 4 h after oral administration in healthy volunteers,[24,25] middle-aged obese subjects,[26] and patients with type 2 diabetes.[27] The ingestion of food does not affect the pharmacokinetic profile of **1**.[24] A single 100 mg oral dose of sitagliptin exhibits a $t_{1/2}$ = 12 h and a mean $AUC_{0-\infty}$ = 8.5 µM·h in healthy volunteers. At steady state, the exposure of sitagliptin increases by ~ 14% compared to exposures after a single dose.[28] Sitagliptin exhibits relatively low reversible plasma protein binding and is widely distributed into tissue. The major elimination pathway for sitagliptin in humans is excretion of parent in urine.[25,29] Consequently, increased exposure (twofold to fourfold) is observed in patients with impaired renal function.[30] In vitro data indicate that **1** is unlikely to be involved in any drug–drug interactions (DDIs) with the cytochrome P450 enzyme system. DDI studies with sitagliptin demonstrated no meaningful alterations of the pharmacokinetic profiles of other co-administered diabetes medications, including metformin, rosiglitazone, and glyburide.[31–33] Conversely, metformin did not meaningfully alter the pharmacokinetic profile of **1**.[31] Finally, a population analysis of Phase 1 and Phase 2 pharmacokinetic studies with sitagliptin revealed no clinically meaningful DDIs in patients treated with 83 different co-administered drugs.[34]

10.5 Efficacy and Safety

The safety and efficacy of sitagliptin (**1**) have been evaluated in multiple clinical studies.[3,4,11,35] In dose escalation studies, a single 100 mg oral dose of **1** was shown to inhibit DPP-4 by > 80% over 24 h in healthy adults.[25] In animals and humans, DPP-4 inhibition of ≥ 80% (uncorrected) affords maximal glucoregulatory effects, suggesting that a once-daily 100 mg dose of sigagliptin is optimal for maximum glucose lowering in humans.[36] The percent inhibition of enzyme in DPP-4 activity assay underestimates the extent of DPP-4 inhibition that occurs in circulating plasma due to dilution of the collected plasma into DPP-4 substrate containing solution. In humans, an uncorrected plasma inhibition of 80% is estimated to correspond to a corrected value of ~ 95–96% inhibition.[36] As expected, based on its mechanism of action, **1** induces a dose-dependent increase in active GLP-1, GIP, and insulin levels as well as a decrease in glucagon levels and glycemic excursion after an oral glucose challenge.[27] In Phase II studies compared to placebo, sitagliptin improves glycemic control in patients with type 2 diabetes by reducing both fasting and postprandial glucose levels with a safety and tolerability profile similar to placebo and a neutral effect on body weight. In Phase III studies, sitagliptin significantly decreases HbA_{1c} levels compared to placebo and increases the proportion of patients that achieve the American Diabetes Association (ADA) goal of < 7% HbA_{1c} levels. In addition, decreases on fasting plasma glucose, 2 h postprandial glucose levels, and improvements in β-cell function have resulted from treatment with sitagliptin.

Sitagliptin 100 mg/day is well tolerated in patients with type 2 diabetes with an overall profile similar to placebo. In dose escalation studies, sitagliptin was well tolerated up to a single dose of 800 mg/day and multiple doses of 400 mg/day for 28 days. A pooled analysis of 12 clinical studies in > 3400 patients up to 2 years in duration with sitagliptin 100 mg/day demonstrated an overall rate of adverse effects (AEs) similar to that of placebo or active comparator and a neutral effect on body weight.[37] Overall, there was a higher rate of incidence of drug-related AEs in the nonexposed group, primarily because of an increased rate of hypoglycemia in studies in which the active comparator was a sulfonylurea. The low incidence of hypoglycemia in sitagliptin treated patients is consistent with its glucose-dependent mechanism of action. Although cardiovascular outcome trials were not included in this pooled analysis, there were no meaningful differences between groups in the incidence rates of cardiac-related or ischemic related AEs.

In 24-week add-on studies with metformin, pioglitazone, glimepiride, or glimapiride/metformin, the addition of sitagliptin 100 mg/day resulted in greater reductions in HbA$_{1c}$ and a greater proportion of patients reaching a target HbA$_{1c}$ of < 7% compared to the respective monotherapies. In addition, a 52 week study that compared patients taking metformin + sitagliptin with metformin + glipizide revealed that ~ 60% of patients in both groups reached HbA$_{1c}$ < 7%.[38] In the group that received glipizide, however, 32% patients had at least one hypoglycemic event compared to only 5% for the sitagliptin group. Furthermore, a 1.5 kg average weight *loss* was reported in the sitagliptin add-on group relative to a 1.1 kg weight *gain* in the glipizide add-on group.

10.6   Syntheses

Sitagliptin **1** was originally synthesized[15,22] by coupling β-amino acid **16** and triazole **17**[39] followed by deprotection of the Boc group in HCl/MeOH. Starting from **13**, asymmetric preparation of α-amino acid **14** was realized by applying Schöllkopf's bis-lactim methodology in two steps. The α-amino acid methyl ester **14** was then transformed into β-amino acid **16** via an Arndt-Eistert homologation through sequential operations: protection, diazo formation-rearrangement, and hydrolysis.

Potts' method[39] was initially used for the preparation of triazole **17**. Even with modifications,[40] however, this synthesis was not suitable for large-scale preparation due to safety hazard issues. Therefore, a practical manufacturing route to **17** was developed.[41] Key intermediate chloromethyloxadiazole **19** was prepared via sequential condensation of 35% aqueous hydrazine with ethyl trifluoroacetate and chloroacetyl chloride followed by dehydration with phosphorus oxychloride in one pot. Treatment of **19** with ethylenediamine in methanol afforded amidine **21**. The use of methanol was crucial for this transformation. In the presence of nucleophilic methanol, active reaction species **20** was generated in situ and subsequently captured by ethylenediamine to afford amidine **21**. Exposure of **21** to aqueous HCl gave the desired triazole **17**, which was directly isolated as its HCl salt by filtration. The overall yield of this manufacturing process is 52%, nearly double that of the original route. Importantly, the use of 1 equivalent of aqueous hydrazine, which is completely consumed in the first step, results in safer operating conditions and cleaner waste streams.

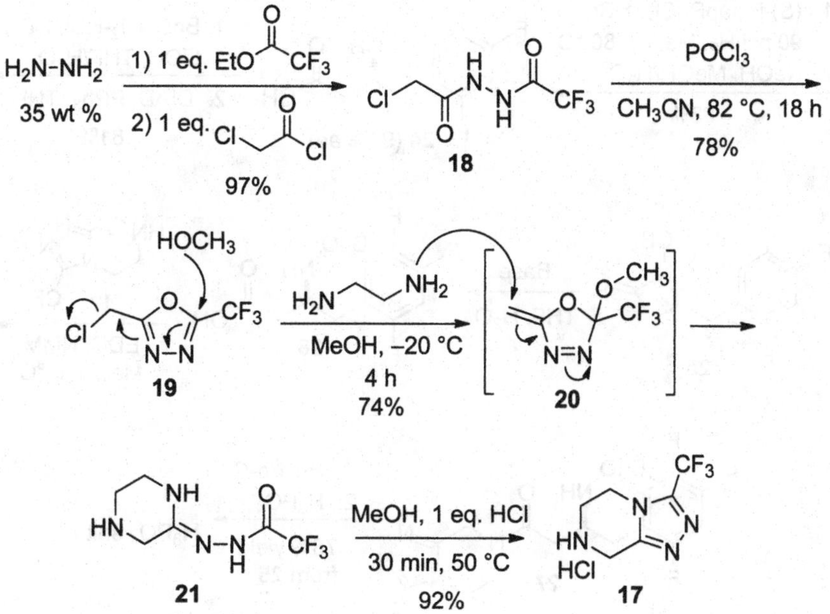

Merck's first-generation large-scale preparation of sitagliptin relied on asymmetric hydrogenation of the requisite β-keto ester **23** to set up the stereochemistry of **1**.[40] Acid **22** was converted to **23** under Masamune's conditions. Enantioselective hydrogenation of **23** was achieved in the presence of (S)-BinapRuCl$_2$–triethylamine complex and a catalytic amount of HBr. It is noteworthy that the use of HBr allowed reduction of the loading of the catalyst to < 0.1 mol% without affecting the enantioselectivity or yield. The isolated acid **24** was further elaborated to lactam **25** in a two-step sequence, EDC coupling followed by a Mitsunobu reaction. Thus the hydroxyl group of **24** was transformed into the corresponding protected amino acid **26** upon hydrolysis of lactam **25**. The N-benzyloxy group of **25** was found to effectively protect the amino group to allow for a selective coupling of **26** with triazole **17** in the presence of EDC and N-methylmorpholine to afford **27** in near quantitative yield. Upon hydrogenation, **1** was isolated in > 99.5% purity as its anhydrous phosphoric acid salt.

The first large-scale synthesis of sitagliptin afforded the desired compound in 45% yield over eight steps from acid **22** and triazole **17**.

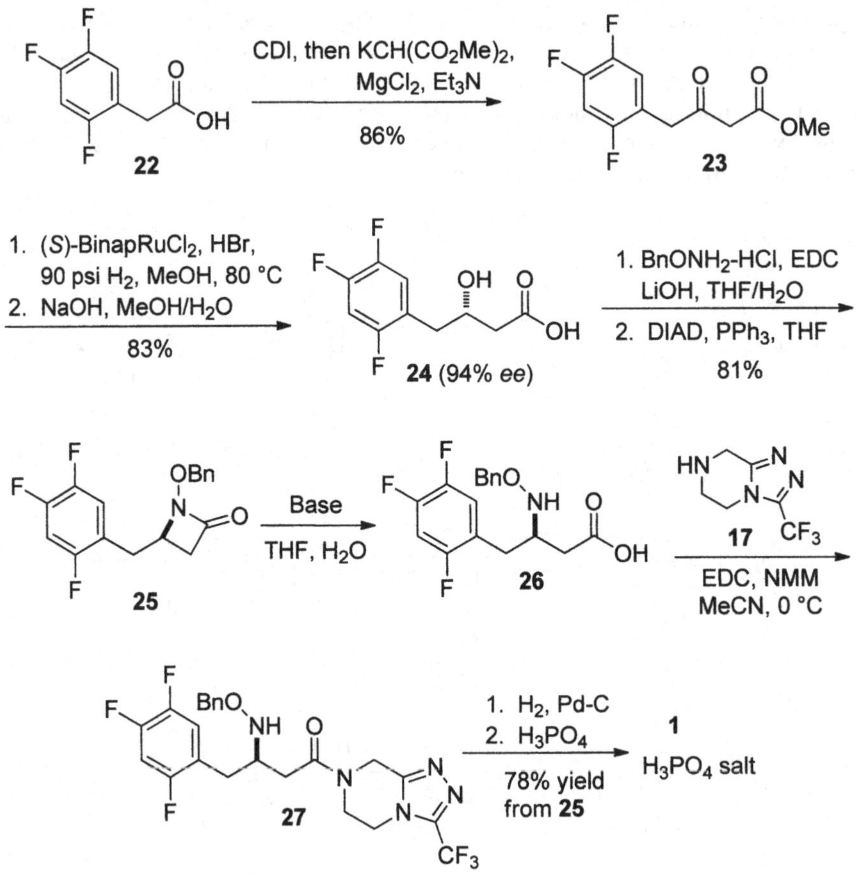

Many aspects of the first-generation synthesis made it suitable for the large-scale preparation of sitagliptin. Nevertheless, the first-generation route relies on an inefficient method of creating the β-amino acid moiety via asymmetric hydrogenation of a β-keto ester intermediate. In particular, the EDC coupling/Mitsunobu sequence used to transform the hydroxyl group of **24** to the masked amino group of **25** provided poor atom economy and was a major contributor to the overall waste output of the synthesis. To overcome this drawback and to achieve a more efficient and environmentally friendly synthesis, the Merck Process Research team focused their efforts on asymmetric hydrogenation of *N*-protected/unprotected enamine substrates to directly set up the desired stereochemistry as well as the functionality of sitagliptin.

The use of a (S)-phenylglycinamide chiral auxiliary (PGA) for asymmetric hydrogenation of enamines was evaluated.[42] Keto ester **23** was condensed with PGA in methanol to afford enamine **28** with 85–90% conversion and the pure Z-enamine isomer was crystallized from the reaction mixture in 80% isolated yield. Various heterogeneous catalysts were screened to hydrogenate **28**. Adam's catalyst (PtO_2) turned out to be the best in terms of diastereoselectivity and conversion. The use of HOAc was essential for the reduction in terms of conversion and diastereoselectivity. Under the optimized conditions, hydrogenation of PGA–enamine **28** in the presence of 5 equivalents of HOAc afforded amine **29** in 90% assay yield and 91% diastereomeric excess. Hydrolysis of **29** followed by EDC coupling with triazole **17** gave amide **30**. Removal of the chiral auxiliary via hydrogenolysis afforded sitagliptin.

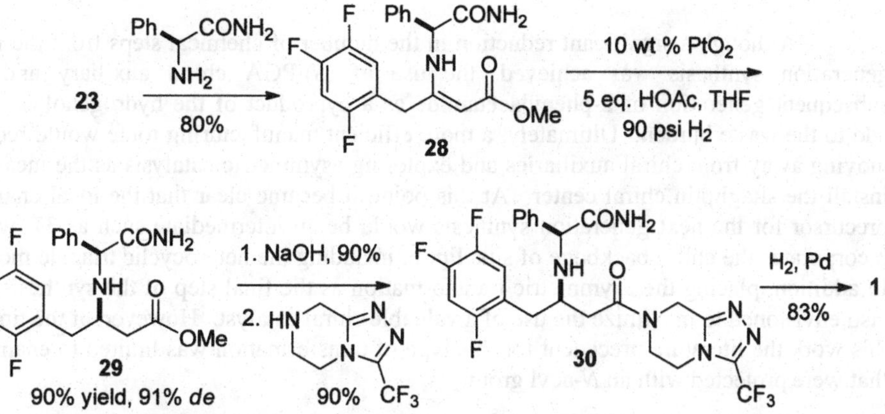

In comparison to the first-generation synthesis, the PGA chiral auxiliary approach resulted in a reduction of three chemical steps. However, some of the functional group manipulation steps could be potentially eliminated by installing the triazole fragment earlier in the synthesis and hydrogenating the corresponding enamine-triazole amide instead of enamine-ester **28** to establish the desired stereochemistry. Thus PGA-0enamine **32** was prepared by heating **31** with (S)-PGA in the presence of a catalytic amount of AcOH to afford the pure Z-enamine isomer in 91% yield. Indeed, hydrogenation (cat. PtO_2, THF/MeOH) afforded the PGA-amine **33** with high selectivity (97.4% de) and 92% assay yield. The PGA group was finally removed under transfer hydrogenolysis conditions to afford sitagliptin (**1**) in 92% assay yield. Sitagliptin was isolated as an L-tartrate salt in 90% recovery and 99.9% enantiomeric excess. The 2-phenylacetamide byproduct resulting from hydrogenolysis cleavage of PGA was completely removed during the crystallization.

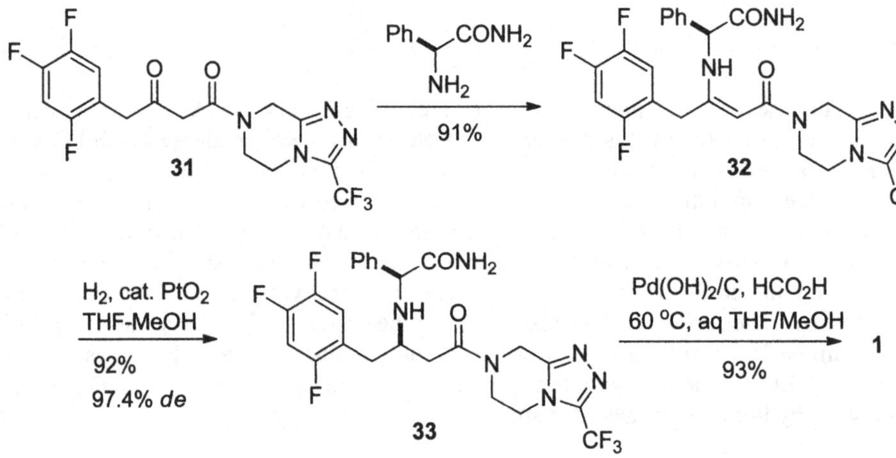

Although a significant reduction in the number of chemical steps from the first-generation synthesis was achieved, the use of (*S*)-PGA chiral auxiliary and the subsequent generation of 2-phenylacetamide as a byproduct of the hydrogenolysis step add to the waste burden. Ultimately, a more efficient manufacturing route would require moving away from chiral auxiliaries and exploring asymmetric catalysis as the means to install the sitagliptin chiral center. At this point, it became clear that the ideal enamine precursor for the next generation synthesis would be an intermediate such as **37**, which incorporates the entire backbone of sitagliptin, including the heterocyclic triazole moiety. In addition, placing the asymmetric transformation as the final step of the synthesis was also envisioned to maximize the use of a valuable chiral catalyst. However, at the time of this work the literature precedent for this type of transformation was limited to enamines that were protected with an *N*-acyl group.

To achieve this ultimate goal, a short, concise synthesis of dehydrositagliptin **37** was developed. The approach used to prepare **37** capitalized on the ability of Meldrum's acid (**34**) to act as an acyl anion equivalent. This process involves activation of **22** by formation of a mixed anhydride with pivaloyl chloride, in the presences of **34**, *i*-Pr$_2$NEt, and a catalytic amount of DMAP to form **35**. The formation of β-keto amide **31** occurred via degradation of **35** to an oxo-ketene intermediate **36** which was trapped with piperazine **17**.[43] Without workup, NH$_4$OAc and methanol were mixed with the crude reaction mixture to furnish dehydrositagliptin **37**, which contains the entire structure of sitagliptin **1** save two hydrogen atoms. Importantly, **37** was prepared in an easily operated one-pot process in 82% overall isolated yield with 99.6 wt % purity through a simple filtration, thereby eliminating the need for aqueous workup and minimizing waste generation.

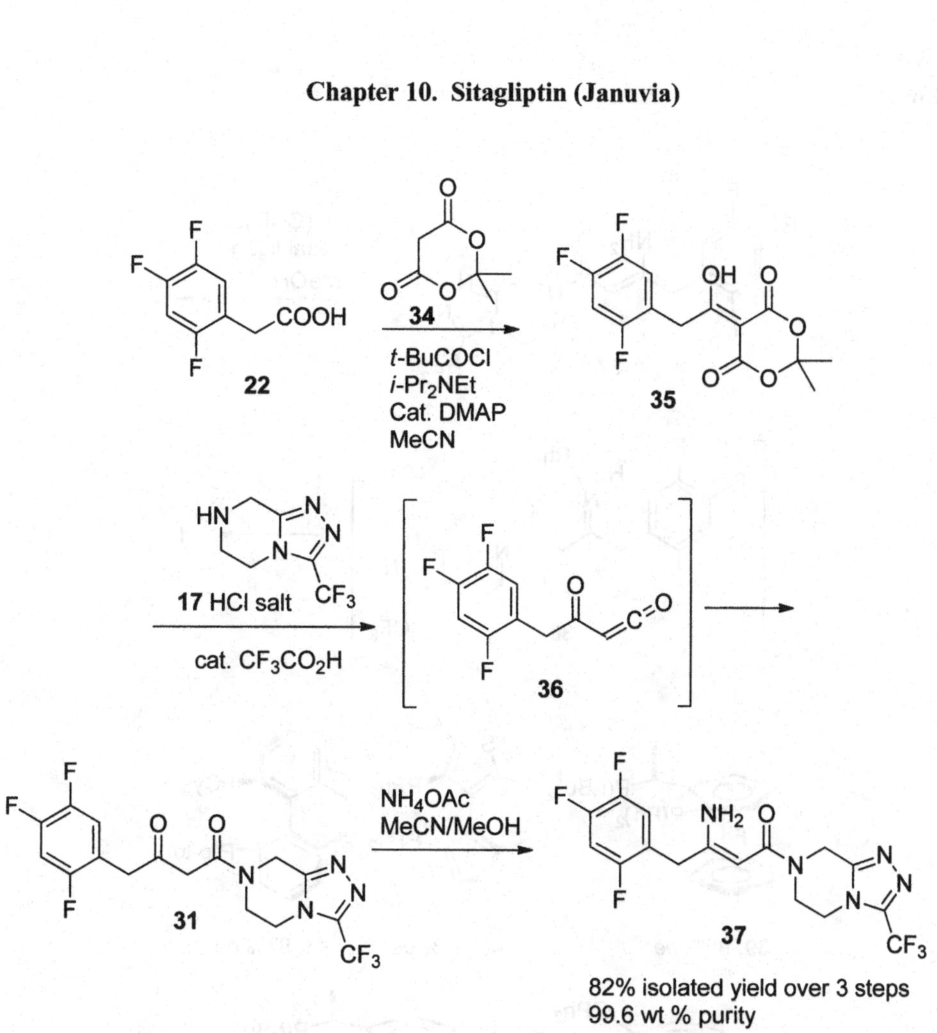

82% isolated yield over 3 steps
99.6 wt % purity

To explore the feasibility of the unprecedented asymmetric hydrogenation of unprotected enamines, a focused pilot screen on substrate **37** with a relatively small set of commercially available chiral bisphosphines in combination with Ir, Ru, and Rh salts was performed.[44] Metal catalysts Ir, Ru, and Rh were selected for this screen due to their demonstrated performance in asymmetric hydrogenations. It was astonishing that the screening results not only showed a trend of enantioselectivities, but also resulted in a very direct hit. While Ir and Ru catalysts gave poor results, [Rh(COD)$_2$OTf], in particular with ferrocenyl-based JOSIPHOS-type catalysts (i.e., **39**), afforded both high conversion and enantioselectivity. Further screening revealed that other ligands not limited to the ferrocenyl structural class could effect this transformation with high enantioselectivity.[45] Using [Rh(COD)Cl]$_2$ as the catalyst, ligands **40–43** provided the highest levels of enantioselectivity for reduction of **37**. In summary, an overall consideration of yield, enantioselectivity, reaction rate, and ligand cost, led to a decision to pursue the [Rh(COD)Cl]$_2$-^{t}Bu-JOSIPHOS **39** combination for further development to deliver a viable hydrogenation process for the commercial manufacture of **1**.

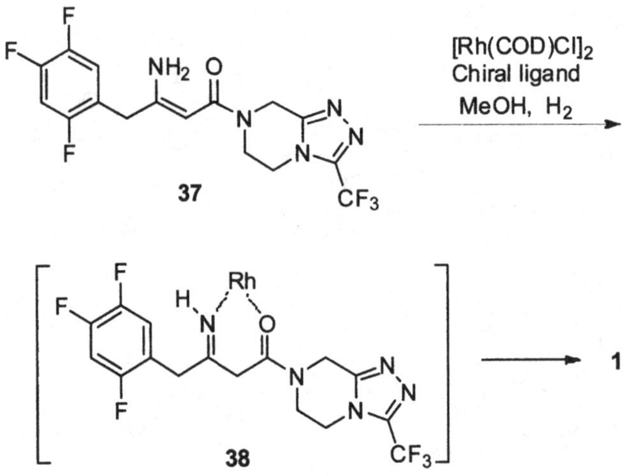

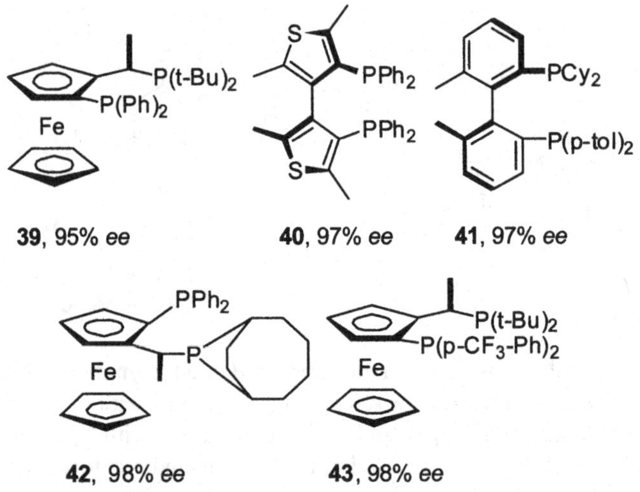

Further exhaustive development showed that hydrogenation of **37** was best performed in methanol. Interestingly, it was found that introducing a small amount of ammonium chloride (0.15–0.3 mol%) was necessary to achieve consistent performance in terms of both enantioselectivity and conversion rate.[46] The effect has been studied in depth in terms of producing consistent hydrogenation results, but the mechanism is not clear. More important, by increasing the pressure to 250 psi, catalyst loading was dramatically reduced to 0.15 mol% without sacrificing yield, enantioselectivity, or reaction rate at 50 °C. The reduction in catalyst loading and, therefore, the cost of the synthesis by simply increasing hydrogen pressure demonstrated the power of asymmetric hydrogenation to set the stereochemistry of **1**.

Mechanistic studies suggested that the hydrogenation proceeded through the imine tautomer **38**. Under optimal conditions, hydrogenation of **37** in the presence of NH_4Cl (0.15 mol%), $[Rh(COD)Cl]_2$ (0.15 mol%), and ^{t}Bu JOSIPHOS **39** (0.155 mol%) in MeOH under 250 psi of hydrogen at 50 °C for 16–18 h proved to be extremely robust and afforded sitagliptin **1** in 98% yield and 95% *ee* reproducibly. Note > 90–95% of the precious rhodium catalyst was recovered by treating the reaction mixture with Ecosorb C-941. The freebase of sitagliptin was then isolated by crystallization in 79% yield from **37** with > 99% *ee* and > 99% purity.

This highly efficient, asymmetric synthesis[47] of sitagliptin (**1**) has been implemented on manufacturing scale. The entire synthesis is carried out with a minimum number of operations: a one-pot process affords crystalline dehydrositigliptin (**37**) in > 99.6 wt %; the highly enantioselective hydrogenation of **37** in the presence of as low as 0.15 mol% ^{t}Bu JOSIPHOS-Rh(I) gives **1** in high yield and > 95% *ee*. After the precious metal catalyst is selectively recovered/removed from the process stream, **1** is isolated as a free base, which is further converted to the final pharmaceutical form, a monohydrate phosphate salt, in > 99.9% purity and > 99.9% *ee*. The use of low Rh(I) loading for the asymmetric hydrogenation in combination with an ease of recovery of the precious rhodium metal made this process highly cost-effective. The overall yield of this process is up to 65%.

The direct enamine hydrogenation route illustrates an example of a synthetic target that was a primary driver for the discovery of a new synthetic transformation—namely the enantioselective hydrogenation of unprotected enamine amides. The efficient route contains all the elements required for a manufacturing process. Furthermore, this straightforward approach to **1** reduces significantly the amount of waste produced in the process. Compared to the first-generation route, the total waste generated per kilogram of sitagliptin produced in this environmentally friendly green process is reduced from 250 kg to 50 kg. Most striking, the amount of aqueous waste produced in the manufacturing process was reduced to zero. The drastic reductions in waste, realized over virtually the entire product lifetime for sitagliptin, coupled with the new chemistry discovered in this efficient process, led to a Presidential Green Chemistry Award, the ICHEME Astra–Zeneca Award, and a Thomas Alva Edison Patent Award for sustainability.[48]

In summary, the discovery, development, and syntheses of sitagliptin (**1**, Januvia) were reviewed herein. Sitagliptin is the first DPP-4 inhibitor approved for the treatment of type 2 diabetes. To date, it has demonstrated an excellent overall clinical profile. Contributions that led to the discovery of **1** have been recognized by a Prix Galien USA Award and a Thomas Alva Edison Patent Award.[49] Furthermore, the strong safety and tolerability profile of sitagliptin make it an attractive candidate for combination therapy with other antihyperglycemic therapies. In 2007, the first fixed-dose-combination therapy of sitagliptin with metformin was also approved by the U.S. FDA and is currently marketed in the United States under the trade name Janumet.[50] Janumet is an important new treatment option that can help patients with type 2 diabetes achieve improved glycemic control compared to existing monotherapies.

10.7   References

1.   Wild, S.; Roglic, G.; Green, A.; Sicree, R.; King, H. *Diabetes Care* **2004**, *27*, 1047–1053.
2.   Stumvoll, M.; Goldstein, B. J.; Haeften, T. W. *Lancet* **2005**, *365*, 1333–1346.
3.   Kulasa, K. M.; Henry, R. R. *Expert Opin. Pharmacother.* **2009**, *10*, 2415–2432.
4.   Herman, G. A.; Stein, P. P.; Thornberry, N. A.; Wagner, J. A. *Clin. Pharm. Ther.* **2007**, *81*, 761–767.
5.   Thornberry, N. A.; Weber, A. E. *Curr. Top. Med. Chem.* **2007**, *7*, 557–568.
6.   Deacon, C. F. *Expert Opin. Invest. Drugs* **2007**, *16*, 533–545.
7.   Karasik, A.; Aschner, P.; Katzeff, H.; Davies, M. J.; Stein, P. P. *Curr. Med. Res. Opin.* **2008**, *24*, 489.
8.   Holst, J. J.; Vilsboll, T.; Deacon, C. F. *Molec. Cell. Endocrinol.* **2009**, *297*, 127.
9.   Kim, W.; Egan, J. M. *Pharm. Rev.* **2008**, *60*, 470.
10.   Havale, S. H.; Pal, M. *Bioorg. Med. Chem.* **2009**, *17*, 1783–1802.
11.   Pei, Z. *Curr. Opin. Drug Disc. Dev.* **2008**, *11*, 512–532.
12.   Ahren, B. *Best Practice Res. Clin. Endocrin. Metab.* **2009**, *23*, 487–498.
13.   Pratley, R. E.; Reusch, J. E. B.; Fleck, P. R.; Wilson, C. A.; Mekki, Q. *Curr. Med. Res. Opin.* **2009**, *25*, 2361–2371.
14.   Tiwari, A. *Curr. Opin. Invest. Drugs* **2009**, *10*, 1091–1104.
15.   Kim, D.; Wang, L.; Beconi, M.; Eiermann, G. J.; Fisher, M. H.; He, H.; Hickey,G. J.; Kowalchick, J. E.; Leiting, B.; Lyons, K.; Marsilio, F.; McCann, M. E.; Patel, R. A.; Petrov, A.; Scapin, G.; Patel, S. B.; Sinha Roy, R.; Wu, J. K.; Wyvratt, M. J.; Zhang, B. B.; Zhu, L.; Thornberry, N. A.; Weber, A. E. *J. Med. Chem.* **2005**, *48*, 141–151.
16.   Lankas, G. R.; Leiting, B.; Sinha Roy, R.; Eiermann, G. J.; Beconi, M. G.; Biftu, T.; Chan, C.-C.; Edmondson, S.; Feeney, W. P.; He, H.; Ippolito, D. E.; Kim, D.; Lyons, K. A.; Ok, H. O.; Patel, R. A.; Petrov, A. N.; Pryor, K. A.; Qian, X.; Reigle, L.; Woods, A.; Wu, J. K.; Zaller, D.; Zhang, X.; Zhu, L.; Weber, A. E.; Thornberry, N. A. *Diabetes* **2005**, *54*, 2988–2994.
17.   Reimer, M. K.; Holst, J. J.; Ahren, B. *Eur. J. Endocrinol.* **2002**, *146*, 717–727.
18.   Pospisilik, J. A.; Martin, J.; Doty, T.; Ehses, J. A.; Pamir, N.; Lynn, F. C.; Piteau, S.; Demuth, H.-U.; McIntosh, C. H. S.; Pederson, R. A. *Diabetes* **2003**, *52*, 741–750.
19.   Mu, J.; Petrov, A.; Eiermann, G. J.; Woods, J.; Zhou, Y.-P.; Li, Z.; Zycband, E.; Feng, Y.; Zhu, L.; Sinha Roy, R.; Howard, A. D.; Li, C.; Thornberry, N. A.; Zhang, B. B. *Eur. J. Pharm.* **2009**, *623*, 148–154.
20.   Edmondson, S. D.; Mastracchio, A.; Beconi, M.; Colwell, L. Jr.; Habulihaz, H.; He, H.; Kumar, S.; Leiting, B.; Lyons, K.; Mao, A.; Marsilio, F.; Patel, R. A.; Wu, J. K.; Zhu, L.; Thornberry, N. A.; Weber, A. E.; Parmee, E. R. *Bioorg. Med. Chem. Lett.* **2004**, *14*, 5151–5155.
21.   Brockunier, L.; He, J.; Colwell Jr., L. F.; Habulihaz, B.; He, H.; Leiting, B.; Lyons, K. A.; Marsilio, F.; Patel, R.; Teffera, Y.; Wu, J. K.; Thornberry, N. A.; Weber, A. E.; Parmee, E. R. *Bioorg. Med. Chem. Lett.* **2004**, *14*, 4763–4766.

22. Xu, J.; Ok, H. O.; Gonzalez, E. J.; Colwell, L. F. Jr.; Habulihaz, B.; He, H.; Leiting, B.; Lyona, K. A.; Marsilio, F.; Patel, R. A.; Wu, J. K.; Thornberry, N. A.; Weber, A. E.; Parmee, E. R. *Bioorg. Med. Chem. Lett.* **2004**, *14*, 4759–4762.

23. Beconi, M. G.; Reed, J. R.; Teffera, Y.; Xia, Y.-Q.; Kochansky, C. J.; Liu, D. Q.; Xu, S.; Elmore, C. S.; Ciccoto, S.; Hora, D. F.; Stearns, R. A.; Vincent, S. H. *Drug Metab. Dispos.* **2007**, *35*, 525–532.

24. Bergman, A.; Ebel, D.; Liu, F.; Stone, J.; Wang, A.; Zeng, W.; Chen, L.; Dilzer, S.; Lasseter, K.; Herman, G.; Wagner, J.; Krishna, R. *Biopharm. Drug Dispos.* **2007**, *28*, 315–322.

25. Herman, G.; Stevens, C.; Van Dyck, K.; Bergman, A.; Yi, B.; De Smet, M.; Snyder, K.; Hilliard, D.; Tanen, M.; Tanaka, W.; Wang, A. Q.; Zeng, W.; Musson, D.; Winchell, G.; Davies, M. J.; Ramael, S.; Gottesdiener, K. M.; Wagner, J. A. *Clin. Pharmacol. Ther.* **2005**, *78*, 675–688.

26. Herman, G.; Bergman, A.; Liu, F.; Stevens, C.; Wang, A. Q.; Zeng, W.; Chen, L.; Snyder, K.; Hilliard, D.; Tanen, M.; Tanaka, W.; Meehan, A. G.; Lasseter, K.; Dilzer, S.; Blum, R.; Wagner, J. A. *J. Clin. Pharmacol.* **2006**, *46*, 876–886.

27. Herman, G.; Bergman, A.; Stevens, C.; Kotey, P.; Yi, B.; Zhao, P.; Dietrich, B.; Golor, G.; Schrodter, A.; Keymeulen, B.; Lasseter, K.; Kipnes, M. S.; Snyder, K.; Hilliard, D.; Tanan, M.; Cilissen, C.; De Smet, M.; Lepeleire, I.; Van Dyck, K.; Wang, A. Q.; Zeng, W.; Davies, M.; Tanaka, W.; Holst, J. J.; Deacon, C. F.; Gottesdiener, K. M.; Wagner, J. A. *J. Clin. Endocrin. Metab.* **2006**, *91*, 4612–4619.

28. Bergman, A.; Stevens, C.; Zhou, C.; Yi, B.; Laethem, M.; De Smet, M.; Snyder, K.; Hilliard, D.; Tanaka, W.; Zeng, W.; Tanan, M.; Wang, A. Q.; Chen, L.; Winchell, G.; Davies, M. J.; Ramael, S.; Wagner, J. A.; Herman, G. A. *Clin. Ther.* **2006**, *28*, 55–72.

29. Vincent, S. H.; Reed, J. R.; Bergman, A. J.; Elmore, C. S.; Zhu, B.; Xu, S.; Ebel, D.; Larson, P.; Zeng, W.; Chen, L.; Dilzer, S.; Lasseter, K.; Gottesdiener, K.; Wagner, J. A.; Herman, G. A. *Drug Metab. Dispos.* **2007**, *35*, 533–538.

30. Bergman, A.; Cote, J.; Yi, B.; Marbury, T.; Swan, S. K.; Smith, W.; Gottesdiener, K.; Wagner, J. Herman, G. A. *Clin. Pharm. Therap.* **2006**, *79*, P75.

31. Herman, G. A.; Bergman, A.; Yi, B.; Kipnes, M. *Curr. Med. Res. Opin.* **2006**, *22*, 1939–1947.

32. Mistry, G. C.; Bergman, A. J.; Luo, W.-L.; Cilissen, C.; Haazen, W.; Davies, M. J.; Gottestiener, K. M.; Wagner, J. A.; Herman, G. A. *J. Clin. Pharm.* **2007**, *47*, 159–164.

33. Mistry, G. C.; Bergman, A. J.; Luo, W.-L.; Zheng, W.; Hreniuk, D.; Zinny, M. A.; Gottestiener, K. M.; Wagner, J. A.; Herman, G. A.; Ruddy, M. *Br. J. Clin. Pharmacol.* **2008**, *66*, 36–42.

34. Herman, G. A.; Bergman, A.; Wagner, J. *Diabetologia* **2006**, *49*, A795.

35. Karasik, A.; Aschner, P.; Katzeff, H.; Davies, M. J.; Stein, P. P. *Curr. Med. Res. Opin.* **2008**, *24*, 489–496.

36. Alba, M.; Sheng, D.; Guan, Y.; Williams-Herman, D.; Larson, P.; Sachs, J. R.; Thornberry, N.; Herman, G.; Kaufman, K. D.; Goldstein, B. J. *Curr. Med. Res. Opin.* **2009**, *25*, 2507–2514.

37. Williams-Herman, D.; Round, E.; Swern, A. S.; Musser, B.; Davies, M. J.;
     Stein, P. P.; Kaufman, K. D.; Amatruda, J. M. *BMC Endocri. Disord.* **2008**, *8*,
     14.
38. Nauck, M. A.; Meininger, G.; Sheng, D.; Terranella, L.; Stein, P. P. *Diabetes
     Obes. Metab.* **2007**, *9*, 194–205.
39. Nelson, P. J.; Potts, K. T. *J. Org. Chem.* **1962**, *27*, 3243–3248.
40. Hansen, K. B.; Balsells, J.; Dreher, S.; Hsiao, Y.; Kubryk, M.; Palucki, M.;
     Rivera, N.; Steinhuebel, D.; Armstrong, J. D. III; Askin, D.; Grabowski, E. J. J.
     Org. Process Res. Dev. **2005**, *9*, 634–639.
41. Balsells, J.; DiMichele, L.; Liu, J.; Kubryk, M.; Hansen, K.; Armstrong, J. D. III
     Org. Lett. **2005**, *7*, 1039–1042.
42. Ikemoto, N.; Tellers, D. M.; Dreher, S. D.; Liu, J.; Huang, A.; Rivera, N. R.;
     Njolito, E.; Hsiao, Y.; McWilliams, J. C.; Williams, J. M.; Armstrong, J. D. III;
     Sun, Y.; Mathre, D. J.; Grabowski, E. J. J.; Tillyer, R. D. *J. Am. Chem. Soc.*
     2004, *126*, 3048–3049.
43. For detailed mechanistic studies, see Xu, F.; Armstrong, J. D. III; Zhou, G. X.;
     Simmons, B.; Hughes, D.; Ge, Z.; Grabowski, E. J. J. *J. Am. Chem. Soc.* **2004**,
     126, 13002–13009.
44. For preliminary communication, see Hsiao, Y.; Rivera, N. R.; Rosner, T.; Krska,
     S. W.; Njolito, E.; Wang, F.; Sun, Y.; Armstrong, J. D. III; Grabowski, E. J. J.;
     Tillyer, R. D.; Spindler, F.; Malan, C. *J. Am. Chem. Soc.* **2004**, *126*, 9918–9919.
45. In collaboration with Solvias AG, Switzerland, extensive screenings were
     carried out to search for the most suitable catalyst for this hydrogenation.
46. Clausen, A. M.; Dziadul, B.; Cappuccio, K. L.; Kaba, M.; Starbuck, C.; Hsiao,
     Y.; Dowling, T. M. *Org. Process Res. Dev.* **2006**, *10*, 723–726.
47. Hansen, K. B.; Hsiao, Y.; Xu, F.; Rivera, N.; Clausen, A.; Kubryk, M.; Krska,
     S.; Rosner, T.; Simmons, B.; Balsells, J.; Ikemoto, N.; Sun, Y.; Spindler, F.;
     Malan, C.; Grabowski, E. J. J.; Armstrong, J. D. III. *J. Am. Chem. Soc.* **2009**,
     131, 8798–8804.
48. This manufacturing process received the Presidential Green Chemistry
     Challenge Award (2006) for alternative synthetic pathways, the IChemE Astra-
     Zeneca Award for excellence in green chemistry and chemical engineering
     (2005), and the Thomas Alva Edison Patent Award (2009) for Merck's U.S. Pat.
     7,468,459.
49. Efforts leading to the discovery and development of sitagliptin led to the receipt
     of the Prix Galien USA Award for Best Pharmaceutical Agent (2007) and the
     Thomas Alva Edison Patent Award (2007) for Merck's U.S. Pat. 6,699,871.
50. Reynolds, J. K.; Neumiller, J. J.; Campbell, R. K. *Expert Opin. Investig. Drugs*
     2008, *17*, 1559–1565.

11

Aliskiren (Tekturna), The First-in-Class Renin Inhibitor for Hypertension

Victor J. Cee

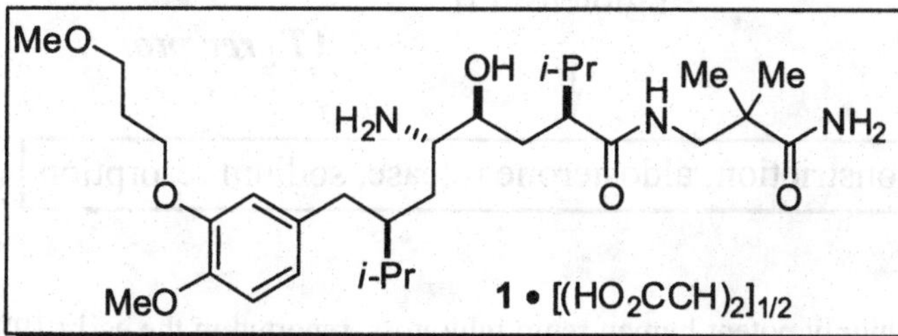

USAN: Aliskiren
Trade name: Tekturna®
Speedel/Novartis
Launched: 2007

11.1 Background

Hypertension is estimated to afflict 1 billion individuals worldwide and is a major risk factor for stroke, coronary artery disease, heart failure, and end-stage renal disease.[1] The discovery of the renin–angiotensin system (RAS) as the master regulator of blood pressure and cardiovascular function provided numerous targets for pharmacologic intervention.[2] Although renin catalyzes the first and rate-limiting step in the activation of the RAS, it was the inhibition of the downstream angiotensin-converting enzyme (ACE) that first established the clinical relevance of this pathway in the treatment of hypertension. The next advance in targeting the RAS for hypertension came from the discovery and development of angiotensin II receptor blockers (ARBs), which provide similar efficacy and lack some of the side effects of ACE inhibitors. With the approval in 2007 of aliskiren, the first-in-class renin inhibitor, clinicians are now able to intervene at all points of the RAS. The impact of aliskiren on the clinical management of hypertension and hypertension-related target organ damage will depend on establishing advantages in efficacy and safety over existing RAS-targeted treatments.

In 1898, the Finnish physiologist Robert Tigerstedt and his Swedish student Per Bergman coined the term *renin* to signify the soluble factor that was responsible for the dramatic rise in blood pressure observed upon injection of kidney extracts into rabbits.[3]

Many years later, it was established that renin, produced in the juxtaglomerular cells of the kidney, is not a direct regulator of blood pressure, but rather an aspartic protease that initiates the production of the vasoconstrictive octapeptide angiotensin II by cleaving the Leu10-Val11 bond of the 57 kD protein angiotensinogen.[4] Since renin is extremely specific for angiotensinogen and the first and rate-limiting enzyme of the RAS, renin inhibition was recognized for decades as an attractive approach for the treatment of hypertension and hypertension-related target organ damage.

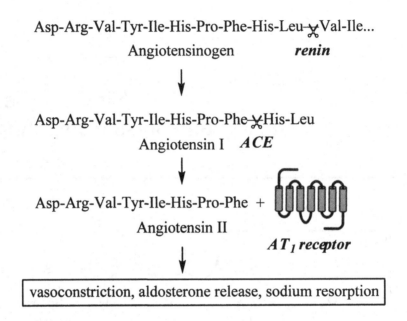

The first highly potent human renin inhibitors, reported in the early 1980s, were analogs of the terminal portion of angiotensinogen containing a nonhydrolyzable amine linkage at the scissile Leu-Val bond. H-142 (2), the most potent in a series of analogs was found to inhibit human renin with an IC_{50} of 10 nM and showed high selectivity against the related aspartyl protease cathepsin D.[5] Shortly after, a research group from Merck reported analogs of pig angiotensinogen incorporating the amino acid statine ((3S,4S)-4-amino-3-hydroxy-6-methylheptanoic acid), based on the observation that statine is a component of the naturally occurring aspartyl protease inhibitor pepstatin. Compound 3 exhibited potent inhibition of human kidney renin (K_i = 1.9 nM) with approximately 70-fold selectivity against cathepsin D.[6] This group was the first to report in vivo proof of concept, showing that an analogue of 3 could completely inhibit the pig kidney renin pressor response in the ganglion-blocked anesthetized rat model when administered at 2 mg/kg i.v. Following the transition-state isostere concept of the statine-based series of inhibitors, extensive medicinal chemistry efforts produced optimized hydroxyl-containing polypeptides, of which remikiren (4, Roche), enalkiren and zankiren (Abbott), ditekiren

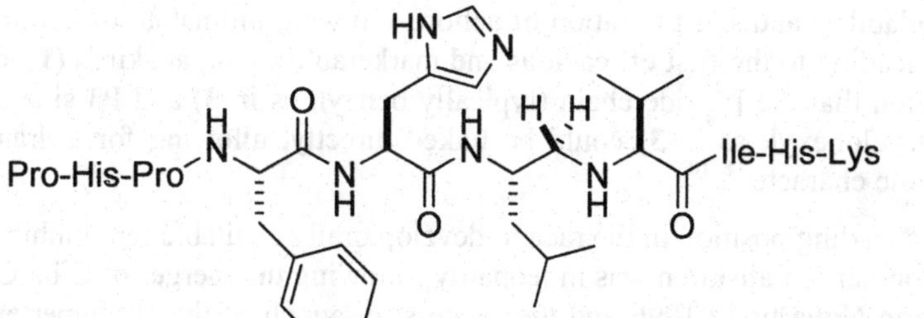

Pro-His-Pro——Ile-His-Lys   H142 (**2**)

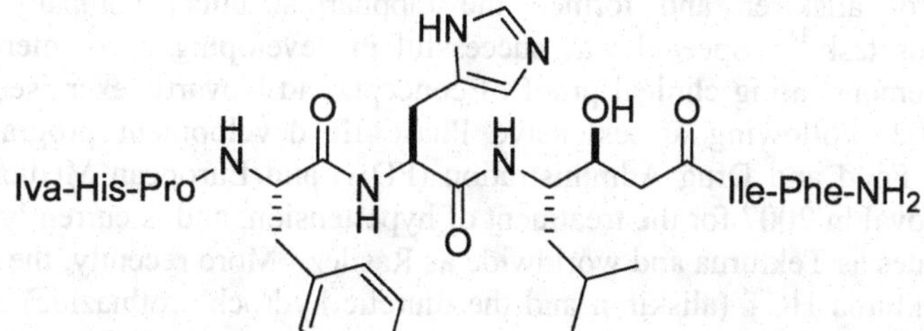

Iva-His-Pro——Ile-Phe-NH₂

3

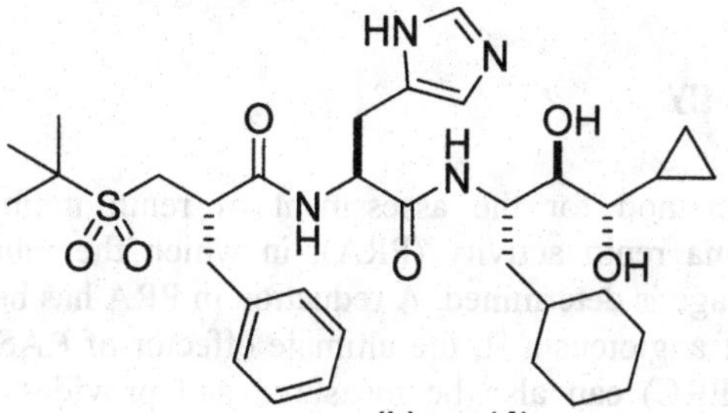

remikiren (**4**)

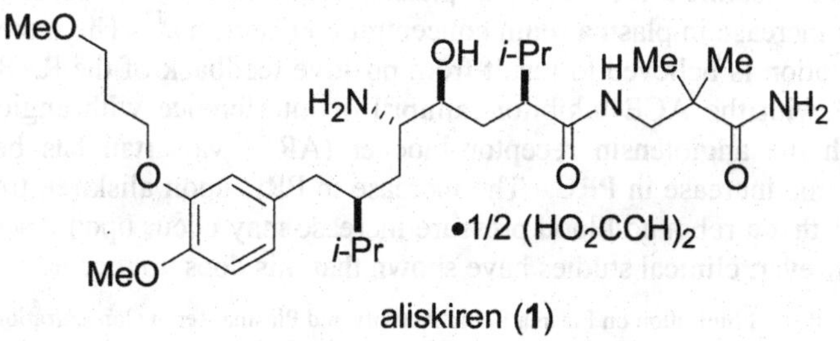

•1/2 (HO₂CCH)₂

aliskiren (**1**)

and terlekiren (Pfizer), SR43845 (Sanofi-Aventis), and FR115906 (Astellas) progressed as far as phase II clinical trials.[7] Unfortunately, these peptidomimetic molecules suffered from poor bioavailability and short duration of action, and were ultimately discontinued. The key advance leading to the first efficacious and marketable agent, aliskiren (1), came from the recognition that the P1 side chain (typically benzyl, as in 3) and P3 side chain (typically methylcyclohexyl, as in 3) could be linked directly, allowing for a dramatic reduction in peptidic character.[8–10]

Despite a leading position in the race to develop orally available renin inhibitors, the clinical development of aliskiren was in jeopardy following the merger of Ciba-Geigy and Sandoz to form Novartis in 1996 and the successful launch of the antihypertensive angiotensin receptor blocker (ARB) valsartan in 1997 (Diovan; Novartis). A group of former Ciba-Geigy employees convinced Novartis to out-license the Phase I/II development of aliskiren and formed the biopharmaceutical company Speedel to accomplish this task.[11] Speedel was successful in developing a commercially viable process and demonstrating clinical proof of concept, and Novartis exercised a call-back option in 2002. Following an extensive Phase III development program, aliskiren received U.S. Food and Drug Administration (FDA) and European Medicines Agency (EMEA) approval in 2007 for the treatment of hypertension, and is currently marketed in the United States as Tekturna and worldwide as Rasilez. More recently, the combination treatments Tekturna HCT (aliskiren and the diuretic hydrochlorothiazide) and Valturna (aliskiren and the ARB valsartan) have received FDA approval for the treatment of hypertension not adequately controlled by monotherapy and as initial therapy in patients likely to require multiple drugs to achieve blood pressure goals.

11.2   Pharmacology

A well established method for the assessment of renin inhibition is the ex vivo measurement of plasma renin activity (PRA), in which the inhibition of the rate of angiotensinogen cleavage is determined. A reduction in PRA has been shown to correlate with reduced levels of angiotensin II, the ultimate effector of RAS activation.[12] Plasma renin concentration (PRC) can also be measured and provides insight into feedback mechanisms in operation on inhibition at different points of the RAS. In a pooled analysis of hypertensive patients receiving 150 and 300 mg aliskiren for 8 weeks, a substantial and sustained reduction in plasma renin activity is observed, despite a compensatory increase in plasma renin concentration (Table 1).[13] The increase in plasma renin concentration is believed to result from positive feedback of the RAS; depletion of angiotensin II with the ACE inhibitor ramipril or interference with angiotensin II/AT$_1$ signaling with the angiotensin receptor blocker (ARB) valsartan has been shown to produce the same increase in PRC. The increase in PRC upon aliskiren treatment led to the hypothesis that a rebound blood pressure increase may occur upon discontinuation of treatment. However, clinical studies have shown that this does not occur.[9]

Table 1. Impact of RAS Inhibition on Plasma Renin Activity and Plasma Renin Concentration After 8 Weeks of Treatment in Hypertensive Patients

Group	PRA (% Change)	PRC (% Change)
Placebo	+11.5	+12.2
Aliskiren, 150 mg	−74.7	+95.9

Aliskiren, 300 mg	−71.5	+146.7
Ramipril, 10 mg	+110.6	+67.9
Valsartan, 320 mg	+159.6	+137.8

11.3 Structure–Activity Relationship (SAR)

The recognition that the P1 and P3 sidechains of first-generation polypeptide inhibitors could be connected via a flexible linker led to a new series of renin inhibitors with reduced peptidic character. Beginning in the early 1990s, research groups at Parke–Davis,[14,15] University of Montreal,[16] Pfizer,[17] and Novartis[18,19] disclosed reasonably potent analogs based on this design principle. In the case of Novartis, this was demonstrated by the synthesis of 6, which was found to be only 100-fold less potent than the earlier clinical candidate CGP038560 (5).

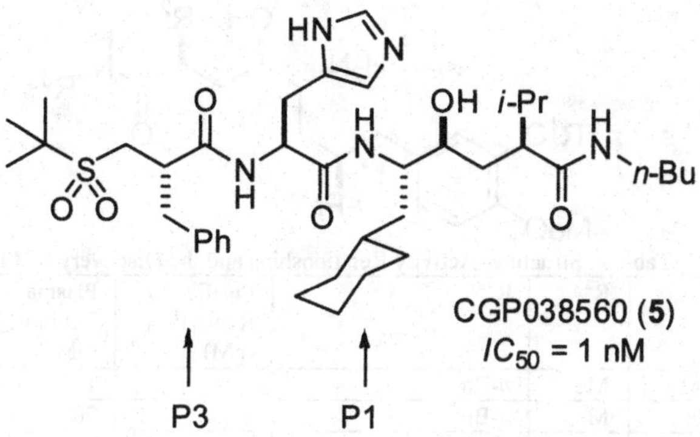

CGP038560 (5)
IC_{50} = 1 nM

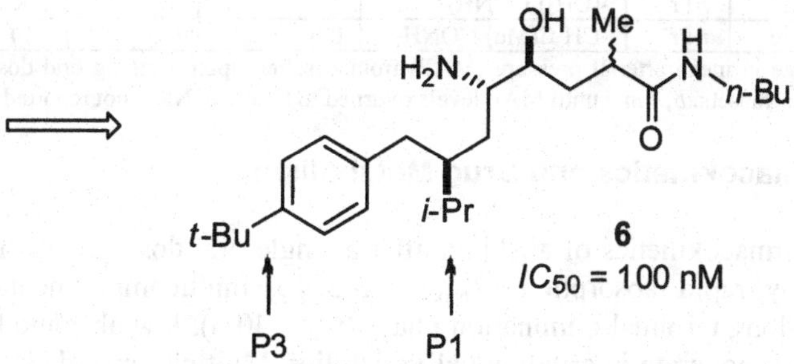

P3   P1

6
IC_{50} = 100 nM

P3   P1

Further optimization of the P3 substituent revealed that 3,4-bis-alkoxy substitution of the phenyl ring improved potency, with the optimal combination 3-(3-methoxypropoxy) and 4-methoxy (7, Table 2). Compound 7 exhibits exceptional renin inhibitory activity in both buffer and plasma and showed moderate in vivo activity in telemetered, sodium-depleted marmosets, with peak reduction of mean arterial pressure of 9 mm Hg and an 8-h duration of action.[20] An X-ray co-crystal structure of compound 7 with recombinant human renin established that the methoxypropoxy sidechain occupies a narrow nonsubstrate pocket, termed S3^{sp}. The importance of terminal methoxy group is evident by comparison of compound **7** with compound **8**, which contains an isosteric *n*-pentoxy group. Compound **8** shows 4- and 70-fold weaker binding to human renin in buffer and plasma, respectively, relative to compound **7**. Researchers at Novartis reported that during the course of extensive optimization of **7**, it was noted that *N*-terminal carboxamides at the P2′ position (R^3) generally provided improved in vivo activity in the sodium-depleted marmoset model, and while carboxamide **9** is somewhat inferior to compound **7** in terms of in vitro activity, the loss in potency could be regained by installation of the larger isopropyl (R^2) P1′ substituent as in **10**. Further optimization of the carboxamide (R^3) revealed that geminal methyl substitution was highly preferred, with aliskiren (**1**) displaying subnanomolar activity in both buffer and plasma. Aliskiren also demonstrated robust (peak ΔMAP = –30 mm Hg) and prolonged (duration > 24 h) activity in the sodium-depleted marmoset model, making it among the most potent renin inhibitors described to date.[21]

Table 2. Structure–Activity Relationships and the Discovery of Aliskiren

Cpd	R^1	R^2	R^3	Purified Renin IC$_{50}$ (nM)	Plasma Renin IC$_{50}$ (nM)	Peak ΔMAP (mm Hg)^a	ΔMAP duration (h)^b
7	-(CH$_2$)$_3$OMe	Me	*n*-Bu	1	1	–9	8
8	-(CH$_2$)$_4$Me	Me	*n*-Bu	4	70	NR	NR
9	-(CH$_2$)$_3$OMe	Me	-(CH$_2$)$_2$CONH$_2$	7	17	NR	NR
10	-(CH$_2$)$_3$OMe	*i*-Pr	-(CH$_2$)$_2$CONH$_2$	1	3	NR	NR
1	-(CH$_2$)$_3$OMe	*i*-Pr	-CH$_2$C(Me)$_2$CONH$_2$	0.6	0.6	–30	>24

a, Maximum change in mean arterial pressure (MAP) from baseline upon 3 mg/kg oral dose in telemetered, sodium-depleted marmosets. *b*, Time until MAP levels returned to baseline. NR = not reported.

11.4   Pharmacokinetics and Drug Metabolism

The human pharmacokinetics of aliskiren after a single oral dose in healthy subjects are characterized by rapid absorption (T_{max} = 1–3 h), multicompartmental elimination kinetics with a long terminal elimination phase ($t_{1/2}$ = 40 h), low absolute bioavailability (F = 2.6%), and moderate interindividual variability. Multiple once-daily oral dosing is characterized by accumulation (accumulation factor = 2) with steady-state reached after 7–8 doses. At steady state, a 300 mg dose of aliskiren produces C_{max} in the range of 200–400 ng/mL (330–660 nM) and C_{trough} in the range of 15–30 ng/mL (25–50 nM). With moderate binding to plasma protein (47–51% binding), approximately half of the measured total drug in plasma is free and unbound. Aliskiren is eliminated primarily via the hepatobiliary route as unchanged drug, with oxidative metabolism (CYP3A4) making a minor contribution, indicating a low exposure to metabolites. Extensive co-administration studies in human subjects have shown that the potential for drug–drug interactions is generally low, with the exception of atorvastatin, ketoconazole, and cyclosporine (increase in aliskiren AUC) and the loop diuretic furosemide (reduction in aliskiren AUC). In vitro experiments have demonstrated that aliskiren is a P-glycoprotein (Pgp) substrate (Michaelis–Menten constant K_m = 2.1 µM), and the increase in aliskiren AUC on co-administration of atorvastatin, ketoconazole and cyclosporine is consistent with the ability of these agents to inhibit Pgp-mediated intestinal efflux. The cause of the decreased aliskiren exposure on co-administration of furosemide is not well understood.[13]

11.5 Efficacy and Safety

The efficacy of aliskiren monotherapy and aliskiren in combination with the diuretic hydrochlorothiazide (HCT), aliskiren in combination with the angiotensin receptor blocker (ARB) valsartan, and aliskiren in combination with the ACE inhibitor ramipril in reducing mean sitting diastolic blood pressure (msDBP) have been reported (Table 3). From a pooled analysis of several placebo-controlled clinical trials, the approved doses of 150 and 300 mg aliskiren provide a placebo-subtracted reduction in msDBP of 3.2 and 5.5 mm Hg,[9] respectively, similar to levels obtained at approved doses of other RAS antagonists such as ARBs and ACE inhibitors, supporting the use of aliskiren as a monotherapy. Combination studies of aliskiren and standard doses of the diuretic HCT[22] as well as aliskiren and a standard dose of the ARB valsartan[23] and aliskiren and a standard dose of ramipril[24] show that additional blood pressure lowering is possible relative to HCT, valsartan, and ramipril alone, supporting the use of these drugs in combination with aliskiren for patients not able to reach blood pressure goals with a single agent.

Table 3. Efficacy of Aliskiren as Monotherapy or in Combination

Aliskiren Dose	Monotherapy or Combination, ΔmsDBP (mm Hg)			
	Monotherapy[a]	+ HCT[b] 6.25/12.5/25 mg	+ valsartan[c] 320 mg	+ ramipril[d] 10 mg
150 mg	−3.2	−1.3/−1.8/−3.3	ND	ND
300 mg	−5.5	ND/−3.8/−4.9	−2.5	−2.1

a Placebo-subtracted change in mean sitting diastolic blood pressure (msDBP) from a pooled analysis of patients with hypertension less than 65 years of age. b HCT (hydrochlorothiazide) monotherapy-subtracted change in msDBP in patients with mild to moderate hypertension after 8 weeks of treatment. c Valsartan monotherapy-subtracted change in msDBP; patients received 160 mg valsartan/150 mg aliskiren for 4 weeks, then were force-titrated to 320 mg valsartan/300 mg aliskiren for 4 additional weeks. d Ramipril monotherapy-subtracted change in msDBP; patients received 5 mg ramipril/150 mg aliskiren for 4 weeks, then were force-titrated to 10 mg ramipril/300 mg aliskiren for 4 additional weeks. ND = not determined.

Aliskiren is presently being evaluated in the ASPIRE HIGHER clinical trial program in more than 35,000 patients in 14 trials to study the effect of renin inhibition on outcomes in patients with conditions such as heart failure and diabetes as well as on organ protection benefit in a variety of disease states.[25] To date, three of the short-term trials with organ protection endpoints have reported findings. In patients with type 2 diabetese, kidney disease, and high blood pressure, the AVOID (Aliskiren in the Evaluation of Proteinuria in Diabetes) study has demonstrated that aliskiren provides additional kidney protection when added to the maximum dose of the ARB losartan.[26] In contrast, the ALLAY (Aliskiren Left Ventricular Assessment of Hypertrophy) trial has shown that in overweight hypertensive patients with evidence of left ventricular hypertrophy, the combination of aliskiren with the ARB losartan does not result in further reduction in left ventricular mass relative to that achieved by either drug alone.[27] Finally, the ALOFT (Aliskiren Observation of Heart Failure Treatment) study has shown that aliskiren added to standard therapy in patients with stable heart failure and hypertension provided improvements in key indicators of heart function.[28] The results of the ongoing ASPIRE HIGHER trials will likely define the ultimate role of aliskiren in the treatment of cardiovascular and renal diseases.

The EMEA filing for aliskiren contained safety information obtained from a large number of clinical studies in a total of 11,566 treated patients. Aliskiren showed a favorable overall adverse event profile (37.7%) compared to placebo (40.2%). Diarrhea was the most common adverse event and was more common in the aliskiren group (2.4%) versus placebo (1.2%). Relative to ACE inhibitors, which are associated with cough due to inhibition of ACE-dependent bradykinin cleavage, aliskiren demonstrated a significantly improved profile (1.0% vs 3.8% for ACE inhibitors), as expected based on the specificity of renin for angiotensinogen.[9]

11.6   Syntheses

The discovery route[20,21,29] to aliskiren targeted three key intermediates, α-aminoaldehyde **11**, the isopropyl-substituted propane unit **12**, and the β-aminopropanamide **13**. The union of **11** and the Grignard reagent derived from **12** establishes the full contiguous carbon backbone of aliskiren and the C_4 stereocenter. Following oxidation-state manipulation, β-aminopropanamide **13** is introduced by a peptide coupling.

The optimized process for the synthesis of aldehyde **11** was reported[29] in 2003 and begins with alkylation of isovanillin **14** with 1,3-dibromopropane followed by sodium methoxide displacement of the remaining bromide to install the methoxypropoxy sidechain (**15**). Reduction and bromination provided the key benzylic bromide **16**. Alkylation of the lithium enolate of acyloxazolidinone **17** provided **18** with high selectivity for the desired diastereomer. This intermediate was transformed into key alkyl bromide **19** in three steps using standard chemistry. Alkylation of the lithiated Schöllkopf auxiliary **20** with bromide **19** proved to be a doubly matched reaction, providing the 2S,5R,2'S product **21** in high yield (85%) and diastereoselectivity (> 98:2 dr). The reaction of ent-**19** with the lithiated Schöllkopf auxiliary **20** was shown to be doubly mismatched, with poor conversion and no selectivity at C_2. The substrate-directed induction from **19** was sufficiently strong to allow the use of an achiral diethoxydihydropyrazine, which provided the alkylated product in 95:5 diastereoselectivity (not shown). In either case, the dihydropiperazine could be cleaved to the corresponding ester with acid and protection of the free amine as the carbamate followed by reduction provided the key aldehyde **11** in high enantiomeric and diastereomeric purity.

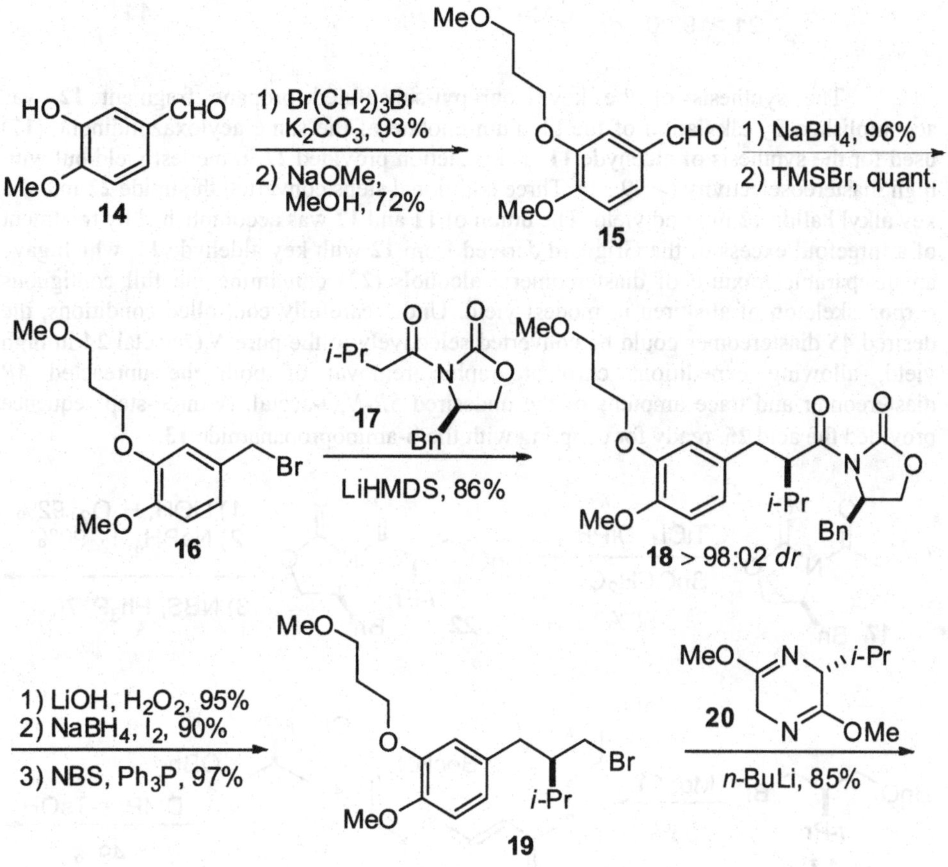

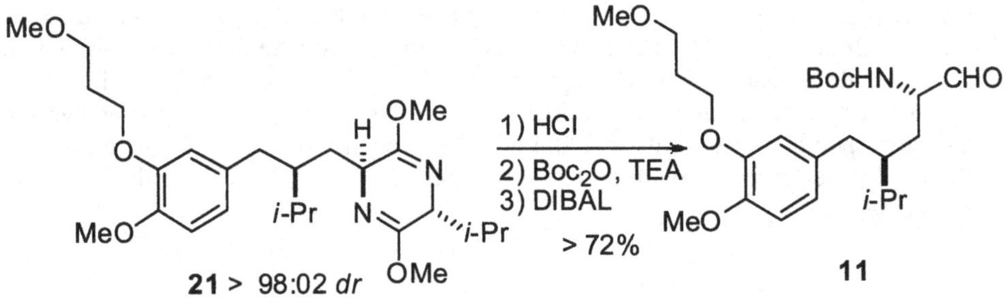

21 > 98:02 *dr*

1) HCl
2) Boc₂O, TEA
3) DIBAL
> 72%

11

The synthesis of the key isopropyl-substituted propane fragment **12** was accomplished by alkylation of the titanium enolate of the same acyloxazolidinone (**17**) used for the synthesis of aldehyde **11**. This reaction provided **22** in modest yield but with high diastereoselectivity (> 99 : 1). Three additional steps converted the imide **22** into the key alkyl halide **12** in good yield. The union of **11** and **12** was accomplished by treatment of a threefold excess of the Grignard derived from **12** with key aldehyde **11**, which gave an inseparable mixture of diastereomeric alcohols (**23**) containing the full contiguous carbon skeleton of aliskiren in modest yield. Under carefully controlled conditions, the desired 4*S* diastereomer could be converted selectively to the pure *N,O*-acetal **24** in high yield, allowing expeditions chromatographic removal of both the unreacted 4*R* diastereomer and trace amounts of the undesired 5*R*-*N,O*-acetal. A three-step sequence provided the acid **25**, ready for coupling with the β-aminopropanamide **13**.

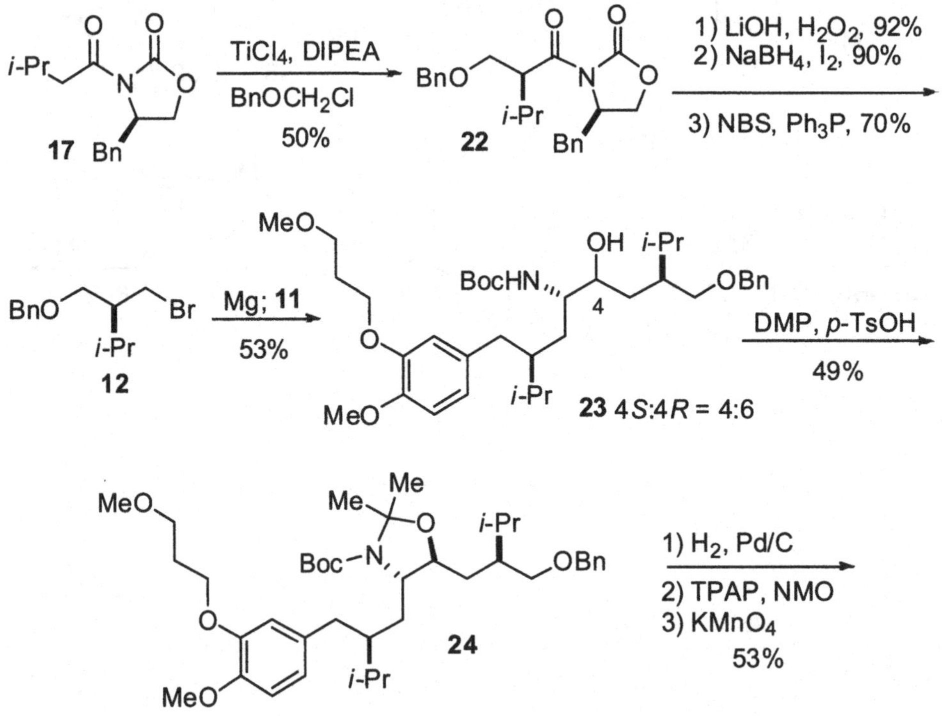

17 Bn

TiCl₄, DIPEA
BnOCH₂Cl
50%

22

1) LiOH, H₂O₂, 92%
2) NaBH₄, I₂, 90%
3) NBS, Ph₃P, 70%

12

Mg; **11**
53%

23 4*S*:4*R* = 4:6

DMP, *p*-TsOH
49%

24

1) H₂, Pd/C
2) TPAP, NMO
3) KMnO₄
53%

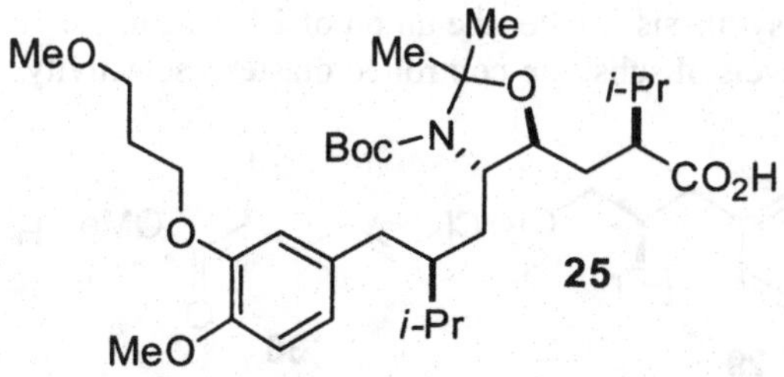

25

The key β-aminopropanamide **13** was synthesized from ethyl 2-cyano-2-methylpropanoate **26**. Nitrile reduction and protection of the resulting amine as its benzyl carbamate provided **27**. High-pressure aminolysis followed by reductive removal of the Cbz group gave **13** as its HCl salt. Compounds **13** and **25** were then coupled using diethyl cyanophosphonate (DEPC) to give protected aliskiren (**28**). A sequential deprotection scheme provided **1** as its HCl salt in good yield. Based on the yields reported to date, the discovery route produces **1** in 3% overall yield with a longest linear sequence of 20 steps.

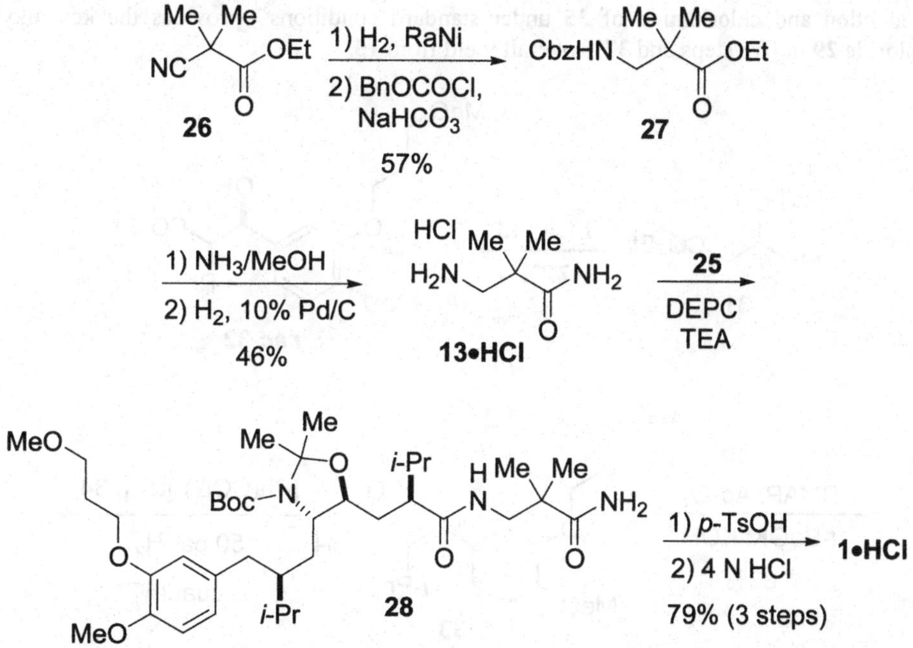

Due to the complexity of aliskiren, the manufacturing process required extensive optimization, and the following discussion focuses on the most likely manufacturing route. Key to the success of the commercial route was the finding that aliskiren could be dissected into roughly equal sized intermediates **29** and **30**, each containing a single isopropyl-substituted stereogenic center, as well as the achiral intermediate **13** used in the medicinal chemistry synthesis.[9] After the union of **29** and **30**, the vicinal aminoalcohol is installed with high levels of substrate-controlled diastereoselectivity.

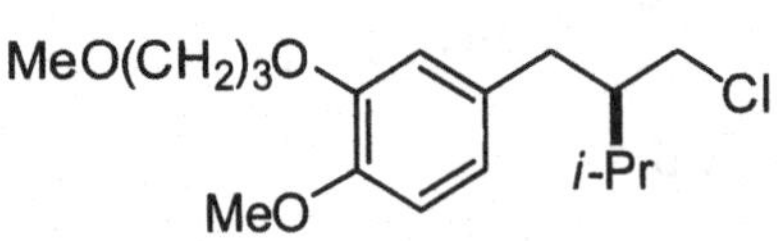

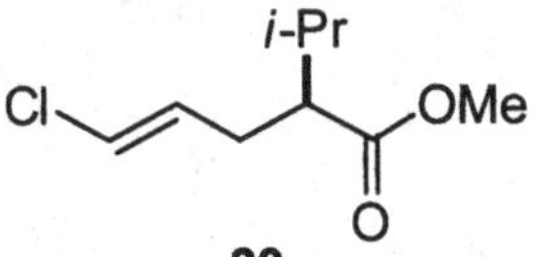

While a number of approaches to the key alkyl chloride intermediate **29** have been reported,[30–32] the approach disclosed in WO 02002500 is particularly efficient, with some steps demonstrated on multikilogram scale.[32] This route begins with an aldol reaction between the lithium enolate of ethyl isovalerate (**31**) and the substituted benzaldehyde (**15**, employed in the medicinal chemistry route) to give **rac-32** with high diastereoselectivity. Acylation followed by elimination installs the trisubstituted olefin and saponification gives the acid **33**. Asymmetric hydrogenation is effected with high efficiency and selectivity under rhodium (I) catalysis using a diphosphanylferrocenyl ligand (**34**) related to the taniaphos series developed by Knochel and co-workers.[33] Reduction and chlorination of **35** under standard conditions[34] provides the key alkyl chloride **29** in five steps and 37% overall yield from **15**.

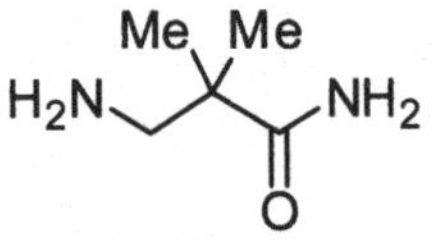

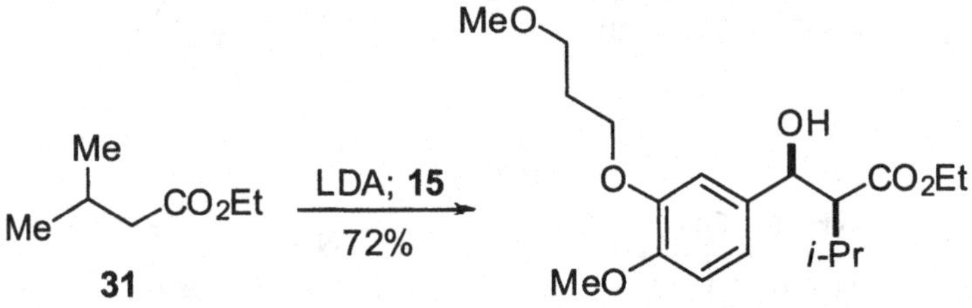

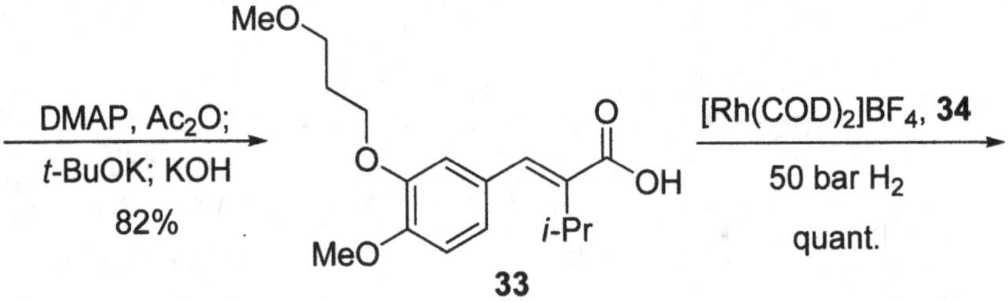

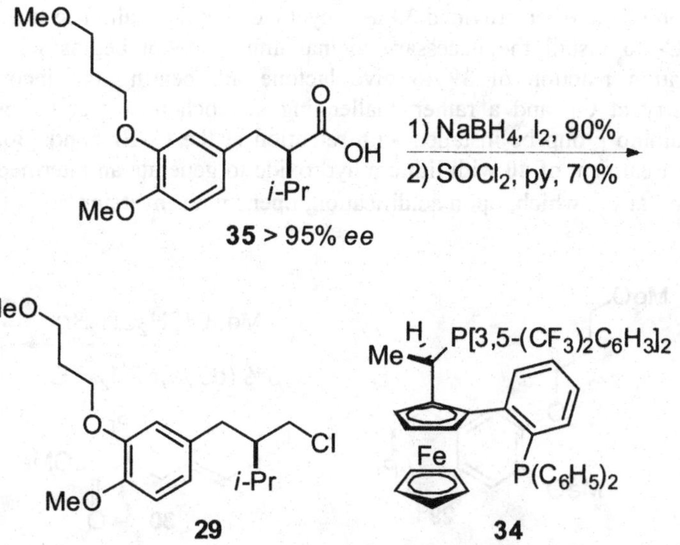
MeO
O
OH
i-Pr
MeO
35 > 95% ee
1) NaBH4, I2, 90%
2) SOCl2, py, 70%
MeO
O
Cl
i-Pr
MeO
29
H
Me
P[3,5-(CF3)2C6H3]2
Fe
P(C6H5)2
34

A number of approaches to the chiral chloropentenoic ester **30** have been disclosed, with enantiomeric purity established either by chiral auxiliaries, diastereomeric salt resolution, or enzymatic resolution.[35,36] The enzymatic resolution reported[36] in WO 0209828 is particularly efficient and has been disclosed as part of the commercial route to aliskiren.[9] The enolate of methyl isovalerate (**36**) is alkylated with (*E*)-1,3-dichloroprop-1-ene to give ***rac*-30**, which is resolved by pig liver esterase (PLE) to give **30** with the desired 2*S*-stereochemistry in high enantiomeric purity. The undesired acid **37** is then recycled in a two-step, one-pot process involving esterification and racemization to return ***rac*-30**.

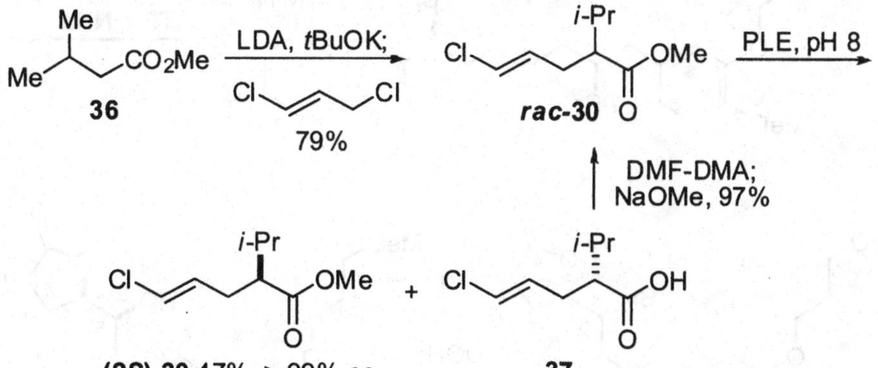
Me
Me
CO2Me
36
LDA, tBuOK;
Cl
Cl
79%
i-Pr
Cl
OMe
O
rac-30
PLE, pH 8
DMF-DMA;
NaOMe, 97%
i-Pr
Cl
OMe
O
(2S)-30 47%, > 99% ee
+
i-Pr
Cl
OH
O
37

The union of alkyl chloride **29** and chloropentenoic ester **30** is accomplished by a nickel-catalyzed Kumada coupling of the Grignard reagent derived from **29** (2.2-fold excess) with **30** to give **38**, containing the full contiguous carbon skeleton of aliskiren.

Saponification of the ester provided **39** as a cyclohexylamine salt. The functionalization at C_4 and C_5 to install the necessary vicinal aminoalcohol begins with a selective halolactonization reaction of **39** to give lactone **40**, bearing the incorrect alcohol stereochemistry at C_4, and a rather challenging stereochemistry at C_5, which would require the amino group be installed with retention at the C_5-Br bond. Both issues are remedied by treatment of **40** with lithium hydroxide to generate an intermediate epoxide with inversion at C_5, which, upon acidification, opens with inversion at C_4 to give the γ-lactone **41**.[37]

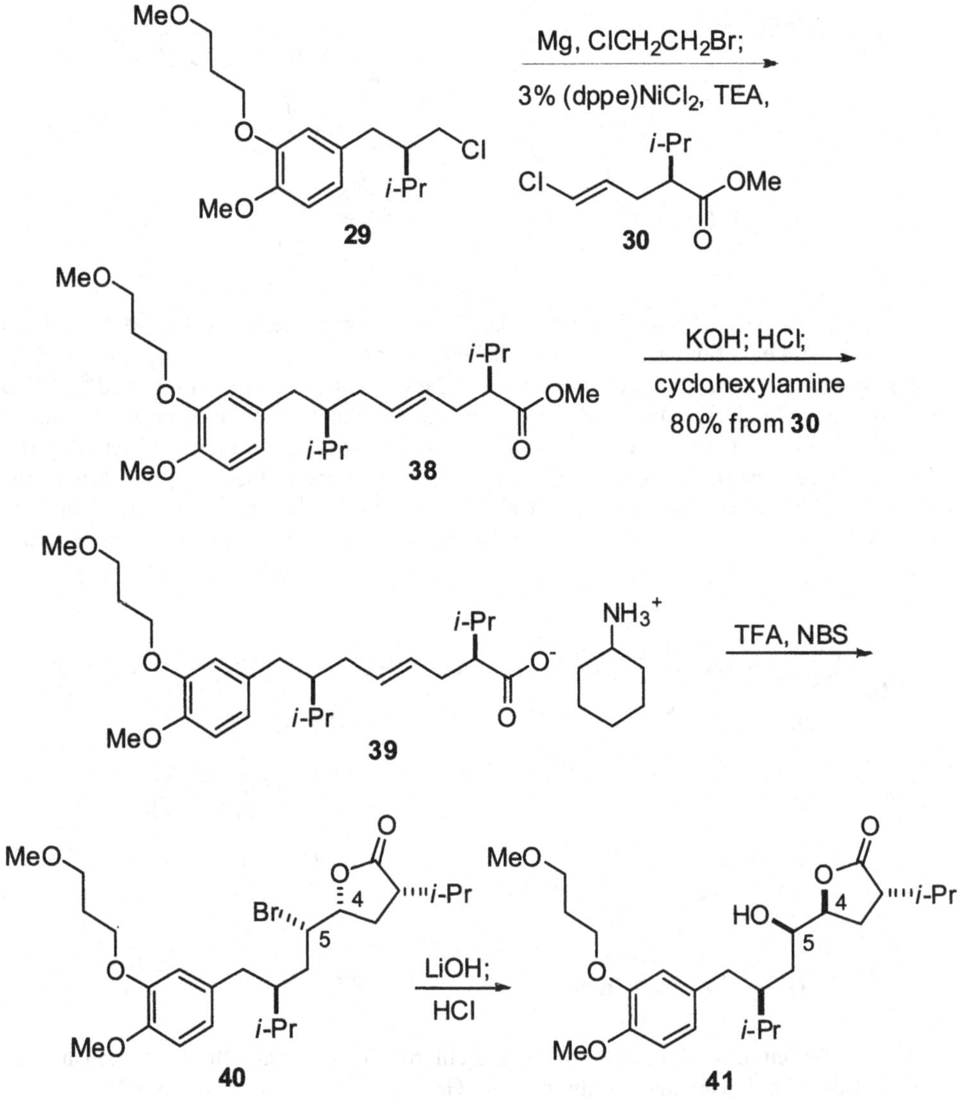

Precedent for the selectivity observed in the halolactonization of **39** is found in the work of Bartlett and co-workers, who showed that 2-substituted pentenoic esters cyclize in a highly stereoselective manner to give *cis*-γ-lactones, presumably by way of intermediate **A**, in which both substituents occupy pseudoequatorial positions.[38] Of note is the finding that the analogous *N,N*-dimethyl amide **42** reacts under similar conditions to give the *trans*-γ-lactone with the correct stereochemistry at C_4.[39] This result is consistent with earlier observations of Yoshida and co-workers,[40] who suggested that avoidance of $A_{1,3}$-strain between the 2-substituent and an *N*-methyl group favors intermediate **B**, with one pseudoequatorial and one pseudoaxial group.

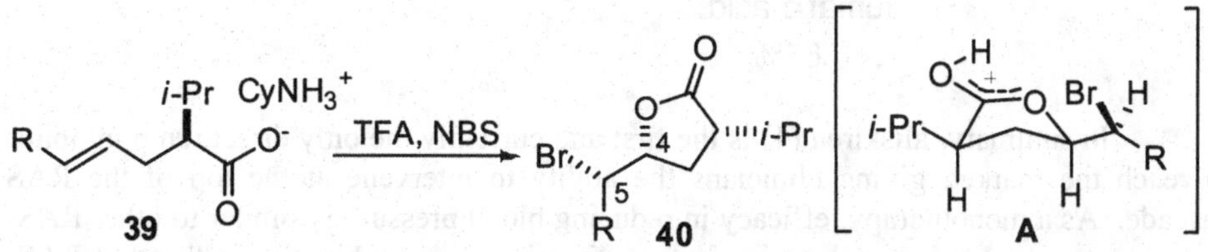

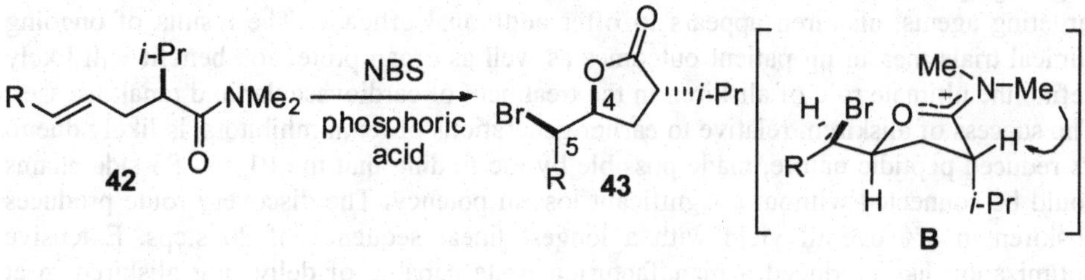

The installation of the amine functionality at C_5 is accomplished by mesylation of **41** followed by azide displacement to give **44** (71% yield over four steps). The β-aminopropanamide **13** is then introduced directly to the lactone under 2-hydroxypyridine catalysis to give the penultimate intermediate **45**. Reduction of the azide and isolation as the hemifumarate salt provides aliskiren hemifumarate (**1**).[37] Based on the information disclosed to date, the synthesis of aliskiren is accomplished with an overall yield of 14% from isovanillin **14** with a longest linear sequence of 15 steps. Given the complexity of aliskiren, this is a remarkably efficient process, and it should be noted that this analysis likely only establishes a lower limit of efficiency, as further optimization of the route on a manufacturing scale is expected.

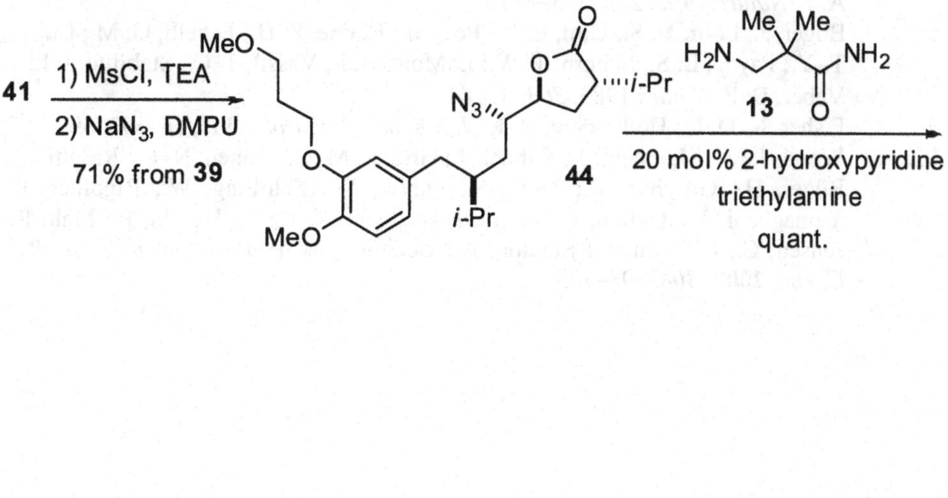

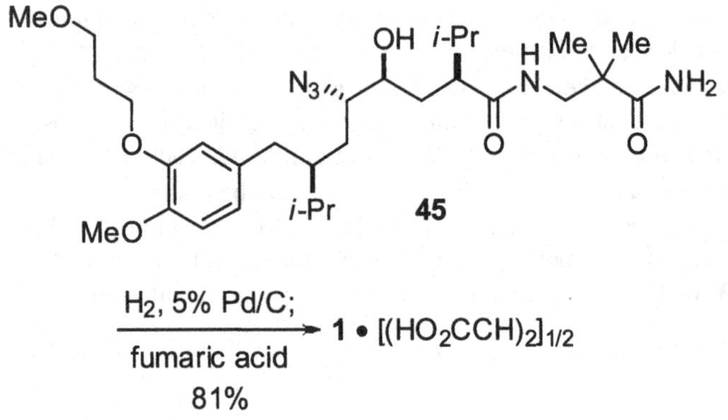

In summary, aliskiren (**1**) is the first and currently the only direct renin inhibitor to reach the market, giving clinicians the ability to intervene at the top of the RAS cascade. As a monotherapy, efficacy in reducing blood pressure is similar to other RAS-targeting agents, but in combination with a diuretic or in combination with other RAS-targeting agents, aliskiren appears to offer additional efficacy. The results of ongoing clinical trials measuring patient outcomes as well as organ protection benefit will likely define the ultimate role of aliskiren in the treatment of cardiovascular and renal diseases. The success of aliskiren, relative to earlier generations of renin inhibitors, is likely due to its reduced peptidic nature, made possible by the finding that the P1 and P3 side-chains could be connected without a significant loss in potency. The discovery route produces aliskiren in 3% overall yield with a longest linear sequence of 20 steps. Extensive optimization has produced a manufacturing route capable of delivering aliskiren in at least 14% overall yield with a longest linear sequence of at most 15 steps.

11.7   References

1.   World Health Organization, International Society of Hypertension Group. *J. Hypertens.* **2003**, *21*, 1983–1992.

2.   Zaman, M. A.; Oparil, S.; Calhoun, D. A. *Nat. Rev. Drug Discov.* **2002**, *1*, 621–636.

3.   Tigerstedt, R.; Bergman, P. *Skand. Arch. Physiol.* **1898**, *8*, 223–271.

4.   Persson, P. B. *J. Physiol.* **2003**, *552*, 667–671.

5.   Szelke, M.; Leckie, B.; Hallett, A.; Jones, D. M.; Sueiras, J.; Atrash, B.; Lever, A. F. *Nature* **1982**, *299*, 555–557.

6.   Boger, J.; Lohr, N. S.; Ulm, E. H.; Poe, M.; Blaine, E. H.; Fanelli, G. M.; Lin, T.-Y.; Payne, L. S.; Schorn, T. W.; LaMont, B. I.; Vassil, T. C.; Stabilito, I. I.; Veber, D. F. *Nature* **1983**, *303*, 81–84.

7.   Fisher, N. D. L.; Hollenberg, N. K. *J. Am. Soc. Nephrol.* **2005**, *16*, 592–599.

8.   Wood, J. M.; Maibaum, J.; Rahuel, J.; Grütter, M. G.; Cohen, N.-C.; Rasetti, V.; Rüger, H.; Göschke, R.; Stutz, S.; Fuhrer, W.; Schilling, W.; Rigollier, P.; Yamaguchi, Y.; Cumin, F.; Baum, H.-P.; Schnell, C. R.; Herold, P.; Mah, R.; Jensen, C.; O'Brien, E.; Stanton, A.; Bedigian, M. P. *Biochem. Biophys. Res. Comm.* **2003**, *308*, 698–705.

9. Jensen, C.; Herold, P.; Brunner, H. R. *Nat. Rev. Drug Disc.* **2008**, *7*, 399–410.
10. Maibaum, J.; Feldman, D. L. *Ann. Rep. Med. Chem.* **2009**, *44*, 105–127.
11. *Drug Discovery Today* **2005**, *10*, 881–883.
12. Nussberger, J.; Wuerzner, G.; Jensen, C.; Brunner, H. R. *Hypertension* **2002**, *39*, e1–8.
13. Vaidyanathan, S.; Jarugula, V.; Dieterich, H. A.; Howard, D.; Dole, W. P. *Clin. Pharmacokinet.* **2008**, *47*, 515–531.
14. Plummer, M.; Hamby, J. M.; Hingorani, G.; Batley, B. L.; Rapundalo, S. T. *Bioorg. Med. Chem. Lett.* **1993**, *3*, 2119–2124.
15. Plummer, M. S.; Shahripour, A.; Kaltenbronn, J. S.; Lunney, E. A.; Steinbaugh, B. A.; Hamby, J. M.; Hamilton, H. W.; Sawyer, T. K.; Humblet, C.; Doherty, A. M.; Taylor, M. D.; Hingorani, G.; Batley, B. L.; Rapundalo, S. T. *J. Med. Chem.* **1995**, *38*, 2893–2905.
16. Hanessian, S.; Raghavan, S. *Bioorg. Med. Chem. Lett.* **1994**, *4*, 1697–1702.
17. Lefker, B. A.; Hada, W. A.; Wright, A. S.; Martin, W. H.; Stock, I. A.; Schulte, G. K.; Pandit, J.; Danley, D. E.; Ammirati, M. J.; Sneddon, S. F. *Bioorg. Med. Chem. Lett.* **1995**, *5*, 2623–2626.
18. Göschke, R.; Cohen, N.-C.; Wood, J. M.; Maibaum, J. *Bioorg. Med. Chem. Lett.* **1997**, *7*, 2735–2740.
19. Rahuel, J.; Rasetti, V.; Maibaum, J.; Rüger, H.; Göschke, R.; Cohen, N.-C.; Stutz, S.; Cumin, F.; Fuhrer, W.; Wood, J. M.; Grütter, M. G. *Chem. Biol.* **2000**, *7*, 493–504.
20. Göschke, R.; Stutz, S.; Rasetti, V.; Cohen, N.-C.; Rahuel, J.; Rigollier, P.; Baum, H.-P.; Forgiarini, P.; Schnell, C. R.; Wagner, T.; Gruetter, M. G.; Fuhrer, W.; Schilling, W.; Cumin, F.; Wood, J. M.; Maibaum, J. *J. Med. Chem.* **2007**, *50*, 4818–4831.
21. Maibaum, J.; Stutz, S.; Göschke, R.; Rigollier, P.; Yamaguchi, Y.; Cumin, F.; Rahuel, J.; Baum, H.-P.; Cohen, N.-C.; Schnell, C. R.; Fuhrer, W.; Gruetter, M. G.; Schilling, W.; Wood, J. M. *J. Med. Chem.* **2007**, *50*, 4832–4844.
22. Villamil, A.; Chrysant, S. G.; Calhoun, D.; Schober, B.; Hsu, H.; Matrisciano-Dimichino, L.; Zhang, J. *J. Hypertens* **2007**, *25*, 217–226.
23. Oparil, S.; Yarows, S. A.; Patel, S.; Fang, H.; Zhang, J.; Satlin, A. *Lancet* **2007**, *370*, 221–229.
24. Uresin, Y.; Taylor, A. A.; Kilo, C.; Tschöpe, D.; Santonastaso, M.; Ibram, G.; Fang, H.; Satlin, A. *J. Renin Angiotensin Aldosterone Syst.* **2007**, *8*, 190–198.
25. Sever, P. S.; Gradman, A. H.; Azzi, M. *J. Renin Angiotensin Aldosterone Syst.* **2009**, *10*, 65–76
26. Parving, H.-H.; Persson, F.; Lewis, J. B.; Lewis, E. J.; Hollenberg, N. K. *New Engl. J. Med.* **2007**, *358*, 2433–2446.
27. Solomon, S. D.; Appelbaum, E.; Manning, W. J. *Circulation* **2009**, *119*, 530–537.
28. McMurray, J. J. V.; Pitt, B.; Latini, R.; Maggioni, A. P.; Solomon, S. D.; Keefe, D. L.; Ford, J.; Verma, A.; Lewsey, J. *Circulation: Heart Failure* **2008**, *1*, 17–24.
29. Göschke, R.; Stutz, S.; Heinzelmann, W.; Maibaum, J. *Helv. Chim. Acta* **2003**, *86*, 2848–2870.

30.   Sandham, D. A.; Taylor, R. J.; Carey, J. S.; Fässler, A. *Tetrahedron Lett.* **2000**, *41*, 10091–10094.
31.   Herold, P.; Stutz, S.; Spindler, F. WO 02002487, 2002.
32.   Herold, P.; Stutz, S. WO 02002500, 2002.
33.   Ireland, T.; Grossheimann, G.; Wieser-Jeunesse, C.; Knochel, P. *Angew. Chem. Int. Ed.* **1999**, *38*, 3212–3215.
34.   Rueger, H.; Stutz, S.; Göschke, R.; Spindler, F.; Maibaum, J. *Tetrahedron Lett.* **2000**, *41*, 10085–10089.
35.   Herold, P.; Stutz, S. WO 01009079, 2001.
36.   Stutz, S.; Herold, P. WO 02092828, 2002.
37.   Herold, P.; Stutz, S.; Spindler, F. WO 02002508, 2002.
38.   Bartlett, P. A.; Holm, K. H.; Morimoto, A. *J. Org. Chem.* **1985**, *50*, 5179–5183.
39.   Herold, P.; Stutz, S. WO 02008172, 2002.
40.   Tamaru, Y.; Mizutani, M.; Furukawa, Y.; Kawamura, S.-I.; Yoshida, Z.-I.; Yanagi, K.; Minobe, M. *J. Am. Chem. Soc.* **1984**, *106*, 1079–1085.

12

Vernakalant (Kynapid): An Investigational Drug for the Treatment of Atrial Fibrillation

David L. Gray

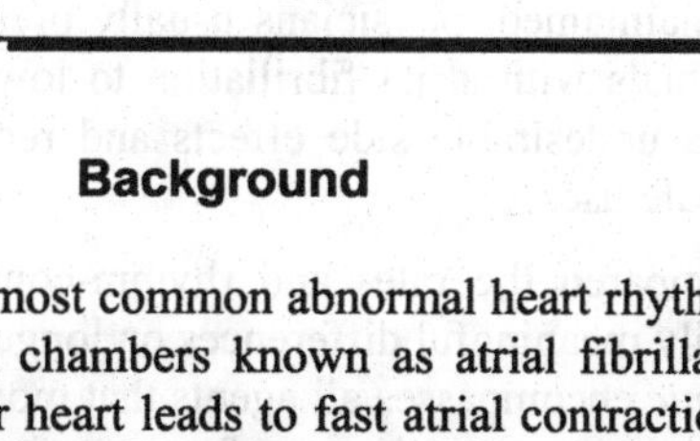

USAN: Vernakalant
Trade name: Kynapid®
Cardiome Pharma
Expected Launch: 2010+

1

12.1   Background

The most common abnormal heart rhythm is an irregular and rapid beating of the smaller atrial chambers known as atrial fibrillation (AF). Aberrant electrical signaling in the upper heart leads to fast atrial contractions that are not synchronized with the ventricles. The incidence of this arrhythmia increases with age, rising to a prevalence of 6–10% after age 50.[1,2] In the United States, it is estimated that 3–5 million people experience AF.[3] There are several distinct underlying pathologies that cause AF, and this condition can be either intermittent (paroxysmal) or persistent. When episodes of atrial fibrillation are perceived by the patient, the arrhythmia is classified as symptomatic AF. In many cases, however, the patient is not specifically aware of the arrhythmia, with diagnosis coming from observation of an abnormal electrocardiogram (EKG) obtained for another purpose. Thus patients present across a spectrum, ranging from paroxysmal asymptomatic AF to persistent symptomatic AF. Paroxysmal episodes of AF can resolve themselves without intervention but often return. People with AF commonly complain of weakness, fatigue, or chest palpitations, and severe episodes can be frightening.[4] The AF arrhythmia itself is not immediately life-threatening, although the rapid atrial beating can precipitate a more serious racing of the ventricular rate. In contrast to atrial rhythm abnormalities like AF and atrial flutter, many ventricular arrhythmias require immediate medical intervention. Over the long term, however, a significantly higher incidence of stroke (4–10× above the normal rate) is a well documented consequence of recurring AF, and the overall cardiovascular mortality risk is doubled for people with this condition.[5,6] The fibrillation can consist of many successive partial atrial contractions which fail to fully move blood through the upper heart chambers. Incomplete exchange of blood in the atria can lead to clot formation within these heart chambers. If these clots dislodge from the atria before they are dissolved, they can enter brain arteries and cause a stroke. It is estimated that 1 of 6 strokes occurs in individuals with atrial fibrillation.[7] Individuals have lived for decades with untreated AF, and the long-term consequences of this arrhythmia are not fully understood; however, it is generally accepted that significant AF will, over time, weaken the heart and cause undesirable remodeling of the cardiovascular muscle.

There is ongoing uncertainty about the best way to mitigate the negative cardiovascular consequences of atrial fibrillation.[8] The correct, synchronous beating of the atria contributes 15–30% to the blood volume ejected from the heart with each contraction of the ventricle. The synchronized action of the atria is important for preloading or swelling the ventricle with sufficient blood for efficient pumping action. Current guidelines suggest that for individuals who can tolerate the AF or are asymptomatic, intermittent episodes of atrial fibrillation are acceptable provided that the potential for thrombotic episodes are controlled with anticoagulation and that these patients are regularly monitored for the development of additional heart abnormalities.[9,10] Where AF episodes are more frequent, severe, or prolonged, a pharmacological intervention is recommended. There are two general approaches to the drug therapy of AF patients.[11,12] One strategy focuses on limiting the abnormal racing of the atria (and ventricles) while allowing for the unsynchronized beating of upper and lower chambers. This intervention strategy is called *rate control therapy*. The other approach seeks to correct the asynchronous nature of the conduction system and to restore a normal sinus heart rhythm (NSR) and is referred to as *rhythm control therapy*. Recently, surgical intervention techniques that attempt to restore NSR by ablating abnormal conductance nodes within the atria have been described for a small number of patients. Regardless of the therapy chosen, if NSR cannot be maintained, physicians usually prescribe chronic anticoagulation therapy for most individuals with atrial fibrillation to lower the risk of stroke.[13,14] Anticoagulation therapy has undesirable side effects and requires vigilant monitoring of blood clotting factors for safe use.

Several large studies have compared the rate- and rhythm-control treatment approaches and have not found statistically meaningful differences on longer-term patient outcomes.[3] The broad term *anti-arrhythmic* encompasses all agents that modify undesired heart rate or rhythm conditions, and these compounds are often grouped into classes according to their primary mode of action using the Vaughn-Williams naming conventions popularized in the 1970s.[15] The rate control strategy relies on using well-characterized and generally safe antiarrhythmics that limit heart rate, including the class II β-blockers (i.e., metoprolol) and the class IV calcium channel blockers (verapamil, diltiazem). These drugs usually do not fully eliminate the underlying arrhythmia but aim to lessen the severity of symptoms and their effect on the heart muscle by preventing overly rapid atrial or ventricular rates. Physicians are familiar with β-blockers and calcium-channel blockers and can use them to simultaneously address other cardiovascular problems through their hypotensive effects.

For effectual rhythm control, the first action is often to restore NSR acutely with a nonsurgical intervention called a cardioversion, by which the patient's heart is reset through the use of electrical current strategically delivered to the heart via external electrode pads. When atrial fibrillation is changed to a normal sinus rhythm, the patient is said to have been converted. The cardioversion process has a good success rate for achieving conversion; however, patients must be anesthetized for the procedure, and the AF often returns.

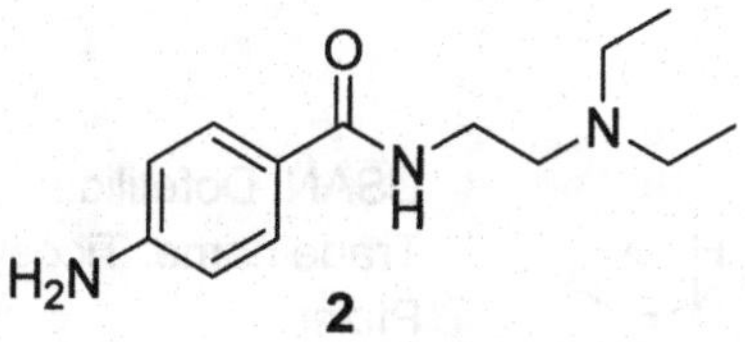

USAN: Procainamide
Trade name: Pronestyl® (IV)
Launched: 1951
Class 1a antiarrythmic

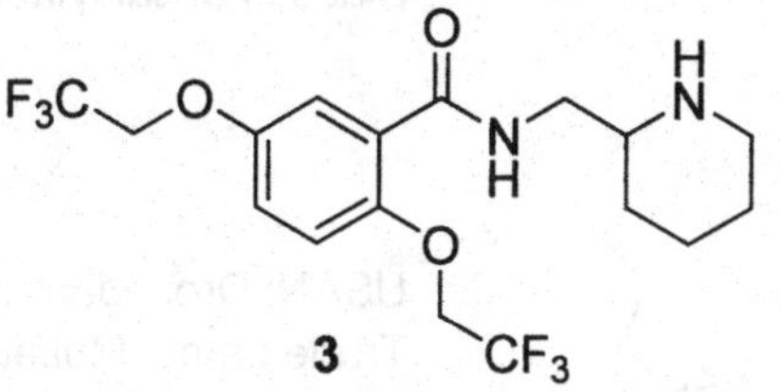

USAN: Flecainide
Trade name: Tambocor®
3M Pharmaceuticals
Launched: 1982
Class Ic antiarrythmic

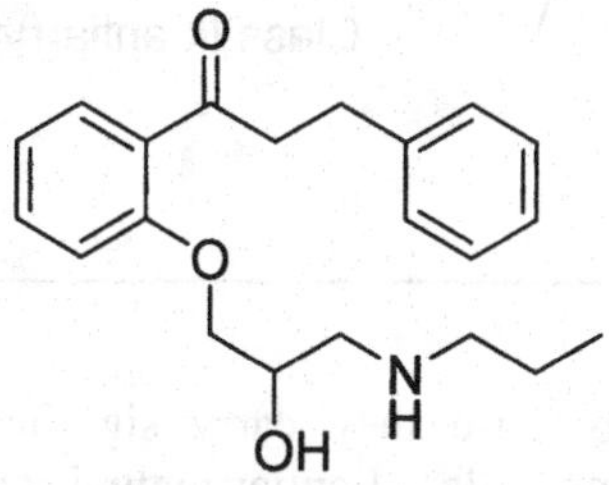

USAN: Propafenone
Trade name: Rhythmol SR®
Abbott
Launched: 1982
Class Ic antiarrythmic

Conversion of AF to NSR can also be accomplished with a subset of antiarrhythmic drugs (including 2–7) that act directly on cardiac muscle cells (myocytes) and antagonize either the sodium channel-mediated propagation currents (procainamide 2, flecainide 3, propafenone 4), or the inwardly rectifying (I_{Kr}) potassium channel currents (ibutilide 5, dofetilide 6). Some of the antiarrhythmics have actions at both potassium and sodium channels (i.e., dronedarone 7 and its close structural progenitor amiodarone). Several compounds from these two mechanistic classes have shown clinical efficacy for restoring NSR in particular groups of AF patients. Of these medications, flecainide, propafenone, ibutilide, dofetilide, and dronedarone are FDA approved for the acute conversion of AF to NSR in certain patient populations.

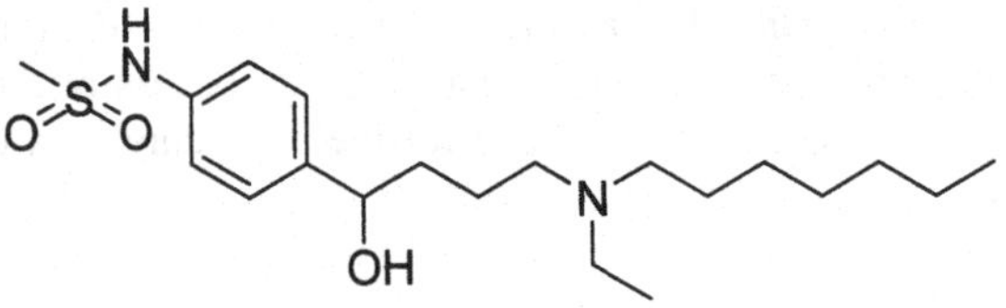

USAN: Ibutilide
Trade name: Corvert®
Pfizer
Launched: 1996
Class III antiarrythmic

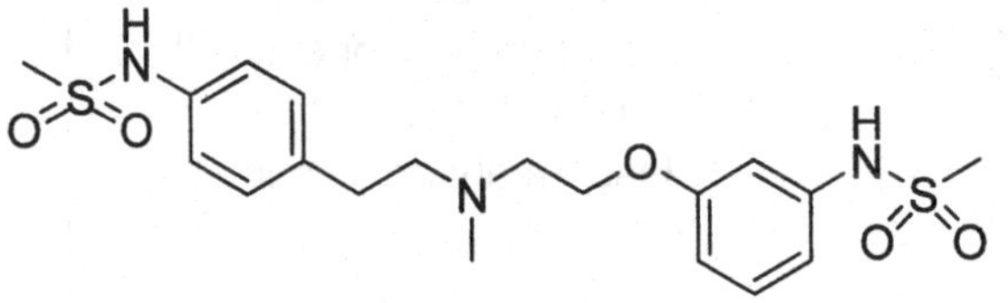

USAN: Dofetilide
Trade name: Ticosyn®
Pfizer
Launched: 2000
Class III antiarrythmic

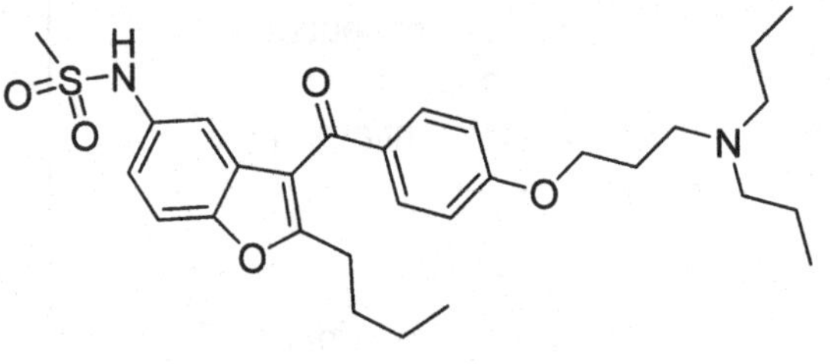

USAN: Dronedarone
Trade name: Multaq®
Sanofi-Aventis
Launched: 2009
Class III antiarrythmic

These agents are used with caution because they carry significant safety liabilities associated with their mechanism of action. Class I antiarrhythmics antagonize sodium channels which raises the threshold for cardiac signal propagation and slows the speed of electrical conduction in the heart muscle. The potency of class I agents increases in a rate-dependent manner; therefore, these compounds can exert a higher degree of signal dampening during fast AF. Class III antiarrhythmics antagonize the rapid inward potassium currents, which prolongs repolarization in cardiac myocytes and renders these cells less sensitive to the aberrant signals that cause AF. Unfortunately, these agents act indiscriminately on all cardiac tissue and can adversely affect the critical ventricular rhythm via their antagonistic actions. A rare, but well-documented arrhythmia associated with Class III (and some Class I) antiarrhythmics is Torsades de Pointes (TdP), a ventricular fibrillation that is fatal if it is not terminated.[16] TdP has been connected to abnormally long delays in a portion of the heart's electrical cycle (specifically, delays in the QT interval on an EKG). Consequently, initial doses of some antiarrhythmics used to treat AF are loaded in a hospital setting in which patients are monitored for very serious cardiovascular events like TdP. The superior AF conversion and maintenance rate of the mixed sodium–potassium channel blocker amiodarone is undermined by a risk of pulmonary toxicity; however, its efficacy supports the notion that activity at multiple ion channels could lead to improved patient outcomes. Even when these agents are initially effective at converting AF and maintaining NSR, about half of successfully treated patients will begin to have reoccurring AF within one year.[17] There remains a high need for treatments that safely convert or manage AF, thereby reducing the need for anticoagulant therapy as well as preventing the damaging effects of chronic atrial rhythm abnormalities.

12.2 Pharmacology

Ion channels open and close to control the exchange of specific ions across cellular membranes and the resulting changes in polarization can regulate or mediate cellular functions. For each major type of ion channel, there are a large number of subtypes that are differentially expressed in various tissues. The localization and expression level of certain sodium, calcium, and potassium channels subtypes allows specific tissues within the heart to respond appropriately to ionic gradients. In humans, the potassium channel subtype Kv1.5 is expressed in the atrial myocytes but is not found in the ventricles.[18,19] The Kv1.5 is a tetrameric homodimer that acts as a delayed rectifier voltage-gated ion channel in cardiac muscle tissue by mediating the slow outflow of potassium ions across the membrane to repolarize these cells. It is also established that Kv1.5 regulates insulin secretion in human atrial myocytes and that the expression of this ion-channel subtype may be down-regulated in people with atrial fibrillation.[20] Blocking repolarization currents that are predominately localized in atrial tissue would theoretically have little or no direct effect on the ventricles but could prolong action potentials in the atria and render that tissue less sensitive to abnormal (re-entrant) signals that cause AF.[21] It has been hypothesized that selectively targeting the Kv1.5 subtype with a selective potassium channel antagonist might be able to normalize atrial rhythm without the corresponding and problematic proarrythmatic effects on the ventricles.[22]

12.3 Structure–Activity Relationship (SAR)

Several companies including Merck,[23,24] Sanofi-Aventis,[26,27] and Proctor & Gamble,[28] published papers that described research programs that targeted selective Kv1.5 blockers. In 2003 Cardiome Pharmaceuticals submitted the first of several patents applications that describe *trans*-cyclohexylamine alkyl ethers (including **8**, **9**, and **1**) and claimed that these compounds showed specificity for inhibition of atrially expressed potassium channel subtypes.[29] Subsequent publications suggest that the initial research effort at Cardiome was actually focused on finding ischemia-selective ventricular antiarrythmetics.[30] As part of this program, they developed an arrhythmia model using anesthetized rats with coronary occlusions that proved to be useful in predicting the atrial antiarrythmetic potential of ion channel blockers. In contrast to the human heart, action potentials in rat ventricles depend on potassium channels like Kv1.5 (which is expressed in rat ventricles). Structure–activity relationships were evolved around compounds that prevented ventricular arrhythmias in this model. Compounds were also tested in HEK cells expressing clones of Nav1.5, Kv1.5, Kv4.2 and Kv2.1 to guide compound design and selection.[31] Promising leads were further advanced into a dog model of atrial fibrillation to confirm their potential to convert AF in a cardiovascular system more similar to the human. Within the cyclohexyl amino ether series, there was an evolution of compounds that emerged as lead candidates at various stages of project progression. In early work, both the *cis*- and *trans*-cyclohexyl amino compounds had some ion channel activity, so initial analogs were often prepared as mixtures and then separated. The ester functionality in **8** was prone to hydrolysis in vivo, leading the medicinal chemistry group to move toward ether linked analogs such as **9**. The authors report evolving toward compounds with selectivity for specific potassium channels (like Kv1.5), but which also had activity for certain sodium channels. Ultimately, vernakalant (**1**) was shown to combine functionally selective antagonism at the Kv1.5 potassium channel with action at the sodium channel Nav1.5. Advanced preclinical studies with vernakalant confirmed its antiarrythmic potential and acceptable pharmacokinetic properties, and this compound was chosen for clinical development.[32]

8   **9**   **10**

12.4   Pharmacokinetics and Drug Metabolism

In human healthy volunteers and patients, vernakalant is cleared both renally and hepatically via cytochrome P450 CYP2D6.[33] The primary metabolic action is *O*-demethylation to the phenolic product **10**, which is cleared and excreted as its glucuronide. Circulating levels of this metabolite (predominately as its glucuronide conjugate) exceed parent drug levels at steady state, but the metabolite is inactive against the targeted sodium and potassium ion channels.

Across several human pharmacokinetic studies, the elimination halfp-life for vernakalant after intravenous administration is 3–4 h. The primary efficacy endpoint in Phase II studies was conversion of recent onset AF to normal sinus rhythm within 90 min. of drug infusion. Robust efficacy by this measure was observed in the 2–5-mg dose range. For conversion of AF patients to NSR, the recommended protocol is a 3-mg dose infused intravenously over 10 min. followed by a 2-mg infusion over 15 minutes if the patient does not convert. More recently, an oral formulation of vernakalant has demonstrated efficacy in converting AF patients to NSR.[34] Using an extended release preparation, a 600-mg oral dose achieves and maintains (for 12 h) plasma drug levels that approximate the 3-mg + 2-mg I.V. dosing regimen. Approximately 15% of the drug is bioavailable via oral administration. Metabolism of **1** by means of CYP2D6 causes modest half-life and clearance differences for polymorphic expressers of this enzyme.

12.5   Efficacy and Safety

Initial clinical work focused on a product that would be used to convert patients who were hospitalized for recent-onset AF as an alternative to a cardioversion procedure. Cardiome partnered with Fujisawa (now Astellas Pharmaceuticals) for the clinical development of a vernakalant (**1**), which produces robust AF conversion rates and none of the ventricular conduction abnormalities associated with less selective agents.[35] By the close of 2009, vernakalant had been in 5 large trials and several smaller trials, including a total of over 1200 AF patients. Four Phase III studies—Atrial Arrhythmia Conversion Trials (ACT 1, 2, 3, and 4) —examined patients with recent onset AF (< 45 days) and the measured the rate of conversion to NSR in various periods after drug administration.[36–38] The conversion rate was 45–63% in these studies, which tended to exclude patients with more compromised heart function or a history of heart failure. While there has been no specific study looking at QT prolongation, monitoring in both patients and in healthy volunteers showed only modest increases in the QT interval with no directly associated incidents of TDP across all trials.

One of the theoretical benefits of atrially selective potassium channel block is minimization of ventricular proarrythmic effects seen with nonselective blockers and the shorter-term studies conducted to date would seem to support this benefit, albeit with data coming from a lower risk population.[39] The most common adverse events reported with vernakalant therapy were altered or lost sense of taste and sneezing. The four ACT trials administered vernakalant intravenously, and after some delays, the companies received an FDA approvable letter for this product and dosing regimen in 2008. Astellas subsequently sold the non-U.S. rights to the intravenous product to Merck, who proceeded to file for European marketing approval in August 2009. Merck also licensed the oral formulation of vernakalant, which has shown statistically meaningful improvement rates for the conversion of recent onset AF out to 90 days after initiating drug therapy. The phase II trials supporting the development of oral vernakalant have examined AF conversion rates as well as the maintenance of converted patients in NSR.[40] In the phase II trials conducted thus far, oral doses have been 300–600 mg twice daily for up to 90 days. Oral dosing at 600 mg b.i.d. for 90 days requires some 20,000 times more drug product relative to a single i.v. infusion of 2–5 mg, which puts a premium on an efficient synthesis to keep manufacturing costs down.

12.6   Synthesis

At the time of this writing, no complete synthesis of vernakalant has appeared in the primary chemistry literature, but there are several Cardiome patents that describe successfully executed routes to this compound.[29,41,42] In 2007, Beatch et al. published a report describing the synthesis and lead evolution of ion channel antagonists that are structurally close to **1**.[31] The authors employ several flexible chemical approaches to explore cyclohexylamine-2-alkylethers like **9**, which they obtained as mixtures of diastereoisomers. Although it is not specifically mentioned, it can be inferred from patents that vernakalant was successfully synthesized using one of these methods and a final stage HPLC separation. Once the advancing candidate (**1**) was identified, the initial medicinal chemistry route was slightly modified and optimized to address the chirality challenges presented by this target. An early, scalable synthesis to enantiomerically pure **1** builds on the original discovery route and was taught in an early patent.[29]

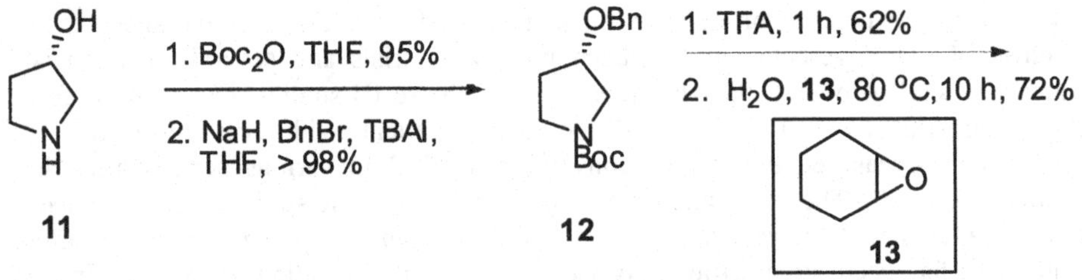

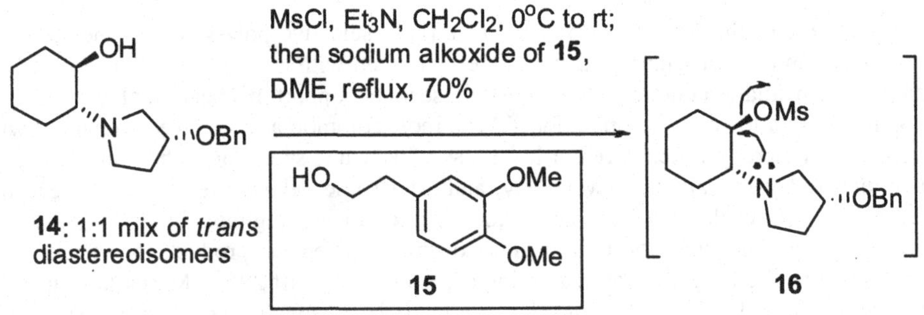

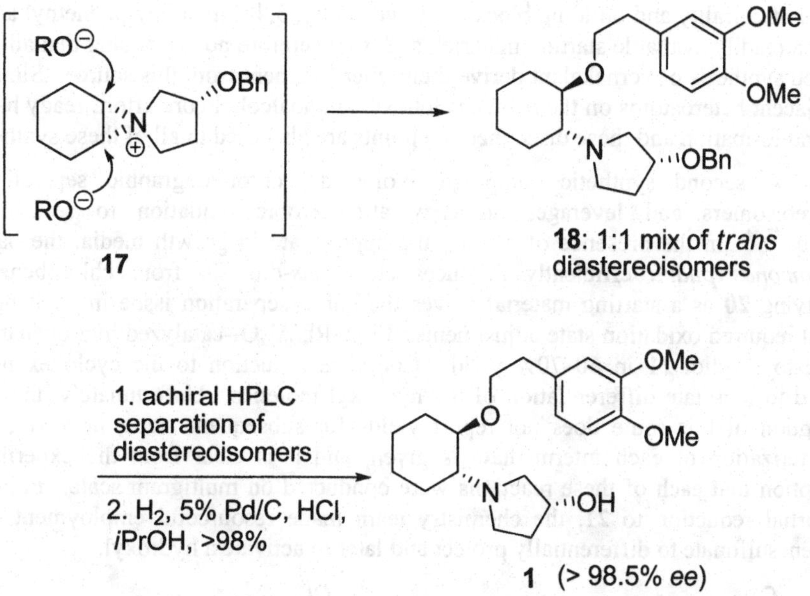

18: 1:1 mix of *trans* diastereoisomers

17

1. achiral HPLC separation of diastereoisomers

2. H₂, 5% Pd/C, HCl, *i*PrOH, >98%

1 (> 98.5% *ee*)

Commercially available (*R*)-pyrrolidin-3-ol (**11**) carries a requisite chiral center, and this compound was *N*-Boc protected and then *O*-benzylated under standard conditions, leading to **12**. After acidic removal of the *N*-Boc, the liberated pyrrolidine amine efficiently opens cyclohexene oxide in a biphasic reaction system to give the adduct **14** with the expected *trans* relationship about the cyclohexane ring. Due to the presence of a chiral center on the pyrrolidine, the epoxide opening gives a 1:1 mixture of diastereoisomers which was carried forward. The cyclohexanol in **14** was activated as its mesylate derivative and subsequently displaced with the pregenerated sodium alkoxide of **15** by refluxing the DME solution for several days. It is interesting that this S_N2 displacement reaction again leads to a *trans* geometry about the cyclohexyl ring. It is reasonable to conclude that the pyrrolidine nitrogen initially displaces the mesylate intramolecularly to generate the symmetrical aziridinium intermediate **17**, which is eventually opened by alkoxide anion (of **15**), leading to the *trans* product.[31] At this stage, the diastereoisomers were separated using achiral preparative HPLC to afford *R,R,R*-(**18**) and *S,S,R* (not shown) single enantiomer compounds with chiral purity in excess of 98.5%. To complete the synthesis of vernakalant, the *O*-benzyl protecting group was quantitatively removed in isopropanol and aqueous HCl catalyzed by 5% Pd/C under a H₂ atmosphere, leading to **1** as a hydrochloride salt. The overall yield on multigram scale was about 11% for the 7 steps in this synthesis. An undesirable aspect of this route is the late-stage achiral HPLC separation, which discards half of the synthesized material, although this approach efficiently translates the absolute chirality of the pyrrolidinol to obviate a chiral resolution. Vernakalant (**1**) contains 3 chiral carbons, so cost-effective installation of these three centers needs to be a key element of any scalable synthesis. Indeed, the progression of the 3 synthetic approaches highlighted here illustrate the never-ending quest to lower active ingredient costs by shifting to more economical sources of chirality and building blocks. Conveniently, 3,4-dimethoxy-phenethyl alcohol (15) is a readily available starting material, and it is therefore not surprising that all of the reported synthesis of vernakalant derive their ether fragment from this source. Similarly, the adjacent heteroatoms on the *trans*-cyclohexyl aminoalcohol core afford ready handles for bond formation and these disconnection points are also used in all of these synthesis.

A second synthetic campaign avoids the chromatographic separation of diastereoisomers and leverages an enzymatic aerobic oxidation to set absolute chirality.[43,44] In the presence of air and the appropriate in growth media, the bacteria *Pseudomonas pudita* efficiently produces chiral *cis*-diol 20 from chlorobenzene.[45] Identifying 20 as a starting material solves the chiral separation issue in exchange for several required oxidation state adjustments. First, Rh/Al$_2$O$_3$-catalyzed hydrogenation of 20 led to *cis*-diol 21 in 60–70% yield. Complete reduction to the cyclohexane was delayed to facilitate differentiation of the hydroxyl moieties. Unfortunately, the patent description of this route does not report yields for subsequent steps; however, a full characterization of each intermediate is given, and it is clear from the experimental description that each of these reactions were conducted on multigram scale. Following the partial reduction to 21, the chemistry team made resourceful employment of the benzenesulfonate to differentially protect and later to activate a hydroxyl.

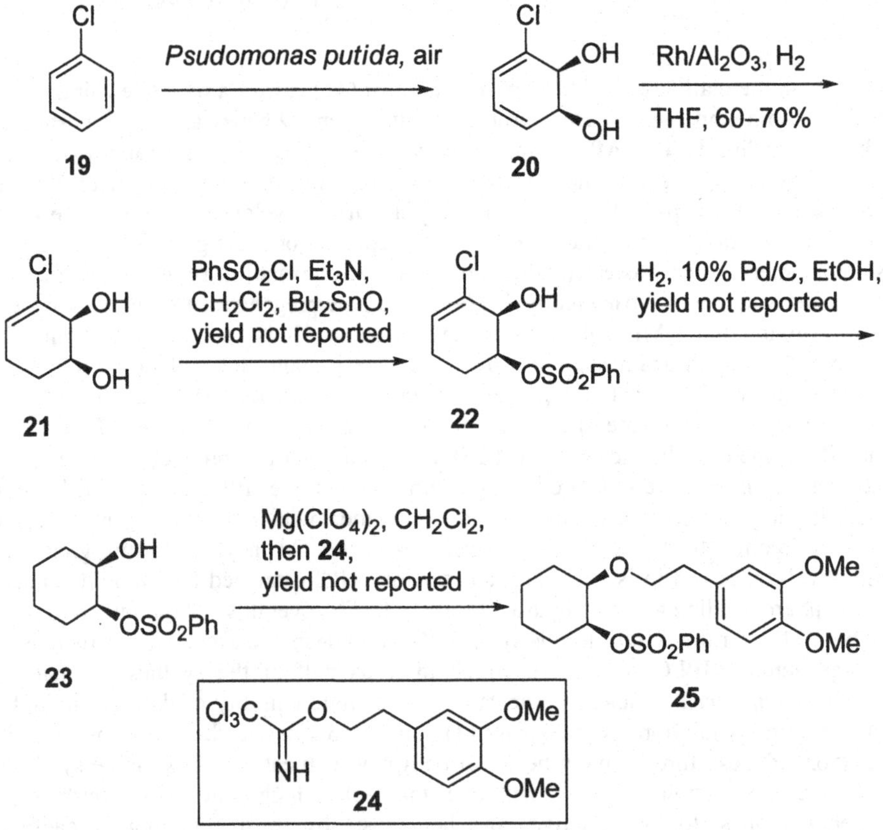

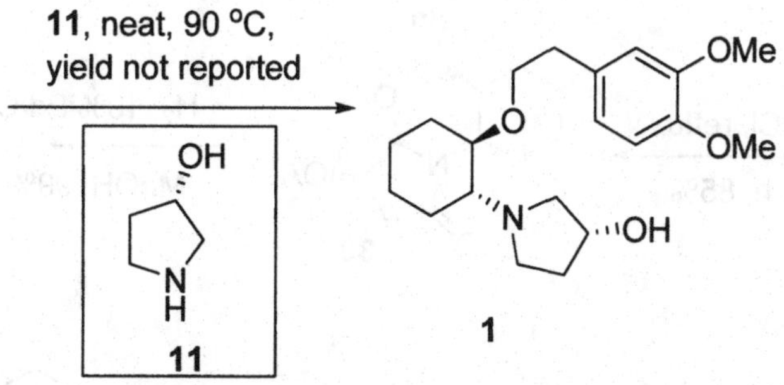

11, neat, 90 °C,
yield not reported

The addition of Bu_2SnO and Et_3N encouraged this latter compound (**21**) to react with benzenesulfonyl chloride selectively on the nonallylic hydroxyl group to distinguish the *trans* alcohols. A second hydrogenation over 10% Pd/C in ethanol led to the monosulfonylated *trans*-cyclohexane diol **23**. The primary hydroxyl group in **15** was activated as its trichloroacitimidate derivative (**24**), which was then combined with **23** using Lewis acid-accelerated conditions to facilitate reaction with the hindered hydroxylic nucleophile. Finally, displacement of the benzenesulfonate within **25** by unprotected (*R*)-pyrrolidin-3-ol delivered vernakalant, which was recrystallized to high chemical and enantiomeric purity.[46]

A more recent refinement of the synthesis takes advantage of two chiral pool materials to streamline the overall synthesis and appears to target elimination of the pyrrolidin-3-ol (**11**) starting material that was employed in the previous schemes.[42,43] As more drug product is required to support a compound's development, costly reagents and starting materials are often eliminated in a targeted fashion to lower manufacturing costs. Although all three of these synthetic processes require a similar number of chemical reactions, there is a clear progression toward more scalable and cost-efficient production of the drug product. The success of the oral form of vernakalant certainly stimulates additional interest in further refining low cost chemical synthesis of this compound.

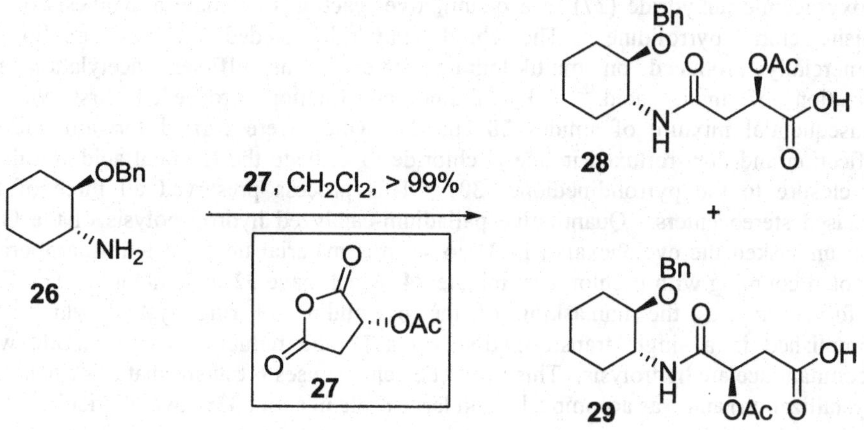

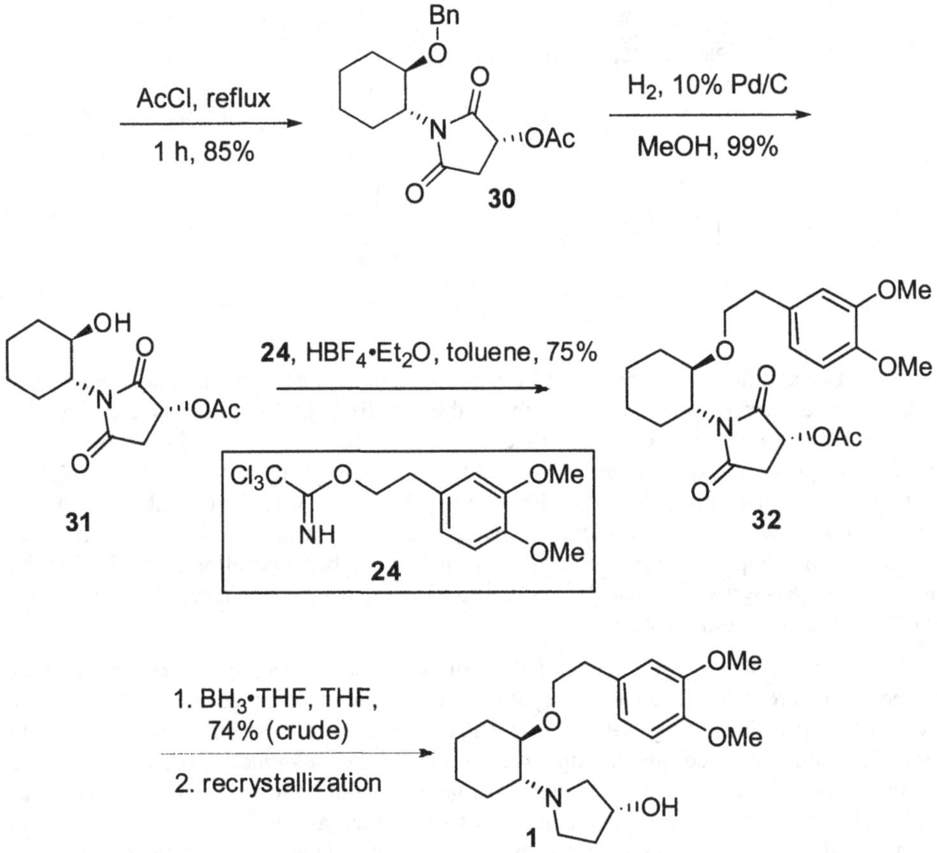

While not a commodity chemical, (1*R*,2*R*)-*trans*-2-benzyloxy-cyclohexylamine (**26**) can be purchased or synthesized.[47]   This amine was combined with (*S*)-2-acetoxysuccinic anhydride (**27**) in a quantitative reaction to initiate a synthesis of the requisite chiral pyrrolidine.   The chiral anhydride needed for this coupling is commercially produced on multikilogram scale via an efficient acetylation and cyclization of malic acid.[48]   Pyrrolidine construction proceeded first via an inconsequential mixture of amides **28** and **29**, which were carried forward without purification and then refluxed in acetyl chloride to activate the terminal acid and drive ring closure to the pyrrolidinedione (**30**).   This process preserved all three of the purchased stereocenters. Quantitative palladium-catalyzed hydrogenolysis of the OBn group unmasked the cyclohexanol in **31** to set this material up for a tetrafluoroborate-promoted coupling with trichloroacetimidate **24**, which gave **32** in 75% yield. The final transformation was the unmasking of the pyrrolidin-3-ol ring system which was accomplished in a single transformation via a borane reduction of the imide with concomitant acetate hydrolysis. This synthetic scheme uses 6 transformations (including recrystallization) and was accomplished on large scale in about 33% overall yield.

In summary, vernakalant (1), a sodium and selective potassium channel blocker, is currently in advanced studies to support its use for the conversion of atrial fibrillation to normal sinus rhythm. Both the oral and the i.v. formulations have demonstrated robust AF conversion efficacy and a favorable ventricular safety profile. At the close of 2009, it is not certain when vernakalant might be launched and marketed in the United States. The FDA and E.U. regulatory agencies have given approvable indications, and the partnering companies are awaiting the outcome of an ongoing Phase III trial before negotiating product labeling. Merck has assumed development of this compound via a partnership announced in 2009 and is preparing for late-stage trials with the oral formulation.

12.7   References

1.   Feinberg, W. M.; Blackshear, J. L.; Laupacis, A.; Kronmal, R.; Hart, R. G. *Arch. Intern. Med.* **1995**, *155*, 469–473.
2.   Podrid, P. J. *Cardiol. Clin.* **1999**, *17*, 173–188, ix–x.
3.   Go A. S.; Hylek E. M., Phillips K. A.; et al. *J. Am. Med. Assoc.* **2001**, *285*, 2370–2375.
4.   Hinton, R. C.; Kistler, J. P.; Friedlich, A. L.; Fisher, C. M. *Am. J. Cardiol.* **1977**, *40*, 509–513.
5.   Flaker, G. C.; Blackshear, J. L.; McBride, R.; Kronmal, R. A.; Halperin, J. L.; Hart, R. G. *J. Am. College Cardiol.* **1992**, *20*, 527–532.
6.   Cabin, H. S.; Clubb, K. S.; Hall, C.; Perlmutter, R. A.; Feinstein, A.R. *Am. J. Cardiol.* **1990**, *65*, 1112–1116.
7.   Wolf, P. A.; Abbott, R. D.; Kannel, W. B. *Stroke* **1991**, *22*, 983–988.
8.   Reiffel, J. A. *Am. J. Cardiol.* **2008**, *102* suppl., 3H–11H.
9.   Fuster V.; Ryden L. E.; Cannom D. S.; et al. *Eur. Heart J.* **2006**, *27*, 1979–2030.
10.   Fuster, V.; Ryden, L. E.; Cannom, D. S.; et al. *Circulation* **2007**, *114*, e257–e354.
11.   Roy, D.; Talajic, M.; Nattel, S.; Wyse, D. G.; Dorian, P.; Lee, K.L.; Bourassa, M. G.; Arnold, J. M.; Buxton, A. E.; Camm, A. J.; Connolly, S. J.; Dubuc, M.; Ducharme, A.; Guerra, P. G.; Hohnloser, S. H.; Lambert, J.; Le Heuzey, J. Y.; O'Hara, G.; Pedersen, O. D.; Rouleau, J. L.; Singh, B. N.; Stevenson, L. W.; Stevenson, W. G.; Thibault, B.; Waldo, A. L. *N. Engl. J. Med.* **2008**, *358*, 2667–2677.
12.   Cain, M. E. *N. Engl. J. Med.* **2002**, *347*, 1825–1833.
13.   Hart, R. G.; Pearce, L. A.; Aguilar, M. I. *Ann. Intern. Med.* **2007**, *146*, 857–867.
14.   Lip, G. Y.; Edwards, S. J. *Thromb. Res.* **2006**, *118*, 321–333.
15.   Vaughan Williams, E. M. In: *Symposium on Cardiac Arrhythmias*, Sandfte E.; Flensted-Jensen E.; Olesen K. H. eds. Sweden, AB ASTRA, Södertälje, **1970**, pp 449–472.
16.   Kowey, P. R.; VanderLugt, J. T.; Luderer, J. R. *Am. J. Cardiol.* **1996**, *78*, 46–52.
17.   Greene, H. L.; Waldo, A. L. *J. Am. Coll. Cardiol.* **2003**, *1*, 20–29.

18.   Mays, D. J.; Foose, J. M.; Philipson, L. H.; Tamkun, M. M. *J. Clin. Invest.* **1995**, *96*, 282–292.

19.   Feng, J.; Wible, B.; Li, G. R.; Wang, Z.; Nattel, S. *Circ. Res.* **1997**, *80*, 572–579.

20.   Wagoner, D. V. R. *Drug. Discov. Today* **2005**, *2*, 291–295.

21.   Stump, G. L.; Wallace, A. A.; Regan, C. P.; Lynch, J. J. *J. Pharmacol. Exp. Ther.* **2005**, *315*, 1362–1367.

22.   Fedida, D.; Wible, B.; Wang, Z.; Fermini, B.; Faust, F.; Nattel, S.; Brown, A. M. *Circ. Res.* **1993**, *73*, 210–216.

23.   Lagrutta, A.; Wang, J. X.; Fermini, B.; Salata, J. J. *J. Exp. Ther.* **2006**, *317*, 1054–1063.

24.   Regan, C. P.; Kiss, L.; Stump, G. L.; McIntyre, C. J.; Beshore, D. C.; Liverton, N. J.; Dinsmore, C. J.; Lynch, J. J. *J. Pharmacol. Exp. Ther.* **2008**, *324*, 322–330.

25.   Nanda, K. K.; Nolt, M. B.; Cato, M. J.; Kane, S. A.; Kiss, L.; Spencer, R. H.; Wang, J. X.; Lynch, J. J.; Regan, C. P.; Stump, G. L.; Li, B.; White, R.; Yeh, S.; Bogusky, M. J.; Bilodeau, M. T.; Dinsmore, C. J.; Lindsley, C. W.; Hartman, G. D.; Wolkenberg, S. E.; and Trotter, B. W. *Bioorg. Med. Chem. Lett.* **2006**, *16*, 5897–5901.

26.   Peukert, S.; Brendel, J.; Pirard, B.; Brueggemann, A.; Below, P.; Kleemann, H.; Werner, H. H.; Schmidt, W. *J. Med. Chem.* **2003**, *46*, 486–498.

27.   Gross, M. F.; Beaudoin, S.; McNaughton-Smith, G.; Amato, G. S.; Castle, N. A.; Huang, C.; Zou, A.; Yu, W. *Bioorg. Med. Chem. Lett.* **2007**, *17*, 2849–2853.

28.   Blass, B. E.; Coburn, K.; Lee, W.; Fairweather, N.; Fluxe, A.; Wu, S. D.; Janusz, J. M.; Murawsky, M.; Fadayel, G. M.; Fang, B.; Hare, M.; Ridgeway, J.; White, R.; Jackson, C.; Djandjighian, L.; Hedges, R.; Wireko, F. C.; Ritter, A. L. *Bioorg. Med. Chem. Lett.* **2006**, *16*, 4629–4632.

29.   Barrett, A. G.; Beatch, G. N.; Choi, L. S. L.; Jung, G.; Liu, Y.; Plouvier, B.; Wall, R.; Zhu, J.; Zolotoy, A. WO2004099137, 2004.

30.   Bain, A. I.; Barrett, T. D.; Beatch, G. N.; Fedida, D.; Hayes, E. S.; Plouvier, B.; Pugsley, M. K.; Walker, M. J. A.; Walker, M. L.; Wall, R. A.; Yong, S. L.; Zolotoy, A. *Drug Dev. Res.* **1997**, *42*, 198–210.

31.   Plouvier, B.; Beatch, G. N.; Jung, G. L.; Zolotoy, A.; Sheng, T.; Clohs, L.; Barrett. T. D.; Fedida, D.; Wang, W. Q.; Zhu, J. J.; Liu, Y.; Abraham, S.; Lynn, L.; Dong, Y.; Wall, Walker, M, J, A. *J. Med. Chem.* **2007**, *50*, 2818–2841.

32.   Orth, P. M. R.; Hesketh, C.; Mak, C. K. H. *Cardio. Res.* **2006**, *70*, 486–496.

33.   Zhongping, L. Mao, L. Z.; Wheeler, J. J.; Clohs, L.; Beatch, G. N.; Keirns, J. *J. Clin. Pharmacol.* **2009**, *49*, 17–29.

34.   Beatch, G. N.; Wheeler, J. J. WO 2008137778, 2008.

35. Roy, D; Rowe, B. H; Steill, I. G.; Coutu, B.; Ip, John H.; Phaneuf, D.; Lee, J.; Vidaillet, H.; Dickinson, G.; Grant, S.; Ezrin, A. M.; Beatch, G. N. *J. Am. Coll. Cardiol.* **2004**, *44*, 2355–2361.
36. Dorian P.; Pinter A.; Mangat I. *J. Cardio. Pharmacol.* **2007**, *50*, 35–40.
37. Kowey P. R.; Roy D.; Pratt C. M. *Circulation* **2007**, *116*, II636–637.
38. Rowe B. H.; Dickinson G.; Mangal B. *Ann. Emerg. Med.* **2008**, *52* Suppl., S48 (Abstract 22).
39. Cheng, W. M. J.; Rybak, I. *Clin. Med. Ther.* **2009**, *1*, 215–230.
40. Ongoing Study by Cardiome Pharma, ClinicalTrials.gov Identifier: NCT00668759
41. Barrett, A. G.; Choi, L. S. L. WO2005016242, 2005.
42. Bain, A. I.; Beatch, G. N.; Walker, M. J.; Plouvier, B.; Sheng, T. ; Longley, C. J.; Yong, S. L.; Zhu J. J.; Zolotoy, A. B.; Wall, R. A.; Zhu, J. J. US2005209307, 2005.
43. Barrett, A. G.; Choi, L. S. L.; Chou, D. T. H.; Hedinger, A.; Jung, G.; Kurz, M.; Moeckli, D.; Passafaro, M. S.; Plouvier, B.; Sheng, T.; Ulmann P. WO2006088525, 2006.
44. Chou, D. T. H.; Jung, G.; Plouvier, B.; Yee, J. G. K. WO 2006138673, 2006.
45. Hudlicky, T.; Luna, H.; Barbieri, G.; Kwart, L. D. *J. Am. Chem. Soc.* **1988**, *110*, 4735–4741.
46. Hashimoto, M.; Eda, Y.; Osanai, Y.; Iwai, T.; Aoki, S. *Chem. Lett.* **1986**, *6*, 893–896.
47. Ditrich, K.; Reuther, U.; Bartsch, M. WO20070912, 2007.
48. Mhaske, S. B.; Argade, N. P. *J. Org. Chem.* **2001**, *66*, 9038–9040.

13

Conivaptan (Vaprisol): Vasopressin V1a and V2 Antagonist for Hyponatremia

Brian A. Lanman

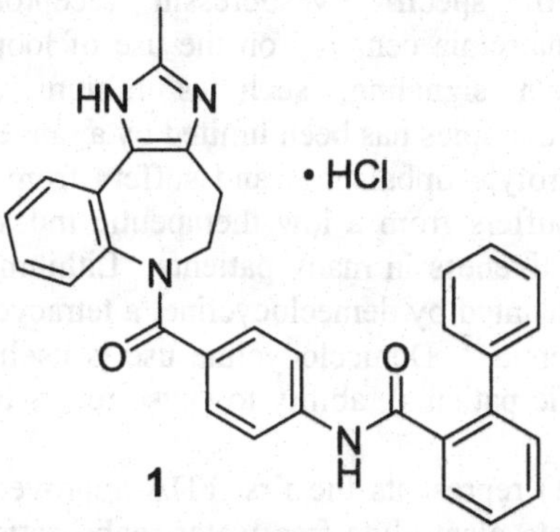

· HCl

USAN: Conivaptan hydrochloride
Trade name: Vaprisol®
Astellas Pharma
Launched: 2006

1

13.1   Background

Hyponatremia, a condition wherein serum sodium concentration is < 135 mmol/L, is the most common electrolyte imbalance among hospitalized patients, occurring in up to 15% of in-patients.[1,2]   In addition to being a potentially life-threatening condition, hyponatremia is an independent predictor of adverse outcomes among patients with heart failure,[3,4] acute ST-elevation myocardial infarction,[5] and cirrhosis.[6]

Hyponatremia is caused by an excess of total body water relative to total body sodium and can result from a number of underlying conditions, including the syndrome of inappropriate antidiuretic hormone secretion (SIADH), cirrhosis, and congestive heart failure (CHF).   In each of these conditions, inappropriate production of arginine vasopressin (AVP) [also known as vasopressin or antidiuretic hormone (ADH)], a neurohormone that regulates renal electrolyte-free water reabsorption, contributes to enhanced renal water retention, leading to decreased serum sodium concentrations.[7] Hyponatremia can be characterized as hypervolemic, euvolemic, or hypovolemic depending on the nature of changes in extracellular fluid (ECF) volume. In hypervolemic hyponatremia, a frequent sideeffect of cirrhosis and CHF, ECF volume is increased; in euvolemic hyponatremia, which most commonly results from SIADH, ECF volume remains normal; and in hypovolemic hyponatremia, as can result from severe cases of diarrhea, ECF volume is decreased.[8,9]

In hyponatremic patients, uncompensated decreases in plasma sodium concentrations result in the shift of water from the extracellular space into brain tissue, where sodium levels are tightly controlled, causing cerebral edema. If hyponatremia develops slowly, loss of solute from the brain can offset brain swelling. However, large or rapid reductions in plasma sodium levels can lead to a range of CNS-related symptoms, such as headache, nausea, vomiting, muscle cramps, lethargy, disorientation, and depressed reflexes. In severe cases, hyponatremia can result in seizures, coma, respiratory arrest, permanent brain damage, and death.[1]

In mild cases of hyponatremia, treatment typically focuses on water restriction (< 800 mL/day); however this approach suffers from poor patient compliance due to thirst brought on by increasing serum osmolality.[1,10] In cases of extreme hyponatremia, infusions of hypertonic saline are used to elevate serum sodium concentrations. Loop diuretics (e.g., furosemide) are often used as an adjunct to such treatment to offset potential volume overload.[1] Hypertonic saline therapy is also suboptimal, as it carries a risk of overly rapid adjustment of plasma sodium levels, which can result in the rapid shift of water from brain tissue to the vascular space, triggering neural demyelination that can result in seizures, coma, quadriplegia, and even death.[1]

Before the introduction of specific vasopressin receptor antagonists, pharmacological treatments for hyponatremia centered on the use of loop diuretics and nonspecific inhibitors of vasopressin signaling, such as lithium carbonate and demeclocycline.[11] The utility of such therapies has been limited by a range of sideeffects. Loop diuretic use can result in electrolyte imbalances and suffers from poor response predictability.[11] Lithium carbonate suffers from a low therapeutic index and a risk of renal damage as well as limited effectiveness in many patients. Lithium carbonate has therefore been nearly completely supplanted by demeclocycline, a tetracycline antibiotic, in the treatment of chronic hyponatremia.[12] Demeclocycline use is itself limited by its nephrotoxicity (particularly in cirrhotic patients), ability to cause reversible uremia, and ability to induce photosensitivity.[1,11]

Conivaptan hydrochloride (1) represents the first FDA-approved agent for the treatment of hyponatremia to modulate electrolyte-free water reabsorption in the renal collecting ducts by antagonizing the action of AVP on vasopressin receptors. As such, 1 represents a novel approach toward the treatment of hyponatremia that could significantly affect the pharmacotherapy of a range of diseases characterized by abnormal water retention.

As of 2009, conivaptan•HCl (1) is one of three vasopressin receptor antagonists approved for use in the treatment of hyponatremia worldwide. The U.S. approval of 1 was preceded by the 2006 approval of mozavaptan hydrochloride (2) in Japan. In 2009, tolvaptan (3) joined 1 as an FDA-approved agent for the treatment of hyponatremia. In this chapter, the pharmacological profile and synthesis of conivaptan hydrochloride (1) is examined in detail.

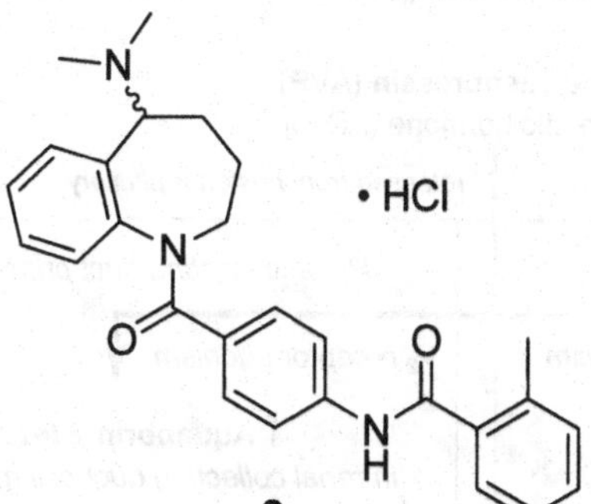

USAN: Mozavaptan hydrochloride
Trade name: Physuline®
Otsuka Pharmaceutical Co.
Launched: 2006 (Japan)

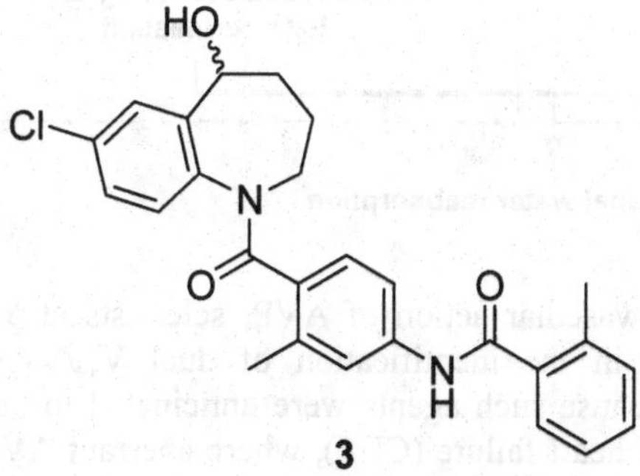

2

USAN: Tolvaptan
Trade name: Samsca®
Otsuka Pharmaceutical Co.
Launched: 2009

3

13.2   Pharmacology

Arginine vasopressin, the antidiuretic hormone in humans, is a nonapeptide hormone produced by the posterior pituitary in response to elevated plasma molality and decreased blood volume.   Vasopressin maintains proper plasma volume and molality by both promoting water reabsorption in the renal collecting ducts and regulating peripheral vasculature tone.   These effects are mediated by two distinct G protein-coupled receptors: the V_{1a} receptor, present in vascular smooth muscle and interstitial cells of the renal medulla, and the V_2 receptor, principally expressed in the renal collecting duct.   The term *vasopressin* stems from the potent vasoconstrictive (pressor) properties of this hormone, which are mediated by agonism of the V_{1a} receptor.   Vasopressin's water-retentive properties, in contrast, arise from agonism of renal V_2 receptors, which results in an increase in the number of water-specific aquaporin-2 channels present in the apical membranes of cells lining the renal collecting ducts.   This increase in the number of water-specific channels significantly enhances the permeability of the renal collecting duct to water, greatly augmenting renal reabsorption of electrolyte-free water.   Although V_2 receptor agonism mediates the most prominent renal response to vasopressin, agonism of renal V_{1a} receptors enhances renal water reabsorption by reducing medullary blood flow, thereby increasing medullary osmolality and augmenting renal water reabsorption (Figure 1).[13]

Figure 1. Effects of vasopressin on renal water reabsorption

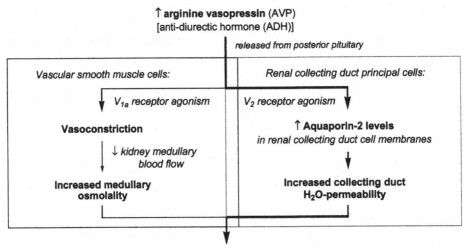

Due to the dual renal and vascular action of AVP, scientists at Yamanouchi Pharmaceuticals became interested in the identification of dual V_{1a}/V_2 vasopressin receptor antagonists, particularly because such agents were anticipated to be of unique utility in the treatment of congestive heart failure (CHF), where aberrant AVP secretion appeared responsible for both the onset of hypervolemic hyponatremia and deleterious increases in vascular resistance.[14] The resulting drug discovery program ultimately lead to the identification of conivaptan•HCl (1).

In CHO cells transfected with human V_{1a} and V_2 vasopressin receptors, 1 inhibits [^{3}H]-AVP binding with K_i's of 4.3 and 1.9 nM, respectively.[15] Compound 1 demonstrates similar activity on rat V_{1a} and V_2 receptors, with K_i's of 0.48 nM and 3.0 nM (Table 1). As a result of significant structural homology between the vasopressin and the oxytocin receptors, 1 and AVP also demonstrate significant oxytocin receptor affinities (rat receptor K_i's of 44.4 nM and 3.4 nM, respectively).[16] As seen in Table 1, the balanced binding affinities of 1 toward rat V_{1a} and V_2 receptors closely parallel those of AVP; in contrast, vasopressin receptor antagonists mozavaptan hydrochloride (2) and tolvaptan (3) demonstrate moderate to significant V_2 receptor selectivity.

Table 1. Rat V_{1a}, V_2 and Oxytocin (OT) Receptor Binding Affinities

Compound	V_{1a} (K_i, nM)[16,17]	V_2 (K_i, nM)[16,17]	OT (K_i, nM)[16]	Selectivity (V_{1a}/V_2)
Arginine vasopressin (AVP)	1.1	3.2	3.4	**0.34**
Conivaptan•HCl (1)	0.48	3.0	44.4	**0.16**
Mozavaptan•HCl (2)	193	42	1550	**4.6**
Tolvaptan (3)	325	1.3	n.a.	**250**

Functionally, **1** inhibits AVP-induced calcium release in hV_{1a}-transfected CHO cells with an IC_{50} of 0.43 nM and AVP-induced cAMP production in hV_2-transfected CHO cells with an IC_{50} of 0.39 nM.[15] In the rat smooth muscle-derived A10 cell line, **1** inhibits V_{1a}-mediated intracellular calcium release with an IC_{50} of 1.2 nM, and in porcine kidney-derived LLC-PK$_1$ cells, **1** inhibits V_2-mediated cAMP production with an IC_{50} of 17.3 nM.[16]

In vivo, i.v. administration of **1** to rats dose-dependently inhibited AVP-induced increases in blood pressure with an ID_{50} of 13 µg/kg. Administration of **1** also significantly increased urine volume (ED_3 of 28 µg/kg)[18] and reduced urine osmolality in a dose-dependent manner.[16] Compound **1** proved orally active in rats, dose-dependently inhibiting AVP-induced increases in diastolic blood pressure and providing dose-dependent diuretic effects that were sustained for 8–10 h after a 3-mg/kg dose.[19] Similar results were obtained in dog pharmacodynamic studies.[20] Notably, daily dosing of rats with **1** (1 or 3 mg/kg for 1 week) resulted in sustained increases in urine volume without sodium excretion, with no evidence of tachyphylaxis.[21]

13.3   Structure–Activity Relationship (SAR)

In work leading to the discovery of the V_2-selective vasopressin antagonist mozavaptan hydrochloride (**2**),[22] researchers at Otsuka Pharmaceuticals reported that des-dimethyl-amino mozavaptan (**4**) displayed comparable affinity for both V_{1a} and V_2 receptors and that substitution of the terminal benzamide moiety with lipophilic groups generally enhanced vasopressin receptor binding affinities. This report prompted Yamanouchi scientists, in their search for a dual V_{1a}/V_2 receptor antagonist, to prepare a range of derivatives of **4**, bearing lipophilic substituents on the terminal benzamide ring. These studies led to the identification of 2-phenybenzamide **5**, which possessed not only similar V_{1a} and V_2 receptor affinities to **4** but also modestly enhanced in vivo activity (as measured by increased in urine volume after oral dosing at 10 mg/kg).[23]

To enhance the water solubility and oral bioavailability of **5**, Yamanouchi scientist next examined the introduction of amino groups at the 4- and 5-positions of the benzazepine ring.[24] Although 1,4-benzodiazepine analogs displayed significantly reduced V_{1a} and V_2 binding affinities, 1,5-benzodiazepine **6** demonstrated enhanced binding affinities toward both receptor isoforms. Substitution of the 5-amino group of **6** with a range of pyridylmethyl, carbamoylmethyl, and aminoalkyl groups in an attempt to further enhance the water solubility of **6** led to the identification of the diethylaminoethyl derivative **7**, which possessed not only similar V_{1a} and V_2 affinity to **6**, but also dramatically enhanced water solubility and oral bioactivity (as reflected in the increase in urine volume).[24]

Restricting the orientation of solubilizing group attached to the 5-position of the benzodiazepine ring through the replacement of the carbon–nitrogen bond with an (E)-alkylmethylidene group was found to further enhance V_{1a} and V_2 binding affinities; however, the resulting compounds were prone to acid- and base-mediated isomerization.[25] To more robustly constrain the orientation of the benzodiazepine 5-substituent, a series of analogs in which this substituent was tethered to the 4-position was prepared.[26] The thiazolobenzazepine ring system (cf., **8**) proved to be a useful template for these studies. Starting from this template, a range of 2-substituted thiazole analogs bearing pyridylmethyl and aminoalkyl solubilizing groups was prepared, revealing 2-aminoalkyl substituted thiazolobenzazepines to be potent binders of both V_{1a} and V_2 (e.g., compound **8**). Further examination of alternatives to the thiazole ring revealed that replacement of the thiazole ring with an imidazole ring uniformly led to enhanced vasopressin receptor binding affinities.[26] Optimization of the chainlength of the aminoalkyl 2-substituent subsequently led to the identification of imidazobenzazepine **9**, which displayed balanced subnanomolar binding affinities for both V_{1a} and V_2 receptors, as well as significantly enhanced oral activity (3.15 mL urine volume increase after oral dosing at 3 mg/kg). Further study of imidazobenzazepine analogs of **9**, however, revealed that replacement of the 2-aminobutyl group with a methyl group (cf., **1**) led to a dramatic increase in oral activity (13.3 mL increase in urine volume after a 3 mg/kg oral dose, presumably as the result of increased oral bioavailability).[26] As a result of its potent V_{1a} and V_2 receptor binding affinities and oral activity, imidazobenzazepine **1** (conivaptan•HCl) became the focus of further pharmacological investigation.

Table 2. Vasopressin Receptor Antagonist Activities In Vitro and In Vivo.[23,24,26]

Entry	Compound	V_{1a} $(K_i, nM)^a$	V_2 $(K_i, nM)^b$	UV $(mL)^c$
1	Mozavaptan•HCl (**2**)	195	9.8	6.65
2	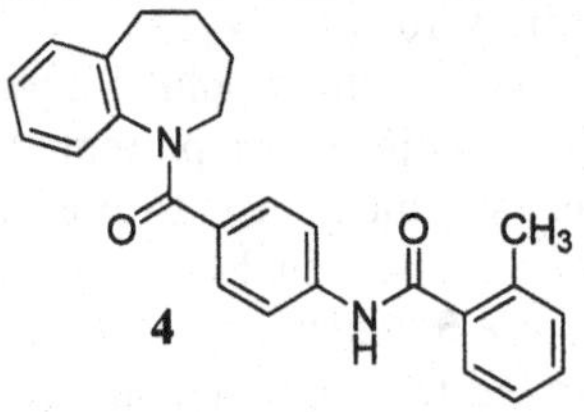 **4**	8.1	7.2	0.38
3	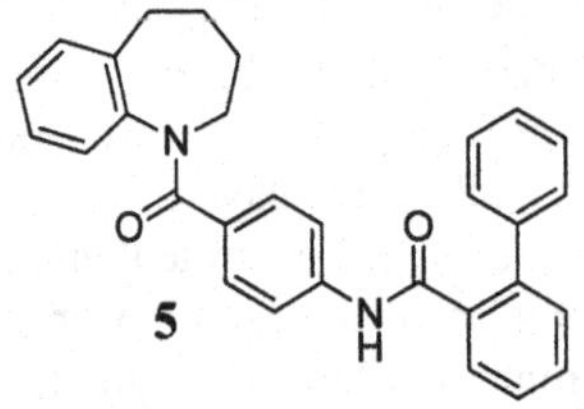 **5**	14	7.6	0.67
4	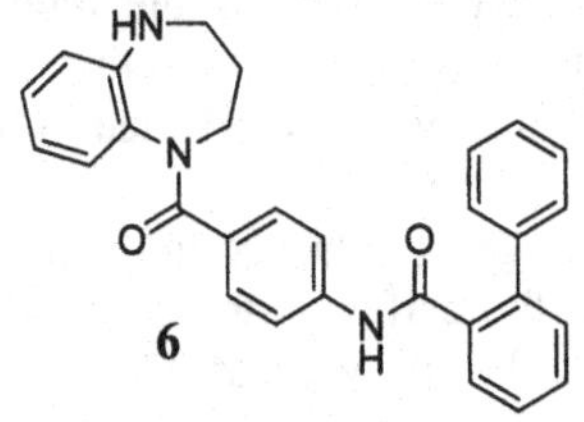 **6**	1.4	1.7	0.76

5	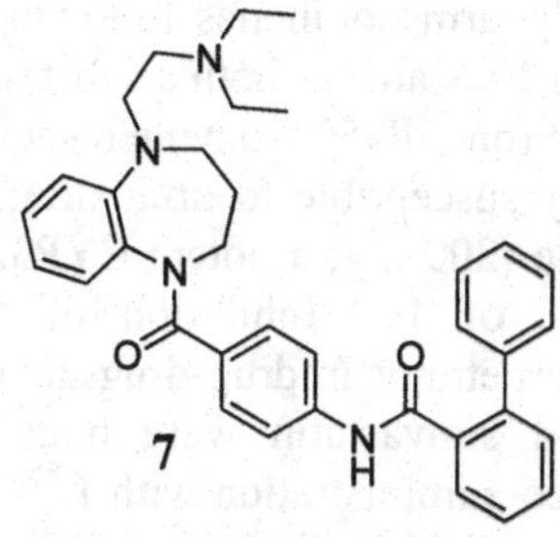 **7**	5.4	5.1	8.13
6	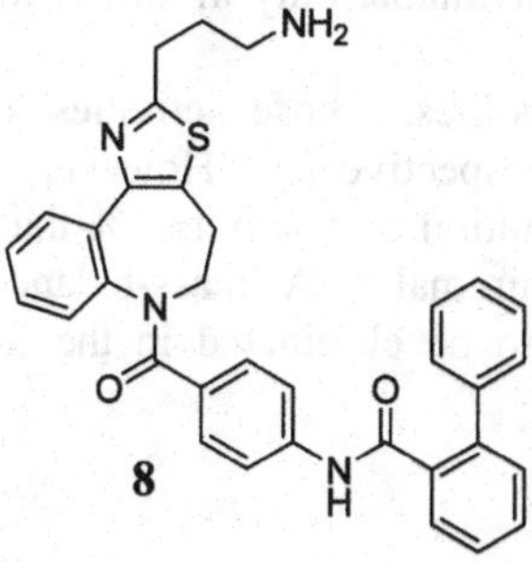 **8**	4.0	1.1	1.80
7	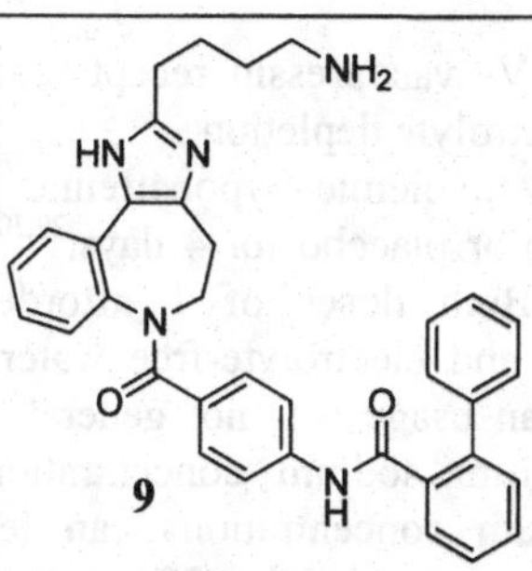 **9**	0.5	0.5	3.15 (3 mg/kg)
8	Conivaptan•HCl (**1**)	0.9	1.5	13.3 (3 mg/kg)

[a] Determined by displacement of [^{3}H]vasopressin from rat liver cell plasma membranes.
[b] Determined by displacement of [^{3}H]vasopressin from rabbit kidney cell plasma membranes.
[c] Mean urine volume (mL) during the 2 h after oral administration (10 mg/kg, unless otherwise noted) of compound to rats.

13.4 Pharmacokinetics and Drug Metabolism

The pharmacokinetic properties of conivaptan•HCl (**1**) in rats and dogs have not been reported in the literature. In humans, peak plasma concentrations of **1** occurred 0.67–2 h after oral administration to healthy males, with conivaptan•HCl (**1**) showing 34–55% oral bioavailability.[27] Intravenous infusion of **1** revealed a mean terminal elimination half-life of 5 h and mean clearance of 15.2 L/h,[28] along with a volume of distribution of 34 L.[27] At plasma concentrations of 10–1000 ng/mL, conivaptan is > 99% bound to human plasma proteins.[28]

Conivaptan•HCl (1) displays nonlinear pharmacokinetics following oral or i.v. dosing.[28] This nonlinearity appears to result from 1's status as both a substrate and potent inhibitor of CYP3A4. CYP3A4 is the sole cytochrome P450 isozyme responsible for the metabolism of 1;[28] and as such, 1 is particularly susceptible to drug–drug interactions. Co-administration of 1 (10 mg) and ketoconazole (200 mg; a potent CYP3A4 inhibitor) resulted in an 11-fold increase in the AUC of 1. Inhibition of CYP3A4 by conivaptan•HCl (1) can also lead to 1 being a perpetrator in drug-drug interactions: The AUC's of CYP3A4 substrates midazolam and simvastatin were both significantly increased (2-fold and 3-fold, respectively) upon co-administration with 1.[28] To minimize the risk of adverse drug–drug interactions, 1 is available only in an i.v. formulation for use in an in-patient hospital setting.

CYP3A4 metabolizes 1 to four metabolites, whose activities at V_{1a} and V_2 receptors are 3–50% and 50–100% that of 1, respectively.[28] However, the combined exposure to these metabolites after i.v. administration of 1 is only 7% that of parent, so their contribution to the clinical effect is minimal. A mass-balance study with radiolabeled 1 revealed that 83% of the dose to be eliminated in the feces, with the remainder being eliminated in the urine.[28]

13.5   Efficacy and Safety

Conivaptan•HCl (1) is a potent dual V_{1a} and V_2 vasopressin receptor antagonist that increases water excretion without significant electrolyte depletion.

In a pivotal Phase III trial, 84 nonhypovolemic hyponatremic patients were randomized to intravenous 1 (40 or 80 mg/day) or placebo for 4 days.[29,30] Dosing was preceded by a 20-mg i.v. loading dose. Both doses of 1 afforded significant improvements in serum sodium concentration and electrolyte-free water excretion as compared with placebo. In addition, conivaptan usage was not generally found to be associated with overly rapid changes in serum sodium concentrations, which is significant, as rapid changes in serum sodium concentrations can lead to neural demyelination. A subsequent study of 251 patients revealed that 20 mg/day i.v. dosing of conivaptan provided similar improvements in serum sodium concentrations and electrolyte-free water excretion.[31]

The hemodynamic effects of 1 were examined in a double-blind, single-dose study in patients with NYHA Class III/IV heart failure.[32] In a study of 142 patients randomized to receive i.v. conivaptan (10, 20, or 40 mg) or placebo, conivaptan•HCl (1) administration resulted in significant reductions in pulmonary capillary wedge pressure and right atrial pressure, along with increased urine output. Cardiac index, systemic and pulmonary vascular resistance, blood pressure, and heart rate displayed no significant changes versus placebo.[32] Ten subsequent Phase II heart failure studies, however, failed to demonstrate that conivaptan usage is associated with improvements in heart failure outcomes, such as length of hospital stay, exercise tolerance, functional status, ejection fraction, or heart failure symptoms compared to placebo.[28]

On the basis of these studies, conivaptan•HCl (1) has been approved by the FDA for the treatment of hospitalized patients with euvolemic and hypervolemic hyponatremia but is not currently indicated for the treatment of congestive heart failure.[28] Due to its aquaretic effects, conivaptan use is contraindicated in patients with hypovolemic hyponatremia, where further reductions in vascular volume could pose serious risks. As conivaptan displays adverse reproductive effects (e.g., decreased fertility), adverse effects on fetal development, and the ability to delay delivery in rats, use during pregnancy warrants considerable caution.[28] The reproductive effects of **1** may be associated with the activity of **1** on rat oxytocin receptors. The most common adverse event occurring in patients treated with **1** were infusion site reactions. Other common adverse events included headache, hypotension, nausea, and constipation.[28]

13.6 Syntheses

The initial synthetic approach to conivaptan•HCl (**1**) employed by the Yamanouchi discovery group[26] commenced with commercially available benzazepinone **10**. Acylation of **10** with *p*-nitrobenzoyl chloride provided benzamide **11**. Subsequent hydrogenation of **11** over palladium on carbon yielded aniline **12**, which was in turn condensed with biphenyl-2-carbonyl chloride to provide bis(amide) **13**. Bis(amide) **13** was subsequently heated with copper(II) bromide in boiling chloroform/ethyl acetate to furnish α-bromoketone **14**. It is interesting that condensation of α-bromoketone **14** with acetamidine hydrochloride in the presence of potassium carbonate in boiling acetonitrile afforded not only the desired imidazobenzazepine product (**1**; 53% yield, 2 steps) but also the related oxazolobenzazepine **15** (7% yield, 2 steps), which presumably resulted from nucleophilic attack of the benzazepinone oxygen on the amidine moiety followed by loss of ammonia. Separation of oxazolobenzazepine byproduct **15** from imidazobenzazepine **1** by silica gel chromatography followed by treatment of the purified imidazobenzazepine free-base with hydrochloric acid then provided conivaptan•HCl (**1**).

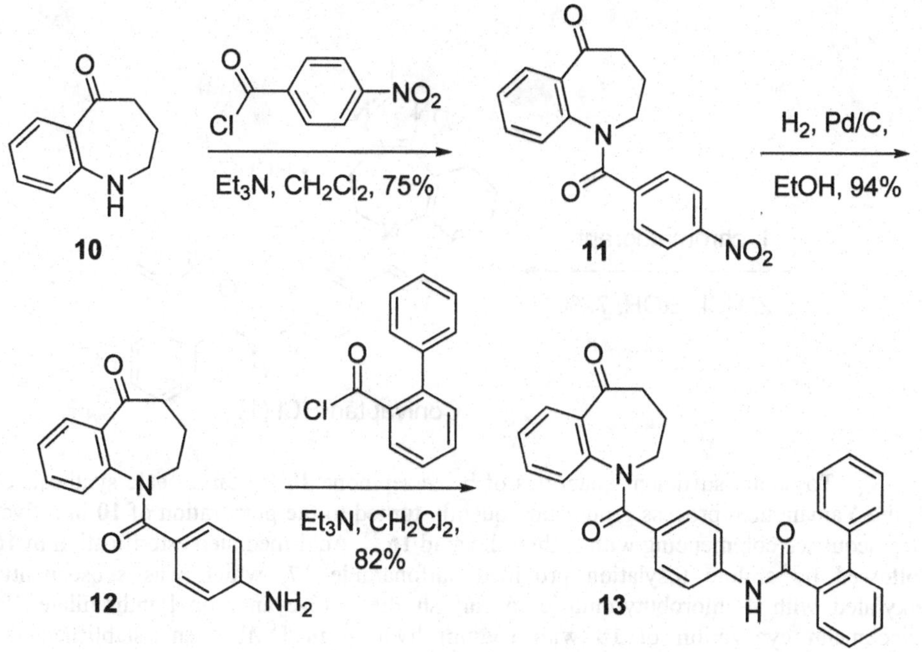

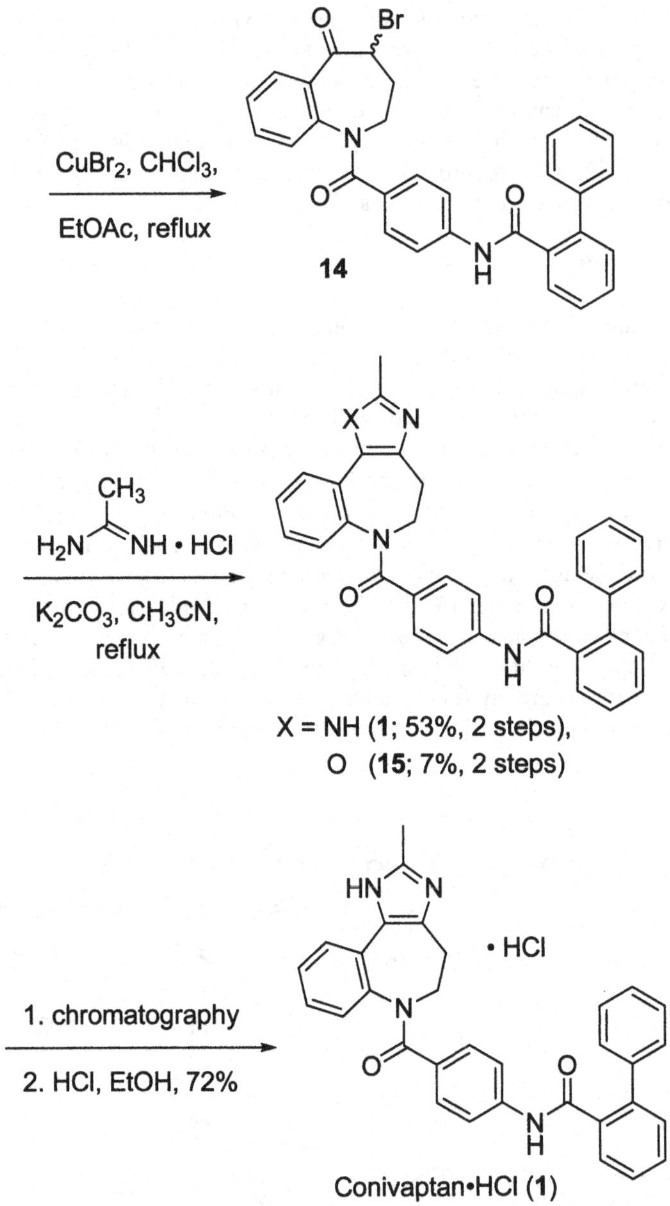

14

CuBr$_2$, CHCl$_3$,
EtOAc, reflux →

CH$_3$
H$_2$N⤴NH • HCl
───────────────
K$_2$CO$_3$, CH$_3$CN,
reflux →

X = NH (**1**; 53%, 2 steps),
O (**15**; 7%, 2 steps)

• HCl

1. chromatography
───────────────
2. HCl, EtOH, 72% →

Conivaptan•HCl (**1**)

To obtain sufficient quantities of benzazepinone **10** for large-scale synthesis of **1**, the Yamanouchi process group subsequently turned to the preparation of **10** in a five-step sequence commencing with anthranilic acid **16**.[33] Acid-mediated esterification of **16** followed by aniline tosylation provided sulfonamide **17**, which was subsequently alkylated with 4-chlorobutyronitrile to furnish the *N*-(3-cyanopropyl)anthranilate **18**. Dieckmann cyclization of **18** with sodium hydride in DMF then established the benzazepinone ring system of **19**. Exposure of **19** to a mixture of boiling acetic acid and concentrated hydrochloric acid effected the sequential hydrolysis and decarboxylation of the nitrile group, along with concomitant cleavage of the tosyl protecting group, to afford benzazepinone **10** in 33% overall yield from **16**.

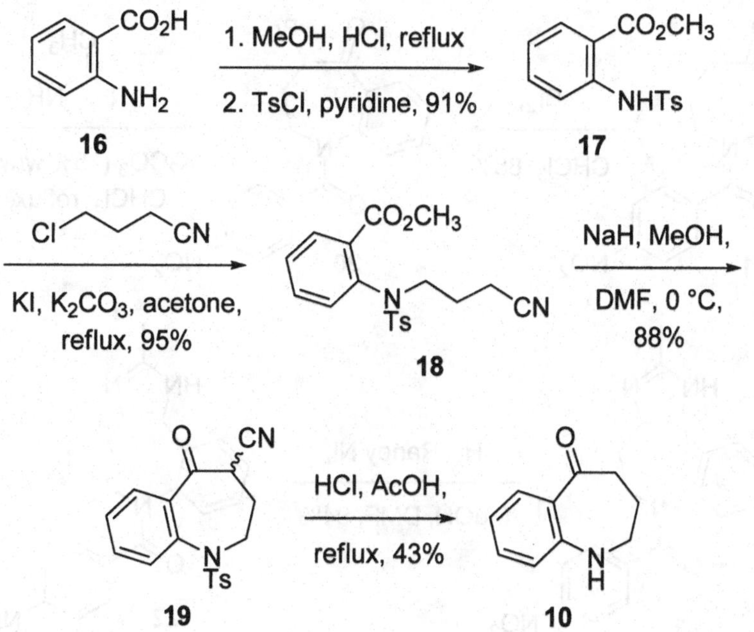

Access to **10** allowed the Yamanouchi process group to turn their attention to two shortcomings of the discovery synthesis: (1) capricious imidazole ring formation (**14→1**), which led to the formation of oxazolobenzazepine byproduct **15** and necessitated the chromatographic purification of **1**, and (2) the installation biphenyl-2-carbonyl chloride (derived from costly biphenyl-2-carboxylic acid) early in the synthetic sequence, which led to considerable waste of this costly material through losses over the subsequent steps of the synthesis.

In the process group's revised synthetic approach to **1**,[33] the order of imidazole ring formation and biphenyl-2-carbonyl chloride introduction were reversed. Thus *p*-nitrobenzamide **11** was initially brominated with bromine in chloroform to provide α-bromoketone **20**. Careful investigation of the condensation of **20** with acetamidine hydrochloride revealed this reaction to be extremely sensitive to the moisture content of the potassium carbonate. Yields of imidazobenzazepine **21** were found to be directly proportional to water content, with anhydrous potassium carbonate affording predominantly the undesired oxazolobenzazepine byproduct in a 1.3:1 mixture with **21**. In contrast, use of potassium carbonate containing 15% (w/w) water provided predominantly the desired imidazobenzazepine **21** in a 10:1 mixture with undesired oxazolobenzazepine. The authors suggest that the sensitivity of the product distribution to water content may reflect shifts in the keto/enol equilibrium of the intermediate α-acetamidino ketone, with increased water content leading to decreased concentrations of the enol tautomer presumably responsible for oxazolobenzazepine formation. Fortunately, crystallization of the 10:1 mixture of imidazobenzazepine **21** and undesired oxazolobenzazepine provided high-purity imidazobenzazepine **21** in 67% yield, obviating the need for chromatographic purification of this intermediate on large scale.

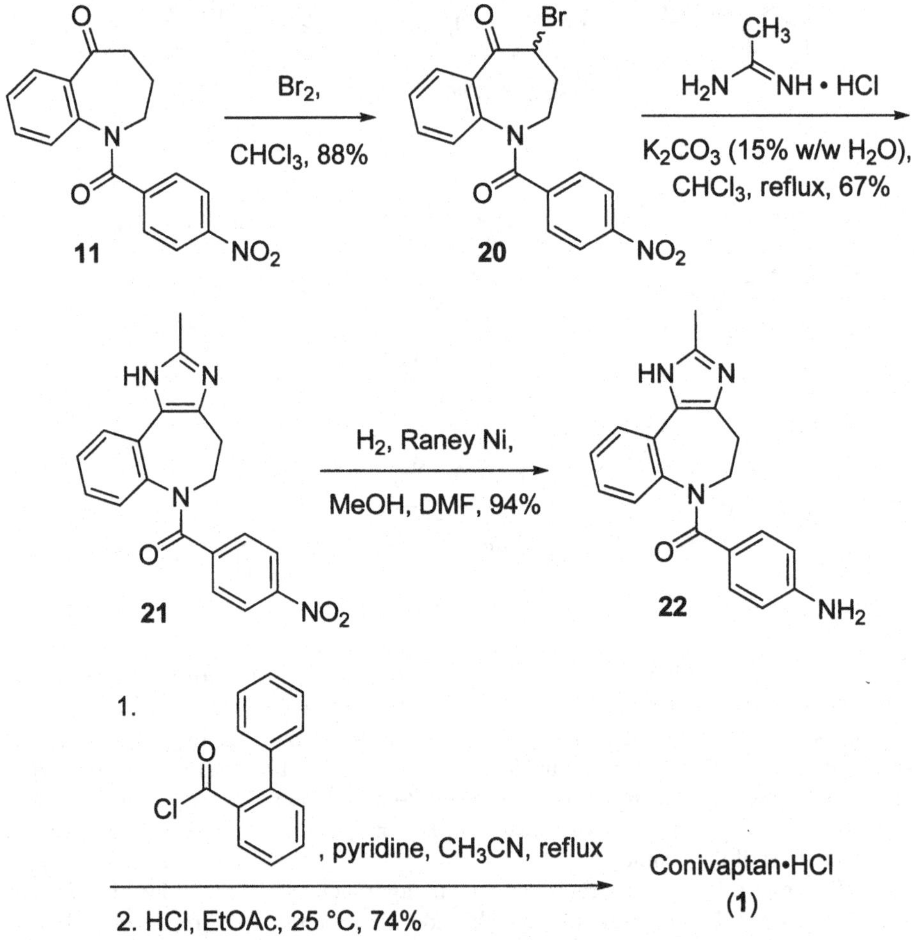

Catalytic hydrogenation of the nitro group of **21** with Raney nickel under hydrogen atmosphere followed by catalyst removal and recrystallization from methanol–water then furnished aniline **22** and set the stage for the ultimate installation of costly biphenyl-2-carbonyl chloride. Selective acylation of the aniline moiety of **22** was readily achieved by treatment with biphenyl-2-carbonyl chloride in a refluxing mixture of pyridine and acetonitrile. Subsequent addition of a solution of hydrogen chloride in ethyl acetate to the cooled reaction mixture resulted in the precipitation of conivaptan•HCl (**1**), which was isolated in 74% yield.

In 2005, the Yamanouchi process group (postacquisition by Astellas) reported further refinements of their initial synthesis of conivaptan•HCl (**1**), leading to a process for the multikilogram synthesis of **1**.[34] Key features of this second-generation process route included (1) improved overall yield, (2) increased synthetic convergence, (3) elimination of the use of chlorinated solvents, and (4) elimination of the Raney nickel hydrogenation.

N-Tosyl benzazepinone **23** served as the starting point for this synthesis. Bromination of **23** with pyridinium hydrobromide perbromide followed by recrystallization from ethanol provided α-bromoketone **24** for investigation of the key imidazole ring-forming reaction. As in the first-generation process synthesis, the use of anhydrous potassium carbonate was found to yield significant quantities (35%) of an oxazolobenzazepine byproduct. The use of either moist potassium carbonate (15% w/w water) or anhydrous potassium carbonate in the presence of 15% (w/w) liquid water largely suppressed this side reaction, however, providing imidazobenzazepine **25** in ~85% yield. It was not surprising that α-bromoketone **24** was subsequently found to be positive in an Ames test, indicating it to be a mutagen; thus attempts were subsequently made to avoid the isolation of this intermediate. Ultimately, bromination of **23** in a mixture of acetic acid and 48% hydrobromic acid followed by partition of the reaction mixture between toluene and water was found to be a suitable alternative for the preparation of α-bromoketone **24**. Acetamidine hydrochloride, potassium carbonate, and 10% (w/w) water were subsequently added to the resulting toluene solution of **24**, and this mixture was heated at 100°C. Aqueous workup of the resulting mixture, followed by recrystallization of the HCl salt of **25** from 2-propanol, provided imidazobenzazepine **25** in 69% yield (2 steps) without intermediate isolation of the α-bromoketone **24**. Compound **25** was next detosylated by incubation with 80% sulfuric acid at 80°C. Due to the high water solubility of the dihydrogen sulfate salt of **26**, imidazobenzazepine **26** was isolated as its free base by extraction into 2-butanone followed by recrystallization from an acetonitrile–water mixture.

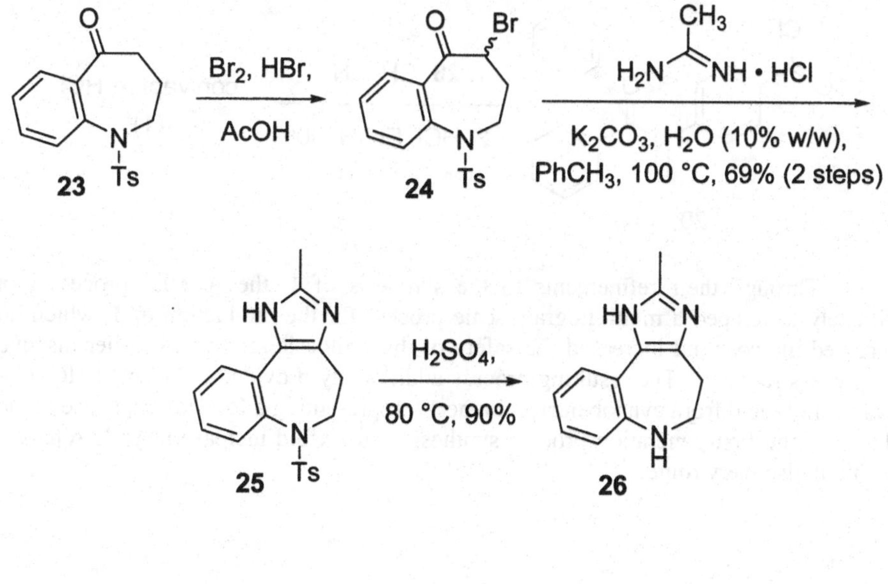

To increase synthetic convergence in the synthesis of **1**, the Astellas process group envisioned the direct acylation of **26** with a preformed 4-biphenyl-2-ylcarboxamidobenzoyl unit (cf. **29**) rather than sequentially elaborating this unit from the amino group of **26**, as had been done in the discovery and first-generation process syntheses. Such a strategy promised to reduce the manufacturing period for **1** by allowing key intermediates **26** and **29** to be prepared simultaneously in separate reactors. It was also hoped that such a strategy would obviate the need to employ a nitrobenzoic acid-derived intermediate, as Raney nickel reduction of such an intermediate would require specialized scale-up facilities.[34]

Initial efforts to prepare benzoic acid **28** from methyl or ethyl 4-aminobenzoate and biphenyl-2-carboxylic acid (**27**) afforded poor yields of **28** (48% and 7%, respectively). However, acylation of 4-aminobenzoic acid with biphenyl-2-carbonyl chloride was found to provide **28** in excellent yield (95%) when DMAP was employed as a base. Selective acylation of the anilinic nitrogen of **26** with benzoic acid **28** was accomplished in analogy with the first-generation process synthesis by conversion of **28** to the corresponding acid chloride (SOCl$_2$, CH$_3$CN) followed by acylation of **26** in acetonitrile. Subsequent addition of ethanolic hydrogen chloride to the reaction mixture resulted in the precipitation of conivaptan•HCl (**1**), which was isolated in 90% yield.

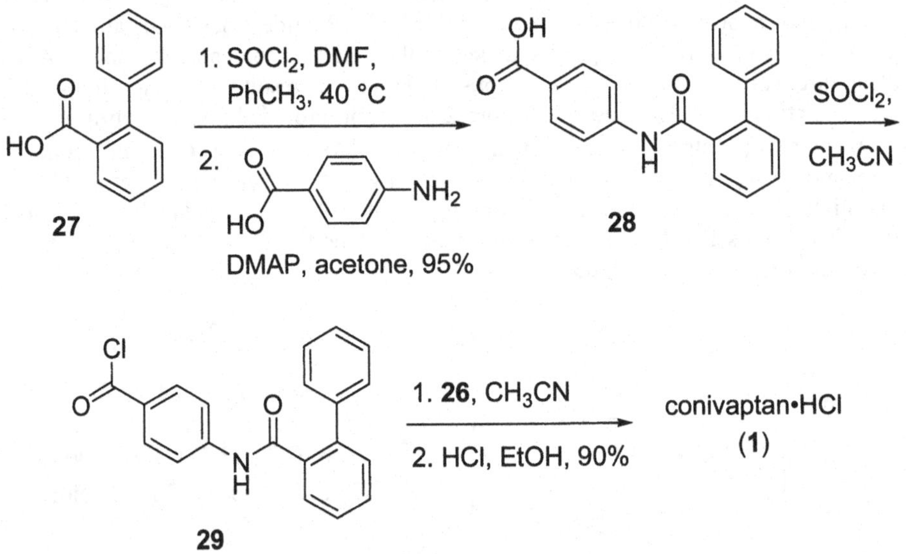

Through their refinements to the synthesis of **1**, the Astellas process group ultimately developed a multikilogram-scale process for the production of **1**, which both decreased the cost and increased the safety of the synthesis relative to earlier discovery and process routes.[34] The resulting process additionally provided conivaptan•HCl (**1**) in 56% overall yield from cyanobenzazepinone **19**, representing a four-fold increase in yield relative to the first-generation process synthesis and sixfold increase in yield relative to the initial discovery route.

In summary, conivaptan•HCl (**1**), an imidazobenzazepine dual antagonist of vascular V_{1a} and renal V_2 vasopressin receptors, represents the first FDA-approved agent for the treatment of euvolemic and hypervolemic hyponatremia to directly inhibit the anti-diuretic effect of AVP at the receptor level. Intravenous infusions of **1** predictably increase electrolyte-free water excretion and raise serum sodium concentrations with minimal risk of overly rapid serum sodium correction and neurological consequences. Although clinical trials of **1** in heart failure patients demonstrate favorable hemodynamic effects, these effects have failed to translate into significant improvements in outcomes such as exercise tolerance or length of hospital stay. Ongoing clinical trials with **1** and related vasopressin receptor antagonists such as tolvaptan (**3**) are continuing to investigate the promise of these agents in the treatment of heart failure, cirrhosis, and other fluid-retentive diseases.

13.7 References

1. Adrogué, H. J.; Madias, N. E. *N. Engl. J. Med.* **2000**, *342*, 1581–1589.
2. Flear, C. T.; Gill, G. V.; Burn, J. *Lancet* **1981**, *2*, 26–31.
3. Gheorghiade1, M.; Abraham, W. T.; Albert, N. M.; Stough, W. G.; Greenberg, B. H.; O'Connor, C. M.; She, L.; Yancy, C. W.; Young, J.; Fonarow, G. C. *Eur. Heart J.* **2007**, *28*, 980–988.
4. Lee, D. S.; Austin, P. C.; Rouleau, J. L.; Liu, P. P.; Naimark, D.; Tu, J. V. *JAMA.* **2003**, *290*, 2581–2587.
5. Goldberg, A.; Hammerman, H.; Petcherski, S.; Zdorovyak, A.; Yalonetsky, S.; Kapeliovich, M.; Agmon, Y.; Markiewicz, W.; Aronson, D. *Am. J. Med.* **2004**, *117*, 242–248.
6. Borroni, G.; Maggi, A.; Sangiovanni, A.; Cazzaniga, M.; Salerno, F. *Dig. Liver Dis.* **2000**, *32*, 605–610.
7. Lehrich, R. W.; Greenburg, A. *J. Am. Soc. Nephrol.* **2008**, *19*, 1054–1058.
8. Lewis, J. L., III. Hyponatremia. In *The Merck Manual for Healthcare Professionals* [Online]; Porter, R. S., Kaplan, J. L., Eds.; Merck Research Laboratories: Whitehouse Station, NJ, 2009. www.merck.com/mmpe/sec12/ch156/ch156d.html, accessed November 2009.
9. Miller, M. *J. Am. Geriatr. Soc.* **2006**, *54*, 345–353.
10. Ghali, J. K. *Cardiology* **2008**, *111*, 147–157.
11. Hline, S. S.; Pham, P.-T. T; Pham, P.-T. T; Aung, M. H.; Pham, P.-M. T; Pham, P.-C. T. *Ther. Clin. Risk. Manag.* **2008**, *4*, 315–326.
12. Forrest, J. N. Jr.; Cox, M.; Hong, C.; Morrison, G.; Bia, M.; Singer, I. *N. Eng. J. Med.* **1978**, *298*, 173–177.
13. Franchini, K. G.; Cowley, A. W. Jr. *Am. J. Physiol.* **1996**, *270*, R1257–R1264.
14. Farhan, A.; Guglin, M; Vaitkevicius, P.; Ghali, J. K. *Drugs* **2007**, *67*, 847–858.
15. Tahara, A.; Saito, M.; Sugimoto, T.; Tomura, Y.; Wada, K.; Kusayama, T.; Tsukada, J.; Ishii, N.; Yatsu, T.; Uchida, W.; Tanaka, A. *Br. J. Pharmacol.* **1998**, *125*, 1463–1470.

16.   Tahara, A.; Tomura, Y.; Wada, K.; Kusayama, T.; Tsukada, J.; Takanashi, M.; Yatsu, T.; Uchida, W.; Tanaka, A. *J. Pharmacol. Exp. Ther.* **1997**, *282*, 301–308.

17.   Yamamura, Y.; Nakamura, S.; Itoh, S.; Hirano, T.; Onogawa, T.; Yamashita, T.; Yamada, Y.; Tsujimae, K.; Aoyama, M.; Kotosai, K.; Ogawa, H.; Yamashita, H.; Kondo, K.; Tominaga, M.; Tsujimoto, G.; Mori, T. *J. Pharmacol. Exp. Ther.* **1998**, *287*, 860–867.

18.   ED$_3$ refers to the dose required to produce a 3-mL increase in urine volume.

19.   Tomura, Y.; Tahara, A.; Tsukada, J.; Yatsu, T.; Uchida, W.; Iizumi, Y.; Honda, K. *Clin. Exp. Pharmacol. Physiol.* **1999**, *26*, 399–403.

20.   Yatsu, T.; Tomura, Y.; Tahara, A.; Wada, K.; Tsukada, J.; Uchida, W.; Tanaka, A.; Takenaka, T. *Eur. J. Pharmacol.* **1997**, *321*, 255–230.

21.   Risvanis, J.; Naitoh, M; Johnston, C. I.; Burrell, L. M. *Eur. J. Pharmacol.* **1999**, *381*, 23–30.

22.   Ogawa, H.; Yamashita, H.; Kondo, K.; Yamamura, Y.; Miyamoto, H.; Kan, K.; Kitano, K.; Tanaka, M.; Nakaya, K.; Nakamura, S.; Mori, T.; Tominaga, M.; Yabuuchi, Y. *J. Med. Chem.* **1996**, *39*, 3547–3555.

23.   Matsuhisa, A.; Tanaka, A.; Kikuchi, K.; Shimada, Y.; Yatsu, T.; Yanagisawa, I. *Chem. Pharm. Bull.* **1997**, *45*, 1870–1874.

24.   Matsuhisa, A.; Koshio, H.; Sakamoto, K.; Taniguchi, N.; Yatsu, T.; Tanaka, A. *Chem. Pharm. Bull.* **1998**, *46*, 1566–1579.

25.   Matsuhisa, A.; Kikuchi, K.; Sakamoto, K.; Yatsu, T.; Tanaka, A. *Chem. Pharm. Bull.* **1999**, *47*, 329–339.

26.   Matsuhisa, A.; Taniguchi, N.; Koshio, H.; Yatsu, T.; Tanaka, A. *Chem. Pharm. Bull.* **2000**, *48*, 21–31.

27.   Burnier, M.; Fricker, A. F.; Hayoz, D.; Nussberger, J.; Brunner, H. R. *Eur. J. Clin. Pharmacol.* **1999**, *55*, 633–637.

28.   Vaprisol (conivaptan hydrochloride injection) [package insert]. Deerfield, IL: Astellas Pharma US, Inc.: **2008**.

29.   Verbalis, J. G.; Bisaha, J. G.; Smith, N. *J. Card. Fail.* **2004**, *10*, S27.

30.   Zeltser, D.; Rosansky, S.; van Rensburg, H.; Verbalis, J. G.; Smith, N. *Am. J.Nephrol.* **2007**, *27*, 447–457.

31.   Verbalis, J. G.; Rosansky, S.; Wagoner, L. E.; Smith, N.; Barve, A.; Andoh, M. *Crit. Care Med.* **2006**, *34*, A64.

32.   Udelson, J. E.; Smith, W. B.; Hendrix, G. H.; Painchaud, C. A.; Ghazzi, M.; Thomas, I.; Ghali, J. K.; Selaru, P.; Chanoine, F.; Pressler, M. L.; Konstam, M. A. *Circulation* **2001**, *104*, 2417–2423.

33.   Tsunoda, T.; Yamazaki, A.; Iwamoto, H.; Sakamoto, S. *Org. Process Res. Dev.* **2003**, *7*, 883–887.

34.   Tsunoda, T.; Yamazaki, A.; Mase, T.; Sakamoto, S. *Org. Process Res. Dev.* **2005**, *9*, 593–598.

14

Rivaroxaban (Xarelto):
A Factor Xa Inhibitor for the Treatment of Thrombotic Events

Ji Zhang and Jason Crawford

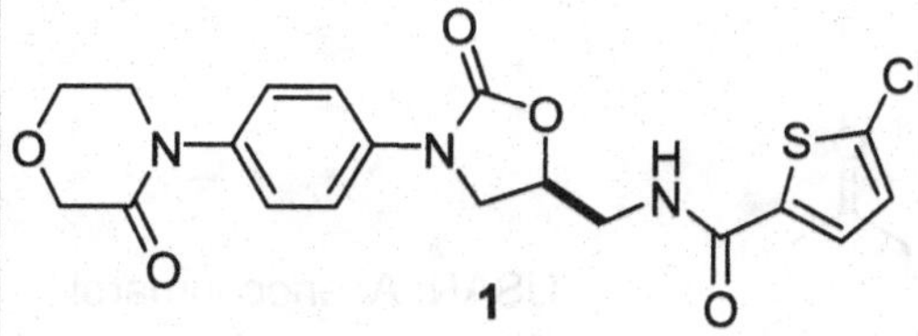

USAN: Rivaroxaban
Trade name: Xarelto®
Bayer/Johnson & Johnson
Launched: 2008(EU)

14.1 Background

Thrombotic (blood clot) events, and subsequent complications, are a leading cause of morbidity and mortality in the general population.[1] In 2005, it was estimated that there were more than 900,000 total venous thromboembolism events in the United States,[2] two thirds of which were acquired in hospital. More than 600,000 of those were nonfatal venous thromboembolism events. Nearly 300,000 were fatal events, including more than 2,200 cases of deep venous thrombosis and 294,000 cases of pulmonary embolism. The majority deaths (93%) were due to sudden fatal pulmonary embolism, or were a consequence of undiagnosed venous thromboembolism. It was estimated that 340,000 patients developed complications from venous thromboembolism, including 336,000 with postthrombotic syndrome and 3,300 with chronic thromboembolic pulmonary hypertension.

The two commonly administered anticoagulant therapies[3] are parenteral heparin, a highly sulfated glycosaminoglycan, and oral warfarin 2, a vitamin K antagonist (which acts by indirectly inhibiting several steps of the coagulation pathway). The major liabilities of these therapies include narrow therapeutic windows, requiring routine and frequent monitoring of patients, and large interindividual and intraindividual variabilities in dose response. It is important to note that warfarin and other coumarin derivatives, such as acenocoumarol **3** (with a short half-life) and phenprocoumon **4** (with a longer half-life) take at least 48 to 72 h to fully develop an anticoagulant effect. The dosing of warfarin is further complicated by the fact that it is known to interact with many commonly used medications and even with certain foods. Thus when any immediate effect is required, heparin and low molecular weight heparins, such as enoxaparin (trade name: Lovenox or Clexane) must be given concomitantly. Unfortunately, these anticoagulant treatments have inconveniences, limitations; and inherent side effects. Individuals exposed to heparins may develop an antibody-based immune response against the drug, which can dramatically complicate the underlying thrombotic disease and its treatment and prognosis.[4] Therefore, the development of safe and highly effective oral anticoagulants could significantly reduce the incidence of venous thromboembolism and related mortality, which will therefore provide meaningful benefits to patients and will lower the economic burden to the healthcare system.[5]

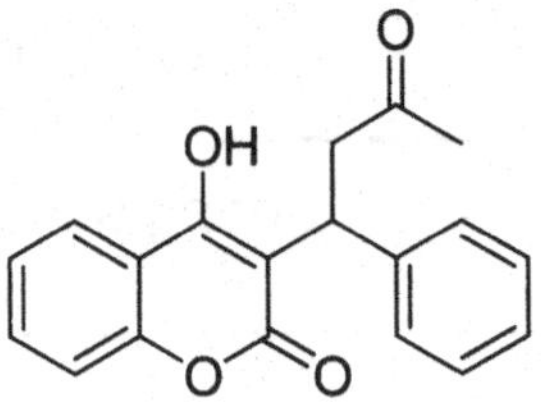

USAN: Warfarin
Trade name: Coumadin®
Jantoven®, Marevan® and Waran®
Launched: 1954

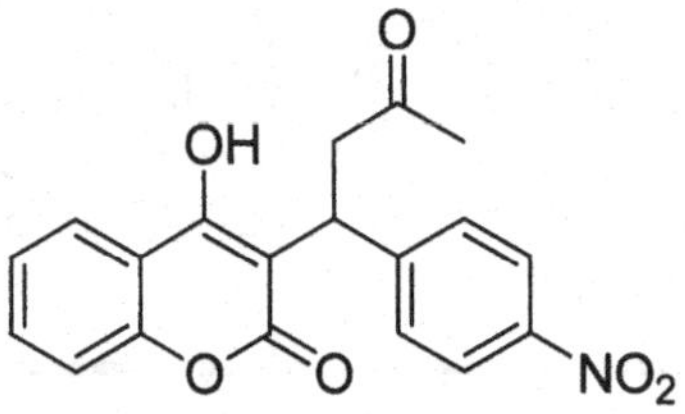

2

USAN: Acenocoumarol
Trade name: Sintrom®
or Sinthrome®
Launched: 1950s

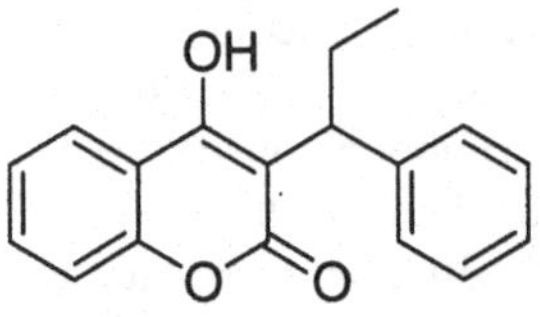

3

USAN: Phenprocoumon
Trade name: Marcoumar®
Marcumar® or Falithrom®
Launched: 1950s

4

To address the unmet medical need, scientists in drug discovery began to focus on factor Xa inhibitors in recent years as targets for therapy.[6] It is known that the activated serine protease Factor Xa plays a vital role in the blood coagulation cascade, as it is activated by both the intrinsic and the extrinsic coagulation pathways. Factor Xa catalyzes the conversion of prothrombin to thrombin through the prothrombinase complex. Thrombin has several thrombotic functions, including the conversion of fibrinogen to fibrin, the activation of platelets, and the feedback activation of other coagulation factors. Overall, these effects provide a feedback loop that serves to amplify the formation of thrombin. The inhibition of Factor Xa would generate antithrombotic effects by decreasing the amplified generation of thrombin, thus diminishing thrombin-mediated activation of both coagulation and platelets, without affecting existing thrombin levels. Thus the existing thrombin levels should be sufficient to ensure primary hemostasis, which would, in principle, provide a favorable safety profile. For these reasons, factor Xa has emerged as a particularly promising target for anticoagulant therapy.[7]

Rivaroxaban **1**, originally named BAY 59-7939 is an orally bioavailable member of a new class of highly potent and novel factor Xa inhibitors (oxazolidinone derivatives) that has been developed by Bayer HealthCare and Ortho-McNeil Pharmaceutical Inc. (Johnson & Johnson). The compound was granted marketing approval in Canada, Germany, and the UK, and was launched under the brand name Xarelto in Europe (2008). In March 2009, the U.S. FDA's Cardiovascular and Renal Drugs Advisory Committee agreed to support the approval of Rivarobaxan. It is noteworthy that if approved, it will mark the first approval of an oral anticoagulant drug since the FDA approved warfarin in 1954. The approval of rivaroxaban marks a significant advance in the treatment of thromboembolic diseases. In this chapter, the pharmacological profile and syntheses of rivaroxaban **1** is profiled in detail.[8]

14.2   Pharmacology

Lead optimization of a series of oxazolidinone derivatives led to the discovery of rivaroxaban **1**.[9] Basic screening tests for this compound demonstrated a highly potent and selective, direct inhibitory action on FXa (IC_{50} = 0.7 nM, K_j = 0.4 nM), excellent in vivo antithrombotic activity and a good pharmacokinetic profile in preliminary studies in animal models (Table 1). The lipophilic chlorothiophene moiety in rivaroxaban is responsible for a decrease in unbound fraction and aqueous solubility and attempts to identify less lipophilic replacements by broad variation were not successful.

The other unique feature of rivaroxaban **1** is the lack of a highly basic group in its active-site binding region, which is an important contributing factor to the oral absorption profile. A high relative bioavailability has been confirmed for rivaroxaban in healthy volunteers.[10] In the clinical study, two major single-center, dose-escalation, placebo-controlled, single-blinded Phase I studies have evaluated the pharmacology of this drug.[11] Both single-dosing and multiple-dosing regimens have been evaluated, and specific pharmacokinetic and pharmacodynamic properties have been elucidated.

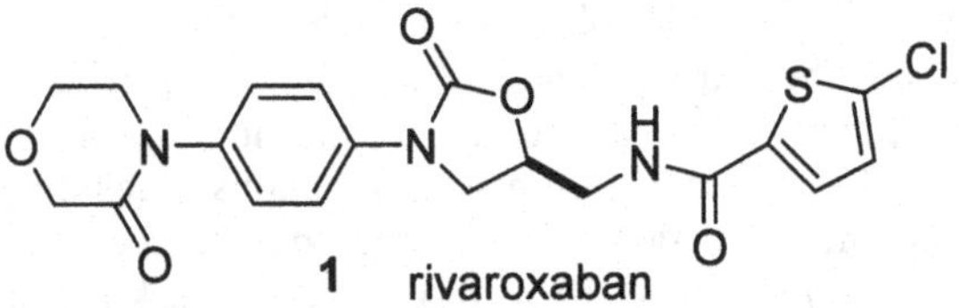

1   rivaroxaban

It was found that rivaroxaban competitively inhibits human FXa and prothrombinase activity (IC_{50} = 2.1 nM). It inhibits endogenous FXa more potently in human and rabbit plasma (IC_{50} = 21 nM) than in rat plasma (IC_{50} = 290 nM). Rivaroxaban has demonstrated anticoagulant effects in human plasma, doubling the prothrombin time (PT) and activated partial thromboplastin time (APTT) at 0.23 and 0.69 µM, respectively. In vivo, rivaroxaban reduced venous thrombosis dose dependently (ED_{50} = 0.1 mg/kg i.v.) in a rat venous stasis model. The pharmacological actions of rivaroxaban have been described in more detail by Perzborn et al.[12]

14.3   Structure–Activity Relationship (SAR)

The SAR around rivaroxaban **1** was exhaustively investigated by Roehrig and co-workers[9] in Bayer and **1** was found to be the most potent member in the series with excellent in vitro activity (IC_{50} 0.7 nM) and good oral bioavailability (60% in male Wistar rats and 60–86% in female Beagle dogs).

Optimization of the lead compound **5** (IC_{50} 120 nM), derived from high throughput screening (HTS), gave isoindolinone **6** (IC_{50} 8 nM) as a potent FXa inhibitor (Figure 1). Unfortunately, in this series, an ideal pharmacokinetic profile could not be achieved due to the low bioavailabilities. Later, it was learned that the 5-chlorothiophene-2-carboxamide moiety was essential for potent FXa inhibition, and when the HTS hits were reevaluated, the oxazolidinone derivatives were identified for further optimization. It was surprising that although **7** (IC_{50} 20 µM) is a very weak FXa inhibitor, the replacement of thiophene moiety in **7** with 5-chlorothiophene provided lead compound **8** (IC_{50} 90 nM), with more than 200-fold improved potency. Based on this promising lead, further SAR optimization was initiated.

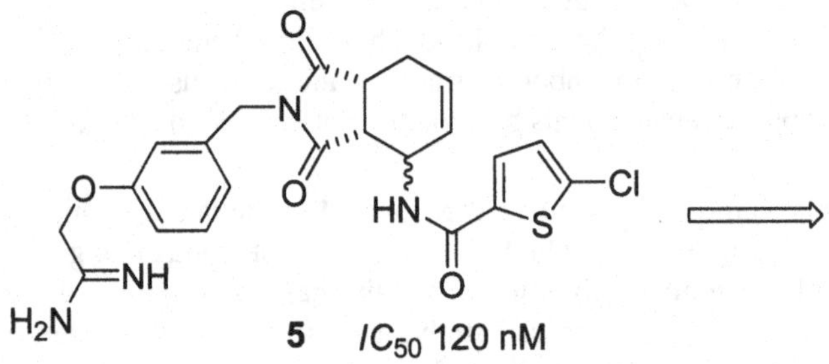

5   *IC*$_{50}$ 120 nM

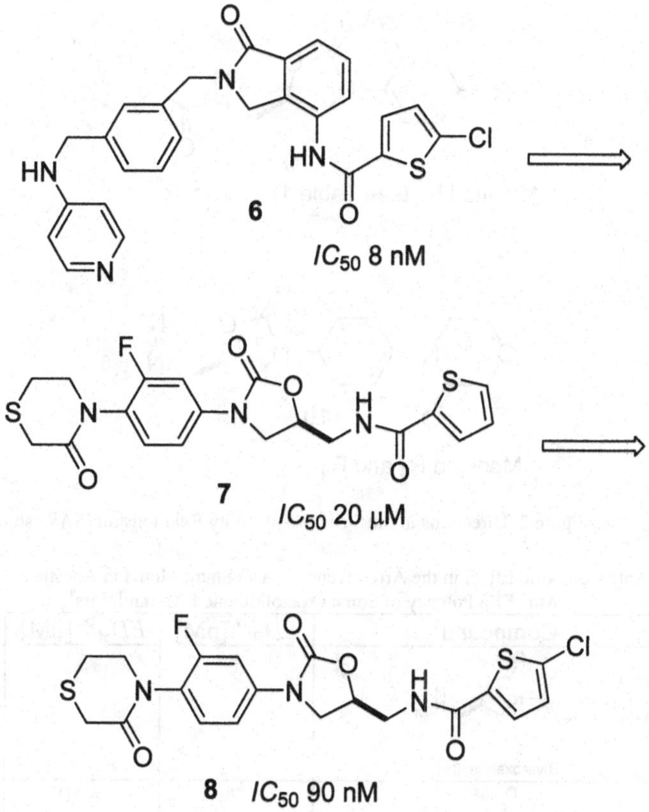

Figure 1. Lead optimization and discovery of rivaroxaban.

From compound **8**, a number of structural analogs were produced, altering the positions outlined in Figure 2. Replacement of R^1 (thiomorpholinone in **8**, 90 nM) with morphilinone or pyrrolidinone served to increase activity to 30–40 nM IC_{50} values for FXa. In contrast, *N,N*-dimethylamine (74 nM) or piperazine (140 nM) were less successful. Based on those results, incorporation of morpholin-3-one (**1**) or pyrrolidinone (**11**) were evaluated, and found to increase activity to the desired range (0.7 nM for **1**, 4 nM for **11**). Replacement of aryl protons (R^2) with fluorine or trifluoromethyl in the optimized morpholinone series did not serve to further increase activity (compounds **12** and **13**). It is interesting that installation of a methyl group at R^3 in the morpholine series substantially decreased activity (1260 nM vs. 43 nM). Unfortunately, in the second targeted SAR studies, which attempted to identify less lipophilic replacements for the 2-chlorothiophene to enhance pharmacokinetic properties, there was little success in the modification of groups R^4 and R^5 in the core structure of **10**, even by broad variation. Thus compound **1**, with subnanomolar activity, was selected for further evaluation.

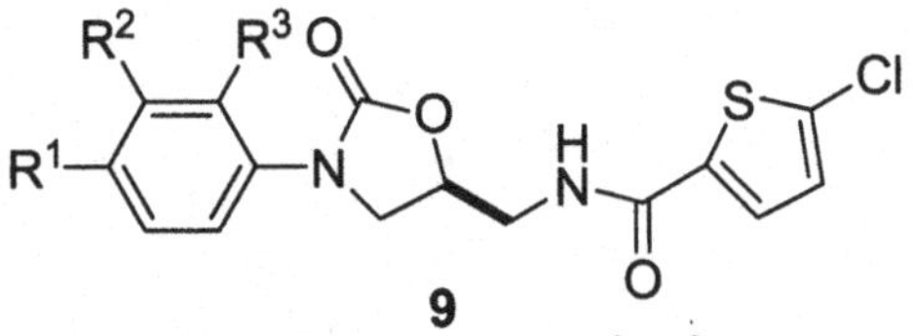

9

Modified R^1 (see Table 1) and R^2, R^3

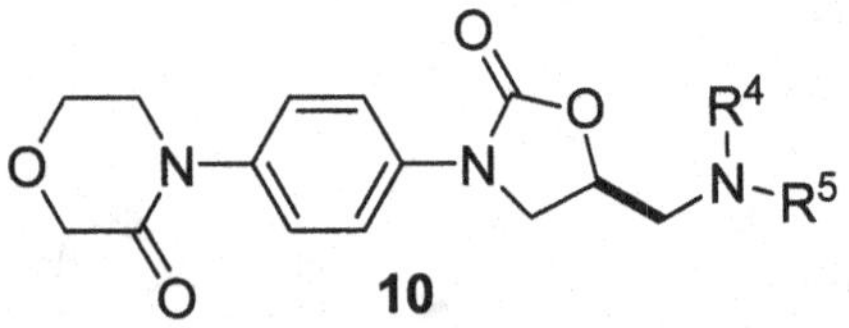

10

Modified R_4 and R_5

Figure 2. Directions in the Structure–Activity Relationship (SAR) studies.

Table 1. In Vivo Antithrombotic Effect in the Arteriovenous (AV) Shunt Model in Anesthetized Rats and In Vitro Anti-FXa Potency of Some Oxazolidinone FXa Inhibitors[9]

Compound	ED_{50}^{iv} [nM]	ED_{50}^{po} [nM]	IC_{50} [nM]
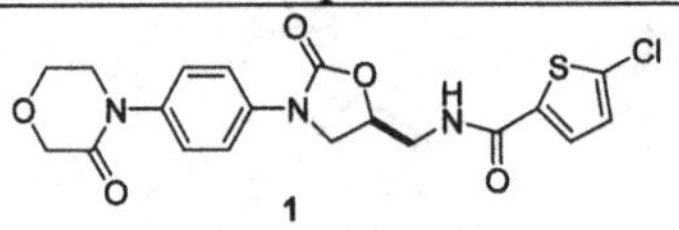 **1** Rivaroxaban	1	5	0.7
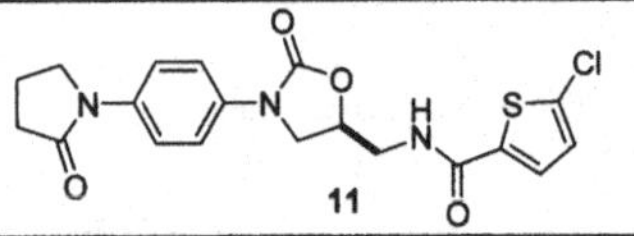 **11**	7	> 30	4.0
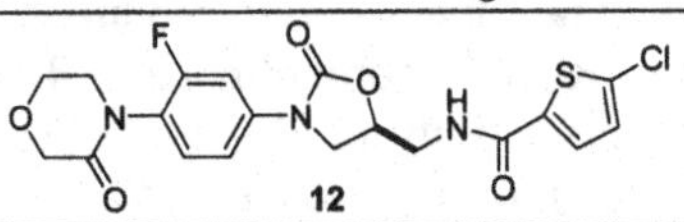 **12**	1	10	1.4
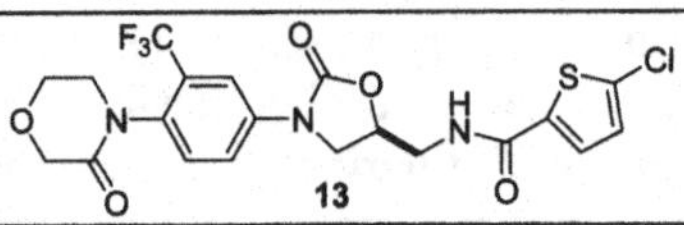 **13**	3	n.d.	1.0

Source: Ref 9.

14.4   Pharmacokinetics and Drug Metabolism

In the single-dose adminstration studies, rivaroxaban solution was orally absorbed and reached peak plasma concentrations in 30 minutes, and when ingested in tablet form, it took two hours to reach peak plasma concentrations. In the multiple daily dose regimens, rivaroxaban took three to four hours to reach peak plasma concentration. The terminal half-life varies from 3.7–9.2 h in the multiple-dose regimen and from 7–17 h in the single-dose regimen.[11]

It was found that renal excretion of rivaroxaban was lower in higher tablet dosages, and this effect was attributed to decreased solubility at higher doses. This trend may have an advantageous effect since it would reduce the risk of unintentional overdosing. Other studies tested the pharmacokinetic properties of rivaroxaban, and one study notably showed that the absorption of the drug was not altered by changes in gastric pH. This is especially important in the setting of polypharmacy in the elderly, which may frequently include co-administration of an antacid or an H2-blocker. These pharmacokinetic parameters alleviate concerns about drug absorption and demonstrate clearly that rivaroxaban is superior to older vitamin K antagonists, which, in contrast, are very sensitive to changes in gastric pH, intestinal motility and concomitant administration of food.

The metabolism and distribution of rivaroxaban were investigated after administration of single intraduodenal and oral (dogs, humans) doses of [^{14}C]-labeled drug.[13] The major compound in plasma was unchanged drug at all time points and across all species. Overall, between 78–95% of the dose administered could be detected as either unchanged drug or metabolites arising from two major metabolic pathways: oxidative degradation of the morpholino moiety and hydrolysis of the amide bond.

Table 2. Pharmacokinetic Properties of Rivaroxaban

Study design	Single-dose Administration	Multiple-dose Administration
area under the curve (AUC)	Dose proportional for tablet and solution	Dose proportional
T_{max}	30 min (for solution) 2 h (for tablets)	3–4 h (for all doses)
$t_{1/2}$	3–4 h (for solution) 7–17 h (for tablets)	Day 0: 3.7–5.8 h Day 7: 5.8–9.2 h

Source: Ref. 11.

14.5   Efficacy and Safety

Two major clinical studies[14] (621 patients who underwent elective total knee replacement, and 706 patients who underwent elective total hip replacement) have evaluated the efficacy and safety of rivaroxaban in the prophylaxis of thrombosis in patients undergoing orthopedic surgery. In these studies, rivaroxaban (2.5–10 mg b.i.d.) compared favorably with enoxaparin (40 mg once daily).

In another Phase III clinical trial,[15] 2531 patients who were to undergo total knee replacement received either oral rivaroxaban (10 mg once daily), beginning 6 to 8 hours after surgery or subcutaneous enoxaparin (40 mg q.d.), beginning 12 h before surgery. It was found that symptomatic events occurred less frequently with rivaroxaban than with enoxaparin ($P = 0.005$). Major bleeding occurred in 0.6% of patients in the rivaroxaban group and 0.5% of patients in the enoxaparin group. The incidence of mainly gastrointestinal drug-related adverse events was 12.0% in the rivaroxaban group and 13.0% in the enoxaparin group.[13]

14.6   Synthesis

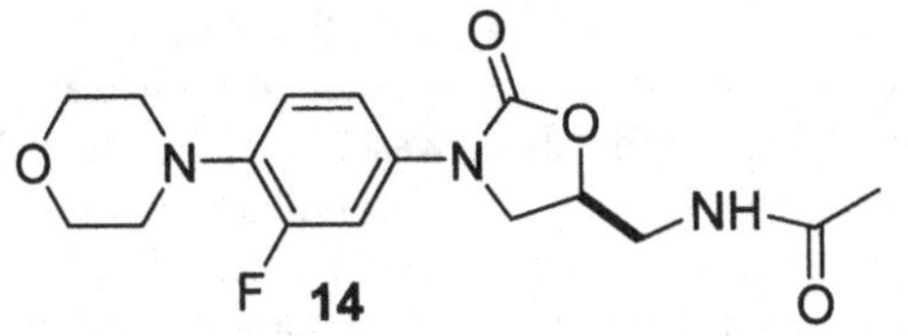

Linezolid (Zyvox®)
Pharmacia/Pfizer
Antibacterial agent
Launched: 2000

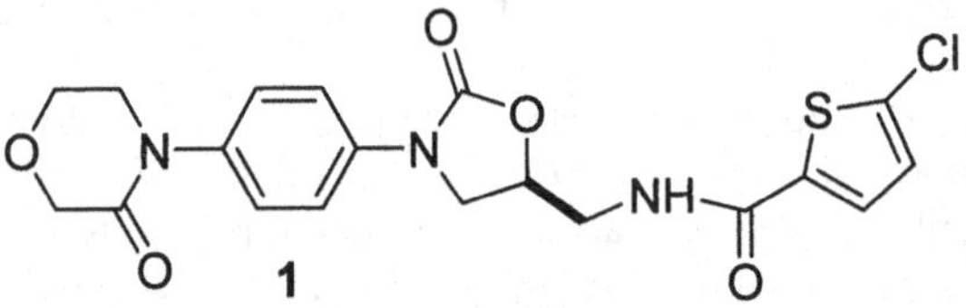

Rivaroxaban
(Xarelto®)
Bayer/J&J
Factor Xa inhibitor
Launched: 2008

Similar to linezolid (Zyvox) 14,[16] the first marketed member of a novel class of oxazolidinone antibacterial agents, the synthesis of rivaroxaban 1 used 4-fluoronitrobenzene 15 as the starting material (Scheme 1). The first synthetic approach to Rivaroxaban, developed by Roehrig and co-workers[9] at Bayer, used the S_NAr displacement of 4-fluoronitrobenzene 16 with morpholin-3-one 15 in the presence of NaH in NMP, to give N-(p-nitrophenyl)morpholinone 17. The nitro substituent was then reduced by catalytic hydrogenation over Pd/C to provide aniline 18. Subsequent coupling of aniline 18 with (S)-2-(phthalimidomethyl)oxirane 19 produced the aminoalcohol adduct 20 in 92% yield, which then cyclizes to the oxazolidinone 21 upon treatment with N,N'-carbonyldiimidazole (CDI) in 87% yield. After deprotection of the N-phthaloyl group in the presence of methylamine in aqueous EtOH (or $NH_2NH_2 \cdot H_2O$ in refluxing MeOH), the resulting primary amine 22 is acylated by 5-chlorothiophene-2-carbonyl chloride to furnish the target imidazolinone compound 1 in 86% yield.

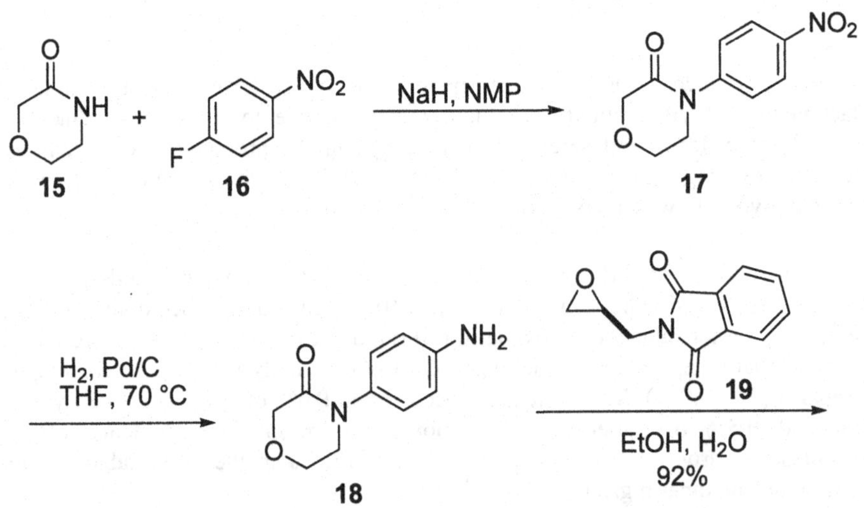

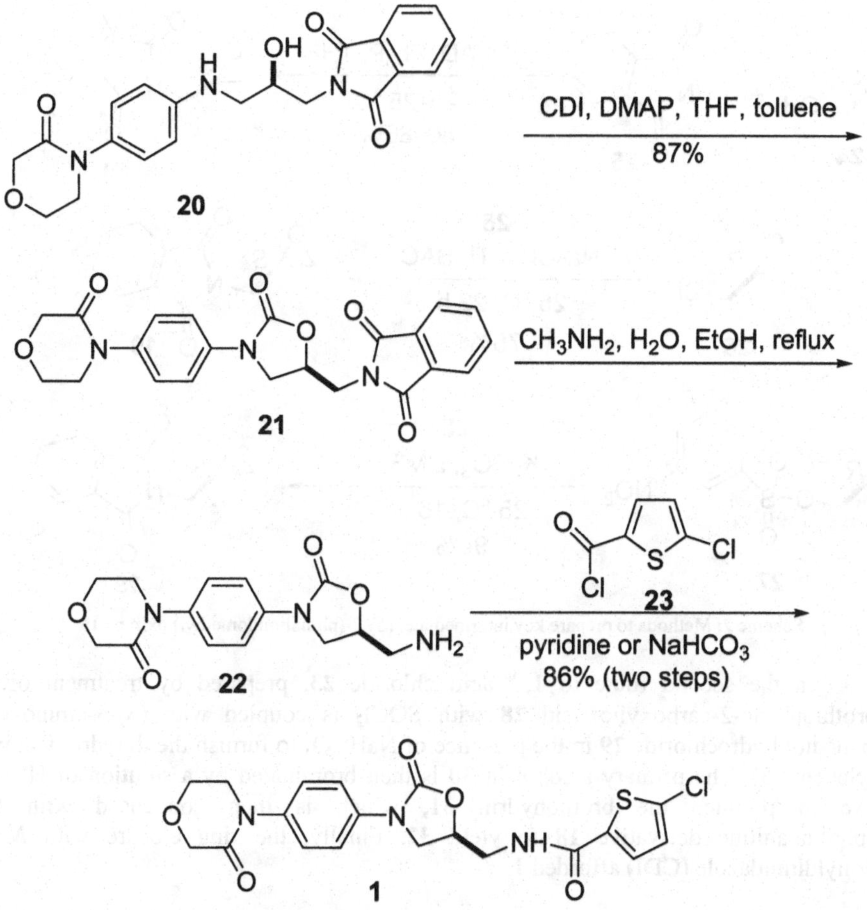

Scheme 1: The first synthetic approach to rivaroxaban

As shown in Scheme 2, (*S*)-2-(phthalimidomethyl)oxirane synthon **19** can be prepared under a variety of conditions, including a condensation of glycidol **24** and phthalimide **25** in THF at room temperature under Mitsunobu reaction conditions (80–86% yield).[17] It was reported that using trimethylbenzylammonium chloride (TMBAC) as phase-transfer catalyst, phthalimide or its potassium salt can be reacted with the inexpensive and commercially available reagent (*S*)-epichlorohydrin, **26**, to give **19** in 72–75% yield.[18] By using easily available optically active (2,3-epoxypropan-1-yl)arylsulfonate **27**, it was possible to prepare **19** in 90% yield.[19]

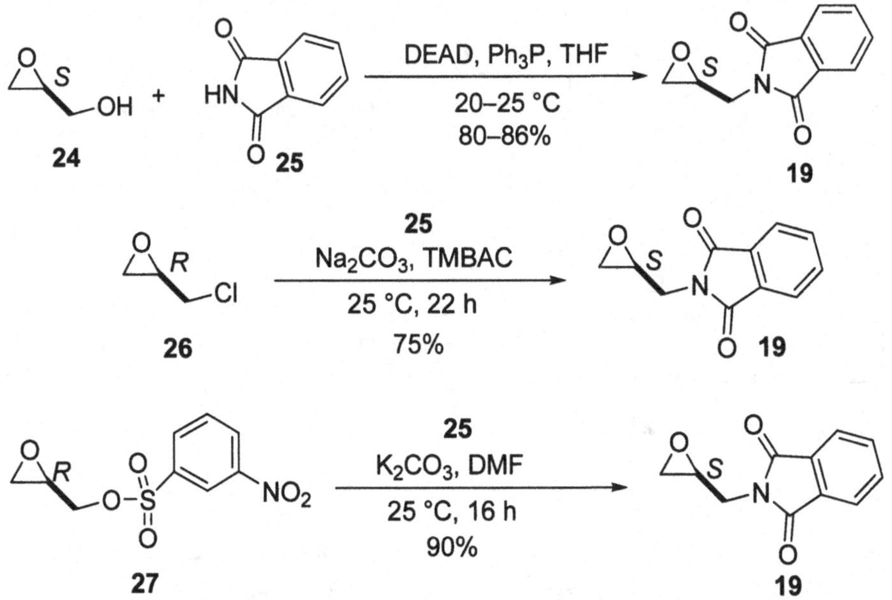

Scheme 2: Methods to prepare key intermediate, (*S*)-2-(phthalimidomethyl) oxirane **19**

In the second route to **1**,[20] acid chloride **23**, prepared by treatment of 5-chlorothiophene-2-carboxylic acid **28** with $SOCl_2$ is coupled with (*S*)-3-amino-1,2-propanediol hydrochloride **29** in the presence of $NaHCO_3$ to furnish the dihydroxy amide **30** (Scheme 3). The primary alcohol in **30** is then brominated by a solution of HBr in HOAc to produce the bromohydrin **31**, which is then condensed with the morpholinoaniline derivative **18** to yield **32**. Finally, the ring closure with *N,N'*-carbonyldiimidazole (CDI) afforded **1**.

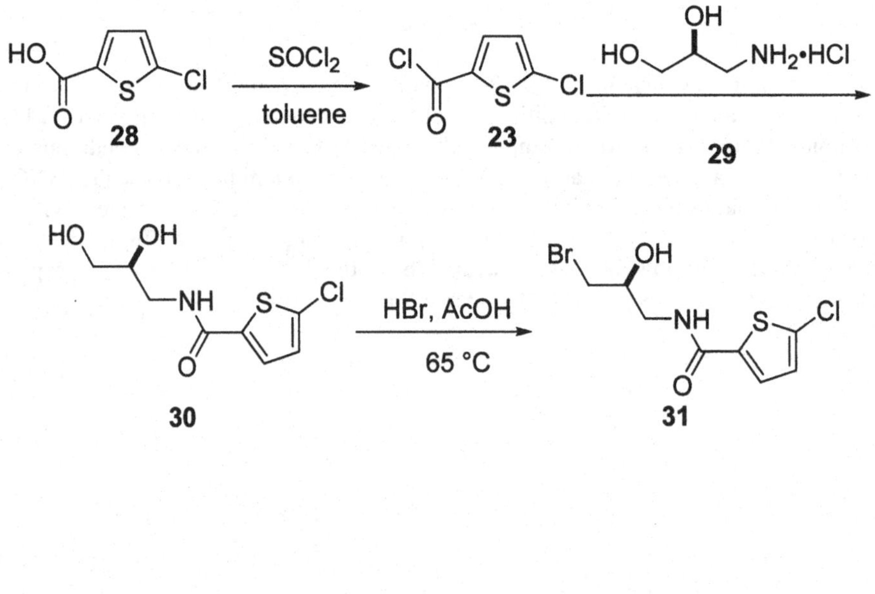

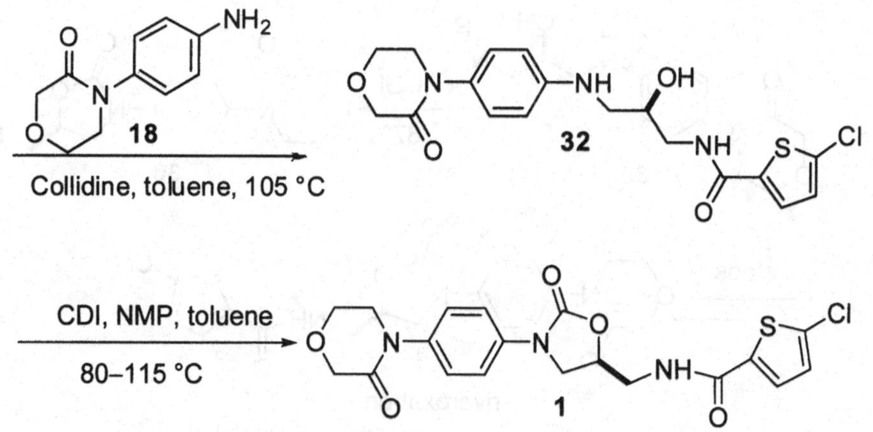

Scheme 3: The second synthetic approach to rivaroxaban

In the scale-up synthesis of linezolid[21] (Pharmacia/Upjohn), the inexpensive (*S*)-epichlorohydrin was used as the chiral source and was used to prepare key intermediate **34**, a crystalline material (Scheme 4). Treatment of carbamate **33** with ^{t}BuOLi in alcohol–DMF, followed by addition of chloride **34**, gave linezolid **14** in 73% isolated yield. In contrast, the Bayer team applied a much simpler method for the formation of oxazolidinone by directly heating the mixture of amino alcohol **35** with *N,N'*-carbonyldiimidazole (CDI) in refluxing THF, using catalytic amounts of DMAP, to provide oxazolidinone **36** (or **21**) in 87% isolated yield.

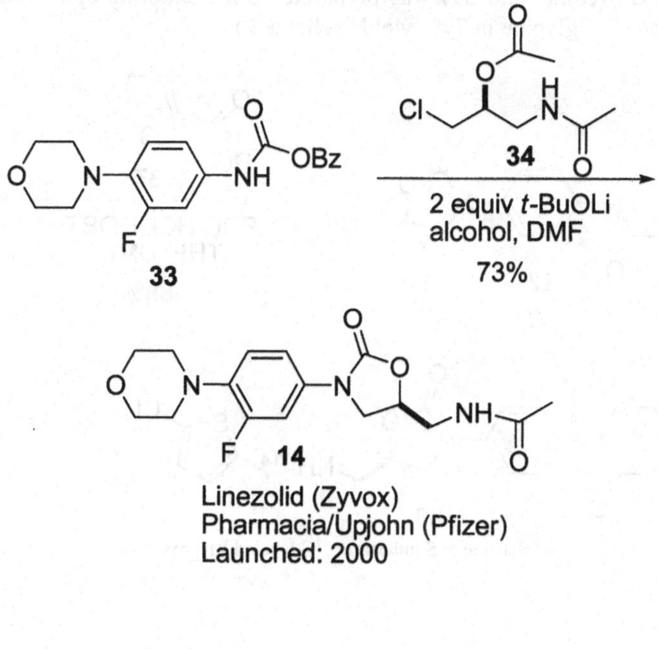

Linezolid (Zyvox)
Pharmacia/Upjohn (Pfizer)
Launched: 2000

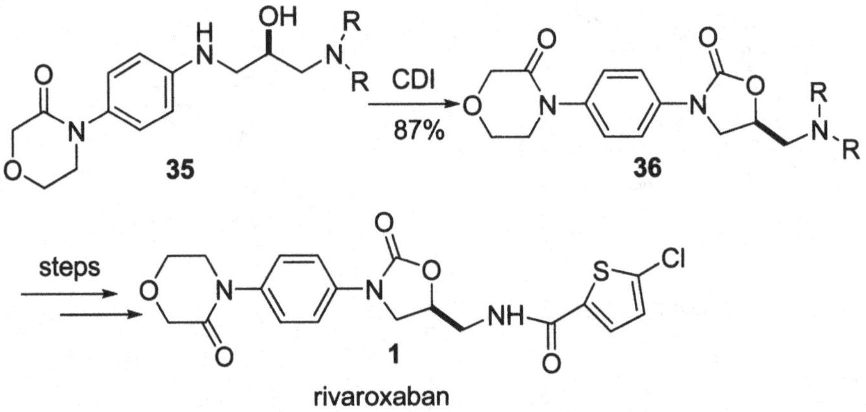

Scheme 4: Different Ring Closure Strategy for Oxazolidinone In The Synthesis of Linezolid and Rivaroaban

During the development of rivaroxaban **1**, Pleiss et al. at Bayer Health Care prepared [^{14}C]-radiolabeled rivaroxaban,[22] which was required for clinical studies of drug absorption, distribution, metabolism, and excretion (ADME studies). The approach taken for the synthesis of ^{14}C labeled rivaroxaban **38** relies on the previously reported synthesis. In the presence of EDC•HCl and HOBT, 4-{4-[5S)-5-(aminomethyl)-2-oxo-1,3-oxazolidin-3-yl]phenyl}-morpholin-3-one **22** was coupled with 5-chloro-2-thiophene [^{14}C]-carboxylic acid **37** and was purified using chiral HPLC to afford the [^{14}C]-radiolabelled rivaroxaban **38** in 85% yield with high chemical and radiochemical purity and with an enantiomeric excess of > 99% *ee* (Scheme 5). Meanwhile, the metabolite M-4 of rivaroxaban (compound **39**) was prepared from 5-chlorothiophenecarboxylic acid chloride **23** and [^{14}C]glycine in 77% yield (Scheme 6).

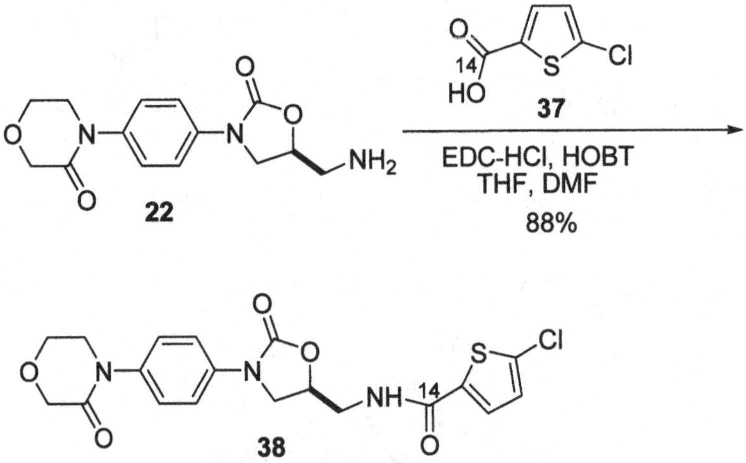

Scheme 5: Synthesis of ^{14}C-labeled Rivaroxaban

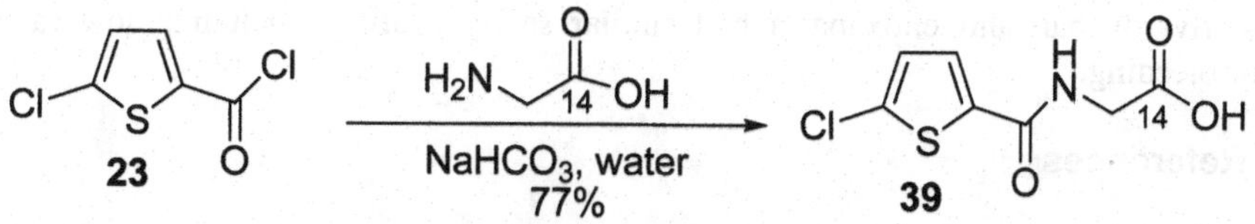

Scheme 6: Synthesis of ^{14}C-labeled rivaroxaban metabolite M-4 (**39**)

14.7. Compounds in Development: Apixaban[23-25] and Otamixaban[26-28]

There are several new factor Xa inhibitors currently under development. One is apixaban **40** (BMS-562247), which is in Phase III clinical trials and, if approved, will be marketed by joint venture of Bristol-Myers Squibb and Pfizer for prevention of venous thromboembolism. The other factor Xa inhibitor being developed is otamixaban **41**, which is being investigated by Sanofi-Aventis in Phase II clinical trials (at the time of writing) as a treatment for acute coronary syndrome.

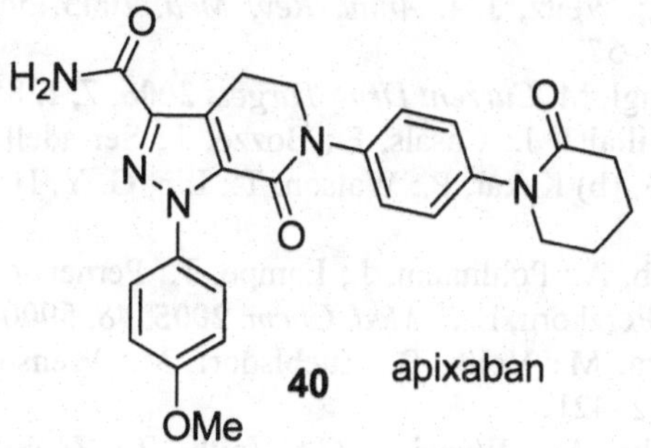

40   apixaban

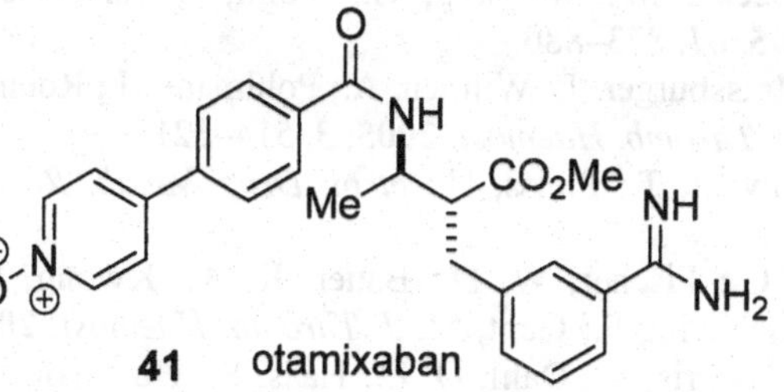

41   otamixaban

In summary, rivaroxaban **1** is an oral direct factor Xa inhibitor and is the first approved factor Xa inhibitor on the European and Canadian market. This class of inhibitors is expected to expand, with new members currently in late-stage clinical development. Rivaroxaban is indicated for the prevention of venous thromboembolic events in patients who have undergone elective total hip or total knee replacement surgery. Rivarobaxan was underwent extensive clinical program that included three Phase III trials of rivaroxaban involving a total of nearly 12,000 patients. The results from these three studies demonstrated the superior efficacy of the factor Xa inhibitor, both in head-to-head comparisons with enoxaparin and when comparing extended-duration (5 weeks) rivaroxaban with short-duration (2 weeks) enoxaparin. In all three trials, rivaroxaban and enoxaparin had similar safety profiles, including low rates of major bleeding.

14.8   References

1.   Merli, G. J. *Am. J. Med.* **2008**, *121*, S2–S9.
2.   Heit, J. A.; Cohen, A. T.; Anderson, F. A. Paper presented at the 47th Ann. Meet. Am. Soc. Hematol., December 10–13, 2005.
3.   (a) Lugassy, G.; Brenner, B.; Samana, M.-M.; Schulman, S.; Cohen, M. *Thrombosis and Anti-Thrombotic Treatment* Martin Dunitz Publishers, **2000**. (b) Iqbal, O.; Aziz, A.; Hoppensteadt, D. A.; Ahmad, S.; Walenga, J. M.; Bakhos, M.; Fareed, J. *Emerging Drugs* **2001**, *6*, 111–135.
4.   (a) Hirsh, J.; Fuster, V. *Circulation* **1994**, *89*, 1449–1468. (b) Hirsh, J.; Fuster, V. *Circulation* **1994**, *89*, 1469–1480.
5.   Eriksson, B. I.; Quinlan, D. J. *Drugs* **2006**, *66*, 1411–1429.
6.   (a) Quan, M. L.; Smallheer, J. M. *Curr. Opin. Drug Discov. Dev.* **2004**, *7*, 460–469. (b) Linkins, L.-A.; Weitz, J. I. *Annu. Rev. Med.* **2005**, *56*, 63–77. (c) Chang, P. *IDrugs* **2004**, *7*, 50–57.
7.   Klauss, V.; Spannagl, M. *Current Drug Targets* **2006**, *7*, 1285–1290.
8.   (a) Escolar, G.; Villalta, J.; Casals, F.; Bozzo, J.; Serradell, N.; Bolós. *Drugs Fut.* **2006**, 31, 484–493. (b) Kakar, P.; Watson, T.; Lip, G. Y. H. *Drugs Today* **2007**, *43*, 129–136.
9.   Roehrig, S.; Straub, A.; Pohlmann, J.; Lampe, T.; Pernerstorfer, J.; Schlemmer, K.-H.; Reinemer, P.; Perzborn, E. *J. Med. Chem.* **2005**, *48*, 5900–5908.
10.   Kubitza, D.; Becka, M.; Voith, B.; Zuehlsdorf, M.; Wensing, G. *Clin. Pharmacol Ther.* **2005**, *78*, 412–421.
11.   Kubitza, D.; Becka, M.; Wensing, G.; Voith, B.; Zuehisdorf, M. *Eur. J. Clin Pharmacol.* **2005**, *61*, 873–880.
12.   Perzborn, E.; Strassburger, J.; Wilmen, A.; Pohlmann, J.; Roehrig, S.; Schlemmer, K. H.; Straub, A. *J. Thromb. Haemost.* **2005**, *3*, 514–521.
13.   Weinz, C.; Schwarz, T.; Pleiss, U.; *et al. Drug Metab. Rev.* **2004**, *36* (Suppl. 1): Abst. 196.
14.   (a) Turpie, A. G.; Fischer, W. D.; Bauer, K. A.; Kwong, L. M.; Irwin, M. W.; Kalebo; P.; Misselwitz, F.; Gent, M. *J. Thromb. Haemost.* **2005**, *3*, 2479–2486. (b) Eriksson, B. I.; Borris, L.; Dahl, O. E.; Haas, S.; Huisman, M. V.; Kakkar, A. K.; Misselwitz, F.; Kalebo, P. *J. Thromb. Haemost.* **2006**, *4*, 121–128.
15.   Lassen, M. R.; Ageno, W.; Borris, L. C.; Lieberman, J. R.; Rosencher, N.; Bandel, T. J.; Misselwitz, F.; Turpie, A. G. G. *N. Engl. J. Med.* **2008**, *358*, 2776–2786.
16.   Li, J. J.; Johnson, D. S.; Sliskovic, D. R.; Roth, B. D, *Contemporary Drug Synthesis*, Wiley & Sons, Hoboken, 83–87, 2004 .
17.   Gutcait, A.; Wang, K.-C.; Liu, H.-W.; Chern, J.-W. *Tetrahedron: Asymmetry* **1996**, *7*, 1641–1648.
18.   Eur. Pat. Appl.; 1403267, 31 Mar **2004**.
19.   PCT Int. Appl., 2004037815, 06 May **2004**.
20.   Thomas, C. R. DE 10300111, EP 1583761, JP 2006513227, WO 2004060887.

21.   Perrault, W. R.; Pearlman, B. A.; Godrej, D. B.; Jeganathan, A.; Yamagata, K.; Chen, J. J.; Lu, C. V.; Herrinton, P. M.; Gadwood, R. C.; Chan, L.; Lyster, M. A.; Maloney, M. T.; Moeslein, J. A.; Greene, M. L.; Barbachyn, M. R. *Org. Proc. Res. Dev.* **2003**, *7*, 533–546.

22.   Pleiss, U.; Grosser, R. *J. Label Compd. Radiopharm.* **2006**, *49*, 929–934.

23.   Bates, S. M.; Weitz, J. I. *Drugs Fut.* **2008**, *33*, 293–301.

24.   Pinto, D. J. P.; Orwat, M. J.; Koch, S.; Rossi, K. A.; Alexander, R. S.; Smallwood, A.; Wong, P. C.; Rendina, A. R.; Luettgen, J. M.; Knabb, R. M.; He, K.; Xin, B.; Wexler, R. R.; Lam, P. Y. S. *J. Med. Chem.* **2007**, *50*, 5339–5356.

25.   Carreiro, J.; Ansell, J. *Expert Opin. Investig. Drugs* **2008**, *17*, 1937–1945.

26.   Nutescu, E. A.; Pater, K. *IDrugs* **2006**, *9*, 854–865.

27.   Cohen, M.; Bhatt, D. L.; Alexander, J. H.; Montalescot, G.; Bode, C.; Henry, T.; Tamby, J.-F.; Saaiman, J.; Simek, S.; De Swart, J. *Circulation* **2007**, *115*, 2642–2651.

28.   Sakai, T.; Kawamoto, Y.; Tomioka, K. *J. Org. Chem.* **2006**, *71*, 4706–4709.

15

Endothelin Antagonists for the Treatment of Pulmonary Arterial Hypertension

David J. Edmonds

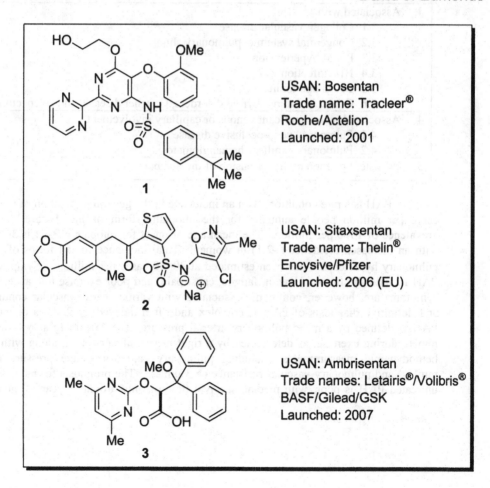

USAN: Bosentan
Trade name: Tracleer®
Roche/Actelion
Launched: 2001

USAN: Sitaxsentan
Trade name: Thelin®
Encysive/Pfizer
Launched: 2006 (EU)

USAN: Ambrisentan
Trade names: Letairis®/Volibris®
BASF/Gilead/GSK
Launched: 2007

15.1   Background

Pulmonary arterial hypertension is a chronic, progressive and, as yet, incurable condition characterized by elevated blood pressure in the pulmonary artery. Untreated, this condition places additional strain on the right ventricle, leading to heart failure and death. Pulmonary hypertension covers a broad group of disorders, which were classified according to similarities in pathology. Pulmonary arterial hypertension (PAH) encompasses the first class of pulmonary hypertension diseases. As shown in Table 1, PAH is further divided onto several subcategories, which reflect the origin of the disease.[1] The term idiopathic PAH is applied to those cases with no clear underlying cause, while similar cases where family history of the disease exists are termed familial. The genetic basis of PAH is still under investigation and has been reviewed recently.[2] Other occurrences may be associated with various underlying conditions, the use of certain drugs (particularly some appetite suppressants and stimulants), or exposure to toxins. PAH associated with pulmonary venous or capillary diseases is classed separately, as is persistent PAH in newborn infants.

Table 1. Clinical Classification of Pulmonary Arterial Hypertension

1.	Idiopathic
2.	Familial
3.	Associated with
	3.1 Collagen vascular disease
	3.2 Congenital systemic–pulmonary shunt
	3.3 Portal hypertension
	3.4 HIV infection
	3.5 Drugs and toxins
	3.6 Other (including thyroid disorders, Gaucher disease, and splenectomy)
4.	Associated with significant venous or capillary involvement
	4.1 Pulmonary veno-occlusive disease
	4.2 Pulmonary capillary hemangiomatosis
5.	Persistent pulmonary hypertension of the newborn

PAH is a rare condition, with an incidence in the general population of around 2 cases per million people annually for the idiopathic form of the disease,[3–5] and a prevalence of 1300 per million. The mean age of onset for idiopathic PAH is 36 years, with an increased incidence (~2:1) in women. Total incidence for all forms of severe pulmonary hypertension has been estimated at 30–50 cases per million.[6] Symptoms of PAH include shortness of breath, fatigue, chest pain, and poor exercise tolerance. Such symptoms are, however, commonly associated with various cardiovascular conditions, and definitive diagnosis of PAH is complex and often delayed by 2 years or more.[3–5] PAH is defined by a mean pulmonary arterial pressure of > 25 mmHg at rest or > 30 mmHg during exercise, as determined by a right heart catheterization, along with other hemodynamic measurements, including pulmonary capillary wedge pressure (≤ 15 mmHg) and pulmonary vascular resistance (> 3 units).[5] The prognosis for patients with untreated PAH is poor, with median survival time of only around 3 years;[3,4] however, with modern treatments, the outlook is considerably improved. When PAH is associated with another condition, such as HIV infection or systemic sclerosis, it is generally a major contributor to mortality in these patient groups.[7,8] The severity of PAH is rated using a WHO scale (Table 2), with many patients placed in class III or IV at diagnosis.[9]

Table 2. World Health Organization Functional Classification of PAH Symptoms

Class	Description
I	PAH without limitation of physical activity; normal activity does not cause undue pain, shortness of breath or dizziness
II	Slight limitation of physical activity; comfortable at rest but ordinary activity causes shortness of breath, fatigue, chest pain or dizziness
III	Marked limitation of physical activity; comfortable at rest but less than ordinary activity causes shortness of breath, fatigue, or chest pain
IV	Inability to carry out any physical activity without symptoms; shortness of breath and fatigue present at rest

The pathogenesis of PAH is also complex and the origins of the disease are poorly understood. There are three characteristic elements that combine to produce the hypertension observed in patients — namely vasoconstriction, localized thrombosis, and vascular-wall remodeling. The latter involves thickening and increased tone of the vascular wall due to hypertrophy and hyperproliferation of smooth muscle and endothelial cells. Imbalances of several endogenous vascular mediators are observed in patients with PAH and are thought to be the result of injury to or dysfunction of pulmonary endothelial cells. Nitric oxide is a powerful vasodilator and anticoagulant and reduced expression of endothelial nitric oxide synthase in the pulmonary arteries has been observed in PAH patients. Decreased excretion of prostacyclin metabolites is also observed, indicating that the concentration of this important vasodilator is also lowered. Production of endothelin-1 in the pulmonary artery is also increased in PAH patients, and increased levels of endothelin-1 are implicated in the progression of the disease.[3–5]

15.2　Treatment of PAH

Historically, the treatment options for patients with pulmonary arterial hypertension were very limited, and the disease was considered untreatable as recently as the 1980s. PAH responds poorly to conventional hypertension treatments, although diurectics and anticoagulants may be used as supportive measures. Calcium channel antagonists, such as nefedipine, are effective in a subset of PAH patients in whom vasoreactivity tests are positive and provide a significant improvement in prognosis for those individuals. Unfortunately, however, only a small proportion of PAH sufferers will respond to calcium channel antagonist therapy. For patients with advanced PAH, surgical intervention was often needed as the disease progressed to right-heart failure. The poor prognosis of PAH made patients good candidates for transplant surgery, and one of the earliest heart–lung transplants was performed on a PAH sufferer. Shortage of organ donors and improvements in transplant surgery now dictate that only those patients with significant heart failure undergo heart–lung transplants, with single or double lung transplants performed preferentially.[3,9]

Targeted drug therapy for PAH began in the 1980s with the use of prostacyclin. Prostacyclin (prostaglandin I_2) is a powerful endogenous vasodilator that also inhibits vascular remodeling. Synthetically derived prostacyclin (**4**), known as epoprostenol, was initially administered as a bridging measure to improve survival chances prior to transplant surgery, but many patients showed sustained improvement to the extent that surgery was no longer needed.[4,9]

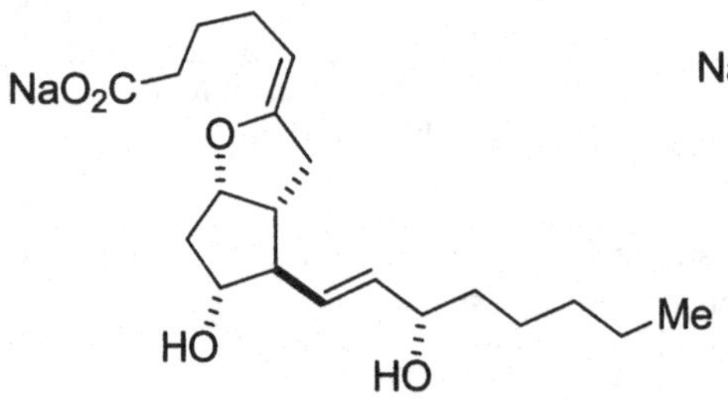

USAN: Epoprostenol sodium (**4**)
Trade name: Flolan®

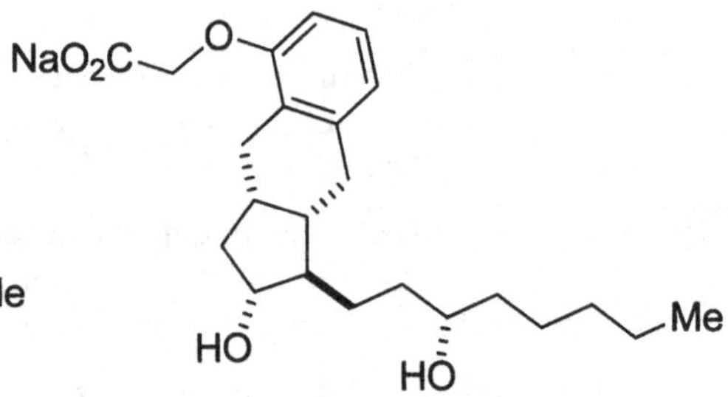

USAN: Treprostniil sodium (**5**)
Trade name: Remodulin®

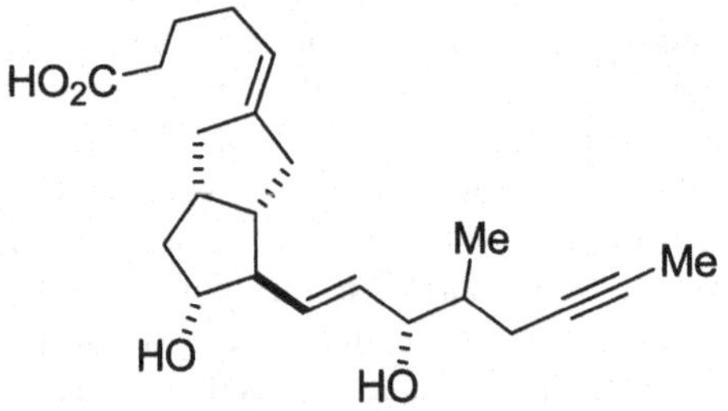

USAN: Iloprost (**6**)
Trade name: Ventavis®

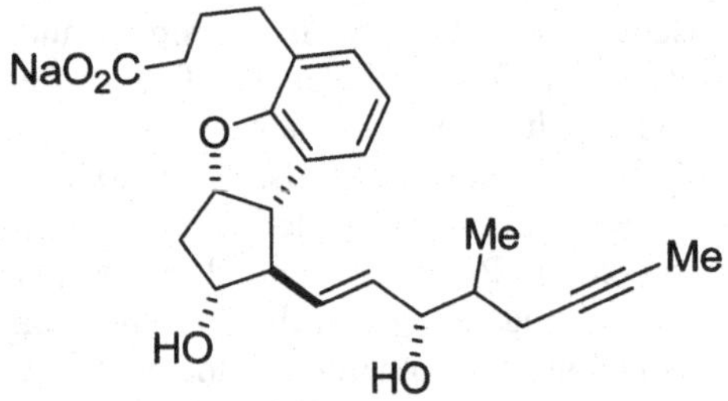

USAN: Beroprost sodium (**7**)

Side effects, such as headache and jaw pain, are observed, but the major drawbacks with epoprostenol therapy relate to its delivery. Epoprostenol has an extremely short half-life in the blood (2–3 min) and therefore must be administered by continuous intravenous infusion via a surgically implanted central vein catheter. This can lead to complications such as local infections, sepsis, or catheter-associated thrombosis. In addition, interruption of therapy due, for example, to pump failure can lead to a life-threatening rebound of symptoms. The compound itself is unstable at room temperature and must be stored in the refrigerator. Despite these severe drawbacks, i.v. epoprosenol remains a useful treatment for patients presenting with WHO class IV PAH. The problems with epoprostenol have led to the development of alternative agents.

Treprostinil (**5**) is a synthetic analog of prostacyclin with a serum half-life of around 3 h. It is also stable at room temperature, making handling easier, and can be administered by subcutaneous infusion using equipment similar to that used for insulin, although pain and inflammation at the infusion site are common. Iloprost (**6**) is another prostacyclin analog used to treat PAH. This compound is rather closer in structure to the natural prostacyclin, but it also offers extended serum stability, with a half-life of around 30 min, along with chemical stability at room temperature. Iloprost is administered by inhalation, which is preferable to intravenous infusion; however, successful treatment requires a nebulizer, and the drug must be administered 6–9 times per day. Both treprosinil and iloprost are used in patients with class III PAH. Beraprost (**7**) has been developed as an orally bioavailable prostacyclin analog, although its short half-life requires 4 times daily dosing. It has been in use in Japan since 2005 and is undergoing trials in the E.U. and the United States.[9,10]

PDE5 inhibitors, already in use for erectile dysfunction, are also promising new agents for PAH. Inhibition of PDE5 leads to increased levels of cGMP, which results in NO mediated vasodilation. Increased cGMP may also lead to increased production of prostacyclin due to its role as an endogenous inhibitor of PDE3. PDE5 expression in the pulmonary vasculature is high, and treatment with PDE5 inhibitors has led to improvement in WHO functional class in PAH patients without inducing systemic hypotension. Sildenafil (**8**) and tadalafil (**9**) are approved for PAH, marketed as Revatio and Adcirca, respectively; vardenafil is also undergoing clinical trials for PAH.[9,11]

USAN: Sildenafil (**8**)
Trade name: Revatio®

USAN: Tadalafil (**9**)
Trade name: Adcirca®

15.3   Endothelin Antagonists

The endothelins are a family of endogenous peptide hormones produced in many tissues, including the vascular endothelium, the kidney, and certain cancerous tissue. Hickey et al. reported the existence of an endothelium-derived vasoconstrictor in 1985 and endothelin-1 (ET-1) was isolated by Yanagisawa et al. in 1988 from porcine endothelial cell culture medium. ET-1 is one of the most potent known vasoconstrictors and was found to be an order of magnitude more potent than angiotensin II. Subsequent studies identified a gene encoding the same peptide in the human genome, along with separate genes for very similar peptides, named endothelin-2 and endothelin-3. Indeed, the endothelin family appears to be highly conserved in mammals, and even shows clear structural similarity with the sarafotoxins, a group of cardiotoxic peptides isolated from snake venom.[12] The biosynthesis of ET-1 begins with the production of preproendothelin-1, a 203 residue peptide. This precursor is cleaved to form big endothelin-1 (or proendothelin). Human big ET-1 is a 38-residue peptide, which is further cleaved to provide ET-1 by endothelin converting enzyme (ECE). The mature ET-1 is a 21-residue peptide that contains two disulfide linkages, one from Cys1 to Cys15 and the other from Cys3 to Cys11, which give the peptide a bicyclic loop structure.[12]

The biological actions of the endothelins are diverse and tissue dependent.[12] The endothelins exert their effects through binding to one of two distinct G protein-coupled receptors, which are approximately 55% homologous. They are expressed in different locations and have different selectivities for the various endothelins. They also have rather divergent effects on the blood vessels. Endothelin receptor type A (ET_A) is found in vascular smooth muscle and fibroblasts and is selective for ET-1, whereas type B (ET_B) is found in the endothelium and smooth muscle cells and has similar affinity for each isoform.[13] The binding of ET-1 to ET_A receptors induces vasoconstriction, which persists even when ET-1 is removed. ET_A receptors also mediate the activation of calcium channels in smooth muscle cells, which perhaps accounts for the persistent effect of ET-1-induced vasoconstriction. In contrast, activation of the ET_B receptors of endothelial cells in healthy pulmonary vasculature results in vasodilation, mediated by the release of prostacyclin and nitrous oxide, inhibition of ET-1 production, and clearance of ET-1 from the circulation. ET_B receptors in the smooth muscle may also mediate vasoconstriction. Both receptor subtypes mediate the potent mitogenic activity of ET-1, which causes proliferation of vascular smooth muscle cells. The situation is complicated by changes to the endothelin system in patients with PAH. In addition to increased levels of ET-1 in the blood, ET_B receptors undergo a functional change. Increased expression in smooth muscle cells increases their contribution to vasoconstriction, and they may also be involved in promoting fibrosis. There is therefore debate as to the relative benefits of dual ET_A/ET_B antagonists as opposed to ET_A-selective agents.[7-9,14,15]

The role of endothelin in PAH, as well as other diseases, led to the early recognition of the therapeutic potential of antagonists, and this has been the focus of significant research in recent years.[8,9,13,16] One such campaign was undertaken by scientists at Roche and led to the discovery of bosentan. Their medicinal chemistry efforts began with a high-throughput screen to identify potential lead matter, a process that yielded pyrimidine sulfonamide **10**, a weak competitive inhibitor of ET-1 binding to both ET_A (IC_{50} = 18 μM) and ET_B (2 μM). This lead was originally prepared as part of a diabetes program and was known to be orally bioavailable but not hypoglycemic.[17,18] Optimization of **10** involved varying the substituents on all four positions of the pyrimidine core, with promising results. A major breakthrough was the discovery of Ro 46-2005 (**11**), a potent mixed antagonist that provided an orally bioavailable tool to assess the pathological role of endothelin receptors in vivo.[19] The series proved flexible in terms of selectivity, with compound **12** proving to be a potent and selective ET_B antagonist, which again contributed to the understanding of the roles of the receptor subtypes.[20] The Roche team's pursuit of a potent mixed antagonist culminated in the discovery of bosentan (**1**), which exhibited IC_{50}s of 8 nM (ET_A) and 150 nM (ET_B). The sulfonamide group of bosentan, which is essential for activity, has a pK_a of 5.5. This acidity contributes to a logD of 1.3 and good aqueous solubility (3 mg/mL at pH 7.4).[18]

10
ET$_A$ IC$_{50}$: 18 μM
ET$_B$ IC$_{50}$: 2 μM

11: Ro 46-2005
ET$_A$ IC$_{50}$: 0.22 μM
ET$_B$ IC$_{50}$: 1 μM

12
ET$_A$ IC$_{50}$: 2.2 μM
ET$_B$ IC$_{50}$: 69 nM

In the clinic, bosentan was found to have bioavailability of around 50%, with ~98% protein binding, a clearance of 2 mL/min/kg and a terminal elimination half-life of 5.4 h. Bosentan is metabolized by cytochrome P450 enzymes (CYPs) and cleared hepatically.[7,15] Major metabolic pathways include O-demethylation and oxidation of the t-butyl group.[21] Bosentan induces CYPs 3A4 and 2C9, the latter resulting in a potential drug–drug interaction (DDI) with the commonly used anticoagulant warfarin, which is metabolized by this enzyme. Other potential DDIs include simvastatin, cyclosporine A, erythromycin, hormonal contraceptives (pregnancy is also contraindicated due to the role of ET-1 in fetal development), and sildenafil. Co-administration of bosentan with sildenafil leads to increased plasma levels of bosentan and reduced sildenafil levels; however, the two drugs have been used in combination therapy without apparent problems.[7,15] In general, bosentan was well tolerated in the clinic. The major adverse effect was hepatotoxicity with increased liver transaminases, believed to be due to inhibition of a bile-salt export pump.[14] Liver enzymes must therefore be monitored monthly during therapy. The liver-enzyme effect has led to termination of therapy in about 3% of patients, however the abnormalities were found to be reversible and resolved after withdrawal of bosentan.[9] Following positive outcomes for patients in the clinic, bosentan was approved in 2001 as the first oral treatment for PAH. It is currently recommended at doses of 62.5 mg or 125 mg twice daily for patients in WHO functional classes II–IV.[9]

Sitaxsentan (**2**) was developed by Encysive as a selective ET$_A$ antagonist, and was the second endothelin antagonist to reach the market. Their lead series included an aminoisoxazole linked to a thiophene via a sulfonamide, a motif ultimately found in sitaxsentan. They examined a broad range of thiophene-2-carboxamide analogs based on this lead series and found that N-arylcarboxamides provided the desired high potency for ET$_A$ while maintaining generally excellent selectivity over ET$_B$. SAR in the N-aryl group appeared to be somewhat flexible, with a wide array of substituents tolerated in both electron rich and electron deficient rings.[22] Unfortunately, their most promising compounds, for example amide **13**, exhibited short half-lives, perhaps due to amide cleavage, and very poor oral bioavailability. Subsequent studies aimed to improve these essential features through replacement of the amide linkage. Imides, ureas, carbamates, and other alternatives all resulted in a dramatic loss of potency against ET_A. Finally, the use of a benzylic ketone linker combined the excellent potency of the amides with good bioavailability. Reoptimization of the aromatic substituents yielded sitaxsentan.[23]

13

ET_A IC_{50}: 3.4 nM
ET_B IC_{50}: 40.4 µM
F = 8%

14

ET_A IC_{50}: 0.04 nM
ET_B IC_{50}: 442 µM
F = 70%

More recent reports suggest an interesting structure-bioavailability relationship in the amide series. In screening for a follow-up candidate, the researchers noted that bioavailability could be conferred on the carboxamides by placing an *ortho* acyl group on the aromatic ring. This observation led to the discovery of compound **14**, which has 70% bioavailability in dogs and excellent potency and selectivity for ET_A. The authors suggest the presence of an intramolecular hydrogen bond between the ketone and amide may contribute to permeability by removing a desolvation penalty.[24]

Sitaxsentan is a potent and selective agent which inhibits ET-1 binding to ET_A receptors (IC_{50} = 1.4 nM), while being essentially inactive at ET_B receptors (IC_{50} = 9.8 µM).[23] In the clinic, it was found to have excellent oral bioavailability (70–100%) and a terminal elimination half-life of 10 h, and is administered as a once daily 100 mg dose. It is highly protein bound in plasma (> 99%) and extensively metabolized in the liver to inactive metabolites, predominantly by CYPs 2C9 and 3A4. Excretion is 50–60% renal, with the balance in the feces.[25] Sitaxsentan inhibits CYP 2C9, and was observed to increase exposure to warfarin by over twofold. The use of cyclosporine A is also contraindicated, but no interactions were observed with sildenafil.[15] Sitaxsentan was well tolerated in trials, with only minor side effects reported. Reversible liver enzyme abnormalities were also observed, but less frequently than with bosentan.[15,25]

The most recently approved endothelin antagonist is ambrisentan, a moderately selective ET_A antagonist originally developed at BASF and Knoll. The program began with screening of the BASF compound collection for small molecules that bind to the ET_A receptor. This screen identified two compounds (**15** and **16**) originally investigated as herbicides, which were found to bind ET_A with good potency (K_i = 250 and 160 nM, respectively) and selectivity (K_i = 3 and 4.7 µM for ET_B). A disadvantage of the lead structures is that they contain two chiral centers, and the research team sought to improve potency while simultaneously simplifying the structure. They were rewarded in this effort with the discovery of darusentan, in which the quaternary center is symmetrically substituted. Further investigations to determine the active enantiomer and optimize the pyrimidine ring yielded ambrisentan, which exhibited potent binding to ET_A (K_i = 1 nM) and around 200-fold selectivity versus ET_B (K_i = 195 nM).[26] This program stands as a clear example of the benefits of high quality lead matter in drug discovery, combined with insightful optimization of molecular structure.

15: R = *i*-Pr
16: R = SMe

17 (Darusentan)

The selectivity of ambrisentan falls in between that observed for bosentan and sitaxsentan, although it should be noted that in each case, binding and inhibition data vary markedly between various assays. While treatment with the highly ET_A-selective sitaxsentan results in lowering of plasma ET-1 levels (presumably due to maintenance of the ET-1 clearance activity of ET_B), use of bosentan or ambrisentan leads to increased plasma ET-1 levels. Ambrisentan had high bioavailability in clinical studies and a long terminal excretion half-life of 15 h, allowing for a once daily dose of 5–10 mg. Ambrisentan is metabolized hepatically by glucuronidation and, to an extent, CYP enzymes, and is a substrate for the P-gp efflux pump. No interactions were observed in studies with sildenafil or warfarin, which is favorable for combination therapy.[7,15]

The advent of orally bioavailable endothelin antagonists to treat PAH has had a marked effect of the therapy of this disease. All three agents have demonstrated significant clinical improvements in trials and on longer term monitoring. Improvements were seen in 6 minute walking distance, a standardized measure of exercise tolerance, as well as time to clinical worsening, hemodynamic parameters, and vascular remodeling. Improvements in WHO functional class have been noted, indicating a tangible benefit in terms of quality of life. Direct comparison of the three available ET-1 antagonists is hampered by a lack of clear comparison data. Similarly, long-term survival following ET-1 antagonist treatment is difficult to establish definitively due to ethical preclusion of long-term placebo-controlled trials.   However, when compared to historical data concerning untreated survival rates, all three agents appear to offer significant improvements.[7–9,14,15]

15.4   Synthesis of Bosentan

The structure of bosentan (**1**) can be viewed as a pyrimidine core bearing four different substituents. In a medicinal chemistry setting, it is important to be able to vary each of the substituents around a core at will to probe SAR. The discovery of bosentan provides an excellent example of a synthetically enabled scaffold, with a single strategy allowing the preparation of a wide range of analogs.[18,27] In this case, the factors favoring diversity generation and process efficiency were coincident and the manufacturing route is little altered from the discovery one. The route began with the introduction of the aryloxy substituent through alkylation of phenol **19** with bromomalonate **18**. The di-ester **20** was then condensed with amidine **21**, incorporating the C2 substituent and forging the core ring system. Pyrimidindione **22** could be converted to dichloride **23** by treatment with POCl$_3$ and the sulfonamide unit introduced via a selective S$_N$Ar reaction. In the discovery route, this was accomplished using the potassium sulfonamide salt of **24** in DMSO at 150 °C, but in the optimized manufacturing route, the S$_N$Ar reaction was carried out in toluene using **24** with an inorganic base and phase transfer catalyst.

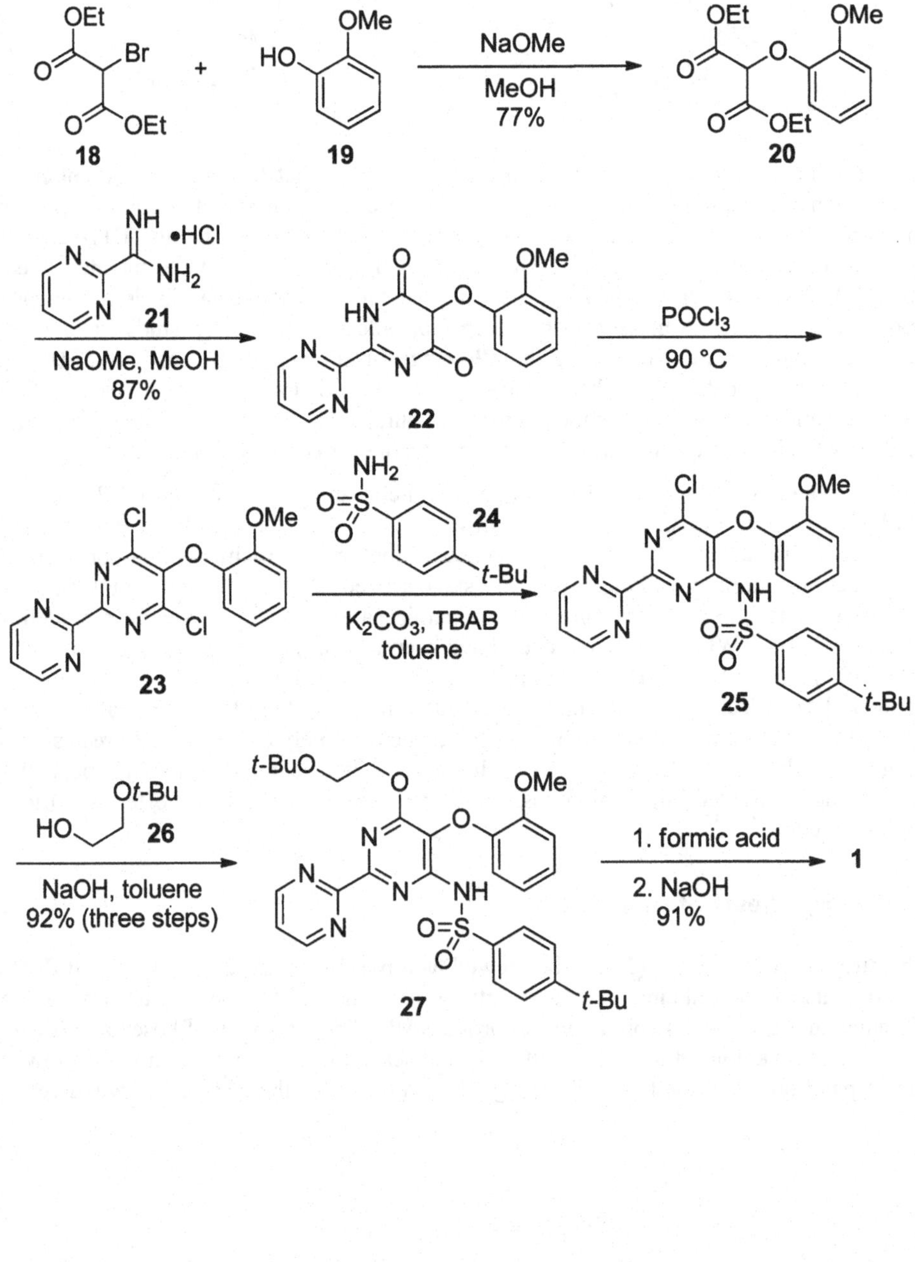

Chloride **25** could be treated with sodium ethylene glycolate in excess ethylene glycol to provide bosentan directly; however, the product was difficult to separate from small quantities of byproducts pyridone **28** and dimer **29**. To address this problem, the process chemistry researchers chose to install the ethylene glycol unit using mono *t*-butyl ether **26**. The addition reaction could then be carried out in toluene, allowing the three-step chlorination and double S_NAr sequence to be telescoped in a single solvent to give bosentan ether **27** in excellent yield. Deprotection was then achieved by heating in formic acid to form bosentan formate, followed by saponification to give bosentan (**1**) in 91% yield after a single recrystallization.[28]

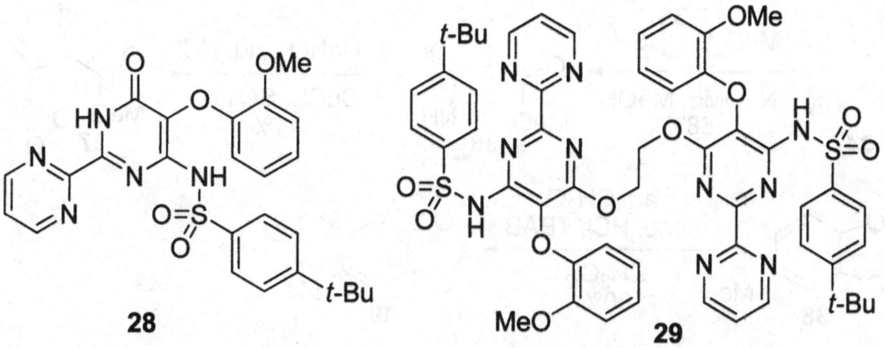

An alternative sequence to avoid dimer formation in the synthesis of bosentan has appeared in the patent literature. The S_NAr reaction with ethylene glycol is carried out first and the primary alcohol protected as the acetate ester **30**. The second displacement with sulfonamide **24** and saponification of the ester protecting group then provide bosentan (**1**).[29] This route is reported to provide the product in high purity and yield, but lacks the advantages of improved throughput afforded by the optimized route.

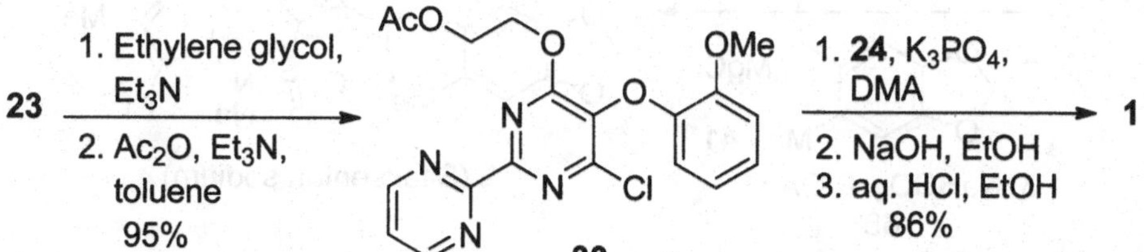

15.5   Synthesis of Sitaxsentan

The structure of sitaxsentan (**2**) is also somewhat modular and was divided into three building blocks—namely the phenyl ring, the central thiophene, and the aminoisoxazole. The isoxazole subunit **33** can be prepared by the reaction of hydroxylamine with butynenitrile (**31**) to form the heterocycle,[30] followed by chlorination with NCS.[22] The thiophene ring can be formed using a Fiesselmann thiophene synthesis starting from chloroacrylonitrile (**34**) and methyl mercaptoacetate (**35**).[31] The sulfonyl chloride group could then be installed via treatment of an intermediate diazonium salt with copper(II) chloride and sulfur dioxide.[32] The final fragment **39** was formed by a selective Blanc chloromethylation of the phenyl ring to install the benzylic chloride directly.[33]

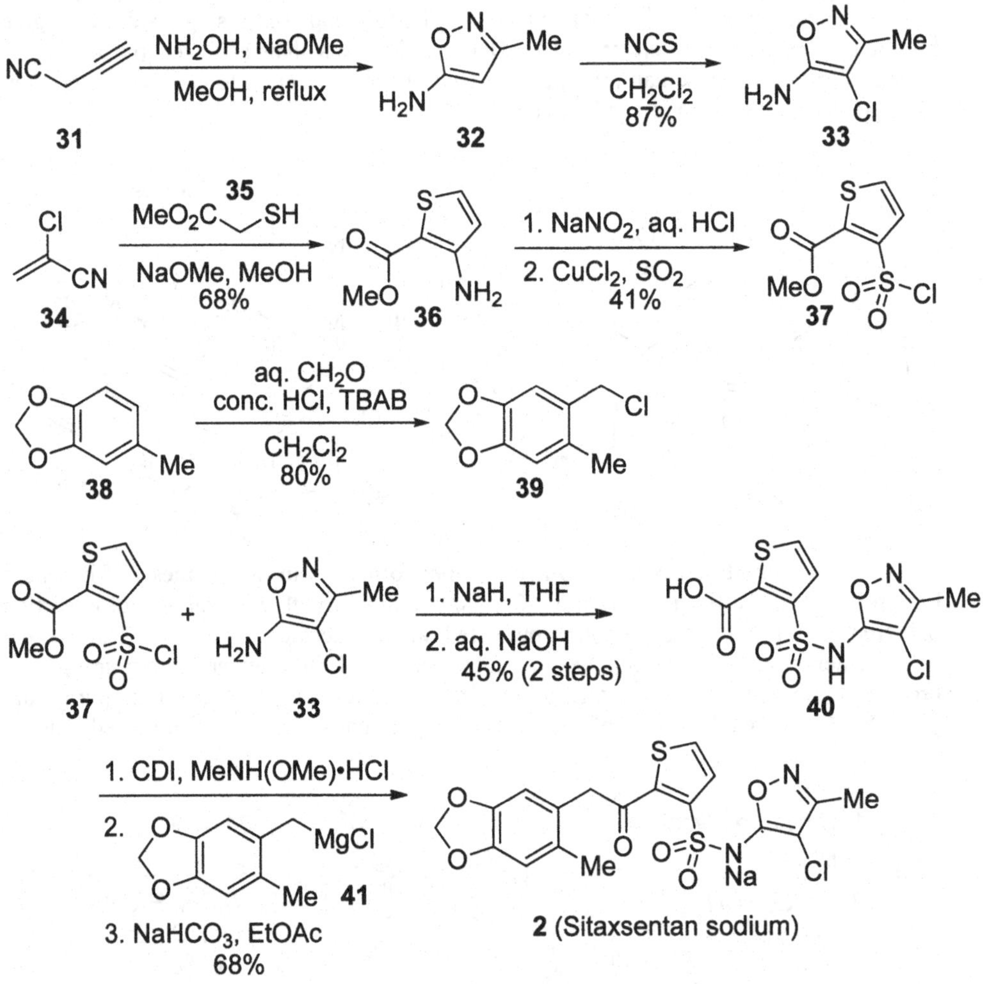

Fragment assembly then began with sulfonamide formation, which required rather forcing conditions via the deprotonated amine, followed by saponification to give thiophene carboxylic acid **40**.[22] This intermediate was a key fragment during the medicinal chemistry campaign and is common to sitaxsentan as well as a wide variety of other analogs, such as **14**. The acid group was converted to the Weinreb amide and treated with the Grignard reagent **41**, formed from chloride **39**.[22] The Grignard formation was challenging, and careful control of the reaction temperature < 8 °C was necessary to prevent Wurtz coupling to give the dimeric byproduct.[33] The fragment assembly sequence provides sitaxsentan (2) in only a few steps, but the difficult bond formation in the final step necessitated a laborious purification protocol to provide the pure sodium salt.[33]

15.6   Synthesis of Ambrisentan

Ambrisentan (3) is distinct from bosentan (1) and sitaxsentan (2) in that it contains a carboxylic acid rather than an acidic sulfonamide. As mentioned above, the medicinal chemistry team sought to simplify the structure of their lead compounds to provide compounds with a single stereogenic center. Their strategy allowed for the rapid construction of analogs from extremely simple building blocks. The discovery route to ambrisentan and related compounds delivered the final compounds as racemates in only four steps.[26,34] Thus a Darzens condensation between methyl chloroacetate (42) and benzophenone (43) provided the key epoxide intermediate 44. In the discovery route, BF$_3$•OEt$_2$ in methanol was used to open the epoxide to give racemic secondary alcohol 45. An S$_N$Ar reaction of the alcohol with pyrimidyl sulfone 46 was used to introduce the pendant heterocycle, and ester hydrolysis under standard conditions furnished ambrisentan (3).   Although yields are not reported for the specific synthesis of ambrisentan, the route was clearly efficient and was used to vary each group systematically. Several symmetrical benzophenone derivatives were used in the first step, the epoxide ring could be opened selectively at the benzylic site with various alcohols or with methyl cuprate, and a wide variety of pyrimidines could be used in the displacement reaction.[26,34]

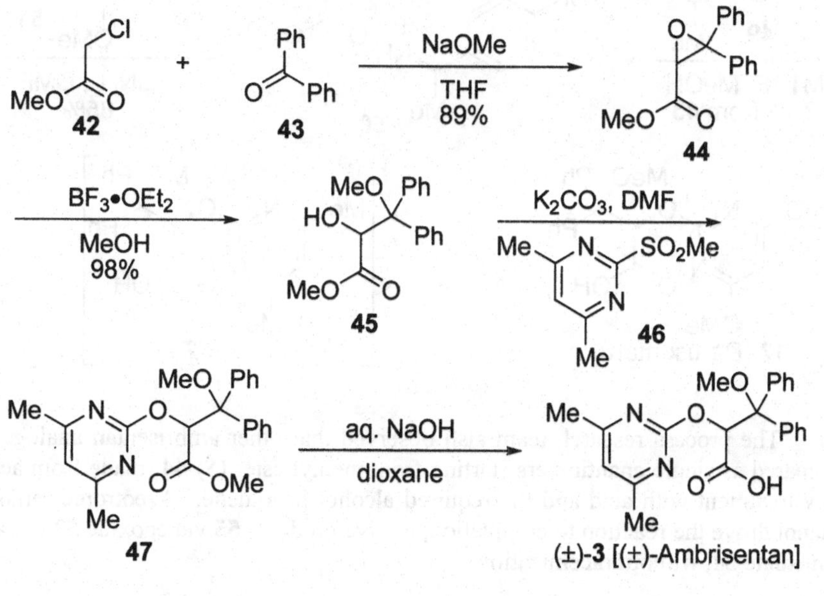

Clearly, access was needed to ambrisentan and related compounds as single enantiomers, and several methods have been described. Epoxide 44 could be prepared in moderate enantiomeric excess by asymmetric epoxidation of the corresponding olefin using a manganese salen catalyst;[34] however, resolution of a key intermediate proved more successful. A resolution strategy will generally result in the loss of 50% of the material, but this disadvantage is mitigated in this case by the very low cost of the starting materials and the short synthetic sequence. Resolution of carboxylic acid **48** proved ideal, as indicated by the process route to darusentan (**17**).[35] In the optimized sequence, ester **45** was prepared as before, but using catalytic *p*-toluenesulfonic acid to effect the epoxide opening. The ester was not isolated, but rather hydrolyzed directly to free acid **48**, which was resolved by crystallization with (*S*)-benzylamine **49**. This sequence gave the ammonium salt **50** of the (*S*)-acid in excellent overall yield from benzophenone. The salt could then be used directly in the arylation reaction, providing darusentan (**17**) in high yield. Again, the synthesis of ambrisentan is not reported specifically; however, the closely related compound **52** was prepared in a similar manner with excellent yields.

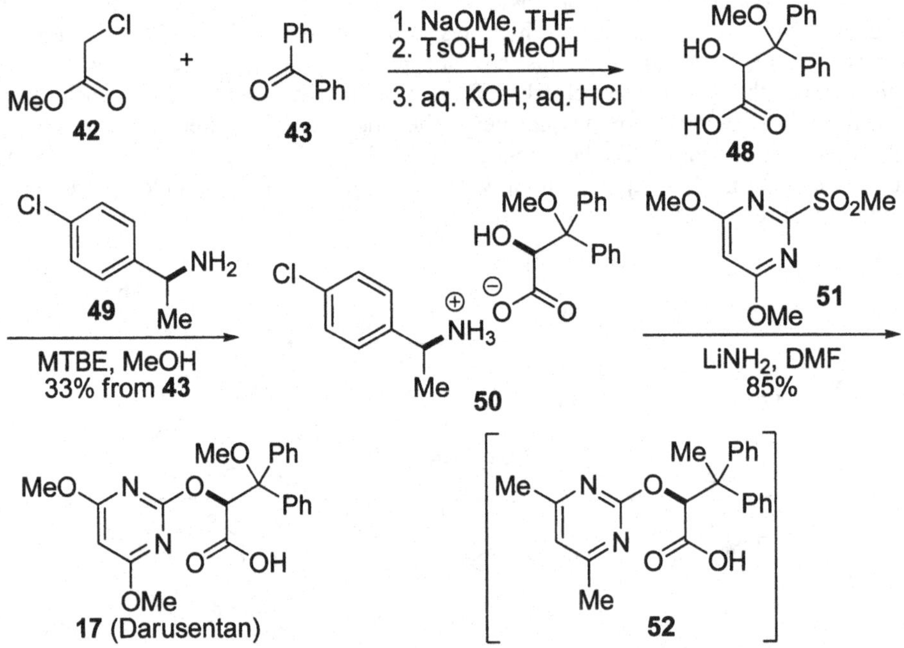

The process research team also observed that other ambrisentan analogs could be prepared as single enantiomers starting from methyl ester (*S*)-**44** (made from acid salt **50**) by treatment with acid and the required alcohol in toluene. Azeotropic removal of methanol drove the reaction to completion, to give products **55** via epoxide **53** or cationic intermediate **54**, without racemization.[35]

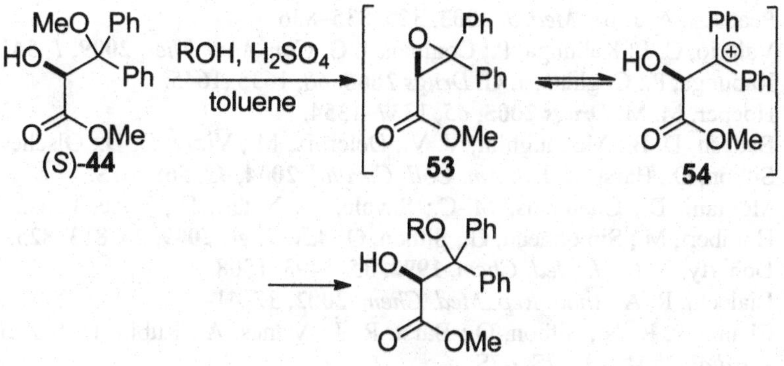

15.7 Conclusion

The scientific efforts in the two decades since the discovery of ET-1 have resulted in the development of much improved therapies for patients with a devastating disease and ET-1 antagonists have become front-line agents in the management of PAH. In each of the cases described here, the development of efficient and versatile synthetic routes allowed for extensive SAR studies, and the quality of the chemistry employed is evident in the similarities between the discovery and process routes to these drugs. In addition to the three agents approved for PAH, darusentan (17) and atrasentan (56) are in the late stages of clinical development for persistent hypertension and hormone-refractory colon cancer, respectively, indicating a continuing interest in this fascinating mechanism.[36,37]

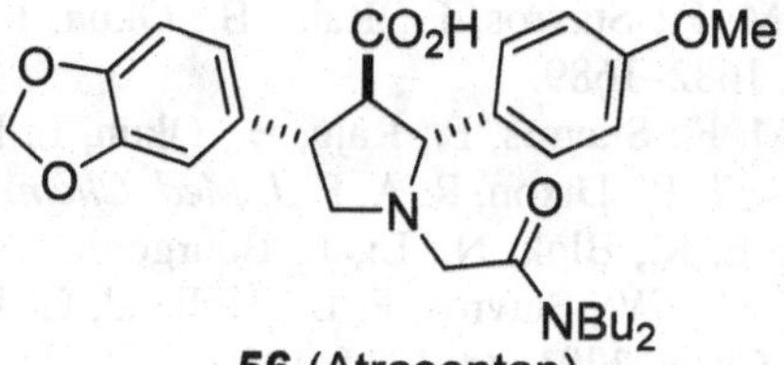

56 (Atrasentan)

15.8 References

1. Simonneau, G.; Galiè, N.; Rubin, L. J.; Langleben, D.; Seeger, W.; Domenighetti, G.; Gibbs, S.; Lebrec, D.; Speich, R.; Beghetti, M.; Rich, S.; Fishman, A. *J. Am. Coll. Cardiol.* **2004**, *43*, 5S–12S.
2. Newman, J. H.; Trembath, R. C.; Morse, J. A.; Grunig, E.; Loyd, J. E.; Adnot, S.; Coccolo, F.; Ventura, C.; Phillips, J. A.; Knowles, J. A.; Janssen, B.; Eickelberg, O.; Eddahibi, S.; Herve, P.; Nichols, W. C.; Elliott, G. *J. Am. Coll. Cardiol.* **2004**, *43*, 33S–39S.
3. Rubin, L. J. *New Engl. J. Med.* **1997**, *336*, 111–117.
4. Gain, S. P.; Rubin, L. J. *Lancet* **1998**, *352*, 719–725.
5. Farber, H. W.; Loscalzo, J. *New Engl. J. Med.* **2004**, *351*, 1655–1665.

6.   Peacock, A. J. *Br. Med. J.* **2003**, *326*, 835–836.
7.   Valerio, C. J.; Kabunga, P.; Coghlan, J. G. *Clin. Med. Ther.* **2009**, *1*, 541–556.
8.   Kabunga, P.; Coghlan, J. G. *Drugs* **2008**, *68*, 1635–1645.
9.   Hoeper, M. M. *Drugs* **2005**, *65*, 1337–1354.
10.   Badesh, D. B.; McLaughlin, V. V.; Delcroix, M.; Vizza, C. D.; Olschewski, H.; Sitbon, O.; Barst, R. J. *J. Am. Coll. Cardiol.* **2004**, *43*, 56S–61S.
11.   Montani, D.; Chaumais, M.-C.; Savale, L.; Natali, D.; Price, L. C.; Jaïs, X.; Humbert, M.; Simonneau, G.; Sitbon, O. *Adv. Ther.* **2009**, *26*, 813–825.
12.   Doherty, A. M. *J. Med. Chem.* **1992**, *35*, 1493–1508.
13.   Bialecki, R. A. *Annu. Rep. Med. Chem.* **2002**, *37*, 41–52.
14.   Channick, R. N.; Sitbon, O.; Barst, R. J.; Manes, A.; Rubin, L. J. *J. Am. Coll. Cardiol.* **2004**, *43*, 62S–67S.
15.   Opitz, C. F.; Ewert, R.; Kirch, W.; Pittrow, D. *Eur. Heart J.* **2008**, *29*, 1936–1948.
16.   Liu, G. *Annu. Rep. Med. Chem.* **2000**, *35*, 73–82.
17.   Burri, K. F.; Breu, V.; Cassal, J.-M.; Clozel, M.; Fischli, W.; Gray, G. A.; Hirth, G.; Löffler, B.-M.; Müller, M.; Neidhart, W.; Ramuz, H.; Trzeciak, A. *Eur. J. Med. Chem.* **1995**, *30*, 385S–389S.
18.   Neidhart, W.; Breu, V.; Bur, D.; Burri, K.; Clozel, M.; Hirth, G.; Müller, M.; Wessel, H. P.; Ramuz, H. *Chimia* **1996**, *50*, 519–524.
19.   Clozel, M.; Breu, V.; Burri, K.; Cassal, J.-M.; Fischli, W.; Gray, G. A.; Hirth, G.; Löffler, B.-M.; Müller, M.; Neidhart, W.; Ramuz, H. *Nature* **1993**, *365*, 759–761.
20.   Breu, V.; Clozel, M.; Burri, K.; Hirth, G.; Neidhart, W.; Ramuz, H. *FEBS Lett.* **1996**, *383*, 37–41.
21.   Mealy, N. E.; Bagès, M. *Drugs Fut.* **2001**, *26*, 1149–1154.
22.   Wu, C.; Chan, M. F.; Stavros, F.; Raju, B.; Okun, I.; Castillo, R. S. *J. Med. Chem.* **1997**, *40*, 1682–1689.
23.   Wu, C.; Chan, M. F.; Stavros, F.; Raju, B.; Okun, I.; Mong, S.; Keller, K. M.; Brock, T.; Kogan, T. P.; Dixon, R. A. F. *J. Med. Chem.* **1997**, *40*, 1690–1697.
24.   Wu, C.; Decker, E. R.; Blok, N.; Li, J.; Bourgoyne, A. R.; Bui, H.; Keller, K. M.; Knowles, V.; Li, W.; Stavros, F. D.; Holland, G. W.; Brock, T. A.; Dixon, R. A. F. *J. Med. Chem.* **2001**, *44*, 1211–1216.
25.   Scott, L. J. *Drugs* **2007**, *67*, 761–770.
26.   Riechers, H.; Albrecht, H.-P.; Amberg, W.; Baumann, E.; Bernard, H.; Böhm, H.-J.; Klinge, D.; Kling, A.; Müller, S.; Raschack, M.; Unger, L.; Walker, N.; Wernet, W. *J. Med. Chem.* **1996**, *39*, 2123–2128.
27.   Burri, K.; Clozel, M.; Fischli, W.; Hirth, G.; Löffler, B.-M.; Neidhart, W.; Ramuz, H. U.S. Pat. 5,292,740, **1994**.
28.   Harrington, P. J.; Khatri, H. N.; DeHoff, B. S.; Guinn, M. A.; Boehler, M. A.; Glasser, K. A. *Org. Process Res. Dev.* **2002**, *6*, 120–124.
29.   Taddel, M.; Naldini, D.; Allegrini, P.; Razzetti, G.; Mantegazza, S. U.S. Pat. 2009/0156811 A1, **2009**.
30.   Haruki, E.; Hirai, Y.; Imoto, E. *Bull. Chem. Soc. Jpn.* **1968**, *41*, 267.
31.   Huddleston, P. R.; Barker, J. M. *Synth. Commun.* **1979**, *9*, 731–734.

32.   Corral, C.; Lissavetzky, J.; Alvarez-Insúa, A. S.; Valdeolmillos, A. M. *Org. Prep. Proc. Int.* **1985**, *17*, 163–167.
33.   Blok, N.; Wu, C.; Keller, K.; Kogan, T. P. U.S. Pat. 5,783,705, **1998**.
34.   Riechers, H.; Klinge, D.; Amberg, W.; Kling, A.; Muller, S.; Baumann, E.; Rheinheimer, J.; Vogelbacher, U. J.; Wernet, W.; Unger, L.; Raschack, M. U.S. Pat. 7,119,097 B2, **2006**.
35.   Jansen, R.; Knopp, M.; Amberg, W.; Koser, S.; Müller, S.; Münster, I.; Pfeiffer, T.; Riechers, H. *Org. Process Res. Dev.* **2001**, *5*, 16–22.
36.   Weber, M. A.; Black, H.; Bakris, G.; Krum, H.; Linas, S.; Weiss, R.; Linseman, J. V.; Wiens, B. L.; Warren, M. S.; Lindholm, L. H. *Lancet* **2009**, *374*, 1423–1431.
37.   Nelson, J. B.; Love, W.; Chin, J. L.; Saad, F.; Schulman, C. C.; Sleep, D. J.; Qian, J.; Steinberg, J.; Carducci, M. *Cancer* **2008**, *113*, 2478–2487.

III

Central Nervous System Diseases

16

Varenicline (Chantix):
An α4β2 Nicotinic Receptor Partial Agonist
for Smoking Cessation

Jotham W. Coe, Frank R. Busch, and Robert A. Singer

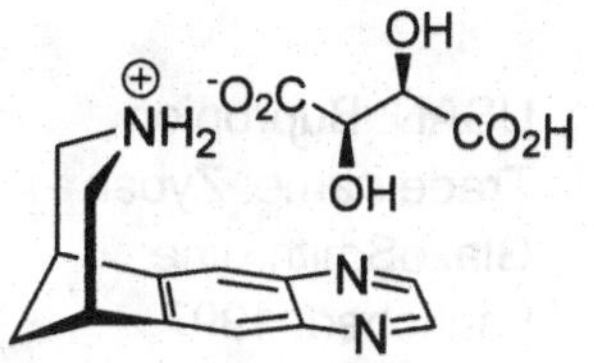

USAN: Varenicline
Trade names: Champix® and Chantix®
Pfizer
Launched: 2006

16.1 Background

Tobacco-related illness is the leading cause of preventable death. If current trends continue, half of the current 1.3 billion smokers worldwide will die from disease directly linked to smoking.[1] Researchers have been measuring the effects of smoking on health since the 1950s.[2] Cumulative results show that half of those who smoke succumb to smoking-related disease and lose an average of 10 years of life.[3] In the United States, scientific and medical awareness of this pandemic emerged in 1964 when the annual Report of the Surgeon General concluded that smoking can cause lung cancer and chronic bronchitis.[4] During the decades that followed, the message from the medical community and health officials has become clearer. Now even passive smoking is considered a health hazard.[5] Still, the challenge has been determining how to aid smokers to quit successfully. Indeed, 70% of smokers want to stop smoking, but > 95% of attempts to quit end in failure owing to the highly addictive nature of smoking and the addictive qualities of nicotine, the active ingredient in tobacco.[6] Nicotine (2) is undisputedly a primary cause of tobacco dependence.

The emergence of treatments for smoking cessation can be traced to smokeless tobacco.[7] In December 1967, the director of pharmaceutical research at AB Leo Läkemedel AB in Sweden, Ove Fernö, received a letter from Stefan Lichtneckert and Claes Lundgren of the University of Lund. Lichtneckert and Lundgren studied the effects of atmospheric pressure on human physiology and noted that some submariners used a smoke-free tobacco product called snus to help them abstain from cigarettes while at sea. Snus was more commonly used by older men in the countryside of Sweden, and they wrote Fernö to suggest that pure nicotine might be a useful smoking cessation aid. Fernö was himself a heavy smoker whose wife disapproved of his smoking. Contemporary work by Murray Jarvik and Michael Russell establishing the connection of nicotine to addiction was not yet fully appreciated, as most at the time ascribed difficulties quitting smoking to its strong habit formation. Fernö, an organic chemist, pursued the delivery of *clean* nicotine, trying several administration forms before deciding on the gum as the safest option. The act of chewing controlled the release of nicotine from the gum giving a self-control element that minimized the risk of intoxication. An ion exchanger incorporated in the gum bound and stabilized the nicotine. Saliva entered upon chewing the gum, releasing nicotinium ions in the oral cavity, but after developing methods to determine nicotine levels in the blood, Fernö discovered that little was being absorbed. Fernö soon realized that added buffer improved absorption.[8]

USAN: Nicotine
Trade name: Nicorette®
Leo Läkemedel AB
Launched: 1978

USAN: Bupropion
Trade name: Zyban®
GlaxoSmithKline
Launched: 1997

Nicotine replacement therapy (NRT) curbs the craving and withdrawal symptoms that smokers experience when quitting.[8] Nicotine Polacrilex, or nicotine gum (e.g., Nicorette) was approved first in Switzerland in 1978 and was introduced in the United States by Merrill Dow in 2 mg strength in 1984 and later in 1992 in 4 mg strength.[9] Today, NRT is available in the United States without a prescription and in many forms, including transdermal patches, lozenges, an inhaler, and a nasal spray. Other forms are available in other markets such as sublingual tablets. NRT on average doubles quit rates vs. placebo in clinical trials. At best, short-term outcomes are ~ 40–50% while placebo rates are ~ 20–25%. Long-term average rates are ~ 20% vs. ~ 10% placebo—all rates depend on the degree of behavioral intervention, which can help smokers achieve higher quit rates. Multiple choices offer individualized management of nicotine exposure to maximize efficacy; however, such choice may complicate the selection of appropriate therapy. Orally, nicotine is poorly bioavailable, with a half-life of only 1–2 h in humans.[10]

Another treatment for smoking cessation, bupropion (3), emerged serendipitously. First marketed in 1985 as Wellbutrin, a safe and well-tolerated antidepressant, it also induced spontaneous smoking reduction and increased smoking cessation rates. Bupropion was reintroduced by GlaxoSmithKline in 1997 as Zyban, the first oral treatment for smoking cessation; it was the first non-nicotine product approved for this purpose.[11] Quit rates similar to those of NRT have been demonstrated in clinical trials with bupropion.[12] Pharmacologic tools such as bupropion add to the means available to smokers who attempt to quit, and they can be further supported by a variety of behavioral modification approaches specifically designed for quit attempts.[13]

The growing worldwide awareness of the health consequences of smoking and the approval of NRT coincided with a significant shift in the perception of nicotine addiction as a treatable neurologic disorder. This shift has encouraged research targeting improved pharmacologic interventions for smoking cessation. The introduction of Zyban was followed 9 years later by that of varenicline (1), the first drug specifically developed to treat nicotine addiction. In this chapter, the pharmacologic profile and syntheses of varenicline (1) are discussed in detail.

16.2 Discovery Chemistry Program

In 1993 Pfizer scientists initiated a project to develop a nicotinic receptor partial agonist to aid smoking cessation. They considered the key aspects of nicotine dependence and addiction that make quitting smoking difficult.[14] Addictive substances have two key features that lead to their dependence and reinforcing effects: In their most addictive routes of administration they achieve relatively rapid exposure in plasma and brain tissues, and exposure is associated with mesolimbic dopamine release in the nucleus accumbens and prefrontal cortex. For example, cocaine is addictive when snorted, but freebased crack cocaine is far more addictive. Another example is heroin, which is diacetylated morphine and serves as a morphine prodrug. Studies have shown that intrathecal (spinal canal) morphine injection achieves peak exposure in the blood in 216 min, whereas the same injection of heroin produces peak *morphine* exposure in 6 min.[15] This rapid rate of rise is a pharmacokinetic characteristic of all substances of abuse, and it highlights the important impact that route of administration has on addictive potential. All dependence-producing and abused substances elevate brain catecholamine levels through pathways consistent with their pharmacology, and all lead to mesolimbic dopamine release in the nucleus accumbens and prefrontal cortex.[16,17]

Nicotine binds at high-affinity nicotinic acetylcholine receptors in the brain (nAChRs). Studies have shown that nAChRs containing β2 subunits in the ventral tegmental area are both necessary and sufficient to sustain the reinforcing effects of nicotine in animals. This relationship was demonstrated in knockout mice lacking β2 subunits by introducing RNA coding for β2 subunits directly into the ventral tegmental area, reinstating a wild-type response to nicotine in models of drug-seeking behavior.[18] Activation of the predominant nicotinic receptors in the ventral tegmental area (> 95 are the α4β2 nAChR subtype) has been shown to cause downstream dopamine release in the nucleus accumbens of the mesolimbic dopamine system and is believed to be the primary mechanism of nicotine addiction. Through this mechanism, the physiologic effects of nicotine from tobacco smoking create reinforcing sensations that support dependence. Abstinence from smoking results in low dopaminergic tone, inducing craving and withdrawal symptoms that trigger the urge to smoke and are a primary cause of relapse in smokers attempting to quit.

Pfizer scientists conceived a medication with a "dual" mechanism of action to treat nicotine dependence from tobacco smoke; they sought an agent that blocked not only the dopamine elevation from smoking but also the dopamine low, or void, during abstinence, thereby addressing both of the sensations that reinforce continued smoking. Accordingly, they targeted nicotinic partial agonists of the neuronal α4β2 nAChR, the primary site of nicotine activity. The hypothesis was that partial activation of this receptor would elicit moderate mesolimbic dopamine levels, blocking the receptor activation and eliminating the pharmacologic reward of smoking. Thus, a partial agonist was uniquely suited as a treatment for nicotine dependence from tobacco smoke.

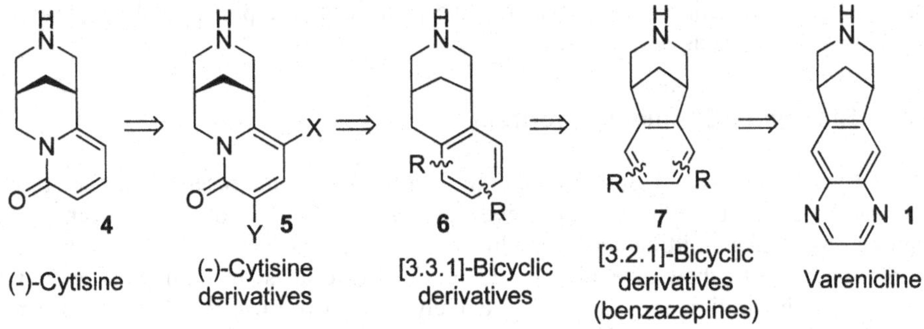

Figure 1

Artificial compound collections and most known natural nicotinic agents were triaged before settling on molecules related to (–)-cytisine (4), a naturally occurring lupin alkaloid found in the seeds of *Laburnum anagyroides* and other plants in the family Leguminosae.[19] Though cytisine was known as early as 1862 and had been the subject of more than a century of chemical and biological inquiry, its nicotinic pharmacology remained uncharacterized until 1994. A potent high-affinity ligand at the α4β2 nAChR, cytisine was identified as a partial agonist by Papke and Heinemann.[20] A literature report suggested that cytisine had potential as a smoking cessation aid, and it is currently marketed in Eastern Europe as Tabex.[21] This application spurred efforts to improve its poor absorption and brain penetration (Figure 1), and further research led to many cytisine derivatives with improved properties (e.g., 5). Novel variations on its [3.3.1]-bicyclic architecture were also pursued (e.g., 6), but these lacked sufficient potency and were often functional antagonists.[22] Improved targets arose after the discovery that [3.2.1]-bicyclic benzazepines (e.g., 7) possessed heretofore unknown nicotinic activity.[23] Development within this series led to the discovery of potent partial agonists that ultimately led to varenicline (1), the active ingredient in Chantix, in 1997. After full development, varenicline was approved for use in the United States by the Food and Drug Administration in 2006.[24,14]

16.3   Pharmacology

Varenicline (1) is a partial agonist of the α4β2 nicotinic acetylcholine receptor, binding with subnanomolar affinity only to α4β2 nAChRs more potently (> 20×) than nicotine. In vitro functional patch clamp studies in human embryonic kidney cells expressing α4β2 nAChRs show that varenicline is a partial agonist with 45% of the maximal efficacy of nicotine. Varenicline demonstrates significantly lower (40–60%) efficacy than that of nicotine in stimulating [3H]-dopamine release from rat brain slices in vitro and in increasing dopamine release from rat nucleus accumbens in vivo. When combined with nicotine in both in vitro and in vivo experiments, varenicline effectively attenuates nicotine-induced dopamine release to the level of varenicline alone, demonstrating its efficacy in antagonizing the effects of concurrent nicotine. In models of nicotine self-administration in rats, varenicline reduces nicotine intake and exhibits lower breakpoints than those of nicotine in drug-seeking self-administration behavior models. These data suggest that varenicline can, to some extent, reproduce the subjective effects of smoking by partially activating α4β2 nAChRs while preventing full activation of these receptors by nicotine. These properties are consistent with the dual mechanism inherent to partial agonism.[25,26]

16.4   Pharmacokinetics and Drug Metabolism

Varenicline (1) is a small, hydrophilic, weak base (log P = 1.1; pKa = 9.9; molecular weight = 211). Studies of its disposition and metabolism showed it to be well absorbed, and after oral administration, 99% of carbon-14-labeled material was recovered in the urine with < 1% in feces in rats, mice, monkeys, and humans. Unchanged varenicline accounted for > 75% of drug-related material in circulation in all species. Drug-related material was observed 1 week post-dose in melanin-containing rat tissues after administration of the carbon-14-labeled compound. N-carbamoyl glucuronidation, N-formylation, a hexose conjugate, and nuclear oxidation metabolites were observed in both excreta and circulation. Protein binding of varenicline is low (f_u ~ 80%; moderate volume of distribution of 1.9 L/kg).

Varenicline is neither a substrate nor an inhibitor of cytochrome P450 enzymes, and it is neither the cause of nor subject to drug interactions via alterations of P450 activities. Nicotine and other tobacco smoke components are subject to P450 metabolism, however. In particular P450 (CYP2A6 and CYP2B6) metabolizes nicotine to cotinine. When quitting smoking, an individual's liver enzyme activity becomes more efficient in response to reduced exposure to these compounds. As a result, other compounds metabolized by P450 enzymes (e.g., warfarin) may need to be monitored or adjusted. Diabetic patients who quit smoking may need to adjust insulin levels, because tobacco-related blood sugar elevation will cease. The plasma half-life of varenicline is 24 h, with peak levels achieved in 4 h and no food effects. Steady state is achieved in 4 days.[27,28]

16.5   Efficacy and Safety

More than 6000 smokers participated in extensive clinical studies of varenicline that revealed its robust efficacy as a pharmacologic aid to smoking cessation. Double-blind, placebo-controlled trials demonstrated superiority over sustained-release bupropion and placebo. Studies in smokers with cardiovascular and chronic obstructive pulmonary disease[29] have demonstrated safety and efficacy. The most common side effects were mild to moderate nausea, sleep disorders, and abnormal dreams that were two to three times placebo rates. Varenicline is neither habit forming nor a scheduled treatment. The careful and methodical clinical evaluation of varenicline led to a U.S. Food and Drug Administration priority review of Pfizer's new drug application for varenicline in November 2005 and subsequent approval in 2006. In 2009, an update to the prescribing information appeared warning of possible neuropsychiatric side effects. A similar warning was added to the bupropion sustained release label. These effects may be associated with nicotine-withdrawal symptoms, and they resolve after ceasing treatment, which is usually associated with a return to smoking. No causal relationship between the medicine and neuropsychiatric events has been demonstrated.[30]

16.6   Syntheses

The Pfizer syntheses of varenicline followed the path of the compound's original discovery from benzazepine (8). The key discovery that benzazepine (8), when suitably protected, converts efficiently to ortho-dinitrated intermediate 9 rather than the meta-isomer (Figure 2) defined both the discovery route and the early process chemistry approaches, as this approach from 8 met the rapid timelines of early development. The brevity and symmetry of these transformations offered a straightforward, late-stage quinoxaline synthesis, and as a result, the primary process chemistry approach to varenicline is essentially identical to the original synthesis. Safety aspects of the original route were thoroughly investigated at numerous stages and rigorously established a safe profile, which was unexpected for a nitration-based sequence. Multiple approaches to 8 proved highly efficient and viable for scale-up, but further research led to yet another route that avoided hazardous intermediates and reagents that required special handling precautions. This synthesis liberated simple, innocuous by-products, alkali metal salts, alcohols, or water, thereby avoiding toxic reagents and making it possible to meet ever-increasing purity and safety specifications.

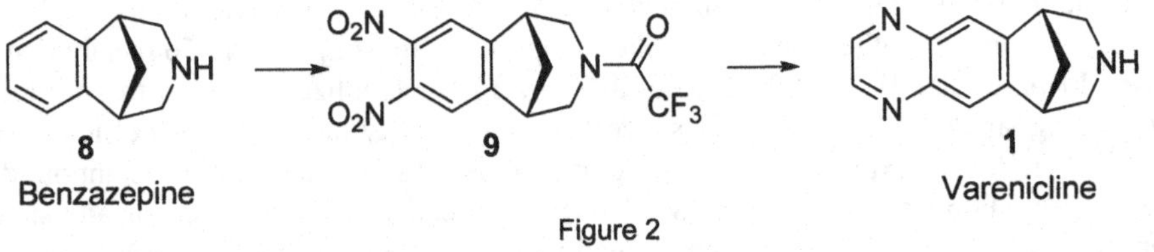

8   **9**   **1**
Benzazepine   Varenicline

Figure 2

The discovery chemistry approach to benzazepine (8) appears in Scheme 1. Benzyne Diels–Alder reaction of 10 was used in the original Wittig/Grignard protocol with cyclopentadiene and magnesium turnings in THF (45–64% yield).[31] Methodological studies improved this step to 89% yield (see Scheme 4). An oxidative cleavage/reductive amination protocol used OsO_4-catalyzed dihydroxylation of benzonorbornadiene (11) via the Van Rheenan procedure followed by oxidative cleavage ($NaIO_4$) of 12 and reductive amination using $NaBH(OAc)_3$ and benzylamine. Hydrogenolysis completed the core benzazepine preparation in 70.8% yield from 11.[32]

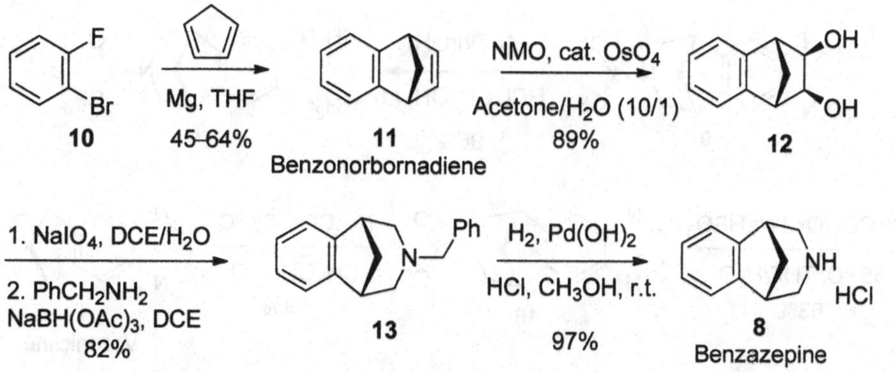

Scheme 1

Conversion of benzazepine (8) to varenicline (1) follows the dinitration path that led to its original discovery (Scheme 2). The bicyclic [3.2.1]-core benzazepine (8) structure holds the amine and aryl functionalities in close proximity. Under electrophilic substitution conditions, electrophilic attack on the aryl ring is inhibited with typical carbamate and amide protection. Even powerful nitrating agents failed to nitrate the aryl ring. Trifluoroacetamide protection was specifically introduced to insulate the nitrogen by removing electron density to avoid the formation of doubly charged cationic intermediates. This protection allowed nitration to proceed. At the time of this discovery, however, a powerful nitrating agent, nitronium triflate, was used specifically to overcome the sluggish nitration of benzazepine derivatives. With trifluoroacetamide protection, nitration of 14 proceeded in excellent yield to give the mononitrated product. Close inspection revealed a small amount of dinitrated material in the crude mixtures, however. On scale-up, with a typical 1.3 eq. of nitronium triflate, the minor dinitrated materials became enriched in the mother liquors from crystallization. This material was isolated and characterized as the *ortho*-dinitration derivative 9. With an excess of 2 eq. of nitronium triflate, dinitrated products were obtained in > 75% yield. Regioselectivity for this conversion likely derives from steric and electronic factors driven by the bicyclic core, leading to 7:1–11:1 preference for the desired *ortho*-dinitro over the *meta*-dinitro regioisomer. Reduction of dinitro compound 9 gives 15, and condensation with glyoxal bisulfite efficiently affords quinoxaline 16. Deprotection generates varenicline (1).[23]

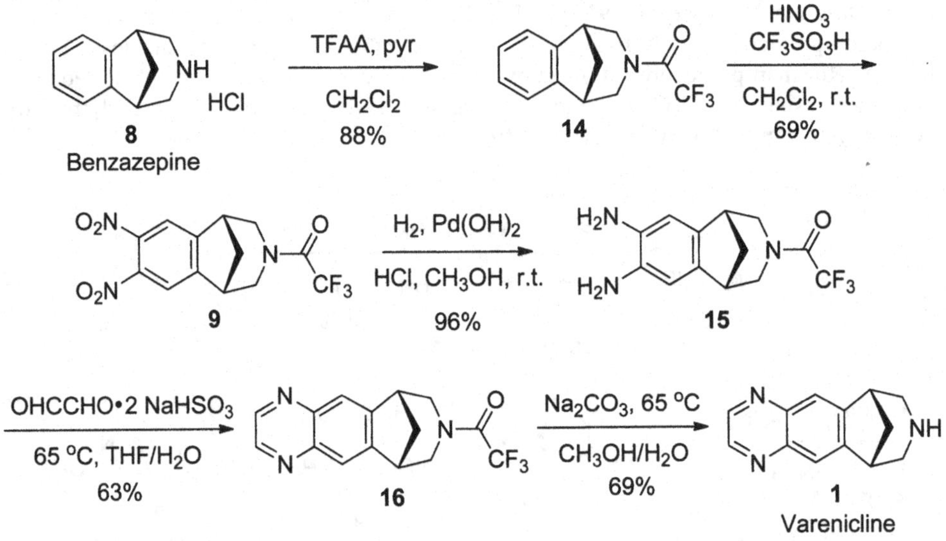

Scheme 2

The process chemistry team carefully considered the benzazepine discovery route to varenicline, as key steps required thorough safety evaluation. Nitrated intermediates are typically avoided in manufacturing routes because they pose dangers in preparation, storage, and handling. Anilines must be fully purged (to ppm levels) from final active pharmaceutical ingredients owing to potential toxicologic concerns. During development, the team implemented extensive characterization and quality controls to ensure purging of nitroaromatic and aniline compounds.[33] Safety assessments of the dinitro intermediates and the nitration reaction conditions demonstrated that the *ortho*-isomer was shock and thermally stable. The reaction conditions themselves were found to pose no safety hazards as long as crude reaction solutions containing the less stable meta-dinitroisomer were not isolated. Isolation of the *ortho*-dinitrated intermediate **9** via crystallization allowed for the controlled disposal of solutions from crude mixtures to reduce potential hazards.

Catalysts and conditions for catalytic reduction were also carefully examined, as reduction of nitrated intermediates also posed dangers and required control strategies. These steps appear late in the quinoxaline synthesis; therefore, full purging of anilines (e.g., **15**) is essential for quality control. After a thorough analysis of all reaction steps, the team concluded that the direct and efficient discovery approach from benzazepine (**8**) to varenicline (**1**) could be safely and efficiently operated in a manufacturing setting with further development.[34]

Specific concerns for process chemistry and manufacture in the discovery synthesis of benzazepine (**8**) included the following:

- Although a commodity chemical, dicyclopentadiene requires cracking at 250–300 °C, necessitating specialized equipment.

- Cyclopentadiene is unstable and requires low-temperature storage and immediate use or dilution for storage in hydrocarbons.
- Benzyne chemistry on scale poses safety hazards requiring considerable evaluation.
- The presence of OsO_4 risked contamination of pilot plant and kilo laboratory equipment, necessitating testing for residual osmium at part per billion levels of all intermediates of dedicated facilities.

The cracking of dicyclopentadiene without dedicated equipment was pursued in the usual manner in the discovery setting without incident, and the team assumed that scale-up would be outsourced to a specialized facility.

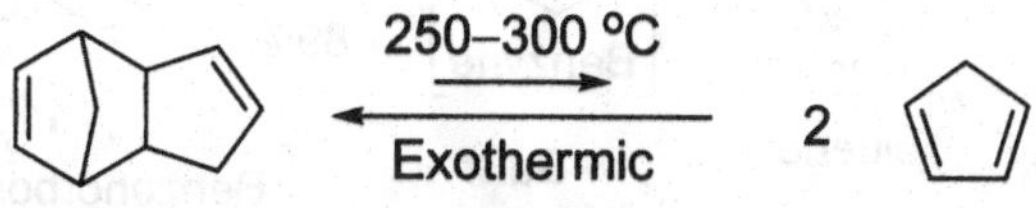

250–300 °C

Exothermic

2

Scheme 3

The instability of cyclopentadiene also required detailed study. Under adiabatic conditions, cyclopentadiene can spontaneously dimerize exothermically and thus should be stored cold for only short duration, especially on kilogram scales (Scheme 3). The impact of container size, volume, temperature, and solvent dilution on storage conditions was carefully studied to assess the potential of a runaway reaction. The published results of this analysis clearly show that hazards increase significantly as the volume of stored cyclopentadiene increases.[35] These generation and storage issues of a simple yet key building block threatened the earliest step in the benzazepine sequence.

The benzyne step also presented scale-up issues. The tremendous potential for benzyne-Diels-Alder chemistry is limited by several factors. Anthranilic acid-based approaches and others that involve gas evolution are specific and involve potentially explosive diazonium intermediates. Precursors are expensive or require lengthy synthesis, and side reactions such as ipso substitution, substrate reduction, and polymerization contribute to poor yields and problematic product isolation. Methods that avoid gas generation included Grignard-type metal exchange or alkali metal amide proton removal processes. Runaway initiation and control issues have discouraged process chemists from implementing these processes, although many halogenated precursors are readily available. More recently, novel benzyne precursors have been used to provide controlled reactions, but these compounds require costly starting materials and require an excess of the Diels-Alder trap for efficiency. These difficulties have decreased general interest in and broad synthetic application of benzyne chemistry.[36] Although the Grignard method originally developed by Wittig in 1958 was effective on a laboratory scale,[31] this method would present additional hazards on a manufacturing scale.[37]

In the course of discovery research, reexamination of the benzyne chemistry led to an easily and safely operated benzyne reaction (Scheme 4).[38] Solvent was found to affect the rate of halogen-metal exchange processes and the basicity of RLi reagents toward cyclopentadiene, a compound with a low pKa (~ 14). Noncoordinating solvents permit halogen-metal exchange at convenient temperatures (0 °C) that support benzyne formation from dibromobenzene in the presence of cyclopentadiene to give 1,4-dihydro-1,4-methano-naphthalene (11) in 89% yield after simple distillation. This reaction solved the benzyne issue on scale, and a number of reactions were operated on multikilogram scales to supply early process campaigns.[34] Cyclopentadiene preparation and usage remained problematic, however.

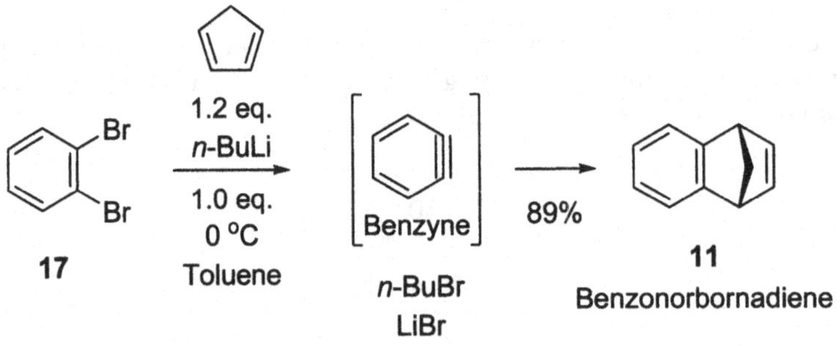

Scheme 4

Oxidative cleavage and reductive amination were shown to be efficient in the discovery laboratory setting (Scheme 1). The efficiency of osmium-catalyzed dihydroxylation of benzonorbornadiene (11) via the Van Rheenan procedure with N-methylmorpholine-N-oxide was confirmed after determining that product precipitated as granular crystals directly from the reaction mixture at high concentration (0.5–1.5 M in 8:1 acetone:H_2O) with rapid stirring. Under these conditions osmium loading was low (0.13–0.26 mol%). Product was readily separated and rinsed with fresh acetone to afford diol 12 in 89% yield. Crystallization of 12 presumably drives the reaction to completion by removing diol from the medium, thus releasing osmium to reenter the catalytic cycle. Geometrically stable dihydroxy benzonorbornadiene osmate esters are believed to resist hydrolysis and sequester osmium, slowing the catalytic cycle. Crystallization of free diol at high concentrations appears to drive the reaction to completion, an example of Le Chatelier's principle. High-purity product was obtained directly from this reaction, and minor byproducts of the benzyne reaction were purged by distillation.[39]

Despite the efficiency and purity of this reaction, however, our engineering group required that residual OsO_x contaminants be purged not only from the synthetic products but also from large-scale multipurpose production equipment. Inductively coupled plasma mass spectrometry accurately determined residual osmium levels and could be used to monitor residuals to as low as 0.04 ppm. Silica gel filtration of reductive amination product 13 (from the next step) effectively removed all residual osmium (< 0.04 ppm). Nevertheless, dedicated equipment or facilities would be required for this step if the synthesis were to be considered for commercial manufacture.[34]

In the discovery synthesis of 8 (Scheme 1), diol 12 was converted to benzyl benzazepine 13 in a two-step procedure. Oxidative cleavage of 12 with $NaIO_4$ (1 eq.) in aqueous dichloroethane gave dialdehyde intermediates that likely exist as bridged hydrates. After being washed with water to remove NaIO$_x$ salts, the extracts were combined with benzylamine (1.05 eq.) in dichloroethane. Intramolecular iminium ion formation presumably occurs, as water separates from this mixture and can be removed with phase separation and filtration through cotton. This mixture was added to a slurry of NaBH(OAc)$_3$ (3.6 eq.) in dichloroethane with aqueous workup, and silica pad filtration gave osmium-free **13** in 82–85% yield. The hydrochloride salt of **13** was hydrogenolyzed to provide benzazepine (**8**) as its HCl salt in 97% yield.

Although the discovery sequence to benzazepine (**8**) is high yielding (64–73%) from dibromobenzene (**17**)—the route would be viable for manufacture only if key challenges were addressed (i.e., management of cyclopentadiene, benzyne, and oxidation). Anticipating that benzonorbornadiene may become a commodity chemical, a tandem ozonolysis process was developed to allow osmium-free conversion of benzonorbornadiene (**11**) to benzazepine (**8**, Scheme 5). Ozonolysis of **8** in MeOH generated methoxyhydroperoxide glycal **18**. Reduction with 5% Pt/C under hydrogen gave methoxy glycal **19**, which was treated with benzylamine and formic acid after which hydrogenation was resumed to produce **13**. Cleavage of the *N*-benzyl group via hydrogenolysis using Pearlman's catalyst with tosic acid gave **8**·as the TsOH salt, which was crystallized in 28% yield overall from **11**, demonstrating a viable manufacturing option for future consideration.[39]

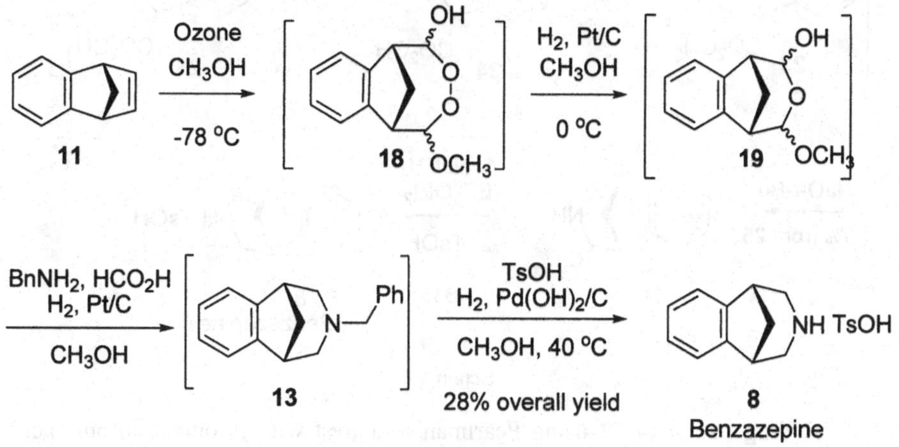

Scheme 5

The dinitration route from benzazepine to varenicline was high yielding, direct, and sufficiently safe for commercial operation. These characteristics defined the process chemistry objective: to enable a benzazepine synthesis for manufacturing. The preferred process route to varenicline evolved with dedicated research, however, as described below.

Process chemists studied an approach to **8** using the readily available indan-1-one-3-carboxylic acid **20** prepared from phenyl succinic anhydride via Friedel–Crafts closure (Scheme 6). After esterification (yielding **21**), homologation of the carbonyl via aminomethylation proceeded in high efficiency to give **25**. In the presence of trimethylsilyl cyanide (1.2 eq.) and catalytic ZnI_2 (0.01 eq.), cyanohydrin **22** was formed on heating in toluene at 50 °C for several hours as a 2:1 mixture favoring the undesired *trans*-diastereoisomer. Batch variability of ZnI_2 plagued this step. Addition of catalytic iodine improved the reproducibility and yield, presumably through the in situ formation of the more reactive and soluble Lewis acid $Zn(I_3)_2$. Added acetonitrile increased the overall solvent polarity and catalytic activity to a level similar to that observed using methylene chloride, giving cyanohydrin **22** in quantitative yield.

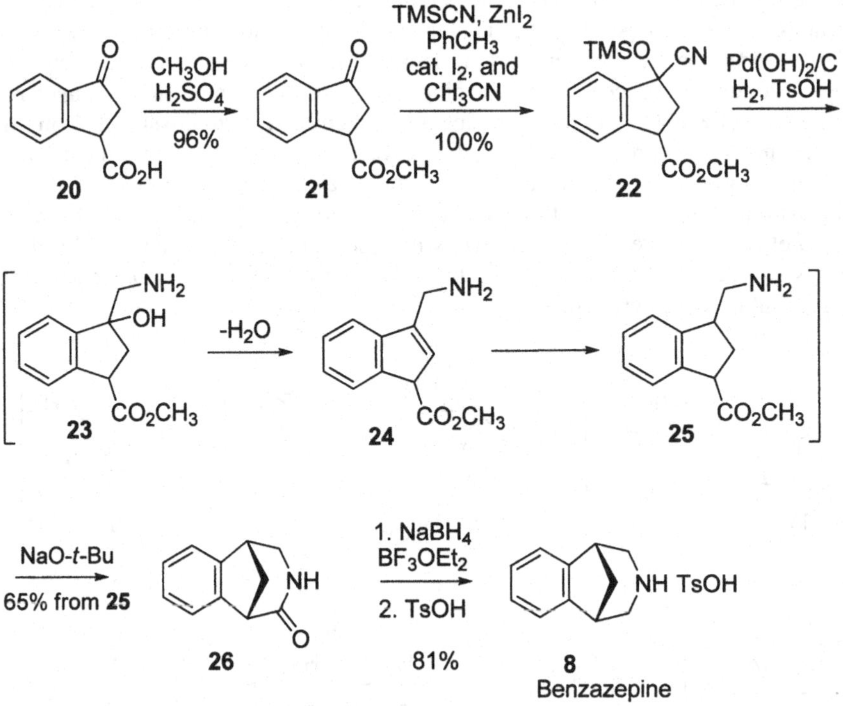

Scheme 6

Hydrogenation of **22** using Pearlman's catalyst with *p*-toluenesulfonic acid or sulfuric acid (> 1 eq.) gave amino-ester **25**, but only after a critical aqueous workup to remove lingering (trace) cyanide—which effectively poisoned the catalyst—if carried forward from the prior step. Acid served to speed nitrile reduction through protonation of intermediate amines **23–25** to prevent catalyst deactivation and to catalyze dehydration of benzylic alcohol **23**. Under these conditions the resulting indene **24** was preferentially reduced, giving **25** in a 10:1 *cis:trans* product ratio. This crude material was converted to a highly crystalline-bridged bicyclic amide **26** under basic conditions, and **26** was efficiently reduced with in situ–generated borane to give benzazepine **8** as its tosylate salt.[40,34]

As with the ozonolysis route, this single-pot approach telescopes via sequential hydrogenation, dehydration, reduction, and basic lactam cyclization chemistry. Although it is possible to remove residual cyanide by aqueous workup, its use was viewed as an unacceptable safety hazard. Given the efficient lactamization of **25**, the highly crystalline state of lactam **26**, and its ready conversion to benzazepine, the task shifted to the optimization of synthetic paths to indane lactam **26** precursors.

One of the most direct annulation strategies to construct the benzazepine core relied on fusion of piperidones and disubstituted arenes (Scheme 7).[41] Enolate coupling to aryl halides was envisioned via both metal-catalyzed approaches and S_NAr couplings of nitro-substituted arenes. In theory the mononitroarene could serve a dual purpose: facilitating couplings and supplying one of two necessary quinoxaline nitrogen atoms in varenicline. Carboxyl-substituted piperidone **28** coupled with **27** in high yield; however, all attempts to close **29** to the indanone **30**—using both S_NAr and metal-mediated methods—failed. Geometric considerations were deemed a likely culprit, but switching the steps (first coupling using palladium-mediated methods) also failed, suggesting that nitro functionality could impede such processes.[42]

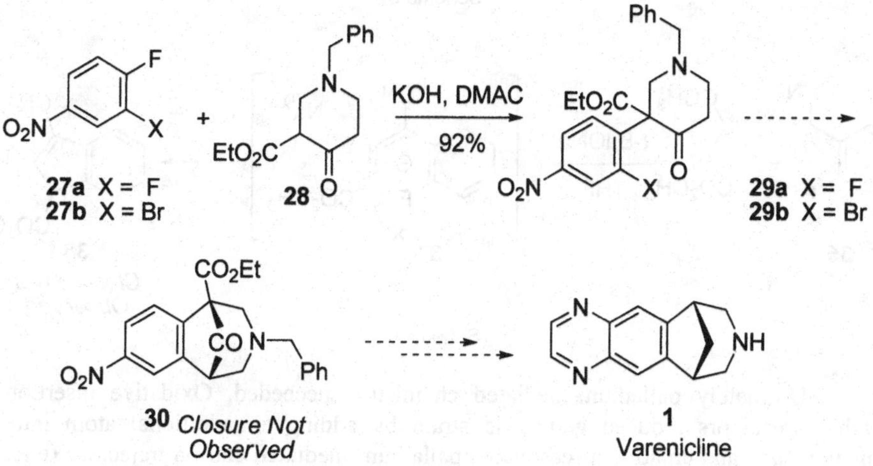

Scheme 7

In a similar fashion, diethyl glutaconate **31** coupled readily with fluoro-nitroarenes **27a** or **27b** to give **32a** or **32b**. Ring closure again failed both via S_NAr (**32a**, X = F) and via palladium catalysis (**32b**, X = Br, Scheme 8). The newly formed allylic anion of **32** presumably delocalized into the nitroaromatic ring, generating highly electron-rich character that inhibited both closure mechanisms. Subsequent investigations relied on nitro-free intermediates.

A brief exploration of annulations with nitro-free fluoro arene **37**—readily prepared from (*E*)-methyl 3-methoxyacrylate (**36**) by a Michael addition elimination strategy with 2-(2-fluorophenyl)acetonitrile (**35**)—again revealed that direct S_NAr cyclization was impeded under all conditions. Extended conjugation may have contributed; however, the necessity for the conjugated sp2-hybridized aryl-propene anion to couple intramolecularly with another sp^2-hybridized carbon atom imposed a highly strained reaction trajectory (Scheme 9).

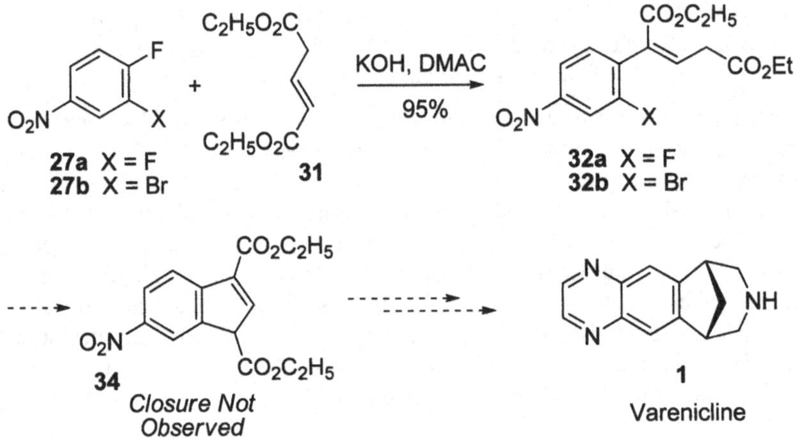

Scheme 8

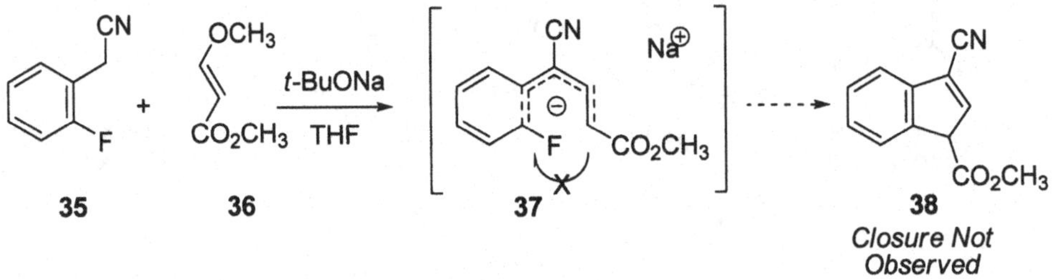

Scheme 9

Ultimately palladium-mediated chemistry succeeded. Oxidative insertion on suitable precursors reduced geometric strain by adding an additional atom into the transition state and creating a reasonable palladium-mediated end-on trajectory (e.g., **41**; Scheme 10). Although the allylic anion could impede oxidative addition in the catalytic cycle through interactions with palladium—as was likely the case with the nitro-substituted systems (see Schemes 7 and 8)—such interactions were likely to favor the *cis*-allylic anion. Initial experiments using catalytic triarylphosphines (e.g., 0.1 eq. PPh$_3$ or diphenylphosphinoferrocene) and various palladium sources (5 mol%) failed under thermal conditions with *t*-BuONa in THF. Delocalization of the allylic anion to the aryl bromide was believed to impede oxidative insertion using palladium, and increasing the electron density on palladium posed one option to overcome this problem. More electron-rich phosphines did indeed facilitate oxidative insertion and proved successful. Indene **38** formed with high efficiency from **40** using 2-dicyclohexyl-biphenylphosphine or tricyclohexylphosphine in 1,2-dimethoxyethane or THF at 60 °C over 2–3 h.[43] Indene **38** was obtained in a one-pot condensation of **39** with acrylate **36** with stoichiometric *t*-BuONa in the presence of the palladium catalyst in 77% yield.

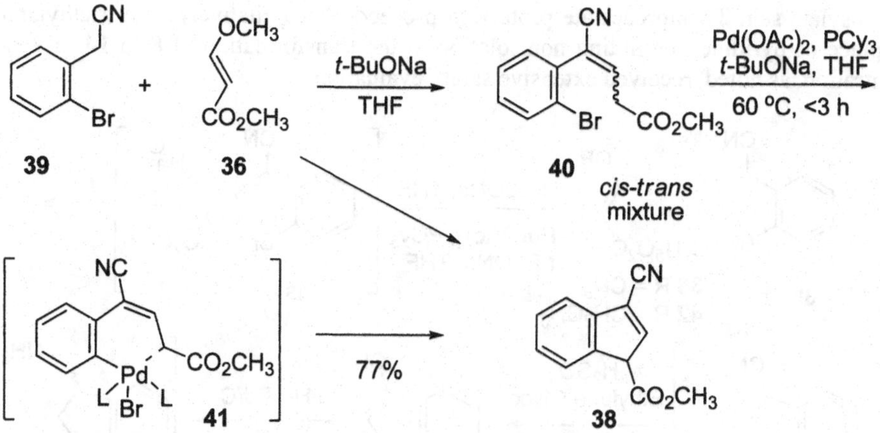

Scheme 10

Reaction efficiency was improved using ethoxy acrylate (42, R = Et; Scheme 11) to avoid methoxide generation. Methoxide is capable of forming bridged dimeric palladium complexes that hinder palladium catalysis, analogous to the process observed with hydroxide. The use of ethoxy acrylate reduced palladium catalyst loading from 5–10 mol% to 0.5–2.0 mol%.

Isolation of neutral indene 38 posed challenges owing to its poor stability within some pH ranges. Furthermore, a purge of phosphines was necessary to enable downstream hydrogenation. Under the reaction conditions the product resides as a single isomer of cyanobenzofulvene olefin sodium salt 44 (determined using single-crystal X-ray diffraction; Scheme 11) but was impractical for routine isolation.[34,43] Ethylene glycol ketene ketal 45 was readily formed in situ by adding ethylene glycol and acid to the reaction mixture after the sodium salt 44 had formed. Sulfuric acid neutralizes the base, catalyzes the dehydration, and forms the ketal. On optimization, ketal 45 precipitated with prolonged stirring at room temperature (12–48 h). Isolation using this method efficiently purged phosphines and other by-products, giving highly pure indene 45 in 75–95% yield from 2-bromophenylacetonitrile (39).

Conversion to benzazepine was achieved under acidic conditions via hydrogenation in alcoholic solvents to form 25, which did not require isolation. Intermediate 25 was directly cyclized under basic conditions to give crystalline lactam 26. The base-mediated ring closure proceeded from the cis/trans mixture of 25 by epimerization of the benzylic methine permitting formation of the cyclic amide from the cis isomer. Finally, conversion of 26 to the desired benzazepine (8) was accomplished by in situ borane reduction and 8 was isolated as the tosylate salt in 81% yield.

This route was efficient and robust, using only inexpensive commodity chemicals without hazardous reagents. By intercepting crystalline intermediates, high states of purity were achieved without chromatography. The complete seven-step synthetic sequence used for manufacturing varenicline involves only filtration processes to remove catalyst and isolate crystalline intermediates in an overall yield of > 50%.[44a] With an efficient route to 8 in hand, the research team made refinements to the synthesis of 1 (Scheme 12). They found it more effective to isolate 8 as the HCl salt rather than as the tosylate salt. Trifluoroacetate protection proceeded quantitatively with triethylamine in place of pyridine, permitting nonisolation in the transformation of **8** to **14** to **9**. The dinitration, as noted, received extensive safety evaluation.

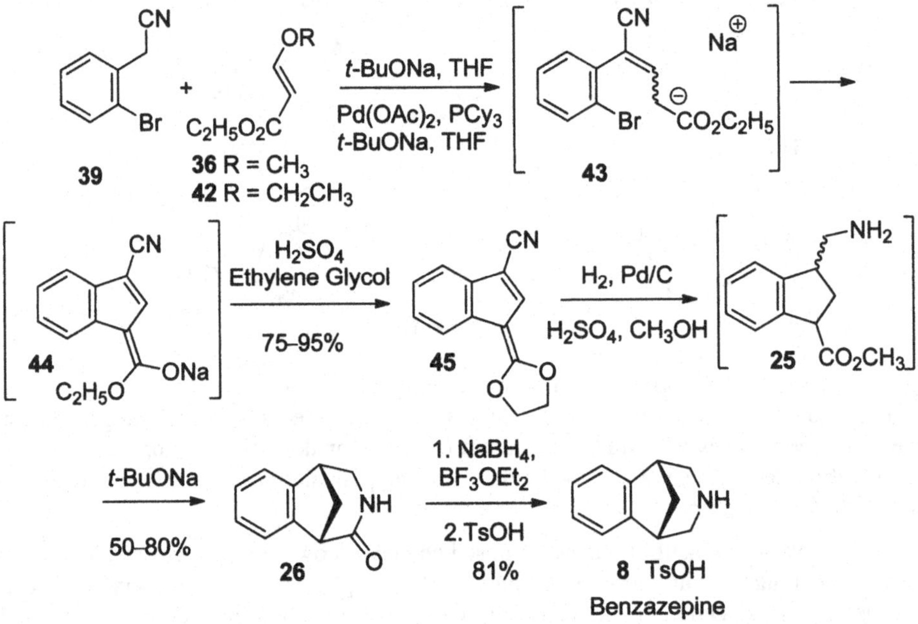

Scheme 11

Dichloromethane, a nongreen solvent atypical in commercial processes, was intentionally selected for its low boiling point (41 °C) and nonflammable properties. The low boiling point provides additional control in the nitration process, limiting the reaction to temperatures with a > 100 °C safety margin between the operating temperature and temperatures at which decomposition of compound **9** begins. After a nitration quench, the crude product is filtered but not dried. Instead it is processed directly to recrystallization to remove the undesired *meta*-dinitro isomer. Purified **9** can then be reduced using palladium on carbon under hydrogenation conditions. Although dianiline **15** is stable in solid form, it decomposes after prolonged standing in solution, thus processes were set up without isolation of **15** using direct addition of aqueous glyoxal to form the desired quinoxaline **16**. Added bicarbonate controlled the pH to avoid unwanted side products in this step.[44,45] The process was completed with rapid and quantitative trifluoroacetamide hydrolysis via treatment with sodium hydroxide, thus combining steps via nonisolated intermediates and telescoping directly into the formation of the *L*-tartrate salt in methanol. The reaction control of the tartrate salt polymorph to the desired form B has been described by Rose and co-workers.[46]

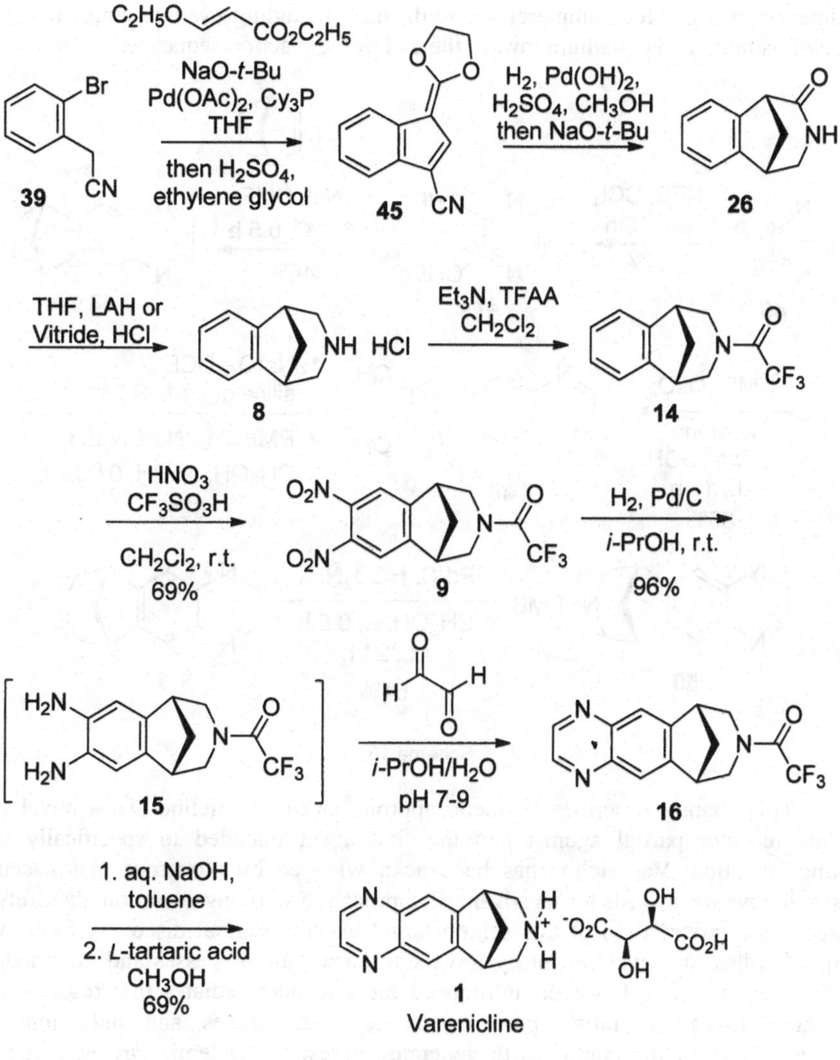

Scheme 12

A Diels-Alder approach to varenicline was recently published by Dr. Reddy's Laboratories. Entry to a key bicyclic intermediate is achieved by an iodide-catalyzed Diels-Alder reaction of tetrabromo dimethyl pyrazine (47) with excess norbornadiene. Dihydroxylation of 48, oxidative cleavage, and reductive amination prepares *N-p*-methoxybenzyl varenicline (50), which is deprotected under transfer hydrogenation conditions to give varenicline (1) in 10% yield for the sequence.[47] This approach continues the theme of building the piperidine of 1 through olefin oxidative cleavage and reductive amination, but by doing so late in the sequence; however, the approach introduces challenges for commercial manufacture, including cyanide containment and the use of osmium and palladium toward the end of the reaction sequence.

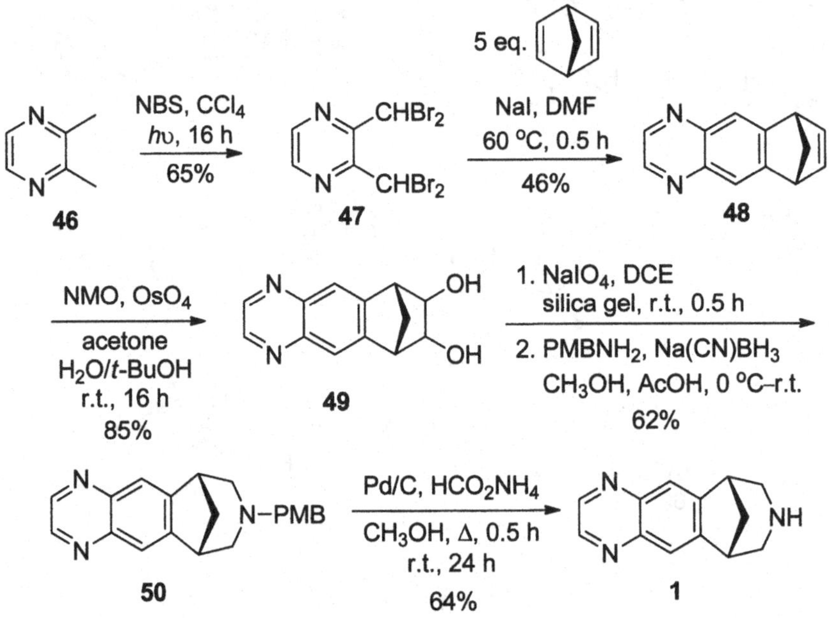

Scheme 13

     This chapter describes synthetic approaches to varenicline (1), a novel α4β2 nicotinic receptor partial agonist and the first agent intended to specifically target smoking cessation. Varenicline has been acknowledged by numerous pharmaceutical industry innovation awards for its advanced approach to smoking cessation, its safety and efficacy.[48] The preparation of this achiral target via the original discovery route posed multiple hurdles for development by traversing hazardous processes and intermediates. The commercial route, however, introduced racemic intermediates that resolve in the final steps while generating pure crystalline intermediates and only innocuous byproducts. Varenicline continues to generate interest in academic circles, and recent synthetic approaches highlight the unique challenges posed by this achiral target.

Acknowledgments

Discovering a new pharmaceutical product requires the input of many scientists. The authors thank all of the dedicated professionals at Pfizer who contributed to the development of varenicline. We'd like to thank Mikael Franzon for his historical perspective and contribution to the discussion on the development of NRT.

16.7   References

1.   WHO, *Report on the Global Tobacco Epidemic, 2009: Implementing Smoke-Free Environments*. Geneva,

http://whqlibdoc.who.int/publications/2009/9789241563918_eng.pdf , 2009, accessed October, 2009.

2. Doll, R.; Hill, A. B. *Br. Med. J.* **1954**, 1451.

3. Doll, R.; Peto, R.; Boreham, J.; Sutherland, I. *Br. Med. J.* **2004**, *328*, 1519.

4. Bayne-Jones, S.; Burdette, W. J.; Cochran, W. G.; Farber, E.; Fieser, L. F.; Furth, J.; Hickam, J. B.; LeMaistre, C.; Schuman, L. M.; Seevers, M. H. *Smoking and Health: Report of the Advisory Committee to the Surgeon General of the Public Health Service.* U.S. Government Printing Office, **1964**, Washington.

5. U.S. Department of Health and Human Services, Centers for Disease Control and Prevention, Coordinating Center for Health Promotion, National Center for Chronic Disease Prevention and Health Promotion, Office on Smoking and Health. *The Health Consequences of Involuntary Exposure to Tobacco Smoke: A Report of the Surgeon General—Executive Summary.* U.S. Government Printing Office, **2006**, Washington.

6. Centers for Disease Control and Prevention, *Early Release of Selected Estimates Based on Data from the January–September 2008 National Health Interview Survey*, U.S. data, **2008**, U.S. Printing Office, Washington.

7. (a) Nordgren P.; Ramström L. *Br. J. Addict.* **1990**, *85*, 1107–1112. (b) Henningfield J. E.; Fagerström K.-O. *Tob. Control* **2001**, *10*, 353–357. (c) Foulds J.; Ramström L.; Burke, M.; Fagerström, K.-O. *Tob. Control* **2003**, *12*, 349–359. (d) Zhu, S.-H.; Wang, J. B.; Hartman, A.; Zhuang, Y.; Gamst, A.; Gibson, J. T.; Gilljam, H.; Galanti, M. R. *Tob. Control* **2009**, *18*, 82–87.

8. Fagerström, K.-O. *J. Behav. Med.* **1982**, *5*, 343.

9. (a) Hjalmarson A. I. M. *J. Am. Med. Assoc.* **1984**, *252*, 2835–2838. (b) Herrera, N.; Franco, R.; Herrera, L.; et al. *Chest* **1995**, *108*, 447–461.

10. (a) Stead, L. F.; Perera, R.; Bullen, C.; Mant, D.; Lancaster, T. *Cochrane Database of Systematic Reviews*, **2008**, *1*, CD000146. DOI: 10.1002/14651858.CD000146.pub3. (b) Fagerström, K.-O.; Axelsson, A.; Sorelius L. *Soc. Res. Nicotine & Tobacco, Newslet.* **2008**, *14*.

11. (a) Hughes, J. R. *Cancer J. Clin.* **2000**, *50*, 143. (b) Fiore, M. C.; Bailey, W. C.; Cohen, S. J.; Dorfman, S. F.; Fox, B. J.; Goldstein, M. G.; Gritz, E. R.; Hasselblad, V.; Heyman, R. B.; Jaén, C. R.; Jorenby, D.; Kottke, T. E.; Lando, H. A.; Mecklenburg, R. E.; Mullen, P. D.; Nett, L. M.; Piper, M.; Robinson, L.; Stitzer, M. L.; Tommasello, A. C.; Villejo, L.; Welsch, S.; Wewers, M. E.; Baker, T. B.; Bennett, G.; Heishman, S. J.; Husten, C.; Kamerow, D.; Melvin, C.; Morgan, G.; Ernestine, M. C.; Orleans, T. *J. Am. Med. Assoc.* **2000**, *283*, 3244. (c) Fiore, M. C.; Bailey, W. C.; Cohen, S. J.; Dorfman, S. F.; Goldstein, M. G.; Gritz, E. R.; Heyman, R. B.; Jaén, C. R.; Kottke, T. E.; Lando, H. A.; Mecklenburg, R. E.; Mullen, P. D.; Nett, L. M.; Robinson, L.; Stitzer, M. L.; Tommasello, A. C.; Villejo, L.; Wewers, M. E. *Treating Tobacco Use and Dependence, U.S.* Department of Health and Human Services, U.S. Public Health Service, **2000**, U.S. Government Printing Office, Washington.

12. Hughes, J. R.; Stead, L. F.; Lancaster, T. *Cochrane Database of Systematic Reviews* **2007**, *1*, CD000031. DOI: 10.1002/14651858.CD000031.pub3.

13. Hughes J. R.; Keely, J.; Naud, S. *Addiction* **2004**, *99*, 29–38.

14.   Coe, J. W.; Rollema, H.; O'Neill, B. T. *Ann. Reports Med. Chem.* **2009**, *44*, 71–101.
15.   Kotob, H. I. M.; Hand, C. W.; Moore, R. A.; Evans, P. J. D.; Wells, J.; Rubin, A. P.; McQuay, H. J. *Anesth. Analg.* **1986**, *65*, 718–722.
16.   Benowitz, N. L. *Ann. Rev. Med.* **1986**, *37*, 21–32.
17.   Goldstein, R. Z.; Volkow, N. D. *Am. J. Psychiatry* **2002**, *159*, 1642–1652.
18.   Maskos, U.; Molles, B. E.; Pons, S.; Besson, M.; Guiard, B. P.; Guilloux, J.-P.; Evrard, A.; Cazala, P.; Cormier, A.; Mameli-Engvall, M.; Dufour, N.; Cloëz-Tayarani, I.; Bemelmans, A.-P.; Mallet, J.; Gardier, A. M.; David, V.; Faure, P.; Granon, S.; Changeux, J.-P. *Nature* **2005**, *436*, 103–107.
19.   Stead, D.; O'Brien, P. *Tetrahedron* **2007**, *63*, 1885–1897.
20.   Papke, R. L.; Heinemann, S. F. *Mol. Pharmacol.* **1994**, *45*, 142–149.
21.   Etter, J.-F. *Arch. Intern. Med.* **2006**, *166*, 1553–1559.
22.   Coe, J. W.; Vetelino, M. G.; Bashore, C. G.; Wirtz, M. C.; Brooks, P. R.; Arnold, E. P.; Lebel, L. A.; Fox, C. B.; Sands, S. B.; Davis, T. I.; Rollema, H.; Schaeffer, E.; Schulz, D. W.; Tingley, F. D. III; O'Neill, B. T. *Bioorg. Med. Chem. Lett.* **2005**, *15*, 2974–2979.
23.   Coe, J. W.; Brooks, P. R.; Vetelino, M. G.; Wirtz, M. C.; Arnold, E. P.; Sands, S. B.; Davis, T. I.; Lebel, L. A.; Fox, C. B.; Shrikhande, A.; Schaeffer, E.; Rollema, H.; Lu, Y.; Mansbach, R. S.; Chambers, L. K.; Rovetti, C. C.; Schulz, D. W.; Tingley, F. D. III; O'Neill, B. T. *J. Med. Chem.* **2005**, *48*, 3474–3477.
24.   Coe, J. W.; Brooks, P. R.; Wirtz, M. C.; Bashore, C. G.; Bianco, K. E.; Vetelino, M. G.; Arnold, E. P.; Lebel, L. A.; Fox, C. B.; Tingley, F. D. III; Schulz, D. W.; Davis, T. I.; Sands, S. B.; Mansbach, R. S.; Rollema, H.; O'Neill, B. T. *Bioorg. Med. Chem. Lett.* **2005**, *15*, 4889–4897.
25.   Rollema, H.; Coe, J. W.; Chambers, L. K.; Hurst, R. S.; Stahl, S. M.; Williams, K. E. *Trends Pharmacol. Sci.* **2007**, *28*, 316–325.
26.   Rollema, H.; Chambers, L. K.; Coe, J. W.; Glowa, J.; Hurst, R. S.; Lebel, L. A.; Lu, Y.; Mansbach, R. S.; Mather, R. J.; Rovetti, C. C.; Sands, S. B.; Schaeffer, E.; Schulz, D. W.; Tingley, F. D. III; Williams, K. E. *Neuropharmacology* **2007**, *52*, 985–994.
27.   Obach, R. S.; Reed-Hagen, A. E.; Krueger, S. S.; Obach, B. J.; O'Connell, T. N.; Zandi, K. S.; Miller, S. A.; Coe, J. W. *Drug Metab. Disp.* **2006**, *34*, 121–30.
28.   Faessel, H. M.; Smith, B. J.; Gibbs, M. A.; Gobey, J. S.; Clark, D. J.; Burstein, A. H. *J. Clin. Pharmacol.* **2006**, *46*, 991–998.
29.   Ebbert et al. *Int. J. Chron. Obstruct. Pulmon. Dis.* **2009**, *4*, 421–430.
30.   Cahill, K.; Stead, L. F.; Lancaster, T. *Cochrane Database Syst. Rev.* **2008**, 3, CD006103. DOI: 10.1002/14651858. CD006103.pub3.
31.   Wittig, G.; Knauss, E. *Chem. Ber.* **1958**, *91*, 895–907.
32.   Brooks, P. R.; Caron, S.; Coe, J. W.; Ng, K. K.; Singer, R.A.; Vazquez, E.; Vetelino, M. G.; Watson Jr., H. H.; Whritenour, D. C.; Wirtz, M. C. *Synthesis* **2004**, *11*, 1755–1758.
33.   Busch, F. R.; Bronk, K. S.; Withbroe, G. J.; Sinay, T. U.S. Pat. 2007/0224690, **2007**.
34.   Coe, J. W.; Watson, Jr., H. A.; Singer, R. A. *Varenicline: Discovery Synthesis and Process Chemistry Developments* in "Process Chemistry in the

Pharmaceutical Industry, Challenges in an Ever Changing Climate," ed. Gadamasetti, K.; Braish, T., Francis & Taylor, **2007**, Boca Raton.

35. am Ende, D. J.; Whritenour, D. C.; Coe, J. W. *Org. Process Res. Dev.* **2007**, *11*, 1141–1146.

36. Riggs, J. C.; Ramirez, A.; Cremeens, M. E.; Bashore, C. G.; Candler, J.; Wirtz, M. C.; Coe, J. W.; Collum, D. B. *J. Am. Chem. Soc.* **2008**, *130*, 3406–3412.

37. Kryka, H.; Hessela, G.; Schmitta, W.; Tefera, N. *Chem. Eng. Sci.* **2007**, *62*, 5198–5200.

38. Coe, J. W.; Wirtz, M. C.; Bashore, C. G.; Candler, J. *Org. Lett.* **2004**, *6*, 1589–1592.

39. Brooks, P. R.; Caron, S.; Coe, J. W.; Ng, K. K.; Singer, R.A.; Vazquez, E.; Vetelino, M. G.; Watson Jr., H. H.; Whritenour, D. C.; Wirtz, M. C. *Synthesis* **2004**, *11*, 1755–1758.

40. O'Donnell, C. J.; Singer, R. A.; Brubaker, J. D.; McKinley, J. D. *J. Org. Chem.* **2004**, *69*, 5756–5759.

41. Satyanarayana, G; Maier, M. E. *Tetrahedron* **2008**, *64*, 356–363.

42. Gallo, E.; Ragaini, F.; Cenini, S.; Demartin, F. *J. Organomet. Chem.* **1999**, *586*, 190–195.

43. Singer, R. A.; McKinley, J. D.; Barbe, G.; Farlow, R. A. *Org. Lett.* **2004**, *6*, 2357–2360.

44. (a) Handfield, Jr., R. E.; Watson, T. J. N.; Johnson, P. J.; Rose, P. R. U.S. Pat. 2007/7285686, **2007**. (b) Busch, F. R.; Withbroe, G. J.; Watson, T. J.; Sinay, T. G.; Hawkins, J. M.; Mustakis, I. G. U.S. Pat. 2008/0275051, **2008**.

45. McCurdy, V. E.; Ende, M. T.; Busch, F. R.; Mustakis, J.; Rose, P.; Berry, M. R. *Pharm. Eng.* **2010**, *30*, accepted.

46. Bogle, D. E.; Rose, P, R.; Williams, G. R. U.S. Pat. 6890927, **2005**.

47. Pasikanti, S.; Reddy, D. S.; Venkatesham, B.; Dubey, P. K.; Iqbal, J.; Das, P. *Tetrahedron Lett.* **2010**, *51*, 151–152.

48. Champix received the 2009 Canadian Prix Galien Innovative Product Award. Chantix was awarded the 2007 United States Prix Galien Award for Best Small Molecule Medicine and the 2006 Scrip for Best Small Molecule Drug Award and was a finalist in the International Prix Galien in 2008.

17

Donepezil, Rivastigmine and Galantamine: Cholinesterase Inhibitors for Alzheimer's Disease

Subas Sakya and Kapil Karki

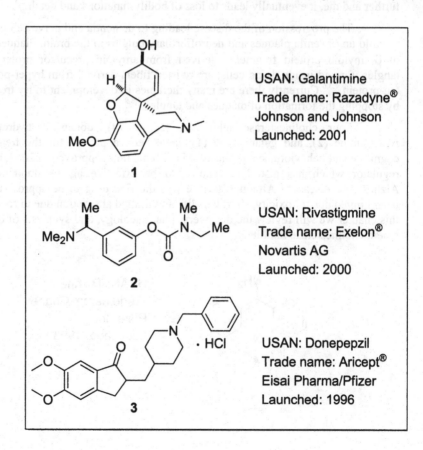

USAN: Galantimine
Trade name: Razadyne®
Johnson and Johnson
Launched: 2001

1

USAN: Rivastigmine
Trade name: Exelon®
Novartis AG
Launched: 2000

2

USAN: Donepepzil
Trade name: Aricept®
Eisai Pharma/Pfizer
Launched: 1996

3

17.1   Background

Alzheimer's disease (AD) is a form of dementia that is associated with memory loss. The first dementia symptom characterization of the disease and the underlying brain pathology, senile plaque deposition, was published by Alois Alzheimer in 1907.[1] Since then the prevalence of the disease has increased worldwide, while the search for a cure only began about 30 years ago. It affects an estimated 5.3 million people in the United States alone. The disease is more prevalent among elderly, with > 10% of people age 65 getting the disease. Prevalence rates increase with age and almost 50% of seniors aged 85 have been reported to have the disease. Worldwide, there are approximately 35 million people with AD, and that number is expected to grow to 107 million by 2050. The direct cost in 2005 associated with the disease was $91 billion in Medicare costs alone for care of beneficiaries with AD and other dementia. The cost is predicted to rise annually for the U.S. and European regions for the care of AD patients, including lost productivity and medical care.[2]

The cause of the disease is not clear but the implication of plaque buildup and a decrease in neuronal function is linked to the severity of the disease. The biggest risk factor for the disease is increasing age, but family history and genetics are also important. The disease progresses slowly in the beginning with gradual loss of cognitive function, resulting in short-term memory loss. At a later stage, the patients experience reduced capacity to perform activities of daily living (ADL), and behavior issues begin to manifest. At advanced stages of the disease, as brain nerve cells begin to deteriorate further and die, it eventually leads to loss of bodily functions and death.[2,3]

The progression of the disease leading to dementia and nerve cell death is linked to build up of senile plaques and neurofibrillary tangles in the brain. Plaques are deposits of β-amyloid protein fragments derived from amyloid precursor protein (APP) and tangles, formed inside dying cells, are twisted fibers derived from hyper-phosphorylated tau-protein.[4,5,6] Currently there are many therapies in development to try treat the disease by stopping the formation of plaques and tangles.[7,8,9]

Four cholinesterase inhibitors, tacrine (4), donepezil hydrochloride (3), rivastigmine (2), and galantamine (1), have been approved for the treatment of the cognitive and behavioral symptoms of AD. Memantine, approved in 2004, is a glutamate regulator which has also been shown to be effective against moderate to severe Alzheimer's disease.[10] Although tacrine was the first drug to be approved, it's use has been limited due to toxicity, short half-life and limited absorption due to food effects.[11] In this chapter we will review the discovery, pharmacology, and synthesis of the three most recent cholinesterase inhibitors.

USAN: Tacrine
Trade name: Cognex®
Pfizer Inc
Launched: 1993

4

17.2   Pharmacology

The hallmark of pathophysiology of Alzheimer's disease, as discovered by Alois Alzheimer 150 years ago, is the prevalence of senile plaques and neurofibrillary tangles in the medial temporal lobe structures and in cortical areas of the brain.[12] In addition, there is general degeneration of the neurons and synapse.[5,6] Studies into the behavioral pharmacology and behavioral neuroscience of AD patients established the importance of the brain neurotransmitter acetylcholine (ACh) in the disposition of AD patients. Early research showed that there is reduced levels of acetylcholine among Alzheimer's patients, which indicated that cognitive function could potentially be improved by increasing acetylcholine levels by blocking acetylcholine esterase (AChE), the enzyme responsible for its degradation.[13,14,15] Butyrylcholine esterase (BuChE), predominantly present in the periphery, also hydrolyzes ACh. In addition, in late stage AD patients, BuChE has been found in the brain and may play a role in the decrease in neuronal activity.[16,17]

Acetylcholine (ACh) acts on the neuronal choline receptors (nAChR) for neuronal activity. In late-stage AD patients, nAChRs are also downregulated, which exacerbates the loss of neuronal function.[18] Thus the cholinergic hypothesis, loss of acetylcholine function in the brains of AD patients resulting in the deterioration of cognitive function, led to active drug discovery efforts. The most successful strategy to elevate acetylcholine has been to inhibit the destruction of acetylcholine by choline esterases, leading to several cholinesterase inhibitors for treatment of AD. Early leads identified were tacrine (4) and physostigmine (19), which had short half-lives and toxicity associated with them.

Among the first to come to the market, donepezil (3) has been shown to inhibit AChE of an electric eel and human red blood cells with a K_i of between 4–9 nM and 2.3 nM, respectively. Using rat brain homogenate derived AChE, the IC_{50} was found to be 5.7 nM compared to an IC_{50} of only 7138 nM against BuChE derived from rat plasma.[19] In vivo activity in rat was demonstrated by doing ex vivo binding studies after dosing 5 mg and 10 mg of the drug orally. The inhibition of rat brain AChE was seen up to 4 h after the 5 mg dose and up to 8 h after the 10 mg dose.[11] In an animal model where rats are treated with scopolamine to generate Nuclueus Basalis Magnocelullaris (NMB) lesions, a passive avoidance response (PAR) test was done with dose titration of the drug in comparison to physostigmine. There was significant improvement in the latency response in the drug-treated group compared to saline treatment group and donepezil performed better than physostigmine.[20]

Rivastigmine (2) has been called a pseudo-irreversible inhibitor of AChE because the drug carbamoylates a serine residue in the active site of the enzyme after binding. However the carbamoylation is transient and the enzyme is reverted back to its original form and activity upon hydrolysis of the adduct. Thus the drug itself is a reversible and noncompetitive inhibitor of both AChE and BuChE. Rivastigmine has an in vitro IC_{50} of 1.2 µM against red blood cell AChE and 1.3 µM against mouse brain AChE, which is about 100 times less potent than physostigmine. However, in vivo inhibition of AChE was reduced by only a factor of 10 compared to physostigmine. AChE is inhibited in a time-dependent manner, and ex vivo inhibition of mice brain showed EC_{50} of 4.2 µmol/kg. Time of peak inhibition of mouse brain AChE in vivo occurred around 1–2 h after subcutaneous (s.c.) dosing. A single dose of 2.0 mg/kg of rivastigmine inhibited the AChE enzyme by ~ 40% for up to 8 h where as physostigmine inhibition lasted only 2 h.[21]

Assessment of the rivastigmine inhibition of AChE in various parts of the brain in mice was also assessed after subcutaneous dosing of drug at 0.5 mg/kg. Rivastigmine was shown to significantly inhibit AChE in the cerebral cortex and hippocampus by as much as 30–40%, compared to AChE inhibition in medulla oblongata and corpus striatum. Similar selective inhibition of AChE was shown when dosed orally as well. Physostigmine on the other hand, inhibited AChE equally in all parts of the brain. This is implicated to have strong correlation to AD since higher deficiences of ACh in the cortex and hippocampus have been shown to have higher deficiencies in cognition and behavioral symptoms among AD patients. The drug was also shown to have good oral absorption by comparing the ED_{50} dosed subcutaneously and orally, which showed it to have a oral to subcutaneous ED_{50} ratio of 1.4. Because of its oral bioavailability, long duration of action, and relatively high therapeutic ratio, the compound was progressed into human testing.[22]

Galantamine (1), a natural product, has a unique dual mode of action, inhibiting AChE and accentuating the affect of ACh in the nAChR via allosteric binding. Galantamine inhibits ACh degradation by competitively inhibiting AChE in the active site. In addition, it has also been shown to preferentially inhibit AChE in the frontal cortex and hippocampus, two regions in the brain where the cholinergic neurotransmission is most affected in AD patients.[23] Galantamine has in vitro IC_{50} of 0.35 μM and 18.6 μM in inhibition of AChE and BuChE, a 53-fold reduced acitivity against BuChE. After a single oral dose (10 mg/kg) in healthy human volunteers, its median maximal RBC AChE inhibition was 40% at 30 min. Among AD patients dosed with galantamine (5 to 15 mg t.i.d. for 2–3 months, the inhibition of the AChE ranged from 20 to 40%. Upon stopping the treatment, no inhibition was seen 30 h after the final dose.[24]

Cholinergic neuronal function is also controlled via nAChRs and loss of the receptors play a crucial role in exacerbating the symptoms and course of AD. A number of subtypes of AChRs (nAChRs) are formed in the brain from a combination of nine known types of alpha-subunits (ACh binding site) and three beta-subunits (structural). Two types of nAChRs, α-7 and α-7 β-2 nAChR subtypes, are thought to affect AD in the brain. Predominant losses of nAChR subtypes also seem to take place in the neocortex and the hippocampus, involved in memory and cognition. In addition, dose-dependent blockade of nAChRs leads to dose-related impairment of cognition and memory. Galantamine has been shown to be a ligand for nAChRs and acts as a partial agonist. Based on its allosteric binding data and its affect on eliciting increased response to the ACh binding to nAChRs, it is thought to act as an allosteric potentiating ligand (APL). Through it functioning as an APL, galantamine has been shown to increase the release of γ-aminobutyric acid (GABA) in hippocampal slices and dopamine in mouse striatal slices. It has also been proven to enhance NMDA receptor whole cell current in cortical neurons. Taken together, these additional APL affects could be the basis for improved behavior in AD patients since GABA, dopamine, and serotonin are involved in mood (e.g., depression, anxiety) and behavioral symptoms (e.g., aggression).[23]

Galantamine has been shown to improve memory in surgical mouse models of memory impairment after dosing it intraperitoneally. In another passive avoidance memory deficiency mouse model, galantamine showed significant passive avoidance improvement compared to tacrine, physostigmin and solvent alone. These animal models suggested that galantamine would improve some cognitive deficit in patients with AD.[25]

17.3 Structure-Activity Relationship (SAR)

17.3.1 Discovery of Donepezil (3)

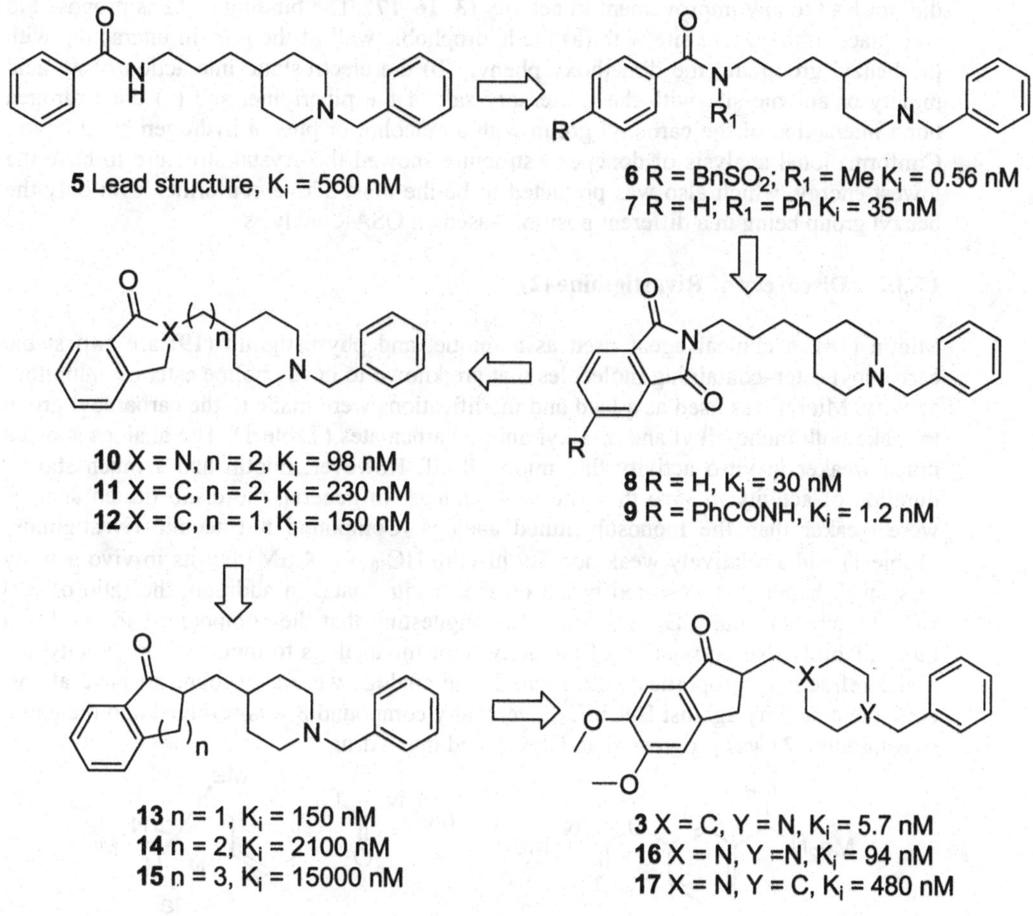

5 Lead structure, K_i = 560 nM

6 R = BnSO$_2$; R$_1$ = Me K_i = 0.56 nM
7 R = H; R$_1$ = Ph K_i = 35 nM

10 X = N, n = 2, K_i = 98 nM
11 X = C, n = 2, K_i = 230 nM
12 X = C, n = 1, K_i = 150 nM

8 R = H, K_i = 30 nM
9 R = PhCONH, K_i = 1.2 nM

13 n = 1, K_i = 150 nM
14 n = 2, K_i = 2100 nM
15 n = 3, K_i = 15000 nM

3 X = C, Y = N, K_i = 5.7 nM
16 X = N, Y =N, K_i = 94 nM
17 X = N, Y = C, K_i = 480 nM

Scheme 1. SAR studies on donepezil.

High-throughput screening of the company sample collection in an AChE inhibition assay resulted in identification of a novel structural class exemplified by lead compound **5**. Subsequent SAR studies were performed by a series of modification to the (a) phenyl group, (b) amide moiety, (c) linker unit, (d) piperidine ring, and (e) benzyl group. Para-substitution on an alkyl substituted amide was found to provide potent analog (**6**, K_i = 0.56 nM vs **7**, K_i = 35 nM), which was shown to be active in vivo using ex vivo studies. Further modification into ring constrained amide analogs led to pthalimide analog **9** with equipotent activity (K_i = 1.2 nM). In addition, it was found that the second carbonyl was not essential (e.g., **10**, K_i = 98 nM). The nitrogen of the amide could be substituted with a carbon and the chain length could be shortened by one carbon (e.g., **12**, K_i = 150 nM) to give almost equipotent compounds between the isoindolone **10** and indanones **11** and **12**. Ring expansion to six- or seven-membered ketones (**14** and **15**) led to poor activity. Finally, 3,4-dimethoxy substitution led to a potent compound **3** (K_i = 5.7 nM) with superior properties. Further modifications of the piperidine rings and the benzyl group did not lead to any improvement in activity (**3**, **16–17**). The binding of **12** is proposed to take place in the active site with (a) the hydrophobic wall of the protein interacting with the benzyl group and the dimethoxy phenyl, (b) the electrostatic interaction of an acid moiety or anionic site with the quaternary salt of the piperidine, and (c) the hydrogen bond interaction of the carbonyl group with an alcohol or phenol hydrogen bond donor. Conformational analysis of donepezil structure showed the crystal structure to have the lowest energy, which also was predicted to be the most active conformer with only the benzyl group being in a different position based on QSAR analysis.[19,26]

17.3.2 Discovery of Rivastigmine (2)

Miotin (**18**), a clinical agent used as a miotic, and physostigmin (**19**) are two stable carbamoyl ester–containing molecules that are known to have choline esterase inhibitory activity. Miotin was used as a lead and modifications were made to the carbamoyl group to make both mono-alkyl and di-alkyl amino carbamates (Table 1). The analogs showed much weaker in vitro activity that miotin itself. However, miotin had a much shorter duration of activity in vivo than the new analogs. In general, the disubstituted analogs were weaker than the monosubstituted analogs. Compound **h** (racemic rivastigmine) (Table 1) had a relatively weak activity in vitro (IC_{50} = ~ 4 μM) but its in vivo activity was much better that expected based on the in vitro data. In addition, the ratio of oral ED_{50} to subcutaneous ED_{50} was only 1.5, suggesting that the compound had good oral bioavailability. No correlation of the activity of the analogs to their hydrophobicity and molar refractivity properties were seen. These analogs were also found to have almost equipotent activity against BuChE.[21] Eventually compound **8** was resolved and the chiral rivastigmine (**2**) was progressed and developed into a drug.

Miotin (**18**)   Physostigmin (**19**)

Table 1. In Vitro and In Vivo Structure–Activity Relationship (SAR) of Carbamoyl Analogs

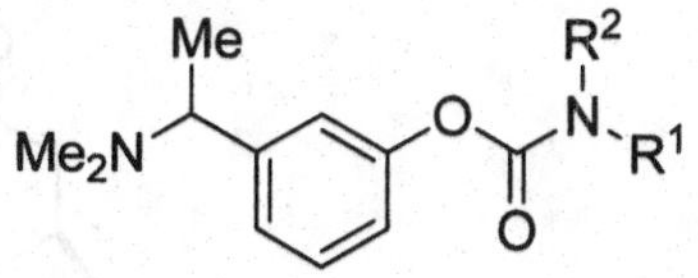

Compound	R^1	R^2	IC_{50} (µM)	ED_{50} (µM)	ED_{50} (p.o.) /ED_{50} (s.c.)
a	H	Me	0.013	0.92	1.3
b	H	Et	0.40	8.47	2.6
c	H	n-Pr	0.11	2.80	4.0
d	H	i-Pr	12.10	40.0	
e	H	Allyl	0.43	6.01	3.8
f	H	c-hexyl	0.093	7.24	3.0
g	Me	Me	0.027	1.14	3.4
h	Me	Et	4.0	4.20	1.5
i	Et	Et	35.0	56.0	1.4

17.3.3 Discovery of Galantamine (3)

Structure–activity relationship studies of galantamine (1) with modifications to the cyclohexane ring, tertiary amino, hydroxyl and methoxy group showed the importance of all these functional groups for activity. Among the active analogs made were desmethoxy derivative 21, carbamate 20, and several quaternary salts of the amine (e.g., 23, 24) along with the N-demethylated analog 22 (Table 2). Among them the N-benzyl ammonium analog 24 and the phenol analog 21 had the most potent in vitro activity. The carbamoyl analog 20 was also shown to have dose-dependent in vivo activity and a sixfold improvement over galantamine in the passive avoidance and forebrain lesioned mice model. In addition, this compound was also shown to have high therapeutic ratio. Several other groups have also published activity on other analogs and intermediates of galantamine.[27-29]

Table 2. Activity of Galantamine Analogs against RBC AChE

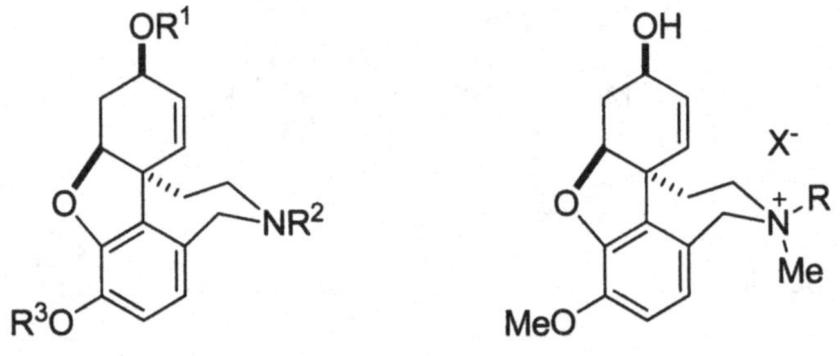

20 R^1 = CONHnBu; R^2 = Me; R^3 = Me   **23** R = Me; X = I
21 R^1 = H; R^2 = Me; R^3 = H   **24** R = Bn; X = Br
22 R^1 = R^2 = H; R^3 = Me

Compound	IC$_{50}$ (μM)
1 (Galantamine)	3.97
20	10.9
21	0.78
22	7.9
23	2.8
24	1.76

17.4   Pharmacokinetics and Drug Metabolism

17.4.1   Donepezil (3):

Pharmacokinetic studies in healthy human volunteers showed that donepezil (**3**) had linear and dose proportional PK after an oral acute dose. The peak plasma concentration (C$_{max}$) occurred at 4.1 ± 1.5 h using a dose range of 0.3–0.6 mg and the terminal disposition half-life was 81.5 ± 22.0 h. Pharmacokinetic properties at oral doses of 2, 4, and 6 mg also showed T$_{max}$ around 3.3–4.7 h and terminal half-life around 54–80 h. Multiple dose studies at 3 and 5 mg doses given daily for 21 days showed steady state plasma concentration (C$_{ss}$) of 18.2 and 13.2 μg/L within 14–21 days. The oral bioavailability was 100% with the maximum concentration reached within 3–4 h. No affect on the rate or extent of absorption was seen if donepezil was given in the morning or evening and food consumption 30 min before drug administration had no influence. The half-life was around ~71 h, allowing for a single daily dose.[20,30]

In vitro plasma protein binding of donepezil is 96%, mainly to albumin (75%). Plasma protein binding competition with furosemide, digoxin and warfarin showed that binding of donepezil was unaffected by these drugs. Multiple dose administration showed the drug to have a volume of distribution of 12 L/kg and the steady state concentration in plasma was reached within 15 days. Comparing the steady state concentration after 5 mg/day administration in healthy (30.2 µg/L) and AD (29.6 µg/L) patients showed no difference.

Donepezil undergoes first-pass hepatic metabolism after an oral dose and is primarily eliminated via renal clearance, with 79% unchanged found in urine, and only 21% found in faeces. In vitro metabolic studies identified CYP3A4 as the primary isoenzyme to metabolize donepezil, with CYP2D6 playing a lesser role. Formal PK studies showed that donepezil was not likely to interfer with metabolism of other drugs. However ketoconazole and quinidine, inhibitors of CYP3A4 and CYP2D6, did inhibit the metabolism of donepezil in the in vitro studies.[11]

17.4.2 Rivastigmine (2)

Rivastigmine (2), dosed at 3 mg orally to healthy adults, was found to be well absorbed with an absolute bioavailability reaching ~36 % with peak plasma concentrations seen after ~1 h. The compound is reported to have a dose-proportional exposure up to the 3 mg dose, which subsequently becomes nonlinear at higher doses.[31] Increasing the dose from 3 to 6 mg b.i.d. results in a threefold increase in the area under the plasma concentration-time curve (AUC). Other studies done in AD patients have found the drug to have nonlinear pharmacokinetics when tested in the dose range of 1–6 mg b.i.d.[32] Administering the drug with food delays absorption and the time to reach maximum concentration by almost 90 min. The maximum concentration is also lowered by 30%, while the AUC increases by 30%. Having food with the drug does seem to alleviate the adverse events seen with the drug while enhancing the bioavailability. Although the half-life of rivastigmine in AD patients is only ~1.5 h, the pharmacodynamic half-life extends to almost 10 h. Its long PD half-life is presumed to be due to the slow hydrolysis and release of the covalent adduct from the AChE active site.[33]

Rivastigmine has a moderate plasma protein binding of 40%, resulting in a large free fraction, and it readily penetrates the blood–brain barrier. It has a volume of distribution of 1.8–2.7 L/kg in healthy adults. The drug is highly metabolized with less than 1% parent compound detected in the feces. The primary inactive metabolite is the phenol resulting from the cleavage of the carbamoyl group by AChE as well as presystemic hydrolysis. The elimination of the metabolite via sulfate conjugation is complete within 24 h and is primarily cleared via renal excretion. Since no liver oxidation is involved in the metabolism of rivastigmine, no drug-drug interactions are expected.[32,33]

17.4.3 Galantamine (1)

Linear pharmacokinetics has been observed over the recommended dose range for galantamine (1) given at 8 to 24 mg/day. A single oral dose of 10 mg given to healthy volunteers shows rapid absorption of the drug resulting in the peak plasma concentrations within 1–2 h. The bioavailability is 100% with low plasma protein binding (~ 18%) and a plasma half-life of ~ 5–6 h.[34,35]

Galantamine dosed repeatedly with oral dosing of 12 and 16 mg b.i.d. gave maximal plasma drug concentrations of 42 to 137 ng/mL and trough plasma concentrations from 19 to 97 ng/mL. After 2 to 6 months of drug therapy, no accumulation of drug was seen. The mean volume of distribution is 175 L, suggesting a very high degree of nonspecific binding.[35]

Food consumption with galantamine delays the absorption rate but has no effect on the extent of absorption in healthy volunteers. Oral dosing of 10 mg drug showed delayed maximum peak plasma concentration by double and the peak plasma concentration was reduced by 25%. The AUC of the drug was unaffected.[36]

Galantamine is primarily metabolized by CYP2D6 and CYP3A4, suggesting co-administration of drugs that inhibit these isoenzymes could elevate its concentration, thus increasing its cholinergic effect and potential side effects. Several major metabolites have been characterized, and all of them were found to be inactive against AChE inhibition. Renal clearance of galantamine accounts for 20–25% of the plasma clearance.[36]

17.5   Efficacy and Safety

17.5.1   Donepezil (1)

Results from several multicenter placebo controlled clinical studies, along with pivotal clinical trials,[37,38,39] done in the United States and Europe with more than 3000 patients have been published.[30] The studies covered multiple doses and periods of therapy ranging from 12 to 52 weeks against patients with mild to moderate AD disease. The outcome looked at the cognitive functions, ADL, behavior, global clinical state and adverse events. Overall, donepezil was found to have a beneficial affect on behavioral and neuropsychiatric symptoms, such as hallucinations, distractability, aberrant motor behavior, and apathy. Several studies of the drug on patients in community or nursing homes with severe AD also showed it to be efficacious.[40] Donepezil is now the only drug approved for the treatment of all stages of AD. Adverse effects were generally mild to moderate with the primary effects on the GI tract. Nausea, vomiting, and diarrhea were the most common and can be attributed to the cholinergic action of the drug.[37]

17.5.2   Rivastigmine (2)

Several clinical trials of rivastigmine demonstrated improvement or maintenance of cognitive function, behavioral function and ability for self-care among patients with mild to moderate AD.[41,42] Open-label trials have shown clinical effects up to 52 weeks.[43] Starting rivastigmine therapy at the earliest stages of the disease seem to fare better than those whose therapy began at the later stages of the disease. Adverse events were mild and found to be associated with GI tract, similar to donepezil.[44,41]

17.5.3   Galantamine (1)

Galantamine clinical studies done with placebo control on AD patients for 3–6 months showed significant benefit on cognitive functions relative to placebo. Other baseline disease severity, apolipoprotein-E4 genotype, age and gender did not affect the beneficial effects on cognitive abilities. Patients who stayed on the treatment for a full 12 months had preserved cognitive functions. Improvement in the basic behavior of the patients in their ability to do basic self-care and do leisure activity also improved significantly compared to the placebo treatment group. The drug-treated patients also had delayed emergence of behavioral symptoms. Galantamine was most effective on reducing symptoms of aberrant behavior, anxiety , disinhibition, and hallucinations.[36,45]

There were no major adverse events (AEs) with galantamine treatment, and most events were mild in severity. Most AEs, including nausea, vomiting, and diarrhea, occurred during escalation of treatment. Some patients lost weight after treatment for 6 months, but this was minimal over a 12-month treatment period.[36]

17.6   Syntheses of Donepezil

In the original discovery route by Sugimoto,[19,46] aldol condensation of 5,6-dimethoxy-1-indanone (25) and 1-benzyl-4-piperidine-carboxaldehyde (26) in the presence of freshly prepared LDA and hexamethylphosphoramide (HMPA) in THF provided exocyclic enone 27 in 62% yield (Scheme 2). Hydrogenation with 10% Pd/C followed by salt formation with hydrochloric acid afforded donepezil hydrochloride (3) in 86% yield.

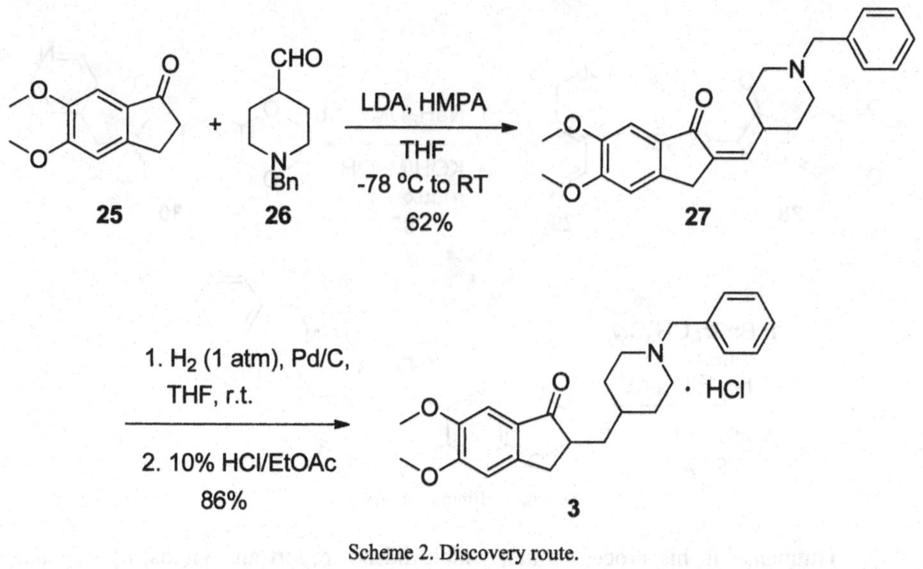

Scheme 2. Discovery route.

Mathad et al.[47] recently disclosed preparation of 3 following a similar process to the discovery route with only modifications at the condensation stage (Scheme 3). They achieved aldol condensation of 25 and 26 using sodium hydroxide as a base under phase transfer conditions to provide the intermediate 27 in excellent yield (88%). Overall yield of this route was 37%, an improvement over the original route.

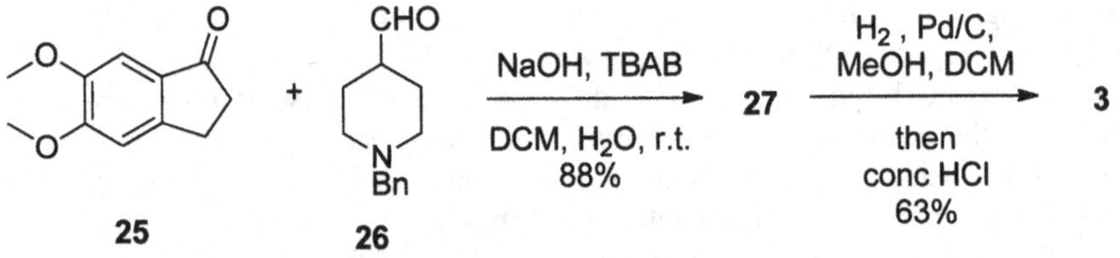

Scheme 3. Mathad et al. modified synthesis.

Iimura[48] disclosed a very different approach to access donepezil (3) as shown in Scheme 4. Alkylation of 5,6-dimethoxy-2-ethoxycarbonyl-1-indanone (28) with 4-pyridylmethyl chloride (29) in the presence of sodium hydride at room temperature followed by decarboxylation provided 30 in 85% yield. After N-benzylation, hydrogenation of the pyridine ring in the presence of PtO₂ followed by hydrochloride salt formation furnished 3 in 91% yield, with an overall yield of 77% from 28. Indanone ester 28 was prepared according a literature procedure[49] by treatment of 5,6-dimethoxy-1-indanone with diethyl carbonate. This route was novel and efficient; however, use of highly reactive and moisture sensitive base (NaH) and expensive catalyst PtO₂ at industrial scale is not desirable.

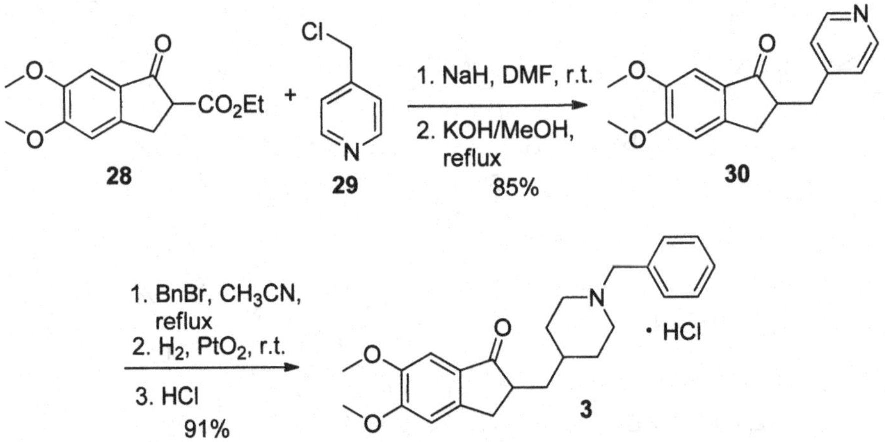

Scheme 4. Iimura synthesis.

Gutman,[50] in his process route, which did not report any yields, hydrogenated the pyridine ring first to access the piperidine moiety and constructed the indanone ring system via an intramolecular Friedel-Crafts acylation (Scheme 5). Hydrogenation of diester 31, obtained from condensation of 4-pyridine carboxaldehyde and dimethyl malonate, followed by benzylation of the piperidine intermediate afforded N-benzylated piperidine 32. Alkylation of 32 with 3,4-dimethoxybenzyl chloride (33) and subsequent hydrolysis gave dicarboxylic acid 34. Subjection of 34 to strong acid resulted in intramolecular Friedel-Crafts acylation and in situ decarboxylation to provide 3.

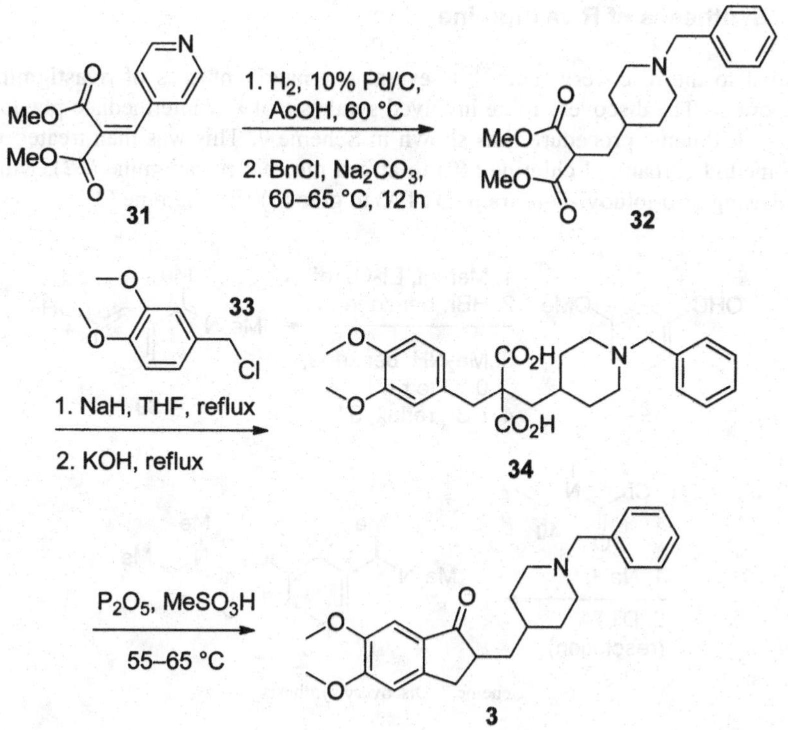

Scheme 5. Gutman et al. synthesis.

A synthesis by Elati et al.[51] involved condensation of 4-pyridinecarboxaldehyde (35) with 25 in the presence of p-toluenesulfonic acid (TsOH) to give the enone 36 in 96% yield. Global hydrogenation of the alkene and pyridine ring in the presence of Pd/C under high-pressure hydrogen gas gave piperidine 37 in 90% yield. Finally, N-benzylation with benzyl bromide provided 3 in 65% yield (Scheme 6).

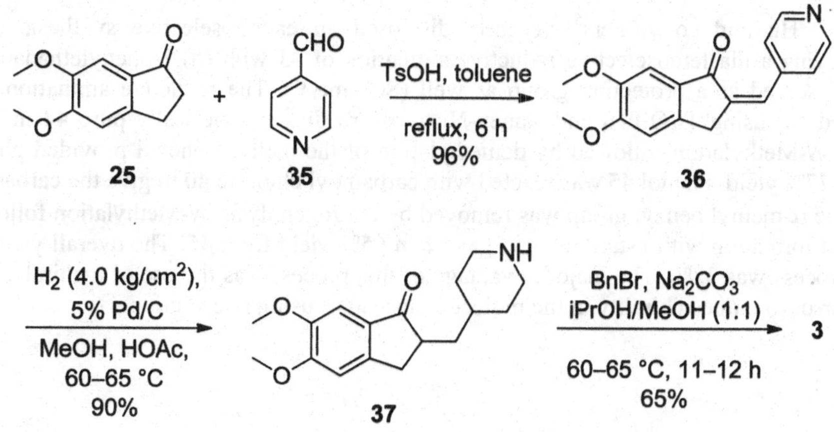

Scheme 6. Elati et al. synthesis.

17.7   Syntheses of Rivastigmine

In addition to the discovery route,[21,52] several alternate syntheses of rivastigmine have been reported. The discovery route involved synthesis of key intermediate phenol **39** by following Stedman's procedure,[53] as shown in Scheme 7. This was then treated with *N*-ethyl-*N*-methyl carbamoyl chloride (**40**) to afford racemic rivastigmine (±**2**), which was resolved using di-*p*-toluoyl-*D*-tartrate (DTTA) to give (*S*)-rivastigmine (**2**).

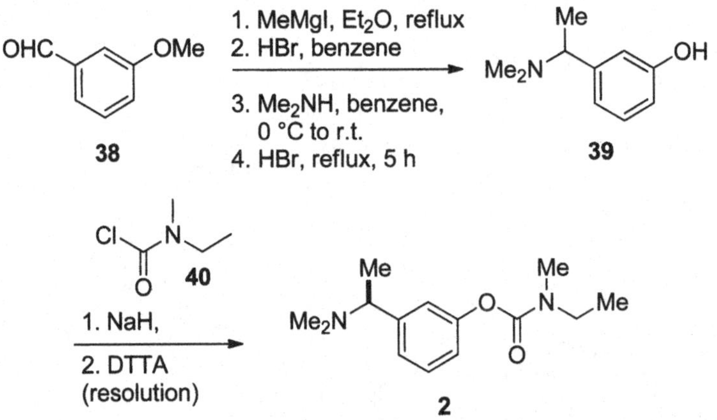

Scheme 7. Discovery synthesis.

Wock-Hardt Ltd.[54] reported a manufacturing process for the preparation of **2** (Scheme 8). In their approach, reductive amination of ketone **41** with dimethylamine hydrochloride using NaCNBH$_3$ afforded amine **42** in 69% yield. The phenol was acylated with *N*-ethyl-*N*-methyl carbamoyl chloride (**40**) using KO*t*-Bu as the base instead of NaH to provide racemic carbamate **2** in 88% yield (98% pure by HPLC). Racemic **2** was further purified by making the oxalate salt, which provided **2** as a colorless crystalline oxalate salt in 100% purity. Resolution with DTTA followed by salt formation with tartaric acid afforded chiral tartrate salt **2**. The overall yield of this process was 20%.

Hu and co-workers[55] recently disclosed an enantioselective synthesis of **2** employing a diastereoselective reductive amination of **43** with (*S*)-1-phenylethanamine, which served as a protecting group as well (Scheme 9). The reductive amination was carried out using Ti(O*i*Pr)$_4$ and Raney-Ni to afford diastereomerically pure **44** in 74% yield. *N*-Methylation followed by demethylation of the methoxyphenyl provided phenol **45** in 77% yield. Phenol **45** was reacted with carbamoyl chloride **40** to give the carbamate and the α-methyl benzyl group was removed by hydrogenolysis. *N*-Methylation followed by salt formation with *L*-tartaric acid gave **2** in 65% yield from **45**. The overall yield for the process was 37%. The major advantage of this process was the use of a chiral amine precursor to induce chirality in the molecule instead of using a resolution.

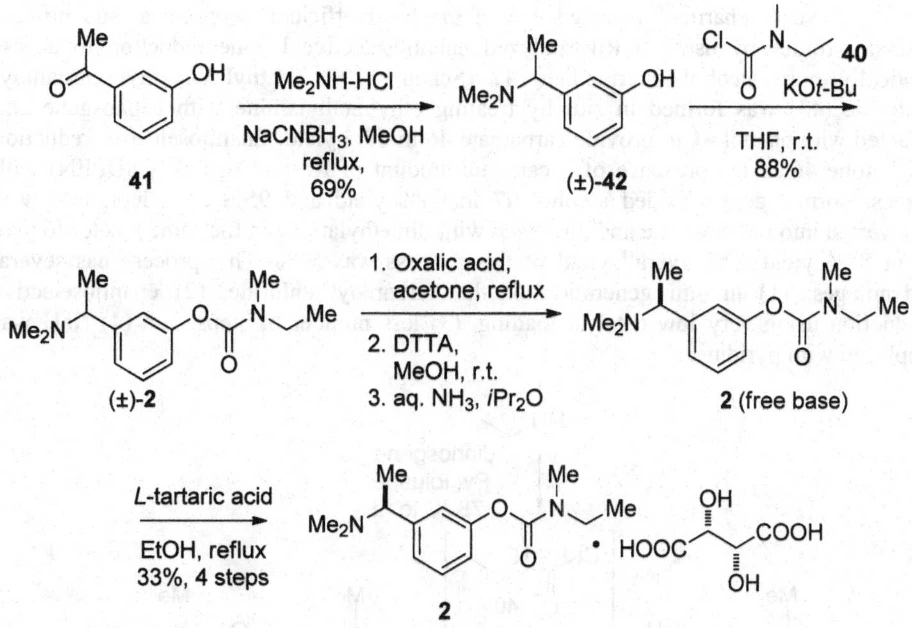

Scheme 8. Wock-Hardt synthesis.

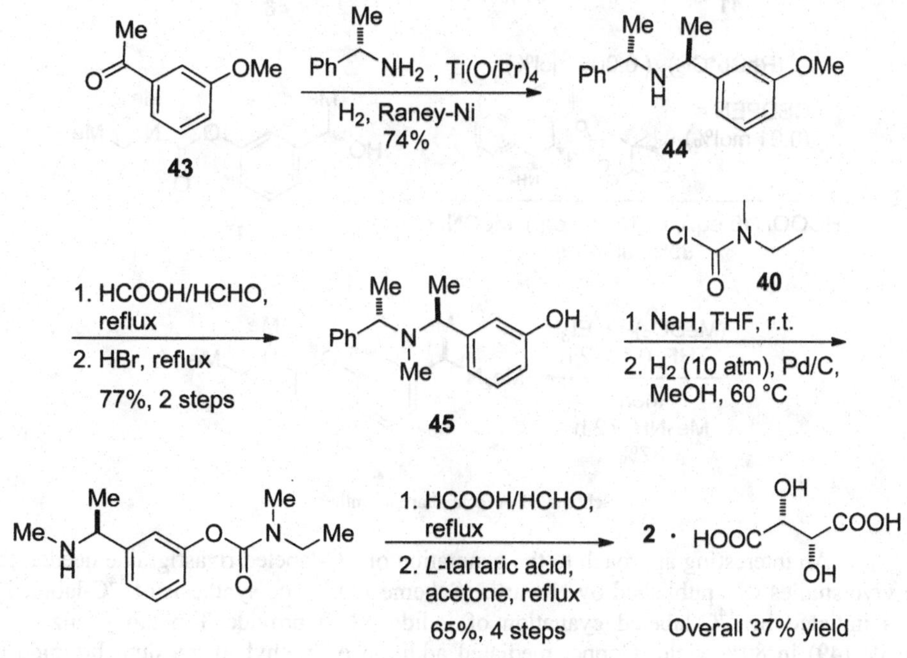

Scheme 9. Hu et al. synthesis.

Avecia Pharma[56] reported one of the most efficient, economical and process friendly routes by using a Rh-catalyzed enantioselective ketone reduction to access optically pure alcohol intermediate **47** (Scheme 10). *N*-Ethyl-*N*-methyl carbamoyl chloride (**40**) was formed in situ by treating ethylmethylamine with triphosgene and reacted with phenol **41** to provide carbamate **46** in 67% yield. Enantioselective reduction of ketone **46** in the presence of a catalytic amount of Rh and ligand (CSDPEN) with excess formic acid afforded alcohol **47** in 95% yield and 95% *ee*. Alcohol **47** was converted into the mesylate and displaced with dimethylamine in the same reactor to give **2** in 87% yield. The overall yield of this process was 56%. This process has several advantages: (1) in situ generation of the carbamoyl chloride, (2) enantioselective reduction using very low catalyst loading, (3) less number of steps, and (4) NaH was replaced with pyridine.

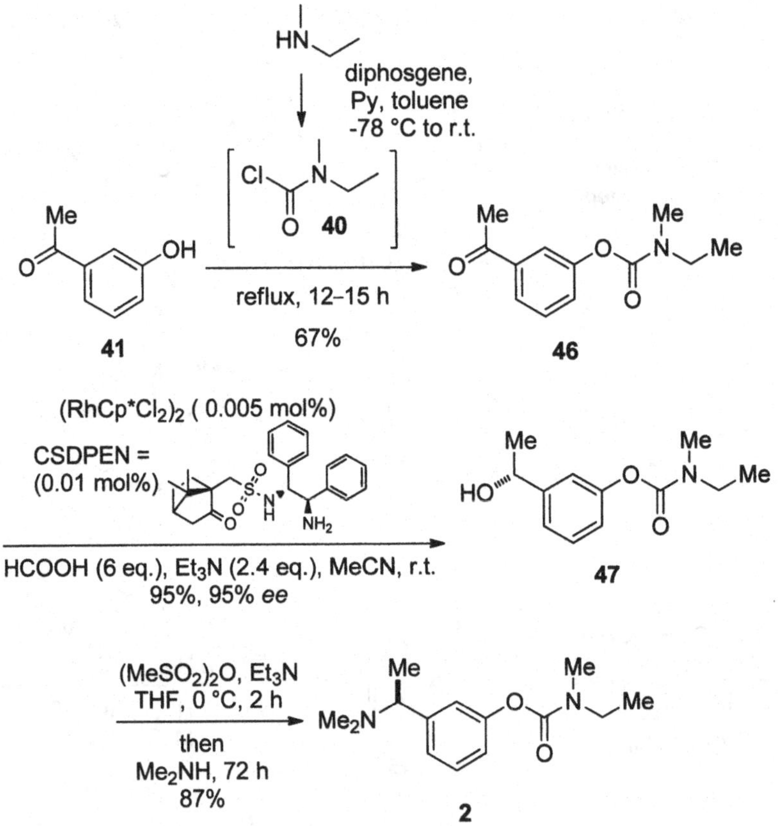

Scheme 10. Avecia pharma synthesis.

An interesting approach to the generation of ^{14}C-labeled rivastigmine needed for in vivo studies was published by Novartis (Scheme 11).[57] The synthesis of ^{14}C-labeled **2** was initiated by ^{14}C-labeled cyanation of iodide **48** to provide 3-methoxybenzo-^{14}C-nitrile (**49**) in 80% yield. Copper-mediated addition of methyl magnesium bromide to nitrile **49** formed an imine that was hydrolyzed with aqueous HCl to the corresponding ketone and then converted into O-methyl oxime **50**. Enantioselective reduction was achieved with $NaBH_4$–$ZrCl_4$ complex and (S)-2-amino-3-methyl-1,1-diphenylbutanol (**51**) to give free amine **52** in 87% yield. This was dimethylated to give **53** and the methoxyphenyl was deprotected to give free phenol, which was converted to ^{14}C-labeled rivastigmine by treating with carbamoyl chloride following established procedures.

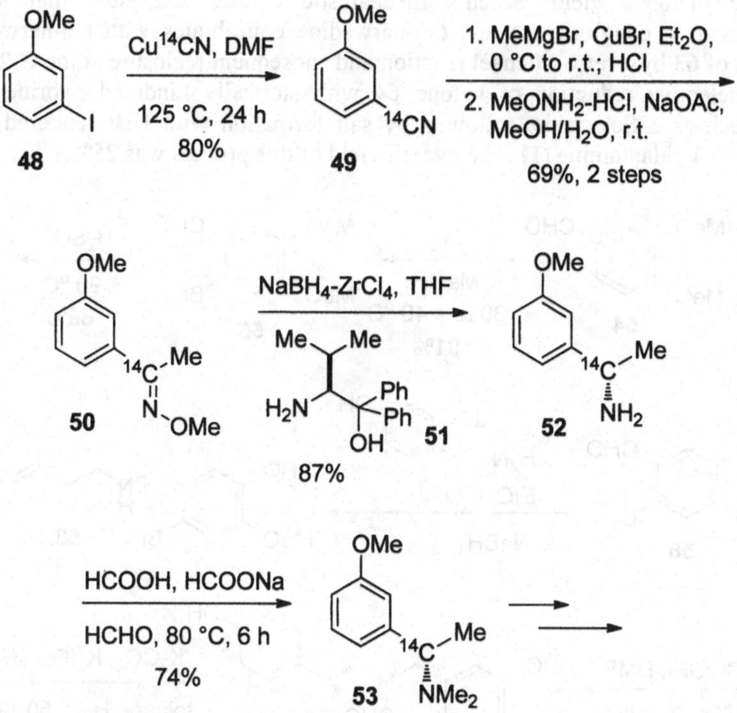

Scheme 11. Synthesis of ^{14}C-labeled rivastigmine precursor

17.8   Syntheses of Galantamine

17.8.1   Racemic Synthesis

An efficient process route[58] has been disclosed on the preparation of galantamine (**1**) on industrial scale (Scheme 12). The route uses a biomimetic approach that involves phenolic oxidative coupling followed by spontaneous ring closure by Michael addition to form a key intermediate natural product (±)-narwedine (**62**). The synthesis was initiated with bromination of aldehyde **54** followed by selective monodemethylation under acidic conditions to give **56**. Reductive amination of aryl adehyde **56** with amine **57** provided **58**, which was N-fomylated to give cyclization precursor **59**. After screening large numbers of phenolic oxidative coupling conditions, it was observed that the highest yields (50–54%) were obtained when **59** was treated with $K_3[Fe(CN)_6]$ at low concentrations. However, for economical reasons, this transformation was carried out at a higher concentration on industrial scale (12 kg of **59**), which resulted in a 40% yield of

60. Intermediate ketone **60** was first protected as propylene glycol ketal **61**, and then the formyl group was reduced to the methyl group with LiAlH$_4$ followed by hydrolysis of the ketal to recover racemic narwedine (**62**) in 80% yield. Racemic narwedine (**62**) was dynamically resolved by crystallization induced chiral transformation into (–)-narwedine (**64**) using seed crystals of (–)-narwedine. This technique was first applied by Barton[59] and later improved by Shieh[60] to get the optically pure enantiomers from racemic narwedine in 85% yield. Shieh's mechanistic studies suggested that narwedine crystallizes as a conglomerate and (+)-narwedine equilibrates with (–)-narwedine via formation of **63** by a retro Michael reaction and subsequent reclosure at the C-2 position. Diastereoselective reduction of ketone **64** with sterically hindered hydride reducing agents, such as *L*-Selectride, followed by salt formation with HBr rendered colorless crystals of (–)-galantamine (**1**). The overall yield of this process was 25%.

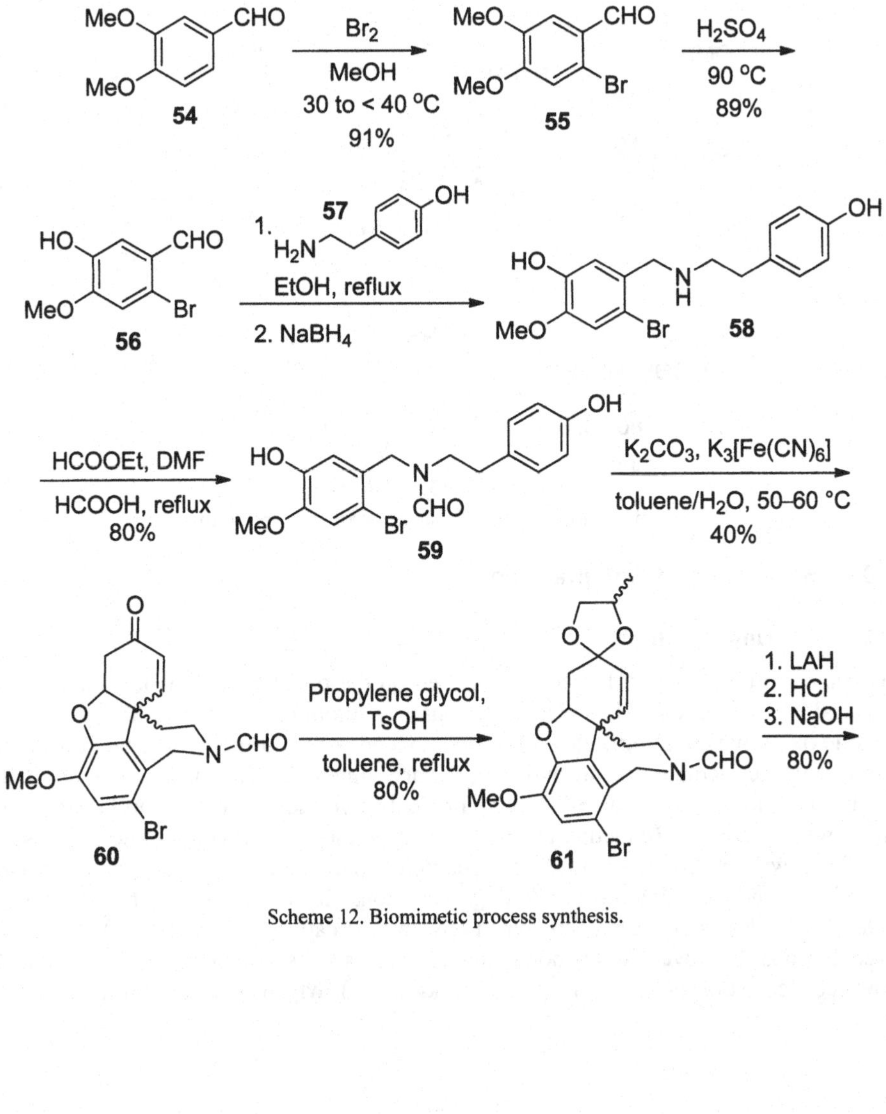

Scheme 12. Biomimetic process synthesis.

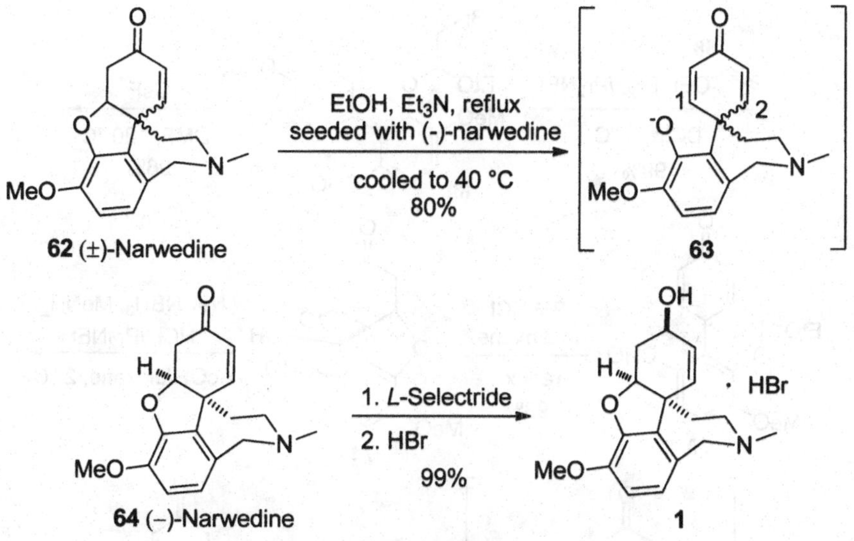

Scheme 12 (cont.).

A recent synthesis improves on the process route to prepare the key intermediate narwedine (64) as shown in Scheme 13. Magnus[61] used a simple *para*-alkylation of a phenol intermediate as the key reaction to form the quaternary C-C bond instead of the problematic phenolic oxidation reaction described in Scheme 12. Cyclization precursor 69 was readily available in high yields from commercially available aryl bromide 65 via Suzuki coupling with 66 to give compound 67, followed by bromine-mediated phenol addition to ethyl vinyl ether. The phenoxide of 69, generated by deprotection of the TBS group, undergoes an intramolecular *para*-alkylation with the alkyl bromide side chain to form cyclized product 70 in 96% yield. Acid-catalyzed hydrolysis of 70 resulted in ring opening of the hemiketal, exposing the phenol, which adds to the enone, giving the furan ring and rearranged hemiacetal 71. Reductive amination of hemiacetal 71 with methyl amine followed by elimination of water with methanesulfonic acid provided (±)-narwedine (64) in 72% yield. The overall yield of 64 from commercially available 65 is 63% yield.

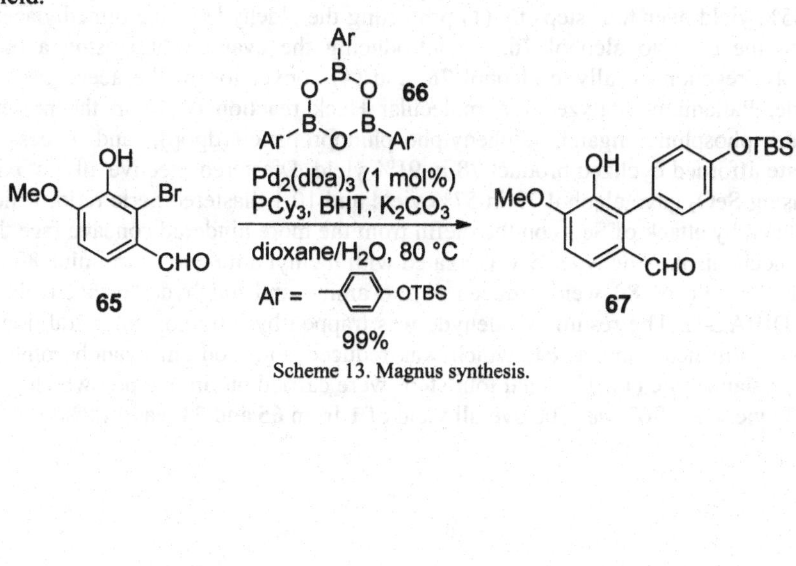

Scheme 13. Magnus synthesis.

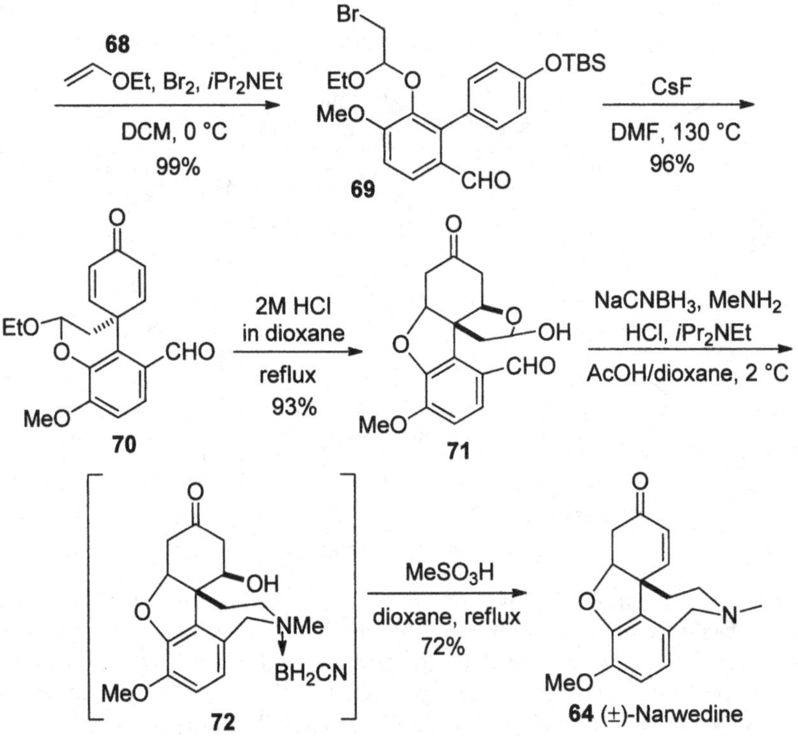

Scheme 13 (cont).

17.8.2   Asymmetric Synthesis of (−)-Galantamine (1)

Many asymmetric total syntheses of (−)-galantamine have been published and reviewed elsewhere.[27] Two of the most efficient and recent syntheses are highlighted below. The Trost group has disclosed three asymmetric syntheses[62–64] of (−)-galantamine. In the third-generation synthesis,[63,64] asymmetric allylic alkylation (AAA) of carbonate 73 with bromide 65 in the presence of catalytic Pd(0) and chiral ligand 74 furnished ether 75 in 72% yield and 88% ee. Intramolecular Heck cyclization precursor 77 was obtained from 75 in 65% yield over four steps by (1) protecting the aldehyde as the dimethylacetal, (2) reducing the ester to alcohol 76, (3) introducing the cyano group using a modified Mitsunobu reaction on allylic alcohol 76, and (4) conversion of the acetal back to the aldehyde. Palladium-catalyzed intramolecular Heck reaction of 77 in the presence of bidentate phosphine ligand, diphenylphosphinopropane (dppp), and excess silver carbonate afforded cyclized product 78 in 91% yield. Diastereoselective allylic oxidation of 78 using SeO$_2$ gave alcohol 79 in 57% yield and 10:1 diastereomeric ratio, which was rationalized by attack of SeO$_2$ on the olefin from the more hindered concave face through an ene mechanism. Aldehyde 79 was treated with methyl amine to form imine 80 and the imine and nitrile of 80 were reduced to the amine and aldehyde, respectively, using excess DIBAL-H. The resulting aldehyde was trapped by intramolecular addition of the amine to form hemi-aminal 81, which was reduced with sodium cyanoborohydride to give (−)-galantamine (1). The final four steps were carried out in one pot, which provided 1 in 62% yield and 96% ee. The overall yield of 1 from 65 and 73 was 14.8%.

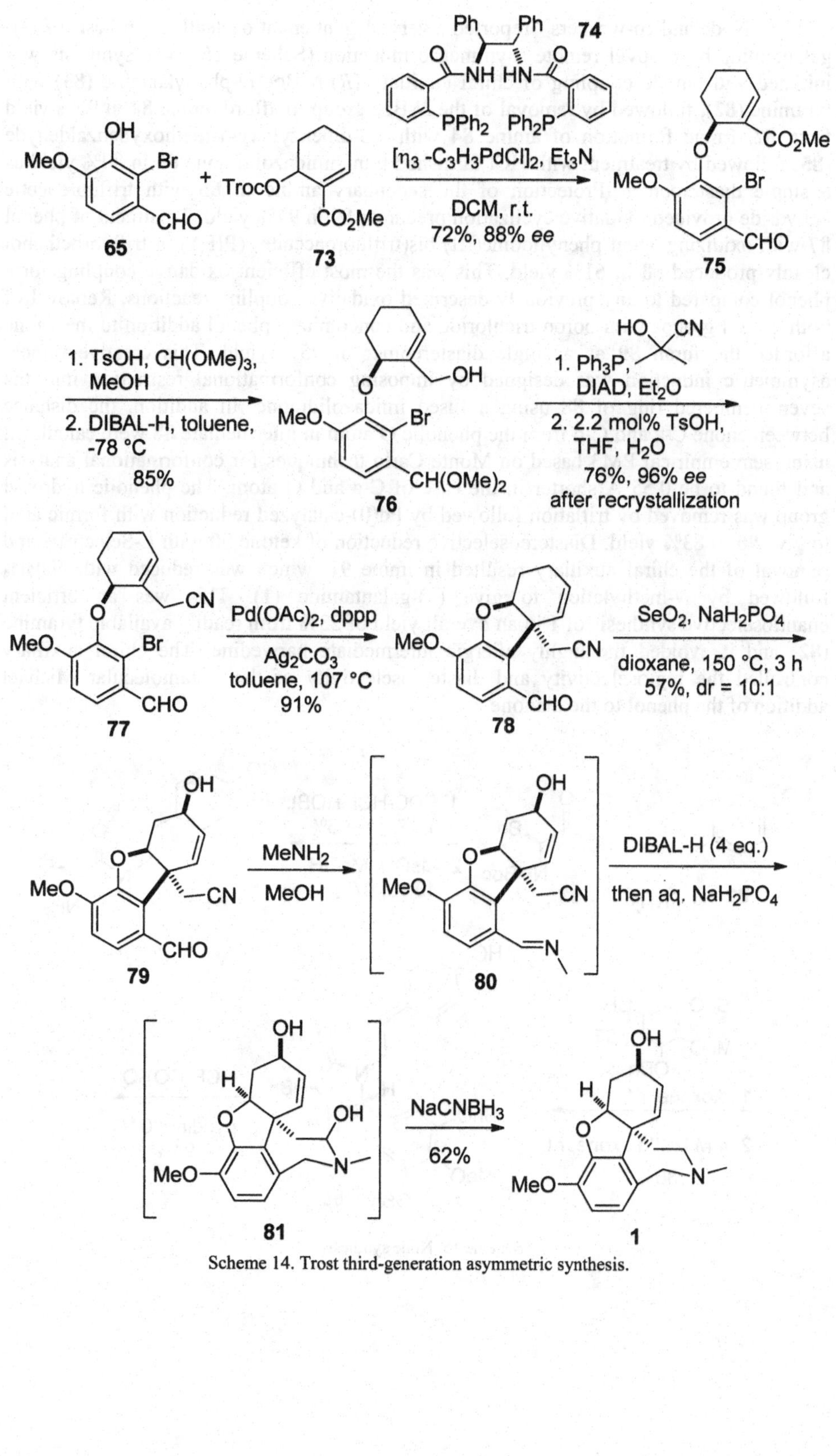

Scheme 14. Trost third-generation asymmetric synthesis.

Node and co-workers[65] reported a very elegant enantioselective synthesis of (–)-galantamine by a novel remote asymmetric induction (Scheme 15). The synthesis was initiated with amide coupling of chiral auxiliary (R)-N-Boc-D-phenylalanine (83) with tyramine (82), followed by removal of the N-Boc group to afford amine 84 in 92% yield from 82. Imine formation of amine 84 with 3,5-dibenzyloxy-4-methoxybenzaldehyde (85) followed by treatment with HCl furnished trans-imidazolidinone 86 in 80% yield as a single diastereomer. Protection of the secondary amine in 86 with trifluoroacetic anhydride provided oxidative cyclization precursor 87 in 94% yield. Treatment of phenol 87 with oxidizing agent phenyliodine(III) bis(trifluoroacetate) (PIFA) in trifluoroethanol cleanly produced 88 in 61% yield. This was the most efficient oxidative coupling for a phenol compared to any previously described oxidative coupling reactions. Removal of both benzyl groups with boron trichloride and concomitant phenol addition to the enone afforded the furan 89 as a single diastereomer in 95% yield. This creative remote asymmetric induction was designed by imposing conformational restriction into the seven-membered ring of 88 using a fused imidazolidinone. In addition, the distance between enone C-a and C-b from the phenolic O atom in intermediate 88 were calculated using semiempirical PM3 based on Monte Carlo techniques for conformational analysis and found to be 0.55 Å shorter in the case of C-a and O atom. The phenolic hydroxyl group was removed by triflation followed by Pd(0)-catalyzed reduction with formic acid to give 90 in 83% yield. Diastereoselective reduction of ketone 90 with L-Selectride and removal of the chiral auxiliary resulted in imine 91, which was reduced with NaBH$_4$ followed by N-methylation to give (–)-galantamine (1). This was an efficient enantioselective synthesis of 1 in an overall yield of 23% from readily available tyramine (82) and it avoided the highly allergic intermediate narwedine. The chiral auxiliary controlled the regioselectivity and diastereoselectivity of the intramolecular Michael addition of the phenol to the dienone.

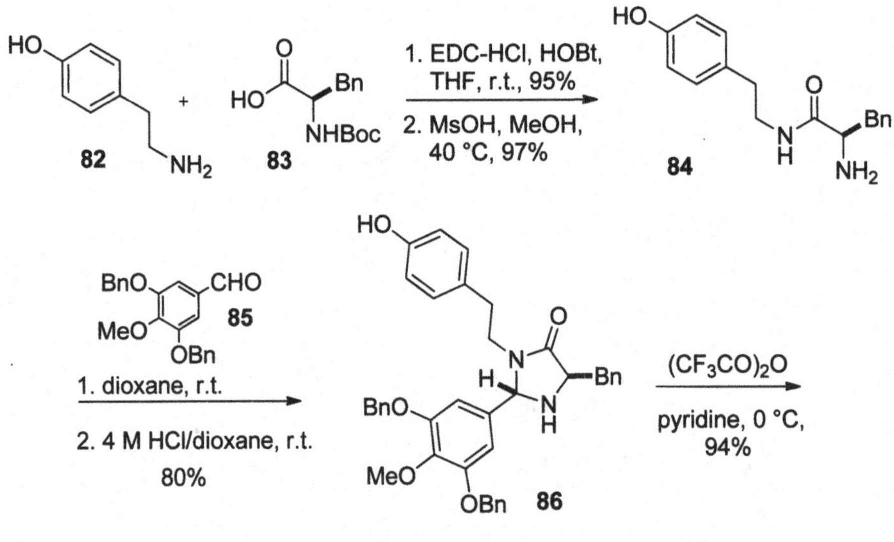

Scheme 15. Node synthesis.

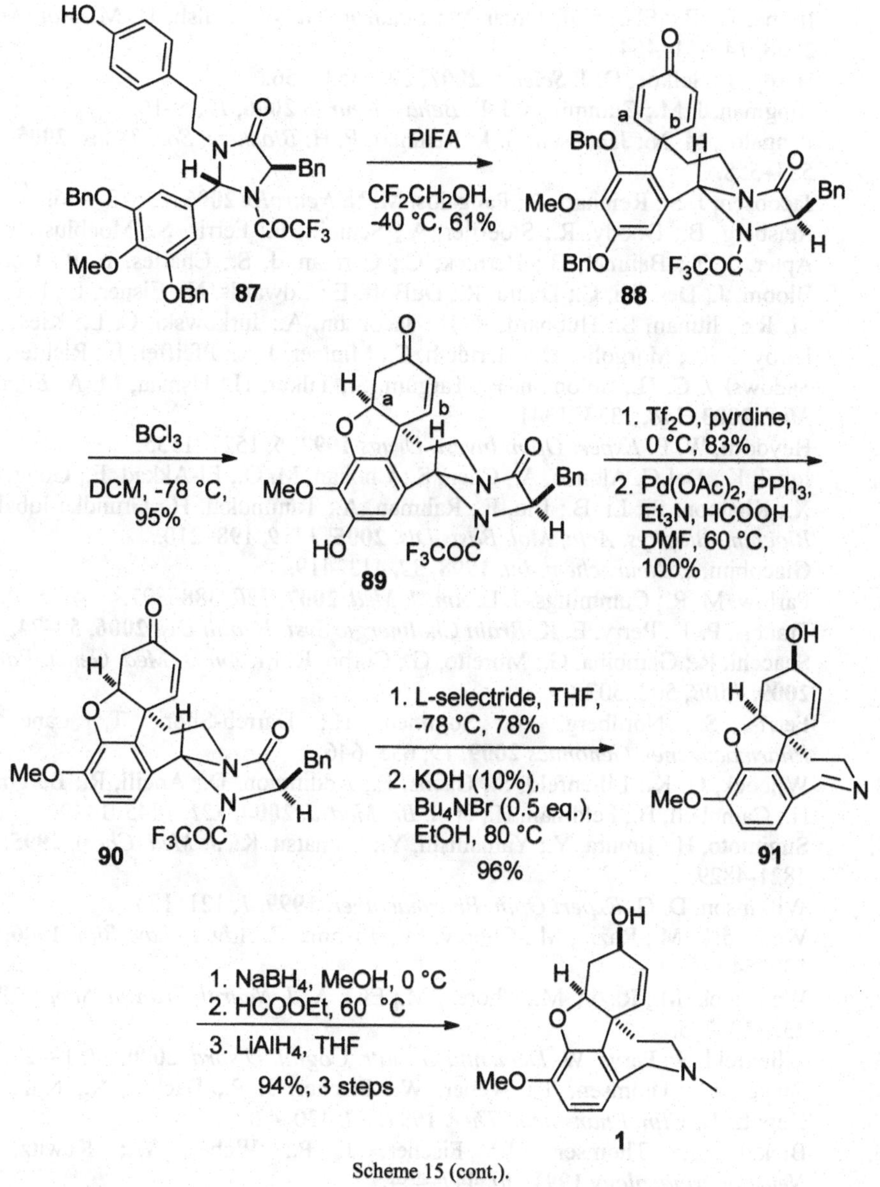

Scheme 15 (cont.).

17.9   References

1.   Alzheimer, A.; Forstl, H.; Levy, R. *Hist. Psychiatry* **1991**, *2*, 71–101.
2.   Anonymous, *Alzheimers Dement.* **2009**, *5*, 234–70.
3.   Katzman, R. *N. Engl. J. Med.* **1986**, *314*, 964–73.
4.   Glenner, G. G.; Wong, C. W. *Biochem. Biophys. Res. Commun.* **1984**, *120*, 885–890.

5.   Irvine, G. B.; El-Agnaf, Omar M.; Shankar, G. M.; Walsh, D. M. *Mol. Med.* **2008**, *14*, 451–464.

6.   Hardy, J.; Selkoe, D. J. *Science* **2002**, *297*, 353–356.

7.   Ringman, J. M.; Cummings, J. L. *Behav. Neurol.* **2006**, *17*, 5–16.

8.   Pangalos, M. N.; Jacobsen, S. J.; Reinhart, P. H. *Biochem. Soc. Trans.* **2005**, *33*, 553–558.

9.   Jacobsen, J. S.; Reinhart, P.; Pangalos, M. N. *NeuroRx* **2005**, *2*, 612–626.

10.   Reisberg, B.; Doody, R.; Stoeffler, A.; Schmitt, F.; Ferris, S.; Moebius, H. J.; Apter, J. T.; Baumel, B.; Bernick, C.; Carman, J. S.; Charles, L. P.; Corey-Bloom, J.; DeCarli, C.; Duara, R.; DuBoff, E.; Edwards, N.; Eisner, L.; Farlow, M. R.; Flitman, S.; Hubbard, R. H.; Jacobson, A.; Jurkowski, C. L.; Kiev, A.; Kirby, L. C.; Margolin, D.; Merideth, C.; Mintzer, J. E.; Pfeiffer, E.; Richter, R.; Sadowsky, C. H.; Solomon, P.; Targum, S.; Tilker, H.; Usman, M. *N. Engl. J. Med.* **2003**, *348*, 1333–1341.

11.   Heydorn, W. E. *Expert Opin. Invest. Drugs* **1997**, *6*, 1527–1535.

12.   Iqbal, K.; Del C. Alonso, A.; Chen, S.; Chohan, M. O.; El-Akkad, E.; Gong, C.-X.; Khatoon, S.; Li, B.; Liu, F.; Rahman, A.; Tanimukai, H.; Grundke-Iqbal, I. *Biochim. Biophys. Acta, Mol. Basis Dis.* **2005**, *1739*, 198–210.

13.   Giacobini, E. *Neurochem. Int.* **1998**, *32*, 413–419.

14.   Farlow, M. R.; Cummings, J. L. *Am. J. Med.* **2007**, *120*, 388–397.

15.   Francis, P. T.; Perry, E. K. *Brain Cholinergic Syst. Health Dis.* **2006**, 59–74.

16.   Scacchi, R.; Gambina, G.; Moretto, G.; Corbo, R. M. *Am. J. Med. Genet. Part B* **2009**, *150B*, 502–507.

17.   Ferris, S.; Nordberg, A.; Soininen, H.; Darreh-Shori, T.; Lane, R. *Pharmacogenet. Genomics* **2009**, *19*, 635–646.

18.   Wilcock, G. K.; Lilienfeld, S.; Gaens, E.; Addington, D.; Ancill, R.; Bergman, H.; Campbell, B.; Feldman, H.; et al. *Br. Med. J.* **2000**, *321*, 1445–1449.

19.   Sugimoto, H.; Iimura, Y.; Yamanishi, Y.; Yamatsu, K., *J. Med. Chem.* **1995**, *38*, 4821-4829.

20.   Wilkinson, D. G. *Expert Opin. Pharmacother.* **1999**, *1*, 121–135.

21.   Weinstock, M.; Razin, M.; Chorev, M.; Tashma, Z. *Adv. Behav. Biol.* **1986**, *29*, 539–549.

22.   Weinstock, M.; Razin, M.; Chorev, M.; Enz, A. *J. Neural. Transm. Suppl.* **1994**, *43*, 219–225.

23.   Lilienfeld, S.; Parys, W. *Dementia Geriatr. Cognit. Disord.* **2000**, *11*, 19–27.

24.   Bickel, U.; Thomsen, T.; Weber, W.; Fischer, J. P.; Bachus, R.; Nitz, M.; Kewitz, H. *Clin. Pharmacol. Ther.* **1991**, *50*, 420–428.

25.   Bickel, U.; Thomsen, T.; Fischer, J. P.; Weber, W.; Kewitz, H. *Neuropharmacology* **1991**, *30*, 447–454.

26.   Cardozo, M. G.; Kawai, T.; Iimura, Y.; Sugimoto, H.; Yamanishi, Y.; Hopfinger, A. J. *J. Med. Chem.* **1992**, *35*, 590–601.

27.   Marco-Contelles, J.; Carreiras, M. D. C.; Rodriguez, C.; Villarroya, M.; Garcia, A. G. *Chem. Rev.* **2006**, *106*, 116–133.

28.   Han, S. Y.; Sweeney, J. E.; Bachman, E. S.; Schweiger, E. J.; Forloni, G.; Coyle, J. T.; Davis, B. M.; Joullie, M. M. *Eur. J. Med. Chem.* **1992**, *27*, 673–687.

29. Jia, P.; Sheng, R.; Zhang, J.; Fang, L.; He, Q.; Yang, B.; Hu, Y. *Eur. J. Med. Chem.* **2009**, *44*, 772–784.

30. Shigeta, M.; Homma, A. *CNS Drug Rev.* **2001**, *7*, 353–368.

31. Polinsky, R. J. *Clin. Ther.* **1998**, *20*, 634–647.

32. Cutler, N. R.; Polinsky, R. J.; Sramek, J. J.; Enz, A.; Jhee, S. S.; Mancione, L.; Hourani, J.; Zolnouni, P. *Acta Neurol. Scand.* **1998**, *97*, 244–250.

33. Williams, B. R.; Nazarians, A.; Gill, M. A. *Clin. Ther.* **2003**, *25*, 1634–1653.

34. Farlow, M. R. *Clin. Pharmacokinet.* **2003**, *42*, 1383–1392.

35. Bickel, U.; Thomsen, T.; Fischer, J. P.; Kewitz, H. *Klin. Pharmakol.***1989**, *2*, 280–283.

36. Lilienfeld, S. *CNS Drug Rev.* **2002**, *8*, 159–176.

37. Tsuno, N. *Expert Rev. Neurother.* **2009**, *9*, 591–598.

38. Rogers, S. L.; Doody, R. S.; Mohs, R. C.; Friedhoff, L. T.; Alter, M.; Apter, J.; Williams, T.; Baumel, B.; Brown, W.; Clark, C.; Cohan, S.; Farlow, M.; Farmer, M.; Folks, D.; Geldmacher, D.; Heiser, J.; Jurkowski, C.; Krishnan, K. R.; Pelchat, R.; Sadowsky, C.; Sano, M.; Strauss, A.; Tune, L.; Webster, J.; Weiner, M.; Stark, S. *Arch. Intern. Med.* **1998**, *158*, 1021–1031.

39. Rogers, S. L.; Friedhoff, L. T. *Eur. Neuropsychopharmacol.* **1998**, *8*, 67–75.

40. Jelic, V.; Haglund, A.; Kowalski, J.; Langworth, S.; Winblad, B. *Dementia Geriatr. Cognit. Disord.* **2008**, *26*, 458–466.

41. Corey-Bloom, J.; Anand, R.; Veach, J. *Int. J. Geriatr. Psychopharmacol.* **1998**, *1*, 55–65.

42. Williams, B. R.; Nazarians, A.; Gillm M. A. *Clin. Ther.* **2003**, *25*, 1634–1653.

43. Farlow, M. R.; Hake, A.; Messina, J.; Hartman, R.; Veach, J.; Anand, R. *Arch. Neurol.* **2001**, *58*, 417–422.

44. Sramek, J. J.; Anand, R.; Wardle, T. S.; Irwin, P.; Hartman, R. D.; Cutler, N. R. *Life Sci.* **1996**, *58* (15), 1201–1207.

45. Raskind, M. A.; Peskind, E. R.; Wessel, T.; Yuan, W.; Allen, F. H., Jr.; Aronson, S. M.; Baumel, B.; Eisner, L.; Brenner, R.; Cheren, S.; Verma, S.; Daniel, D. G.; DePriest, M.; Ferguson, J. M.; England, D.; Farmer, M. V.; Frey, J.; Flitman, S. S.; Harrell, L. E.; Holub, R.; Jacobson, A.; Olivera, G. F.; Ownby, R. L.; Jenkyn, L. R.; Landbloom, R.; Leibowitz, M. T.; Zolnouni, P. P.; Lyketosos, C.; Mintzer, J. E.; Nakra, R.; Pahl, J. J.; Potkin, S. G.; Richardson, B. C.; Richter, R. W.; Rymer, M. M.; Saur, D. P.; Daffner, K. R.; Scinto, L.; Stoukides, J.; Targum, S. D.; Thein, S. G., Jr.; Thien, S. G.; Tomlinson, J. R. *Neurology* **2000**, *54*, 2261–2268.

46. Sugimoto, H.; Tsuchiya, Y.; Higurashi, K.; Karibe, N.; Iimura, Y.; Sasaki, A.; Yamanashi, Y.; Ogura, H.; Araki, S.; Takashi, K.; Atsuhiko, K.; Michiko, K.; Kiyomi, Y. EP 296560, **1988**.

47. Niphade, N.; Mali, A.; Jagtap, K.; Ojha, R. C.; Vankawala, P. J.; Mathad, V. T. *Org. Process Res. Dev.* **2008**, *12*, 731–735.

48. Iimura, Y. WO 9936405, **1999**.

49. Peglion, J. L.; Vian, J.; Vilaine, J. P.; Villeneuve, N.; Janiak, P.; Bidouard, J. P. EP 534859, **1993**.

50. Gutman, L. A.; Shkolnik, E.; Tishin, B.; Nisnevich, G.; Zaltzman, I. WO 2000009483, **2000**.

51.   Elati, C.; Kolla, N.; Chalamala, S. R.; Vankawala, P.; Sundaram, V.; Vurimidi, H.; Mathad, V. *Synth. Commun.* **2006**, *36*, 169–174.
52.   Weinstock, R. M.; Chorev, M.; Tashma, Z. Phenyl carbamates. EP 193926, **1986**.
53.   Stedman, E.; Stedman, E. *J. Chem. Soc.* **1929**, 609–617.
54.   Jaweed, M. S. M.; Upadhye, B. K.; Rai, V. C.; Zia, H. WO 2007026373, **2007**.
55.   Hu, M.; Zhang, F.-L.; Xie, M.-H. *Synth. Commun.* **2009**, *39*, 1527–1533.
56.   Fieldhouse, R. WO 2005058804, **2005**.
57.   Ciszewska, G.; Pfefferkorn, H.; Tang, Y. S.; Jones, L.; Tarapata, R.; Sunay, U. B. *J. Labelled Compd. Radiopharm.* **1997**, *39*, 651–668.
58.   Kueenburg, B.; Czollner, L.; Froehlich, J.; Jordis, U. *Org. Process Res. Dev.* **1999**, *3*, 425–431.
59.   Barton, D. H. R.; Kirby, G. W. *J. Chem. Soc.* **1962**, 806–817.
60.   Shieh, W.-C.; Carlson, J. A. *J. Org. Chem.* **1994**, *59*, 5463–5465.
61.   Magnus, P.; Sane, N.; Fauber, B. P.; Lynch, V. *J. Am. Chem. Soc.* **2009**, *131*, 16045–16047.
62.   Trost, B. M.; Toste, F. D. *J. Am. Chem. Soc.* **2000**, *122*, 11262–11263.
63.   Trost, B. M.; Tang, W. *Angew. Chem., Int. Ed.* **2002**, *41*, 2795–2797.
64.   Trost, B. M.; Tang, W.; Toste, F. D. *J. Am. Chem. Soc.* **2005**, *127*, 14785–14803.
65.   Kodama, S.; Hamashima, Y.; Nishide, K.; Node, M. *Angew. Chem., Int. Ed.* **2004**, *43*, 2659–2661.

18

Aprepitant (Emend): A NK₁ Receptor Antagonist for the Treatment of Postchemotherapy Emesis

John A. Lowe III

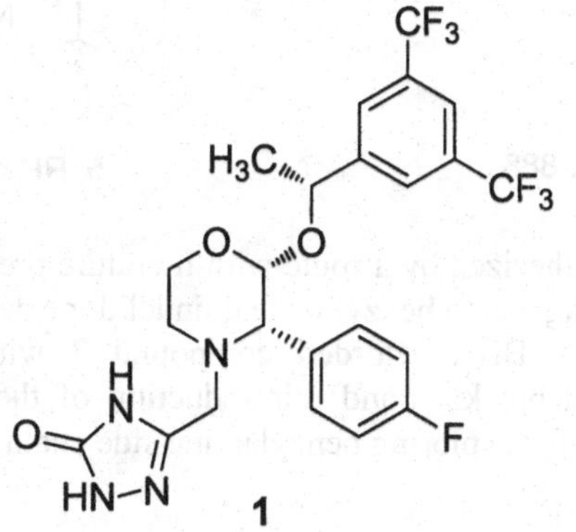

USAN: Aprepitant
Trade name: Emend®
Merck
Launched: 2003

18.1 Background

Substance P is an undecapeptide hormone originally isolated in 1931 as a contractile substance from intestine smooth muscle. Following its structural characterization in 1970, it was found to play a role in coordinating the inflammatory and stress/pain response to injury. Substance P was subsequently found to effect its in vivo activity through agonist stimulation of the NK_1 neurokinin receptor.[1] The peptide structure of substance P served as the starting point for a considerable effort to find NK_1 receptor antagonists throughout the 1970s and 1980s. However, the discovery of the first nonpeptide antagonist, CP-96,345 (2), in 1988 by the Pfizer group was the key event that opened the door to new therapeutics based on this target.[2] Subsequent efforts at Pfizer and other companies led to numerous structurally diverse NK_1 receptor antagonists, exemplified by 2–5. Based on the early pharmacology of substance P, initial clinical trials were directed toward inflammatory endpoints such as asthma and migraine, but were unsuccessful. The subsequent discovery of potent anti-emetic activity in animal models with some of the early compounds led to successful clinical trials for aprepitant (**1**), Merck's lead clinical candidate, in postchemotherapy emesis, establishing this as its clinical indication. We first review the medicinal and synthetic chemistry that led from CP-96,345 (**2**) to aprepitant (**1**), and then review the latter in more detail.

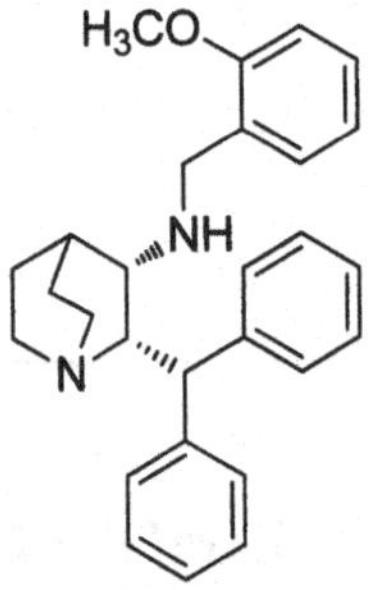

2, CP-96,345 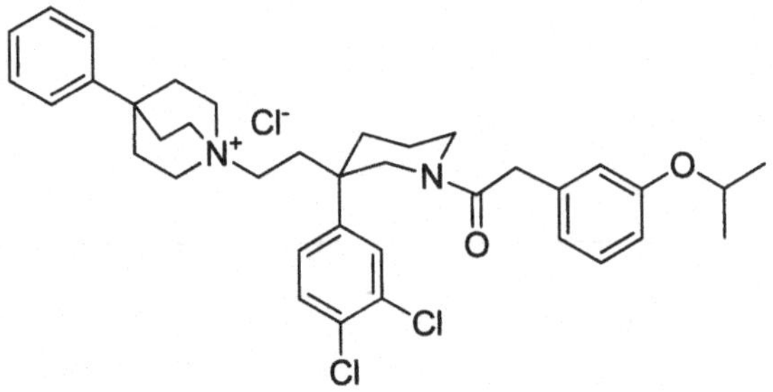   **3**, SR-140,333

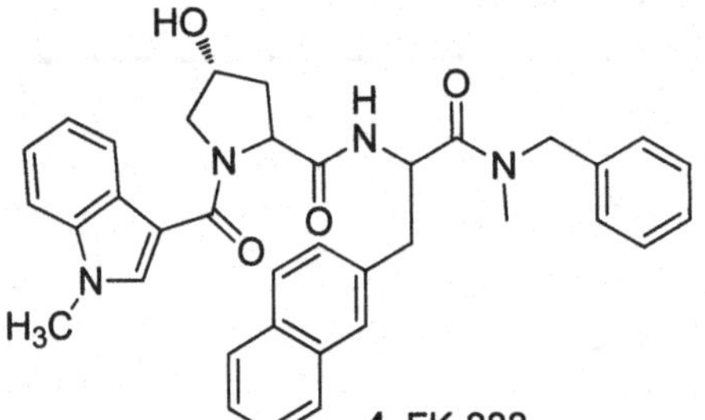

4, FK-888 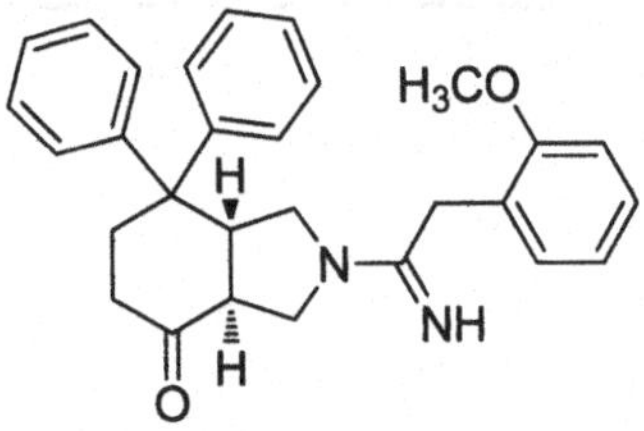   **5**, RP-67,580

CP-96,345 (**2**) was synthesized by a route with literature precedent based on the 1,4-addition of aryl Grignard reagents to benzylidene quinuclidone derivatives such as **6**.[3] The hindered borane reagent 9-BBN afforded compound **7** with the desired *cis*-stereochemistry. Subsequent deblocking and reintroduction of the benzyl side chain furnished analogs depicted by **8** for exploring benzylamine side chain SAR.

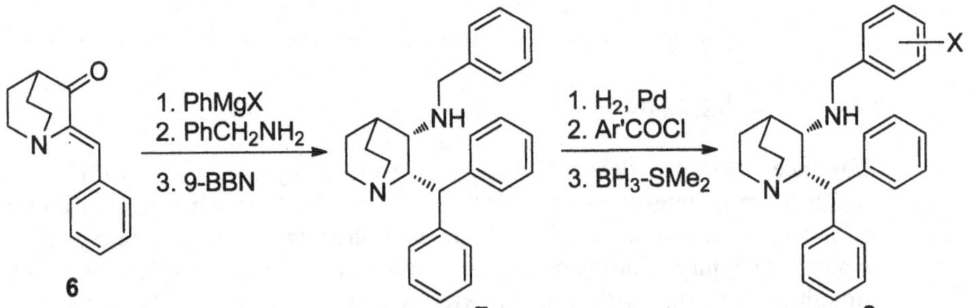

The synthesis of CP-96,345 (**2**) in optically active form, which relies on a chiral isocyanate-based resolution, is shown in Scheme 1.

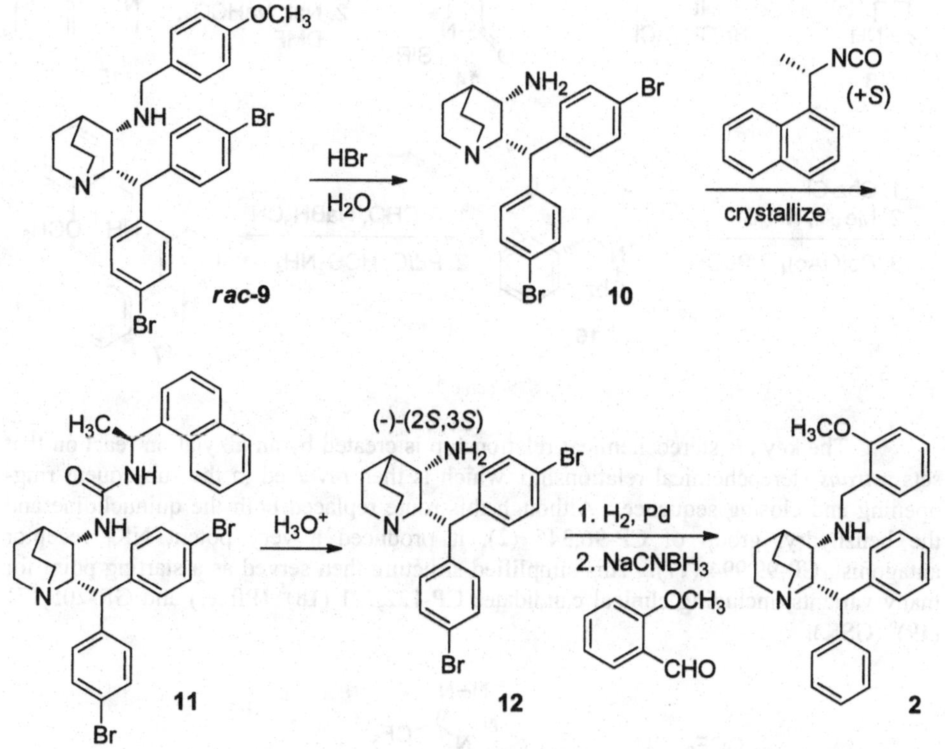

Scheme 1.

The next stage in optimizing the structure came from its simplification to a piperidine by Desai and co-workers. They discovered an efficient synthetic route to the piperidine template **17** using the readily available 4-phenyl azetidin-2-one **13**, as shown in Scheme 2.[4]

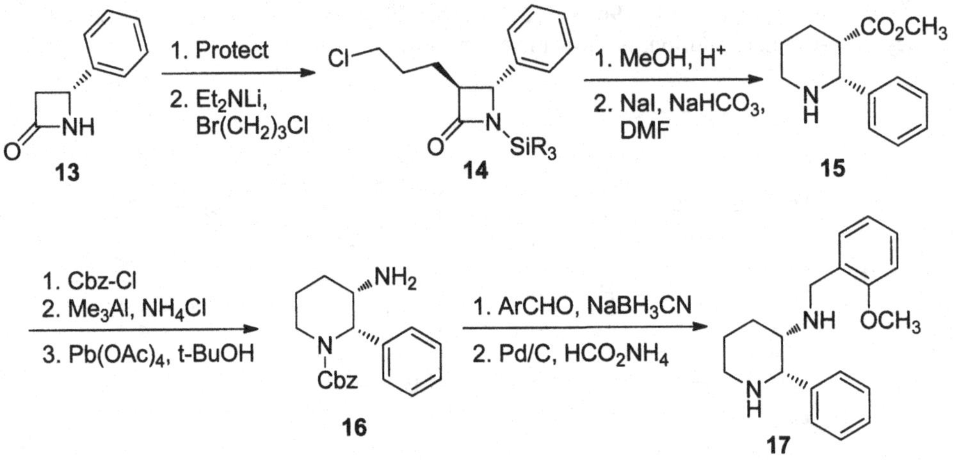

Scheme 2.

The key *cis* stereochemical relationship is created by an alkylation reaction that sets a *trans* stereochemical relationship, which is then reversed in the subsequent ring-opening and closing sequence. Although this route replaced both the quinuclidine and the benzhydryl group of CP-96,345 (**2**), it produced a very potent NK$_1$ receptor antagonist, CP-99,994 (**17**). This simplified structure then served as a starting point for many variants, including clinical candidates CP-122,721 (**18**)[5] (Pfizer) and GR-205,171 (**19**)[6] (GSK).

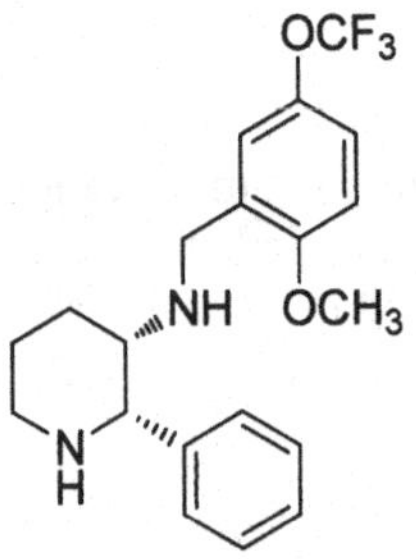

18, CP-122,721 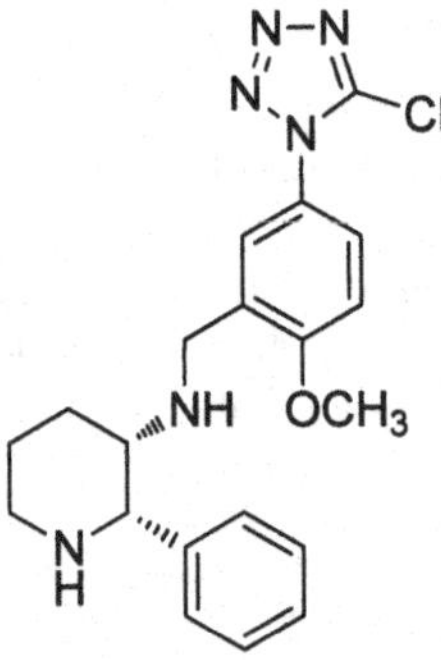   **19, GR-205,171** 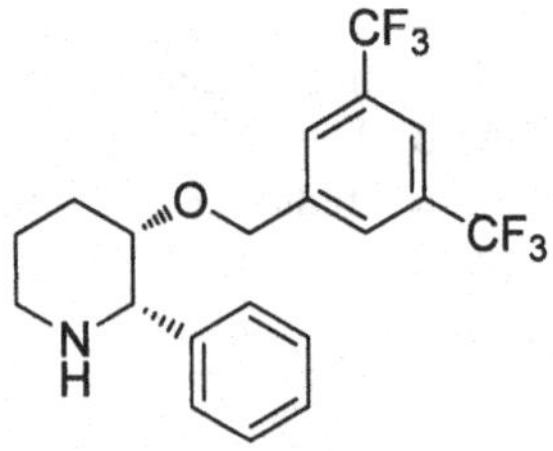   **20, L-733,060**

At Merck, the next alteration in the structure was to replace the benzylamine side chain with a benzyl ether, shown above in L-733,060 (**20**)[7] and then to replace the piperidine with a morpholine ring, which ultimately led to the discovery of aprepitant (**1**). These changes afforded patentability, and also reduced off-target activity at ion channels relative to CP-96,345 and CP-99,994.

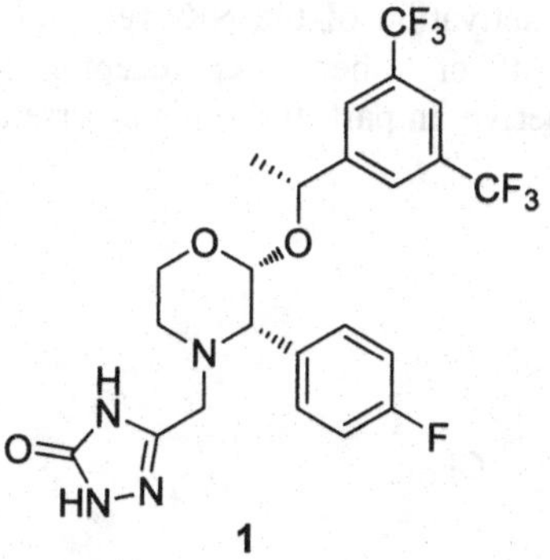

It is worth noting that a different set of substituents and a different substitution pattern afford optimal NK_1 receptor antagonist activity in the Merck series as compared with the Pfizer series. It was subsequently discovered that the change from a benzhydryl quinuclidine to a phenyl piperidine changes the way each class of compounds binds to the NK_1 receptor, with the phenyl piperidines moving farther down in the binding site. In the case of the benzyl ether side chain, this movement to a new binding site necessitates a new substituent pattern, from 2-OMe to 3,5-diCF$_3$, to afford optimal receptor binding.

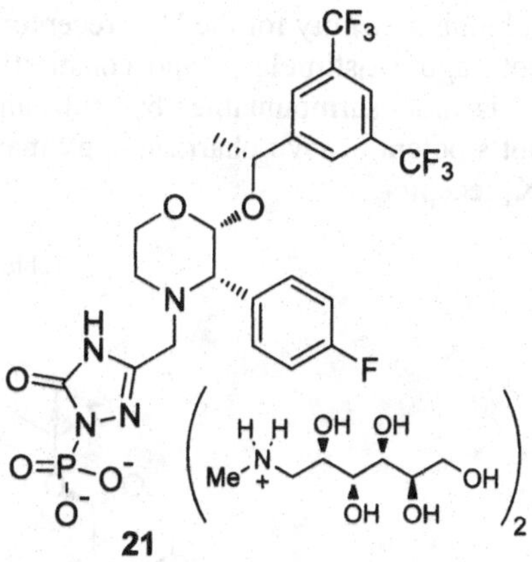

Finally, a more soluble form of aprepitant was recently developed for intravenous administration, using a phosphate group that is readily hydrolyzed in vivo to afford the parent drug. Fosaprepitant (**21**) shows equivalent activity to aprepitant in vivo after p.o. or i.v. administration.[9]

18.2   In Vitro Pharmacology and Structure-Activity Relationships (SAR)

The initial in vitro activity of several of the standard NK_1 receptor antagonists compared with aprepitant is given in Table 1. The three clinical candidates from Pfizer, GSK, and Merck stand out as the most potent compounds in this regard.

Table 1.

Entry	Compound	hNK$_1$ K$_i$ (nM)^a
1	**2**	0.4
2	**17**	0.5
3	**18**	0.14
4	**19**	0.08
5	**20**	0.87
6	**1**	0.09

^{a}K$_i$ value at the human NK$_1$ receptor.

The SAR of the aprepitant series is summarized in Table 2, which illustrates that, once the triazolinone substituent was optimized,[10] many close-in compounds were found to bind as well as **1**.[8] These compounds will be differentiated later. In addition to its potent binding affinity for the NK_1 receptor, aprepitant is an inverse agonist—that is, it blocks both agonist-stimulated and constitutive activation of the NK_1 receptor, and its blockade is not surmountable by substance P or other NK_1 receptor agonists. Aprepitant's potent in vivo pharmacology may derive, in part, from this inverse agonism at the NK_1 receptor.[8]

Table 2.

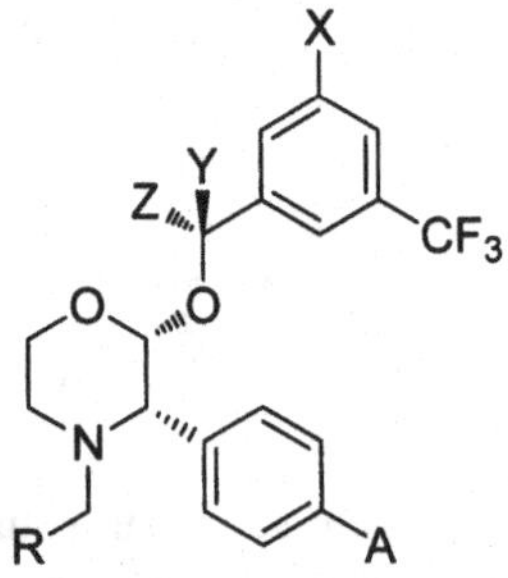

Entry	Compound	A	X	Y	Z	R	hNK$_1$ K$_i$ (nM)^a
1	1	F	CF$_3$	H	CH$_3$	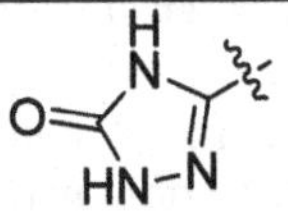	0.09
2	22	H	CF$_3$	H	H	CO$_2$CH$_3$	1.6
3	23	H	CF$_3$	H	H	CO$_2$H	66
4	24	H	CF$_3$	H	H	CONH$_2$	1.1
5	25	H	CF$_3$	H	H	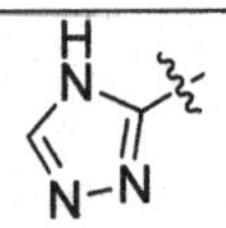	0.13
6	26	H	CF$_3$	H	H	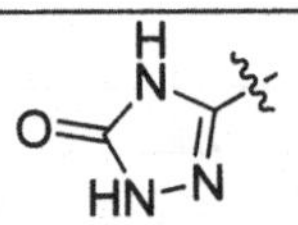	0.09
7	27	H	CF$_3$	H	CH$_3$	"	0.88
8	28	H	CF$_3$	CH$_3$	H	"	0.09
9	29	H	F	H	CH$_3$	"	0.10
10	30	H	H	H	CH$_3$	"	0.27
11	31	F	CF$_3$	H	H	"	0.07

^{a}K$_i$ value at the human NK$_1$ receptor.

18.3 In Vivo Pharmacology

To prioritize the best compound from the preceding collection in Tables 1 and 2, in vivo pharmacology was designed to examine activity both acutely (1–4 h postdose) and chronically (24 h postdose). Two animal models were reported: resiniferatoxin-induced systemic vascular leakage (SYVAL) and GR-73632-induced foot tapping in gerbils (Foot Tap). Vascular leakage or extravasation resulting from an injury is triggered by NK_1 receptor–mediated leakiness in the vascular endothelium, which allows circulating white blood cells to extravasate, or enter into, the site of an injury to initiate the inflammatory process. In the SYVAL test, Evans blue dye is used in place of white blood cells as a readily measurable endpoint. As shown in Table 3, aprepitant (1) shows good activity at 1 h postdose, along with several of the other compounds tested, but is superior at 24 h postdose. This result indicates 1 has the best pharmacokinetics and longest half-life/duration of action.[8]

Table 3.

Entry	Compound	ID_{50} (p.o., 1 h, mg/kg)^a	ID_{90} (p.o., 24 h, mg/kg)^a
1	18	0.010	>10
2	19	0.007	>10
3	26	0.006	5.4
4	28	0.010	2.3
5	31	0.008	2.3
6	1	0.008	1.8

^{a}ID is the inhibitory dose for the SYVAL assay.

The Foot Tap test was developed as an indication of CNS activity, in that a compound must penetrate the CNS to be active. In the test, a peptidic NK_1 receptor agonist, GR-73632 is administered by i.c.v. injection through a hole in the skull into the ventricles of the brain to induce the foot-tapping behavior. Once again, aprepitant (1) proved to have the longest duration of action by virtue of its lowest ID_{50} value at 24 h postdose, as shown in Table 4.[8]

Table 4.

Entry	Compound	IC_{50} (i.v., mg/kg)^a		
		$t = 0$ h	$t = 4$ h	$t = 24$ h
1	18	0.03	0.24	5.37
2	19	0.04	0.12	>10
3	26	0.85	nd	2.88
4	28	0.16	0.04	1.11
5	31	0.30	0.07	1.24
6	1	0.36	0.04	0.33

^{a}IC_{50} is the 50% inhibitory concentration for the Foot Tap assay.

Finally, the ability of aprepitant to block emesis, a therapeutically relevant endpoint for the ultimate clinical application of the drug, as discussed below, was tested in the presence of several emetogens in the ferret. This species is believed to be a valid animal model for cancer chemotherapy–induced emesis in humans. Aprepitant (1) effectively blocked retching and vomiting due to cisplatin, apomorphine, and morphine at a dose of 3.0 mg/kg p.o. and 0.3 mg/kg i.v. (in the case of cisplatin).[8]

18.4   Pharmacokinetics and Drug Metabolism

In the ferret, aprepitant (1) shows a plasma half-life of 9.7 h after a 0.5 mg/kg i.v. dose. The volume of distribution is 1.3 L/kg, giving a clearance value of 1.5 mL/min/kg, and confirming the slow clearance of the compound and long duration of action seen in the in vivo studies reported above. After a 1 mg/kg p.o. dose, T_{max} is reached in 3.3 h, and the oral bioavailability is 45%.[11] Brain penetration was assessed in this study using C-14-labeled aprepitant, which showed a brain to plasma ratio of about 0.8 after an oral dose of 3 mg/kg. Although several metabolites of aprepitant were detected, the study concluded that the activity of aprepitant was due to the parent drug.

In human volunteers, aprepitant showed between 59% and 67% oral bioavailability, and there was no effect of food on drug absorption.[12] The compound is also a substrate and a moderate inhibitor of cytochrome P450 3A4 and an inducer of 2C9, but shows only modest 3A4 inhibition after an oral dose and modest induction of these enzymes over time in the clinic.[13]

In addition, Merck used positron emission technology (PET) studies to determine the NK_1 receptor occupancy of aprepitant in human brain. Using an F-18-labeled selective NK_1 receptor ligand and then measuring its displacement after oral doses of aprepitant, their study showed that aprepitant achieved >90% NK_1 receptor occupancy at doses of 100 mg or higher.[14]

A recent study showed that 115 mg fosaprepitant (21) provides an equivalent amount of circulating aprepitant to a 150 mg dose of the parent drug, thus serving as a parenteral alternative to the oral form of the parent.[15]

18.5   Efficacy and Safety

In the initial study of aprepitant (1) for the treatment of emesis in cancer chemotherapy patients, it was used with a combination of granisetron, a 5-HT$_3$ receptor antagonist, and dexamethasone, the standard therapy at that time.[16] Acute emesis (occurring within 24 h) and delayed emesis (occurring on days 2 to 5) were assessed with aprepitant given as a single 400 mg dose before cisplatin or given in addition as two 300 mg doses on days 2 and 3, compared with placebo. Aprepitant showed significant superiority to placebo in both the acute and (especially in the) delayed phases of emesis. In a subsequent study, aprepitant was given alone and in combination with granisetron and dexamethasone, as above, but was compared with these agents alone rather than with placebo.[17] Aprepitant alone was effective in 43% and 57% of patients, respectively, in treating acute and delayed emesis following cisplatin therapy. In combination with granisetron and dexamethasone, it was effective in 80% and 63% of the patients in the acute and delayed groups and was far more effective than the combination of granisetron and dexamethasone alone. Consequently, aprepitant is now used in this regimen as a combination with a 5-HT₃ receptor antagonist and a corticosteroid for the treatment of postchemotherapy emesis. Fosaprepitant (21) is currently being studied as an alternative to oral aprepitant therapy, given its bioequivalence as noted above.[18]

18.6   Synthesis of Aprepitant (1) and Fosaprepitant (21)

The initial synthesis of aprepitant (1), which relies on a Tebbe olefination and reduction to install a methyl group on the benzyl ether side chain, is shown in Scheme 3.[8,19] The initial steps are from a literature-precedented synthesis of p-fluorophenyl glycine based on conversion of chiral oxazolidinone 33 to azide 34. Formation of morpholinone intermediate 36 proceeds via benzylation and reaction with 1,2-dibromoethane.

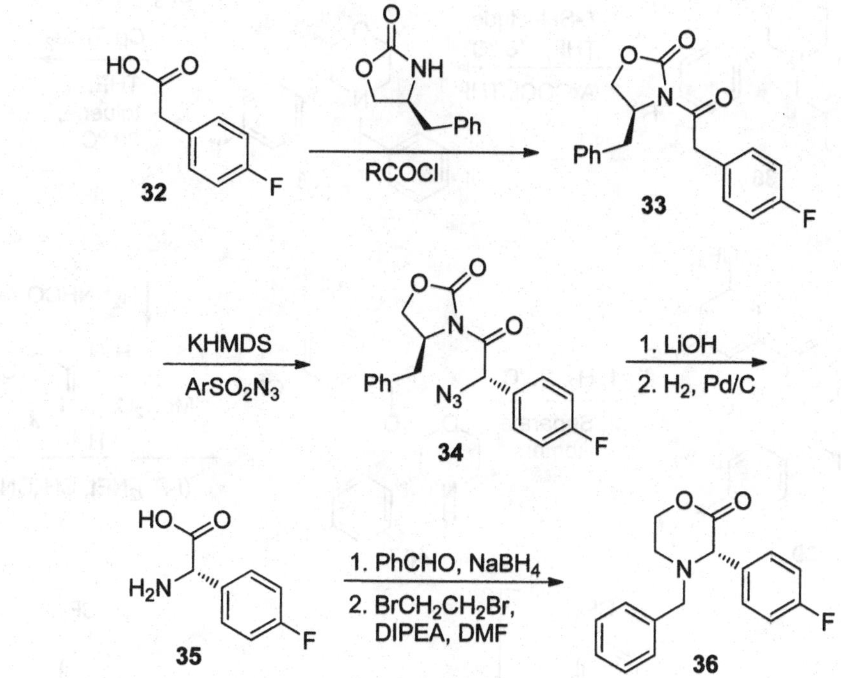

Scheme 3.

        A classical resolution using dibenzoyl tartrate affords an even simpler synthesis of the chiral phenyl glycine intermediate 35.

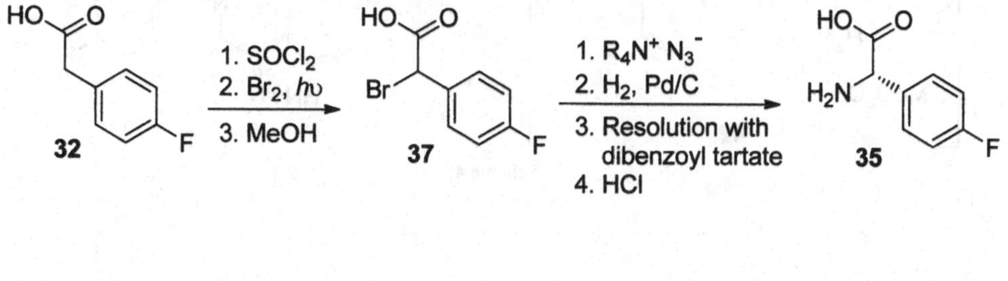

To install the ether side chain, the morpholinone is reduced to the desired *cis* stereoisomer with the hindered reagent *L*-Selectride, and the acetal is acylated to afford intermediate **38** (Scheme 4). The required methyl group is then added via a modified Tebbe olefination procedure to provide **39**, and the resulting olefin is reduced and the benzyl group is removed under hydrogenation conditions in a single step to give **40**. The triazolinone ring was initially installed using a two-step procedure based on alkylation of morpholine **40** with intermediate **41** to form intermediate **42**, followed by ring closure in refluxing xylenes to furnish aprepitant (**1**).

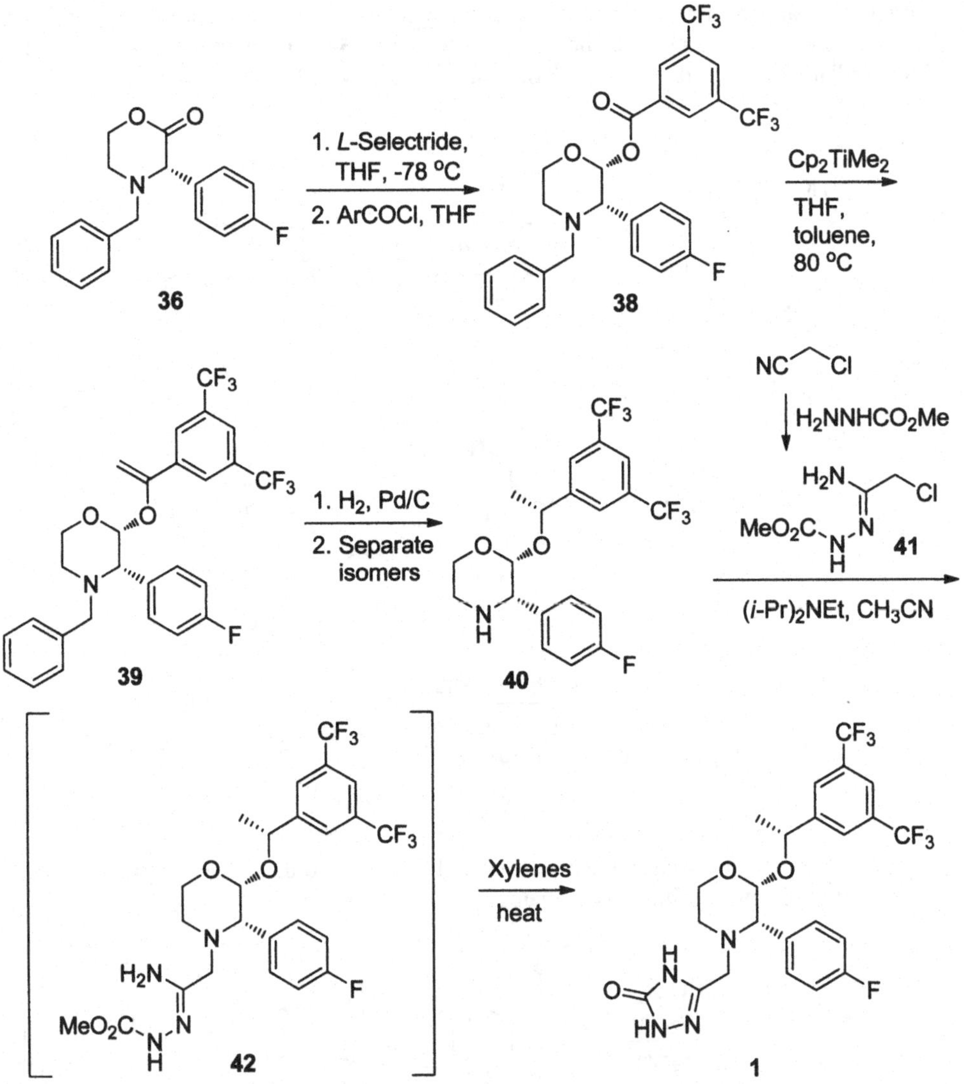

Scheme 4.

Merck developed a simpler approach to the final installation of the triazolinone side chain via addition of **45** to key morpholine intermediate **40** (Scheme 5).[20] In this case, **45** is prepared either directly from intermediate **43** using chloroacetate *ortho*-ester, or stepwise using the benzyl-protected acid chloride of glycolic acid to give intermediate **44**, followed by debenzylation and activation of the alcohol as the chloride.

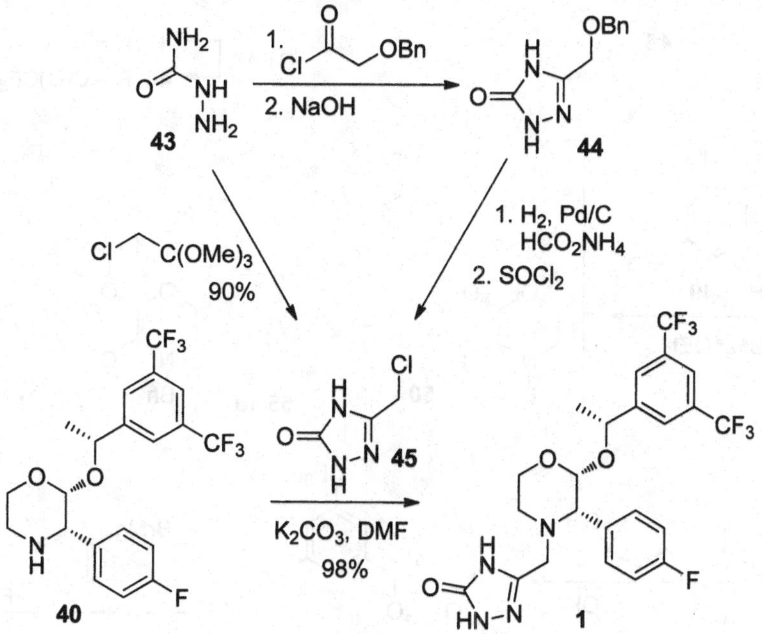

Scheme 5.

After the identification of aprepitant as a clinical candidate, Merck invested considerable process research toward an improved synthesis of aprepitant, which culminated in the elegant manufacturing process shown in Scheme 6.[21,22] The key step relies on displacement of a trifluoroacetate from intermediate **48** by the optically active alcohol intermediate **49**. The synthesis of **49** was accomplished via an oxazaborolidine-catalyzed borane reduction of the corresponding acetophenone. Although the displacement resulted in an almost equal mixture of the two diastereomers **50** and **51**, the desired diastereomer **50** could be recovered in high yield by base-catalyzed equilibration of the mixture and crystallization. Addition of *p*-fluorophenyl magnesium bromide followed by hydrogenolysis afforded the key intermediate **40**, which can be readily converted to **1** as detailed in the previous synthesis.

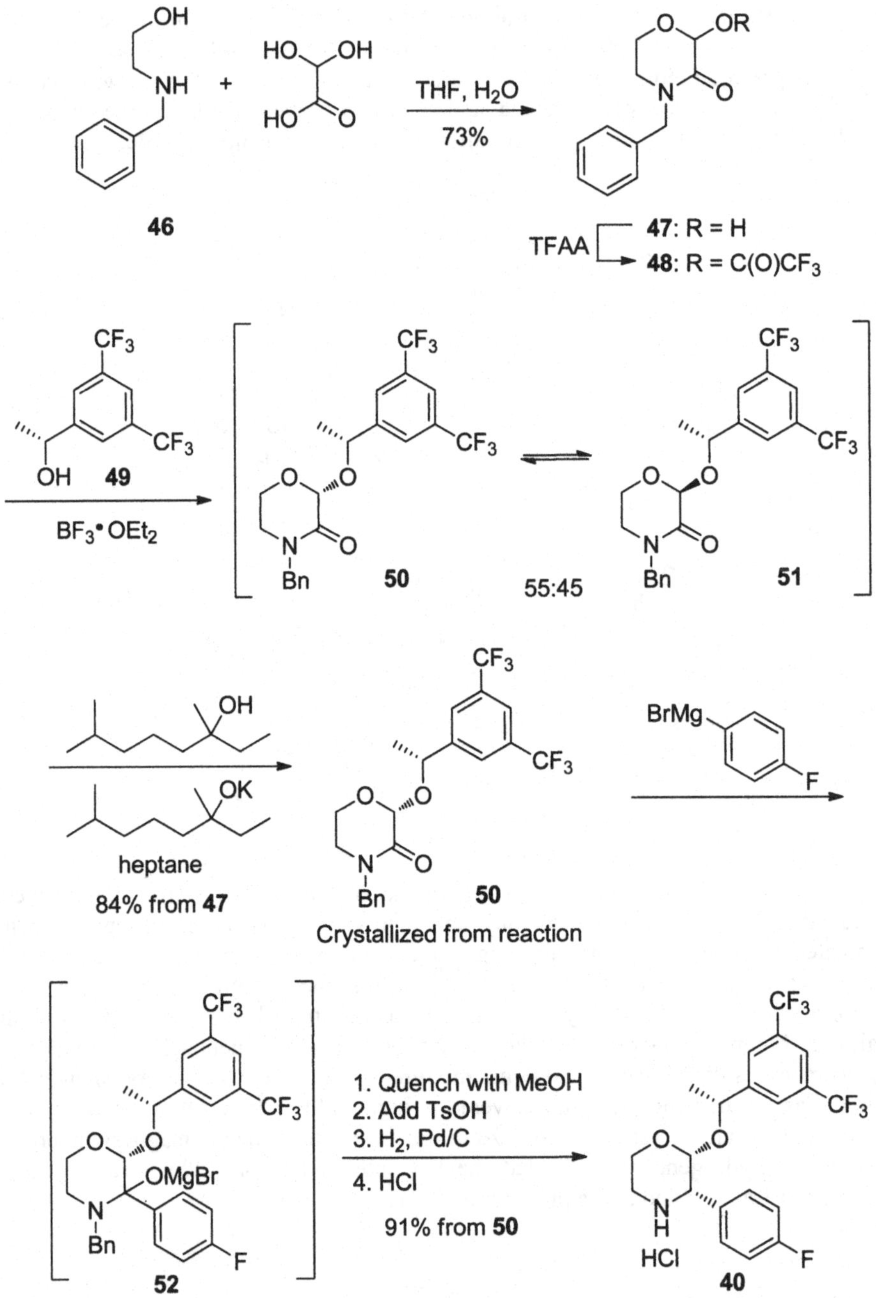

Scheme 6.

Dr. Reddy's Laboratories has also disclosed a synthetic route to aprepitant, which uses an oxidation/reduction sequence as the key step to afford the desired *cis-*

stereochemistry of the morpholine intermediate **40**. Merck had previously reported that the alkylation of the hydroxy-morpholine intermediate with the corresponding secondary alkyl halide to install the benzyl ether side chain was unsuccessful, affording only acyclic by-products. But in the synthesis in Scheme 7, the alkylation is carried out on the *trans*-stereoisomer of the hydroxy-morpholine **57**, which is evidently less hindered and therefore permits successful alkylation. The oxidation (**59** to **60**)/reduction (**60** to **40**) sequence is then necessary to install the correct *cis*-stereochemistry.[23]

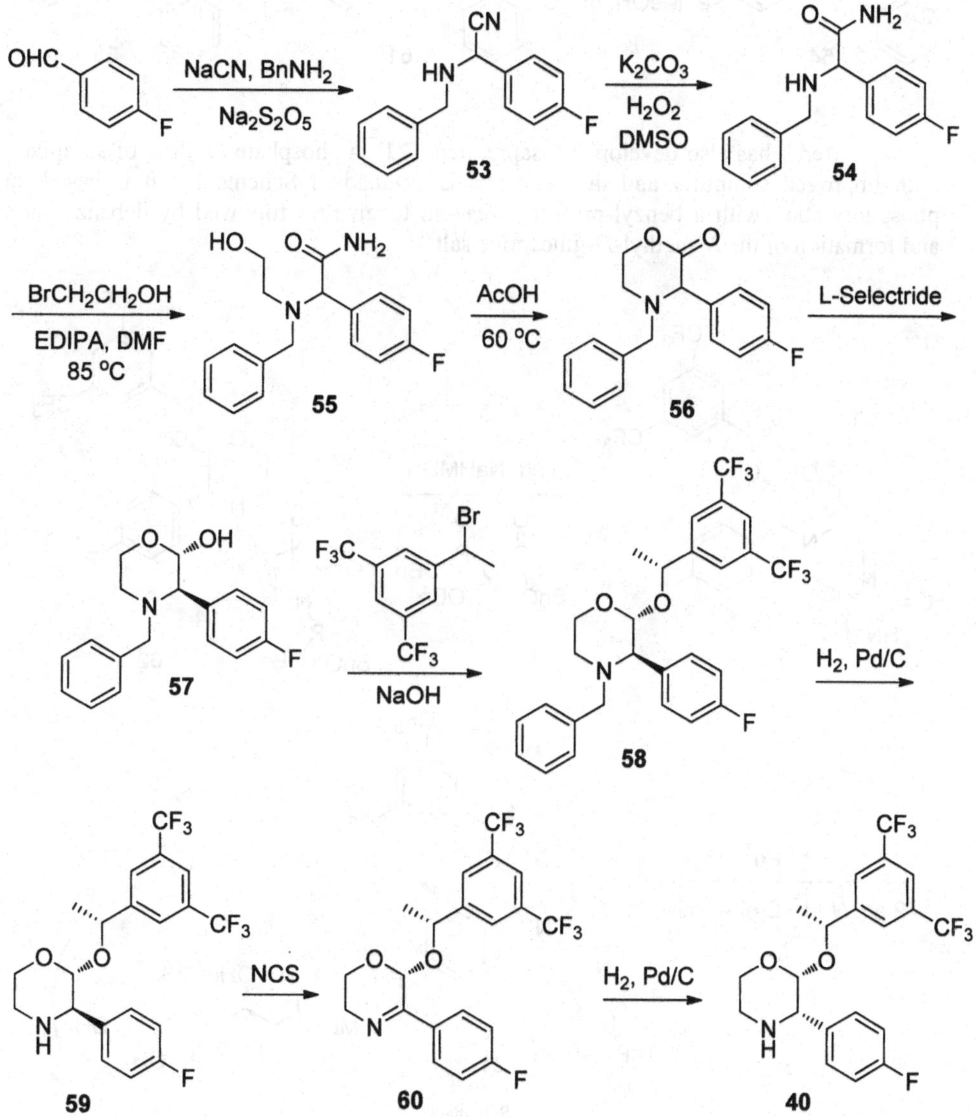

Scheme 7.

To prepare aprepitant in optically active form, this group reported a resolution-based scheme using di-*p*-toluoyl-tartaric acid (*D*-(+)-DPTTA) to afford the key morpholinone intermediate **36**.[24]

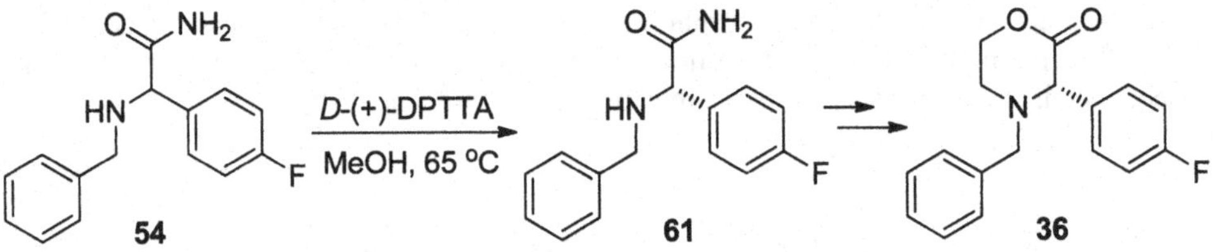

Merck has also developed fosaprepitant (**21**), a phosphate prodrug of aprepitant with improved solubility and the synthesis is detailed in Scheme 8. It is based on phosphorylation with a benzyl-protected reagent to give **62** followed by debenzylation and formation of the *N*-methyl-*D*-glucamine salt.[9]

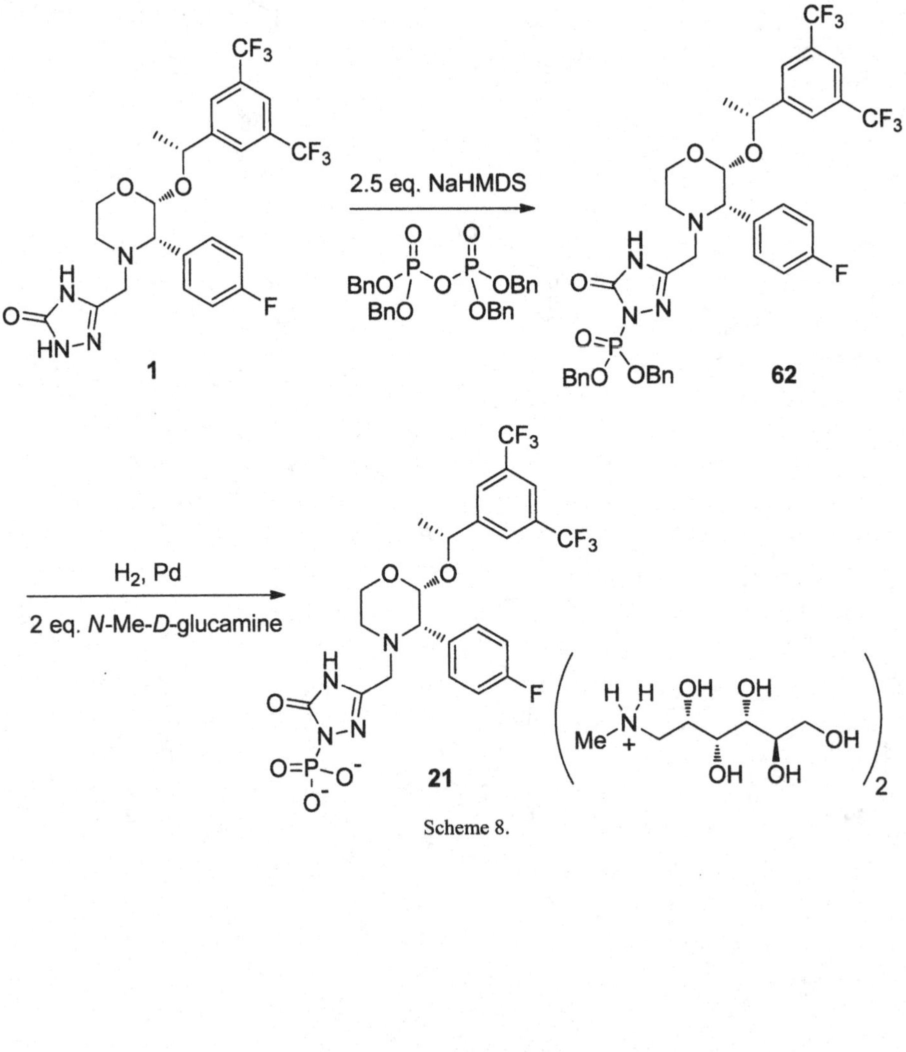

Scheme 8.

While the original synthesis of **1** features a clever use of the Tebbe olefination reaction, the process routes use elegant methods to set the ring stereochemistry through displacement reactions (and equilibration), followed by reduction, reactions that are much more scalable. This difference reflects the contrasting need for SAR development in the medicinal chemistry work vs. the need for scalability in the process work.

18.7   References

1.   Hokfelt, T.; Pernow, B. et al. *J. Intern. Med.* **2001**, *249*, 27–40.
2.   Snider, R. M.; Constantine, J. W.; Lowe, J. A., III; Longo, K. P.; Lebel, W. S.; Woody, H. A.; Drozda, S. E.; Desai, M. C.; Vinick, F. J.; Spencer, R. W.; et al. *Science* **1991**, *251*, 435–437.
3.   Lowe, J. A., III; Drozda, S. E.; Snider, R. M.; Longo, K. P.; Zorn, S. H.; Morrone, J.; Jackson, E. R.; McLean, S.; Bryce, D. K.; Bordner, J.; et al. *J. Med. Chem.* **1992**, *35*, 2591–2600.
4.   Desai, M. C.; Lefkowitz, S. L.; Thadeio, P. F.; Longo, K. P.; Snider, R. M. *J. Med. Chem.* **1992**, *35*, 4911–4913.
5.   Rosen, T. J.; Coffman, K. J.; McLean, S.; Crawford, R. T.; Bryce, D. K.; Gohda, Y.; Tsuchiya, M.; Nagahisa, A.; Nakane, M.; Lowe, J. A., III. *Bioorg. Med. Chem. Lett.* **1998**, *8*, 281–284.
6.   Gardner, C. J.; Armour, D. R.; Beattie, D. T.; Gale, J. D.; Hawcock, A. B.; Kilpatrick, G. J.; Twissell, D. J.; Ward, P. *Regul. Pept.* **1996**, *65*, 45–53.
7.   Harrison, T.; Williams, B. J.; Swain, C. J.; Ball, R. G. *Bioorg. Med. Chem. Lett.* **1994**, *4*, 2545–2550.
8.   Hale, J. J.; Mills, S. G.; MacCoss, M.; Finke, P. E.; Cascieri, M. A.; Sadowski, S.; Ber, E.; Chicchi, G. G.; Kurtz, M.; Metzger, J.; Eiermann, G.; Tsou, N. N.; Tattersall, F. D.; Rupniak, N. M.; Williams, A. R.; Rycroft, W.; Hargreaves, R.; MacIntyre, D. E. *J. Med. Chem.* **1998**, *41*, 4607–4614.
9.   Hale, J. J.; Mills, S. G.; MacCoss, M.; Dorn, C. P.; Finke, P. E.; Budhu, R. J.; Reamer, R. A.; Huskey, S. E.; Luffer-Atlas, D.; Dean, B. J.; McGowan, E. M.; Feeney, W. P.; Chiu, S. H.; Cascieri, M. A.; Chicchi, G. G.; Kurtz, M. M.; Sadowski, S.; Ber, E.; Tattersall, F. D.; Rupniak, N. M.; Williams, A. R.; Rycroft, W.; Hargreaves, R.; Metzger, J. M.; MacIntyre, D. E. *J. Med. Chem.* **2000**, *43*, 1234–1241.
10.   Hale, J. J.; Mills, S. G.; MacCoss, M.; Shah, S. K.; Qi, H.; Mathre, D. J.; Cascieri, M. A.; Sadowski, S.; Strader, C. D.; MacIntyre, D. E.; Metzger, J. M. *J. Med. Chem.* **1996**, *39*, 1760–1762.
11.   Huskey, S. E.; Dean, B. J.; Bakhtiar, R.; Sanchez, R. I.; Tattersall, F. D.; Rycroft, W.; Hargreaves, R.; Watt, A. P.; Chicchi, G. G.; Keohane, C.; Hora, D. F.; Chiu, S. H. *Drug Metab. Dispos.* **2003**, *31*, 785–791.
12.   Majumdar, A. K.; Howard, L.; Goldberg, M. R.; Hickey, L.; Constanzer, M.; Rothenberg, P. L.; Crumley, T. M.; Panebianco, D.; Bradstreet, T. E.; Bergman, A. J.; Waldman, S. A.; Greenberg, H. E.; Butler, K.; Knops, A.; De Lepeleire, I.; Michiels, N.; Petty, K. J. *J. Clin. Pharmacol.* **2006**, *46*, 291–300.
13.   Shadle, C. R.; Lee, Y.; Majumdar, A. K.; Petty, K. J.; Gargano, C.; Bradstreet, T. E.; Evans, J. K.; Blum, R. A. *J. Clin. Pharmacol.* **2004**, *44*, 215–223.

14.   Bergstrom, M.; Hargreaves, R. J.; Burns, H. D.; Goldberg, M. R.; Sciberras, D.;
        Reines, S. A.; Petty, K. J.; Ogren, M.; Antoni, G.; Langstrom, B.; Eskola, O.;
        Scheinin, M.; Solin, O.; Majumdar, A. K.; Constanzer, M. L.; Battisti, W. P.;
        Bradstreet, T. E.; Gargano, C.; Hietala, J. *Biol. Psychiatry* **2004**, *55*, 1007–1012.
15.   Navari, R. M. *Expert Opin. Investig. Drugs* **2007**, *16*, 1977–1985.
16.   Navari, R. M.; Reinhardt, R. R.; Gralla, R. J.; Kris, M. G.; Hesketh, P. J.;
        Khojasteh, A.; Kindler, H.; Grote, T. H.; Pendergrass, K.; Grunberg, S. M.;
        Carides, A. D.; Gertz, B. J. *N. Engl. J. Med.* **1999**, *340*, 190–195.
17.   Campos, D.; Pereira, J. R.; Reinhardt, R. R.; Carracedo, C.; Poli, S.; Vogel, C.;
        Martinez-Cedillo, J.; Erazo, A.; Wittreich, J.; Eriksson, L. O.; Carides, A. D.;
        Gertz, B. J. *J. Clin. Oncol.* **2001**, *19*, 1759–1767.
18.   Navari, R. M. *Expert Rev. Anticancer Ther.* **2008**, *8*, 1733–1742.
19.   Sorbera, L. A.; Castaner, J.; Bayes, M.; Silvestre, J. *Drugs Fut.* **2002**, *27*, 211–
        222.
20.   Cowden, C. J.; Wilson, R. D.; Bishop, B. C.; Cottrell, I. F.; Davies, A. J.;
        Dolling, U.-H. *Tetrahedron Lett.* **2000**, *41*, 8661–8664.
21.   Brands, K. M. J.; Payack, J. F.; Rosen, J. D.; Nelson, T. D.; Candelario, A.;
        Huffman, M. A.; Zhao, M. M.; Li, J.; Craig, B.; Song, Z. J.; Tschaen, D. M.;
        Hansen, K.; Devine, P. N.; Pye, P. J.; Rossen, K.; Dormer, P. G.; Reamer, R. A.;
        Welch, C. J.; Mathre, D. J.; Tsou, N. N.; McNamara, J. M.; Reider, P. J. *J. Am.
        Chem. Soc.* **2003**, *125*, 2129–2135.
22.   Nelson, T. D. in *Strategies and Tactics in Organic Synthesis*. Vol. 6, Ed.
        Harmata, M., Elsevier: Amsterdam, **2005**, pp. 321–351.
23.   Vankawala, P. J.; Elati, R. R. C.; Kolla, N. K.; Chlamala, S. R.; Gangula, S. WO
        2007/044829, **2007**.
24.   Kolla, N.; Elati, C. R.; Arunagiri, M.; Gangula, S.; Vankawala, P. J.;
        Anjaneyulu, Y.; Bhattacharya, A.; Venkatraman, S.; Mathad, V. T. *Org. Process
        Res. Dev.* **2007**, *11*, 455–457.

19

Armodafinil (Nuvigil): A Psychostimulant for the Treatment of Narcolepsy

Ji Zhang and Jason Crawford

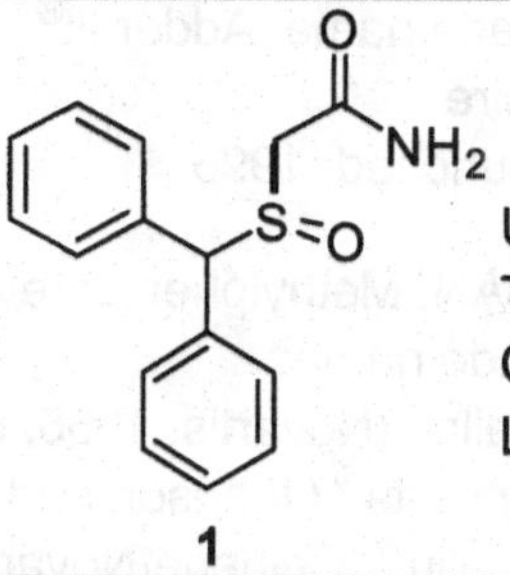

USAN: Armodafinil
Trade name: Nuvigil®
Cephalon
Launched: 2007(USA)

1

19.1   Background

Narcolepsy is a chronic disorder, characterized by excessive daytime sleepiness, and is estimated to affect as many as three million people worldwide.[1] The excessive drowsiness associated with the disabling neurological disorder often interferes with normal daytime activities, and may present a significant safety risk if the patient is driving or operating mechanical equipment, which could lead to accident-related injury or loss of life. Excessive drowsiness has also been seen as inappropriate in the workplace or in social activities, creating embarrassment or physical harm. Mood changes, including increased irritability and impaired cognitive function in the areas of concentration and memory are reported to be related to the disorder.[2]

The treatment of narcolepsy with psychostimulants such as amphetamine **2** (Adderall), and methylphenidate **3** (Ritalin) has been reported.[3] However, these are schedule 2 DEA-controlled substances and have a potential risk of abuse, overdose, and dependence, which present substantial barriers to widespread use.[4] As a result, there has been a significant effort to identify novel therapeutic agents for the treatment of narcolepsy.[5] Armodafinil **1** (Nuvigil) is a unique psychostimulant drug that was approved by the FDA on June 15, 2007, for the treatment of shift work sleep disorder and excessive daytime sleepiness associated with obstructive sleep apnea. Armodafinil **1** is the active component, (–)-(R)-enantiomer, of the racemic drug modafinil **4** (Provigil).[6]

Modafinil **4** and adrafinil **5** were originally discovered by scientists working for the French pharmaceutical company Lafon in the late 1970s.[7] Adrafinil **5** was first offered as an experimental treatment for narcolepsy in France in 1984. It was found that the amide modafinil **4** is the primary metabolite of the hydroxylamide adrafinil **5** and has similar biological activity. Modafinil was approved in France in 1994 under the name modiodal, and was subsequently approved in the United States in 1998 as was Provigil. Unlike other CNS stimulants, Modafinil has been reported to have little abuse liability. However, the compound was not as potent, as a CNS stimulant, as methamphetamine. Nevertheless, studies have shown that the drug promotes vigilance and wakefulness without the central and peripheral side effects associated with other psychostimulants. Armodafinil is the R- and longer lasting isomer of the racemic modafinil. It was found that armodafinil is eliminated approximately three times more slowly than the S-isomer of racemic modafinil,[7b] thus making single-oral daily dosing feasible.

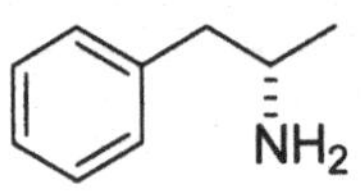

USAN: Amphetamine
Trade name: Adderall®
Shire
Launched: 1996

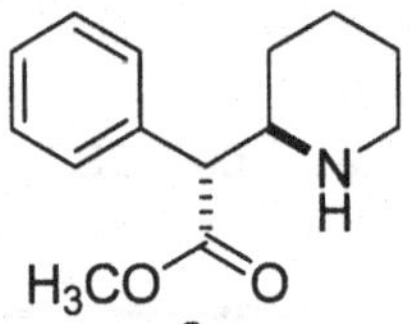

2

USAN: Methylphenidate
Trade name:
Ritalin® (Novartis, 1955, racemate)
Concerta® (Johnson and Johnson, 2000)
Focalin® (Celgene/Novartis, 2001)
Daytrana® (Shire, 2006)

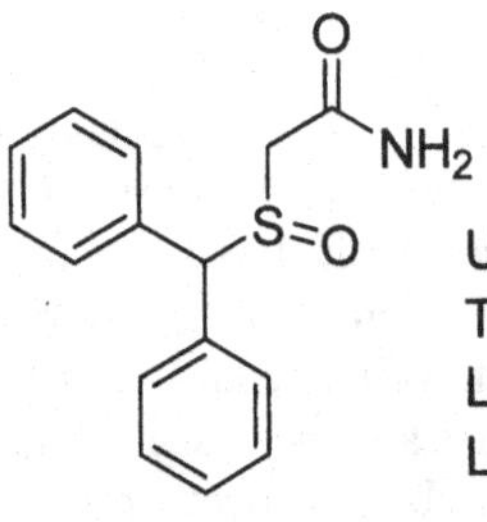

3

USAN: Modafinil
Trade name: Provigil®
Lafon Laboratories
Launched: 1998(USA)

4

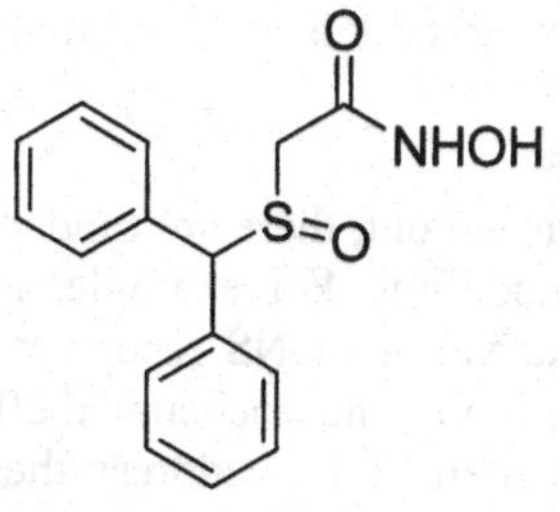

USAN: Adrafinil
Trade name: Olmifon®
Lafon Laboratories
Launched: 1984 (France)

5

Provigil, also known as a stay-awake pill, has been tested by the military for sustained operations in some countries and can keep people awake for days (most military operation tests were on the order of 48–72 h).[8] For example, the current practice in the French military is to place modafinil pills in the ejection seat of fighter planes and in rescue boats.[9] To help manage fatigue due to sleep loss during extended military operations, modafinil has been approved for use on certain Air Force missions in the United States.[10] One study of helicopter pilots suggested that 600 mg modafinil given in three doses can be used to keep pilots alert and maintain their accuracy at predeprivation level for 40 h without sleep (although behavioral and administrative fatigue countermeasures should be the first-line approach for sustaining aircrew performance). It was reported that modafinil has also been shown to be effective in the treatment of depression,[11] cocaine addiction,[12] Parkinson disease,[13] schizophrenia, and cancer-related fatigue.[14] Although modafinil is considered to be effective in the treatment of attention-deficit hyperactivity disorder (ADHD),[15] the FDA did not approve its use for pediatric indication.

The approval of armodafinil marked a significant advance in the treatment of narcolepsy. In this chapter, the pharmacological profile and syntheses of modafinil 4 and armodafinil 1 are profiled in detail.[16]

19.2   Pharmacology

Preclinical animal studies with modafinil demonstrated that stimulant effects are distinct from those of amphetamine and may not involve the dopamine system in the brain.[17a–c] The reported nondopaminergic effects of modafinil include activation of $\alpha 1$ adrenergic receptors, enhancement of serotonin (5-HT) function, inhibition of GABA release,[17d] and stimulation of glutamate and histamine release.[17e] In 1994, Mignot reported that modafinil inhibits dopamine transporter (DAT) binding, with an IC_{50} value of 3.2 µM,[18] whereas Madras showed that modafinil occupies DAT and norepinephrine transporters (NET) in living primate brain.[19] Consistent with these data, modafinil administration increases extracellular levels of dopamine in the brain as measured by in vivo microdialysis. It was found that wake-promoting actions are absent in DAT-knock-out mice. Based on the available evidence, it seems that modafinil interacts with multiple molecular targets in the brain, including DAT proteins.

Although several studies were carried out, there are fundamental unresolved issues regarding the pharmacology of modafinil. For example, few investigations have screened the activity of modafinil at various CNS receptors and transporters, and no studies have attempted to correlate in vivo neurochemical effects of modafinil with ongoing behaviors. A recent study attempted to address these issues by first examining the activity of modafinil against a range of receptors and transporters using binding assays, including transfected cells expressing cloned human G protein-coupled receptors or monoamine transporters.[20a] Transporter-mediated uptake and release were examined in rat brain synaptosomes and the effects of modafinil on motor activity and neurochemistry were determined in rats undergoing in vivo microdialysis in nucleus accumbens. It was found that modafinil displayed measurable potency only at dopamine transporters, inhibiting tritiated dopamine ($[^3H]DA$) uptake, with an IC_{50} of 4.0 μM. Accordingly, modafinil pretreatment antagonized methamphetamine-induced release of the DAT substrate $[^3H]$1-methyl-4-phenylpyridinium. Intravenous modafinil (20 and 60 mg/kg) produced dose-dependent increases in motor activity and extracellular DA, without affecting serotonin. The interaction of modafinil with the DAT site was characterized in vitro,[20b] and the effects of modafinil administration were compared with those of the indirect DA agonist, such as (+)-methamphetamine (METH). The findings show that modafinil interacts with DAT proteins as an uptake blocker, and this action is involved with stimulant properties of the drug. It is suggested that nondopaminergic mechanisms may also contribute to the pharmacology of modafinil.

19.3   Pharmacokinetics[21] and Drug Metabolism[22]

The pharmacokinetic properties of armodafinil were evaluated in humans using both single and multiple oral dose administration. In doses ranging from 50 to 400 mg, systemic exposure is shown to be doseproportional with no change in kinetics over 12 weeks of dosing. Steady-state appears to be reached within 7 days of dosing, and at that point, systemic exposure is 1.8 times greater than the exposure seen in a single dose. When compared with modafinil (100 mg), the concentration time profile of armodafinil (50 mg) is essentially equivalent.

The pharmacokinetics of modafinil are not affected, to a clinically significant extent, by volunteer age or food intake, but both the maximum plasma concentration and the elimination half-life of the drug are increased in patients with hepatic or renal impairment. It was found that peak plasma concentrations of modafinil were reached 2.3 h after a single 200 mg oral dose in healthy volunteers. Over the dose range 200 to 600 mg, the pharmacokinetics of modafinil were linear and dose dependent. Orally administered modafinil is extensively biotransformed in the liver to the inactive metabolites modafinil acid 6 (major metabolite) and modafinil sulphone 7 (minor metabolite), before being eliminated primarily in the urine. It was found the elimination half-life is 9 to 14 h. It is important that stereospecific pharmacokinetics of modafinil was demonstrated. The *S*-modafinil enantiomer was eliminated at a threefold faster rate than *R*-modafinil.

Figure 1: Metabolites of modafinil.

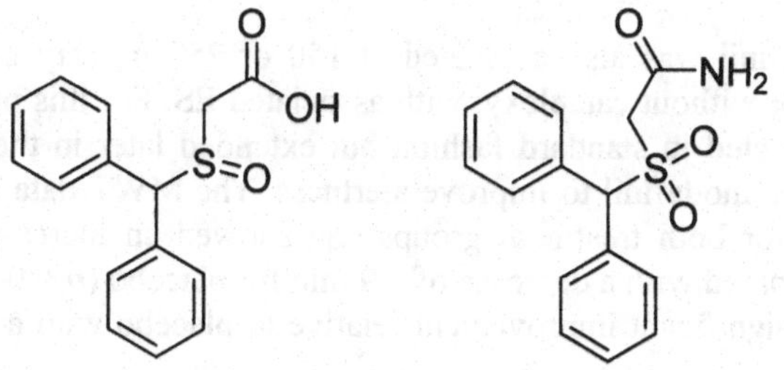

6   7

In metabolism and distribution studies, armodafinil's primary metabolic pathway is via a non-CYP-related amide hydrolysis. As a result, concomitant medications having CYP-interactions are not likely to have deleterious effects (or drug–drug interactions) on the pharmacokinetic profile of armodafinil. The details of the pharmacokinetics of armodafinil are listed in Table 1.

Table 1. Pharmacokinetic properties of armodafinil

Study Design	Single-Dose Administration (n = 93)	Multiple-Dose Administration (n = 34) (Day 7)
C_{max} (mg/L)	1.3 ± 0.4	1.8 ± 0.4
T_{max} (h)	1.5 (0.5–6.0)	2.0 (0.5–6.0)
$t_{1/2\beta}$ (h)	13.8 ± 3.3	15.3 ± 3.0
AUC_{∞} (mg/L.h)	24.1 ± 6.9	NA

Source: Ref. 23.

19.4   Efficacy and Safety[16]

The armodafinil development program was very similar to the development program previously undertaken for modafinil (or racemic API). Four double-blind and two opened-label clinical trials (1090 patients enrolled with 645 receiving active treatment with armodafinil and 445 receiving placebo) have evaluated the efficacy and safety of armodafinil for the treatment of excessive sleepiness (ES) associated with obstructive sleep apnea (OSA), shift-work disorder (SWD), and narcolepsy.[24]

Armodafinil was initially evaluated in healthy volunteers undergoing acute sleep loss using the maintenance of wakefulness test (MWT) and psychomotor vigilance testing (PVT).[25] In this single-dose, double-blind, parallel group trial, armodafinil (100, 150, 200 or 300 mg), modafinil (200 mg) or placebo was administered. Wakefulness, measured by the MWT was improved for all doses of armodafinil and modafinil ($p < 0.0001$) along with improved PVT attention scores ($p < 0.0001$). Both the effect size and duration for armodafinil were observed at later time points and were related to dose. It was suggested that armodafinil 150 and 250 mg would have effects similar to those of modafinil 200 and 400 mg, respectively.

Armodafinil was also evaluated at 150 or 250 mg/day in 196 narcolepsy patients,[24] with or without cataplexy, with associated ES. For this particular trial, the MWT was conducted in standard fashion but extended later in the day to examine the potential for armodafinil to improve alertness. The MWT data at 09:00–15:00 h were combined for both treatment groups and showed an increase of 1.9 min for armodafinil compared with a decrease of 1.9 min for placebo ($p < 0.01$). MWT sleep latency showed significant improvement relative to placebo with a difference of 2.8 min. ($p < 0.05$).

19.5   Synthesis

19.5.1 Synthesis of Modafinil, the Racemic API

The racemic API, modafinil, can be synthesized via several approaches. For example, treatment of α-phenyl benzenemethanethiol **8** with methyl chloroacetate **9** at 100 °C for 4 h gave the methyl ester of benzhydrylsulfanylacetic acid, **10**. Treatment of **10** with ammonia produced amide **11**. The subsequent thioether oxidation was easily carried out using H_2O_2 to deliver modafinil, **1** (Scheme 1).[26]

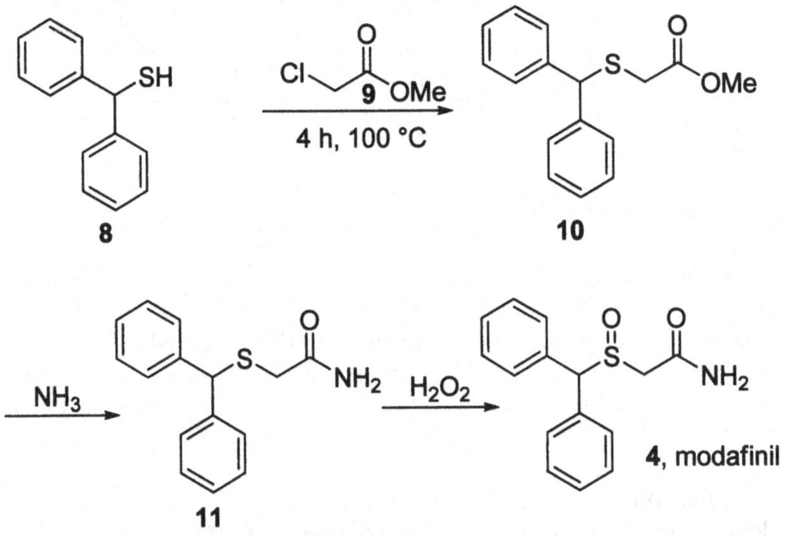

Scheme 1

In another route (Scheme 2), mixing benzhydrol **12** with thioglycolic acid **13** in trifluoroacetic acid afforded benzhydrylsulfanylacetic acid **14** in 99% yield. The treatment of acid **14** with thionyl chloride, followed by concentrated ammonium hydroxide gave acetamide **11** in 87% yield. The oxidation of the thioether moiety with 30% H_2O_2 in acetic acid produced racemic modafinil **1** in 67% yield.[27] The overall yield of the three-step route is approximately 57%.

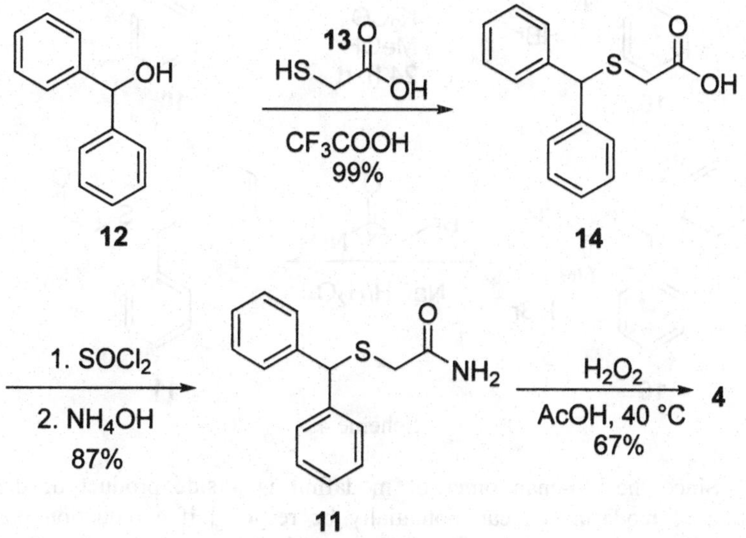

Scheme 2

Alternatively, benzhydrylsulfanylacetic acid **14** was prepared from benzhydryl bromide **15** and thioglycolic acid, **13** in excellent yield (Scheme 3). Similarly, the reaction of benzhydryl chloride and thioglycolic acid, **13** also provided benzhydrylsulfanylacetic acid **14**.[28]

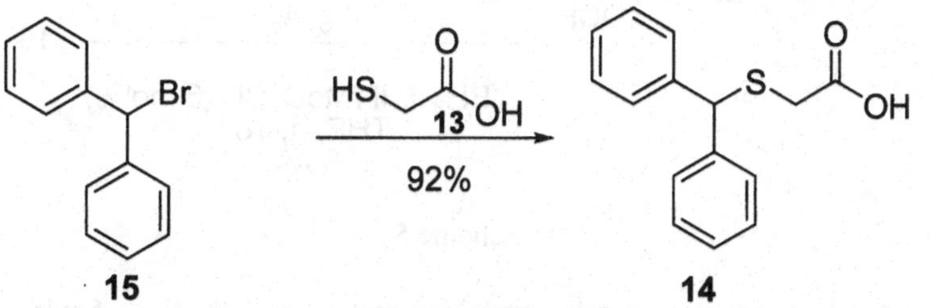

Scheme 3

To avoid the use of volatile, odiferous reagents like thioglycolic acid, an improved route (Scheme 4) employed benzhydryl thiuronium bromide **16** (which was isolated in 99% yield from the reaction of benzhydrol **12** and thiourea in HBr/water).[29] Treatment of **16**, as the HBr salt, with methyl chloroacetate **9** in the presence of K_2CO_3 and MeOH generated the methyl ester of benzhydrylsulfanylacetic acid, **10**. Similarly, the reaction of the HBr salt **16** with 2-bromoacetamide in the presence of NaOH gave acetamide **11**.[30]

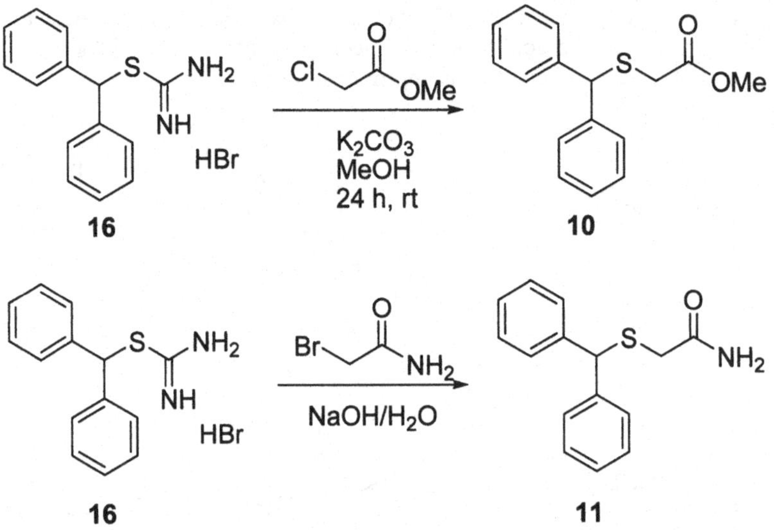

Scheme 4

Since the (S)-enantiomer of modafinil is a side product of the chiral separation of modafinil, it can potentially be recycled if a reduction method is employed. Two novel methods were developed by Fernandes and Romão.[31] Using PhSiH$_3$ and 5% MoO$_2$Cl$_2$ in refluxing THF or BH$_3$·THF/ MoO$_2$Cl$_2$ (5 mol%) at room temperature, treatment of sulfoxide **17** gave methyl ester of benzhydrylsulfanylacetic acid, **10** in 96% and 85% yield, respectively (Scheme 5).

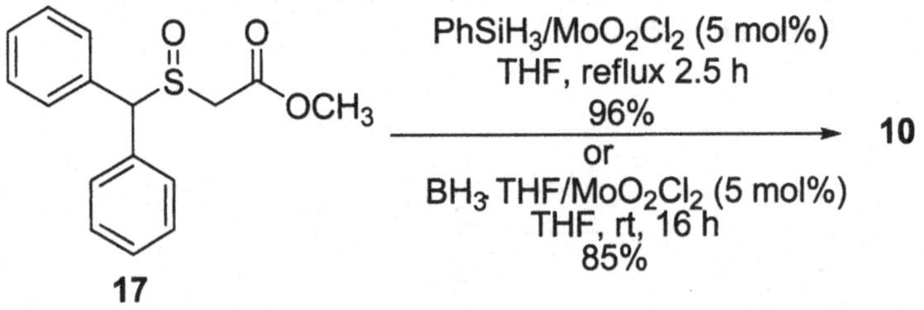

Scheme 5

A robust, process friendly, transition metal-free reduction of sulfoxide was developed recently by Guillen and co-workers.[32] Using 4.5 equiv KI with 7.0 equiv acetyl chloride, several modafinil derivatives were reduced easily under mild conditions, giving the corresponding sulfides in excellent yield (Table 2). It was found that decreasing the amount of KI or increasing the reaction temperature led to an increased formation of benzhydrol **12** as a side product. By using a saturated solution of sulfoxide and KI, in which the solid reagents slowly dissolve as the reaction proceeds, the concentration of sulfoxide and KI is kept constant throughout most of the reaction, and the acetyl chloride is always present in large excess relative to the dissolved sulfoxide, thus suppressing the formation of benzhydrol **12**.

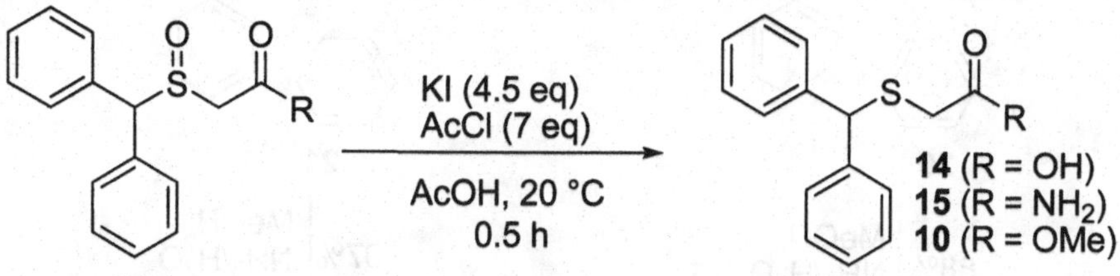

Table 2. Reduction of Modafinil Derivatives Using KI/AcCl System

Compound	R	Reaction Yield (%)
14	OH	88
15	NH$_2$	87
10	OMe	95

Source: Ref. 32.

19.5.1. The Synthesis of Enantiomers of Modafinil and the Asymmetric Synthesis of Armodafinil

The resolution of the racemic mixture of modafinil acid **6** using thiazolidinethione **19** as the chiral auxiliary was achieved in 88% yield (Scheme 6) in the presence of DCC.[33] Two diastereomeric intermediates **20** and **21** were easily separated by silica gel column chromatography and the absolute stereochemistry was assigned based on the single X-ray crystallographic analysis. Finally, the addition of ammonia to diastereomeric thiazolidinethione **20** yielded armodafinil **1**.

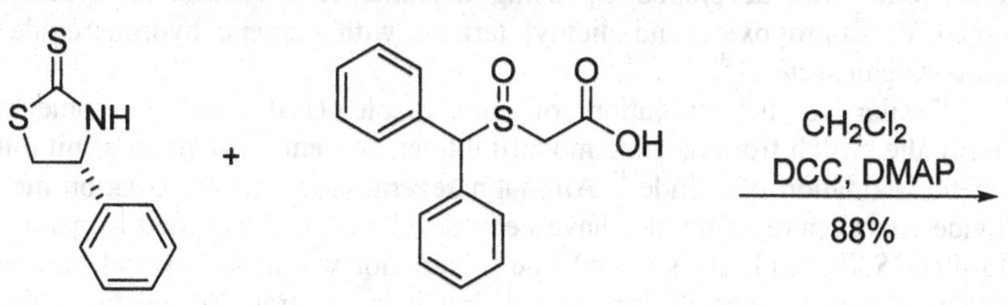

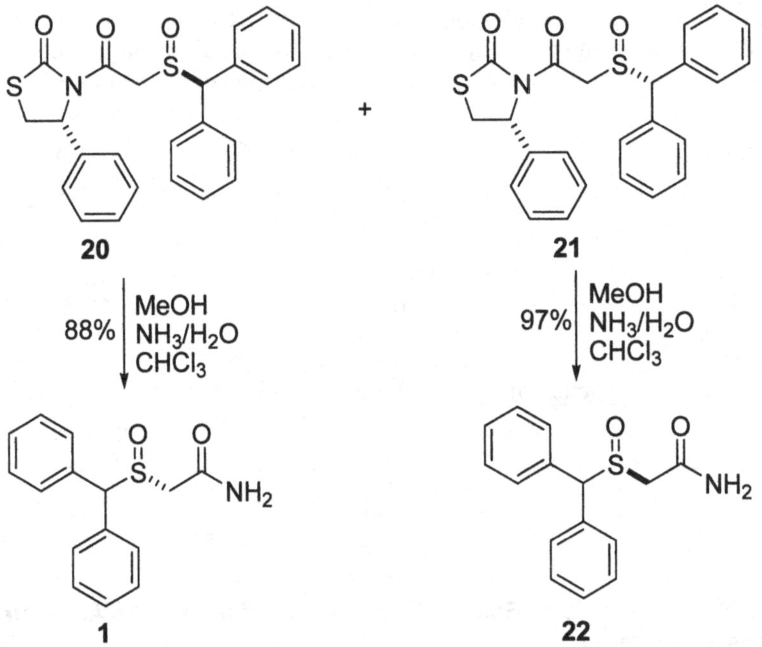

Scheme 6

Three synthetic approaches were used to provide armodafinil during the process development by Cephalon/Novasep.[34] Since the racemic modafinil is commercially available, the resolution via preferential crystallization of modafinic acid **6** was employed for phase I clinical trials and was subsequently replaced by large-scale chiral chromatography. Meanwhile, an economical enantioselective synthetic route was developed by using asymmetric oxidation catalyzed by a titanium (IV) isopropoxide and diethyl tartrate with cumene hydroperoxide (the Sharpless/Kagan system).[36a]

Similar to the evolution of omeprazole (Prilosec) to esomeprazole (Nexium), the switch from racemic modafinil to enantioenriched armodafinil utilized asymmetric oxidation of sulfide.[35] Although several asymmetric oxidation methods to provide enantiopure sulfoxides have been developed, the modified Kagan system [(Ti(Oi-Pr)$_4$/(S,S)-DET] was selected due to superior yields and optical purities (% ee).[36b] The Kagan method is very useful, but it is substrate dependant (Table 3). Several sulfide derivatives of modafinil were screened to determine a direction for optimization. It was found the sulfide amide **11** provided excellent optical purity and further optimization.

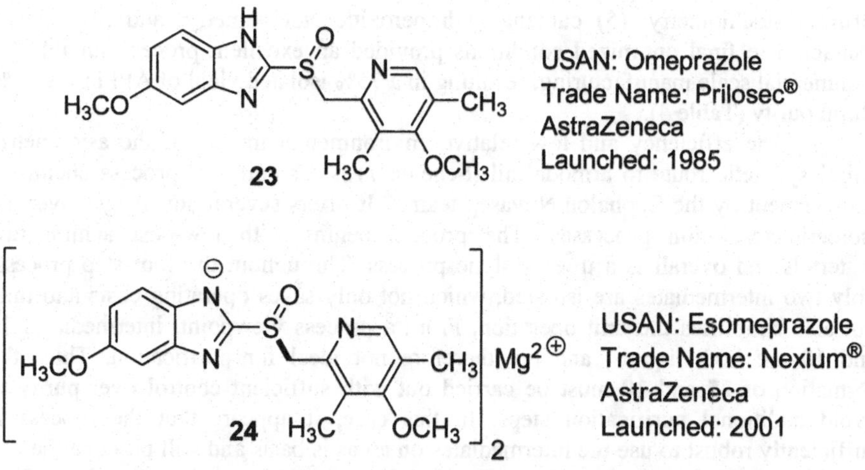
USAN: Omeprazole
Trade Name: Prilosec®
AstraZeneca
Launched: 1985
23
USAN: Esomeprazole
Trade Name: Nexium®
AstraZeneca
Launched: 2001
24

Figure 1. Omeprazole **23** (racemic) and esomeprazole **24** (chiral).

Table 3. Initial Substrate Screening for Asymmetric Oxidation Using the Kagan Method

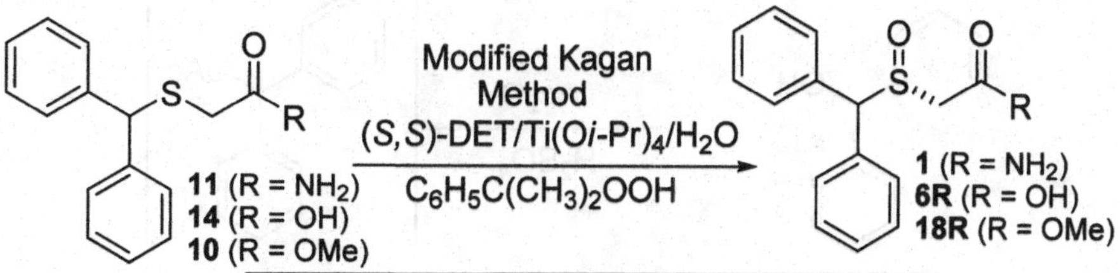
Modified Kagan
Method
(S,S)-DET/Ti(Oi-Pr)4/H2O
C6H5C(CH3)2OOH
11 (R = NH2)
14 (R = OH)
10 (R = OMe)
1 (R = NH2)
6R (R = OH)
18R (R = OMe)

Entry	R	Reaction Yield (%)	% *ee*
1	OMe	50	65
2	OH	ND	0
3	NH_2	70	> 98

Source: Ref. 34.

Table 4. Optimization of Asymmetric Oxidation of **11** (to Generate **1**)[34]

Reaction Solvent	HPLC Purity of **1** (area%)	*ee* of **1** (%)	Reaction Yield (%)
Toluene	> 99	93.0	92
Ethyl acetate	**> 99**	**99.5**	**75**
CH_2Cl_2	> 98.5	98.0	61
CH_3CN	> 98.5	99.3	70
THF	> 99	99.7	50
Acetone	> 99	99.6	45

Source: Ref. 34.

To optimize the asymmetric oxidation conditions for yield and chiral purity, several reaction parameters were evaluated carefully, including (1) the choice of solvent, (2) water stochiometry, (3) Ti catalyst stochiometry, (4) (*S*,*S*)-DET (diethyl tartrate) stochiometry, (5) cumene hydroperoxide stochiometry, and (6) catalyst contact. The final optimized conditions provided an excellent process suitable for commercial scale manufacturing, resulting in a 75% isolated yield of API in > 99.5% chiral purity (Table 4).

The efficiency and low relative environmental impact of the asymmetric chiral synthetic route to armodafinil (Scheme 7) is a significant process chemistry achievement by the Cephalon/Novasep team.[34] It offers several advantages over the isomeric resolution processes: The process begins with low-cost achiral raw materials and overall is a true catalytic process. Throughout the four-step process, only two intermediates are isolated, which not only saves operating costs and time but also simplifies the unit operation. From a process viewpoint, intermediates 25 and 10 are both liquids, and are therefore not ideal for purification. Thus, the formation of 25 and 10 must be carried out with sufficient control over purity to avoid additional purification steps. In this case, it appears that the process is sufficiently robust to use the intermediates on an as is basis and still produce the key intermediate 11 as a pure solid compound. In addition, the armodafinil isolated from the asymmetric oxidation is typically > 99% chemical purity and > 99.5% chiral purity, meeting the specification in every way for the API.

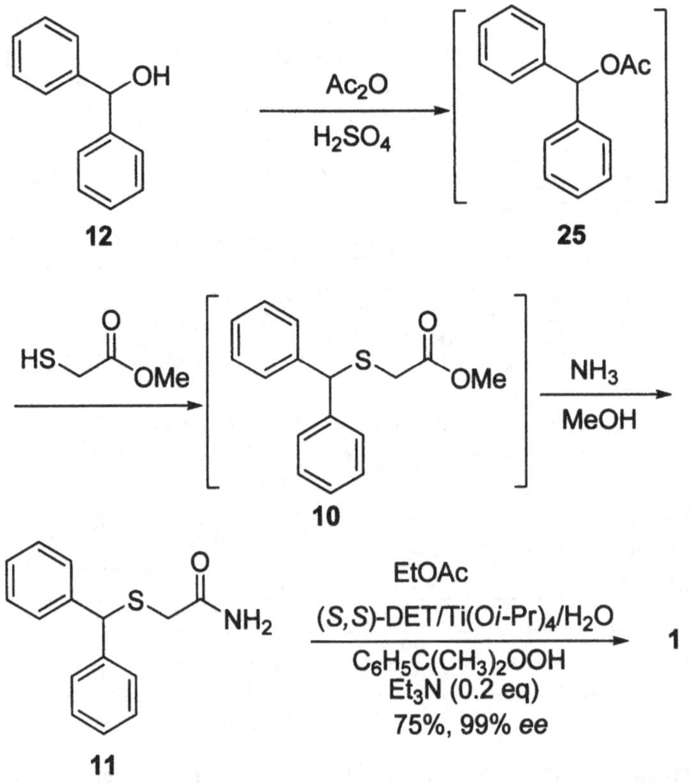

Scheme 7: The Novasep/Cephalon commercial manufacturing route for armodafinil.

In summary, armodafinil is one of the few drugs that has been specifically developed for the treatment of narcolepsy. Armodafinil is the (R)-enantiomer of the wake-promoting compound (racemic modafinil), with a considerably longer half-life of 10–15 h. The data from phase III trials support the efficacy and safety of armodafinil in the treatment of excessive daytime sleepiness associated with narcolepsy. This drug demonstrates objective efficacy as well as subjective reports of improvement in measures of EDS. The efficacy profile for armodafinil has demonstrated that it has longer lasting wake-promoting effects than modafinil. Regarding other clinical indications, there is a possibility that armodafinil will be demonstrated to have beneficial effects on EDS associated with other co-morbidities and more clinical trials will be required in the future.

19.6 References

1.   Becker, P. M.; Schwartz, J. R.; Feldman, N. T.; Hughes, R. J. *Psychopharmacology* **2004**, *171*, 133–139.
2.   (a) Millman, R. P. *Pediatrics*, **2005**, 115, 1774–1786. (b) Shah, N.; Roux, F.; Mohsenin, V. *Treat. Respir. Med.*, **2006**, *5*, 235–244.
3.   (a) Howell, L. L.; Kimmel, H. L. *Handbook of Contemporary Neuropharmacology*, John Wiley & Sons, **2007**, 2, 567–611. (b) Tafti, M.; Dauvilliers, Y. *Pharmacogenomics* **2003**, *4*, 23–33.
4.   Kollins, S. H. *Curr. Med. Res. Opin.* **2008**, *24*, 1345–1357.
5.   (a) Abad, V. C.; Guilleminault, C. *Expert Opin. Emerg. Dr.* **2004**, *9*, 281–291. (b) Billiard, M. *Neuropsychiatr. Dis. Treat.* **2008**, *4*, 557–566.
6.   (a) Kumar, R. *Drugs* **2008**, *68*, 1803–1839. (b) Ballas, C. A.; Kim, D.; Baldassano, C. F.; Hoeh, N. *Expert Rev. Neurother.* **2003**, *2*, 449–457.
7.   (a) L. Lafon U.S. Pat., 4177290, 1979; Ger. Offen, 2809625, 1978; and (b) Wong, Y. N.; Simcoe, D.; Harman, L. N.; et al. *J. Clin. Pharmacol.* **1999**, *39*, 30–40.
8.   "UK Army Tested Stay Awake Pills" BBC News, October 26, **2006**.
9.   (a) Lagarde, D. *Ann. Pharm. Fr.* **2007**, *65*, 258–264. (b) Buguet, A.; Moroz, D. E.; Radomski, M. W. *Avia. Space Envir. Md.* **2003**, *74*, 659–663.
10.   (a) Caldwell, J. A.; Caldwell, J. L. *Avia Space Envir Md.* **2005**, *76*, C39–51. (b) Eliyahu, U.; Berlin, S.; Hadad, E.; Heled, Y.; Moran, D. S. *Mil. Med.* **2007**, *172*, 383–387.
11.   Orr, K.; Taylor, D. *CNS Drugs* **2007**, *21*, 239–257.
12.   (a) Haney, M. *Addict. Biol.* **2009**, *14*, 9–21. (b) Martinez-Raga, J.; Knecht, C.; Cepeda, S. *Cur. Drug Abuse Rev.* **2008**, *1*, 213–221.
13.   van Vliet, S. A. M.; Blezer, E. L. A.; Jongsma, M. J.; Vanwersch, R. A. P.; Olivier, B.; Philippens, I. H. C. H. M. *Brain Res.* **2008**, *1189*, 219–228..
14.   Cooper, M. R.; Bird, H. M.; Steinberg, M, *Ann. Pharmacother.* **2009**, *43*, 721–725.
15.   Turner, D. *Expert Rev. Neurother.* **2006**, 6, 455–468.
16.   (a) Lankford, D. A, *Expert Opin. Investig. Drugs*, **2008**, *17*, 565–573. (b) Nishino, S.; Okuro, M, *Drugs Today* **2008**, 44, 395–414. (c) Laffont, F., *Drugs Today* **1996**, *32*, 339–347.

17.   (a) Duteil, J.; Rambert, F. A.; Pessonnier, J.; Herman, J. F.; Jean, F.;
        Gombert, R.; Assous, E, *Eur. J. Pharmacol.* **1990**, *180*, 49–58. (b) Simon,
        P.; Hemet, C.; Ramassamy, C.; Costentin, J. *Eur. Neuropsychopharmacol.*
        1995, *5*, 509–514. (c) McClellan, K. J.; Spencer, C. M. *CNS Drugs*, **1998**, 9,
        311–324. (d) Ferraro, L.; Tanganelli, S.; O'Connor, W. T.; Antonelli, T.;
        Rambert, F.; Fuxe, K. *Neurosci. Lett.* **1999**, *220*, 5–8. (e) Ishizuka, T.;
        Sakamoto, Y.; Sakurai, T.; Yamatodani, A. *Neurosci. Lett.* **2003**, *339*, 143–
        146.

18.   Mignot, E.; Nishino, S.; Guilleminault, C.; Dement, W. C. *Sleep* **1994**, *17*,
        436–437.

19.   Madras, B. K.; Xie, Z.; Lin, Z.; jassen, A.; Panas, H.; Lynch, L.; Johnson,
        R.; Livni, E.; Spencer, T. J.; Bonab, A. A. *J. Pharmacol. Exp. Ther.* **2006**,
        319, 561–560.

20.   (a) Zolkowska, D.; Jain, R.; Rothman, R. B.; Partilla, J. S.; Roth, B. L.;
        Setola, V.; Prisinzano, T. E.; Baumann, M. H. *Pharmacol. Exp.Ther.* **2009**,
        329, 738–746. (b) Madras, B. K.; Xie, Z.; Lin, Z.; Jassen, A.; Panas, H.;
        Lynch, L.; Johnson, R.; Livni, E.; Spencer, T. J.; Bonab, A. A.; Miller, G.
        M.; Fischman, A. J. *Pharmacol. Exp.Ther.* **2006**, *319*, 561–569.

21.   Keating, G. M.; Raffin, M. *CNS Drugs*, **2005**, *19*, 785–803.

22.   Robertson, P. J.; Hellriegel, E. T. *Clin. Pharmacokinet.* **2003**, *42*, 123–137.

23.   Darwish, M.; Kirby, M.; Hellriegel, E. T.; Yang, R.; Robertson, P. J. *Clin
        Drug Invest.* **2009**, *29*, 87–100.

24.   Harsh, J. R.; Haydak, R.; Rosenberg, R.; et al. *Curr. Med. Res. Opin.* **2006**,
        22, 761–774.

25.   Boyd, B.; Castaner, J. *Drugs Fut.* **2006**, *31*, 17–21.

26.   Faming Zhuanli Shenqing Gongkai Shuomingshu, Appl.: CN 2005-
        10049330 20050310, 2006.

27.   Prisinzano, T.; Podobinski, J.; Tidgewell, K.; Luo, M.; Swenson, D.
        Tetrahedron Asymm. **2004**, *15*, 1053–1058.

28.   Faming Zhuanli Shenqing Gongkai Shuomingshu, Appl.: CN 2006-
        10155494 20061227, 2007.

29.   Liang, S. PCT Int. Appl. 2005042479, 2005.

30.   Eur. Pat. Appl. 1260501, 2002.

31.   (a) Fernandes, A. C.; Romão, C. C. *Tetrahedron* **2006**, *62*, 9650–9654. (b)
        Fernandes, A. C.; Romão, C. C. *Tetrahedron Lett.* **2007**, *48*, 9176–9179.

32.   Ternois, J.; Guillen, F.; Piacenza, G.; Rose, S.; Plaquevent, J.-C.; Coquerel,
        G. *Org. Proc. Res. Dev.* **2008**, 12, 614–617.

33.   Osorio-Lozada, A.; Prisinzano, T.; Olivo, H. F. *Tetrahedron Asymm.* **2004**,
        15, 3811–3815.

34.   Hauck, W.; Adam, P.; Bobier, C.; Landmesser, N. *Chirality* **2008**, *20*, 896–
        899.

35.   (a) Kagan, H. B.; Luukas, T. O. In *Transition Metals for Organic Synthesus*
        (2nd Edition), 2004, 479–495. Eds. Beller, M.; Bolm, C. Wiley-VCH
        Verlag. (b) Kagan, H. B. In *Catalytic Asymmetric Synthesis*, (2nd Edition),
        2000, 329–356, Ed. Ojima, I. Wiley-VCH.

36. (a) Philippe, P.; Kagan, H. B. *J. Am. Chem. Soc.* **1984**, *25*, 1049–1052. (b) Ternois, J.; Guillen, F.; Plaquevent, J.-C.; Coquerel, G. *Tetrahedron Asymm.* **2007**, *18*, 2959–2964. (c) Olivo, H.; Osorio-Lozada, A.; Peeples, T. *Tetrahedron Asymm.* **2005**, *16*, 3507–3511.

V

Miscellaneous

20

Raloxifene, Evista: A Selective Estrogen Receptor Modulator (SERM)

Marta Piñeiro-Núñez

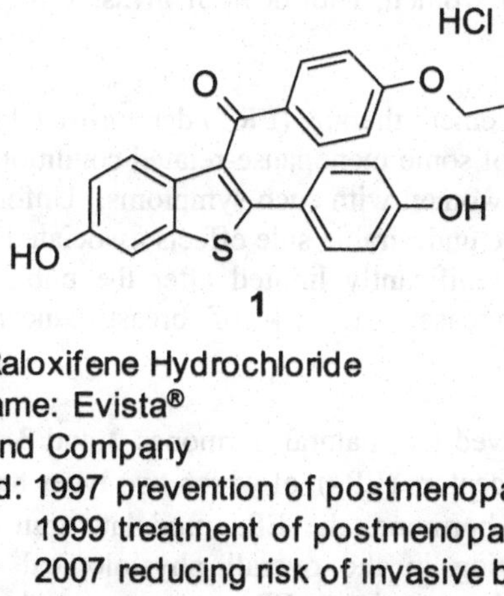

USAN: Raloxifene Hydrochloride
Trade Name: Evista®
Eli Lilly and Company
Launched: 1997 prevention of postmenopausal osteoporosis
1999 treatment of postmenopausal osteoporosis
2007 reducing risk of invasive breast cancer

20.1 Background

It has been long known that endogenous estrogens such as 17β-estradiol (2) and estrone (3) play essential roles in reproductive endocrinology, as they are the main hormones involved in the development and maintenance of the female sex organs and mammary glands. More recently, estrogen regulation has been recognized as playing a pivotal role in the growth and function of a number of other tissues, both in males and females, such as the skeleton, cardiovascular system and central nervous system. When a sharp decrease in the amount of circulating natural estrogens occurs during menopause, there are a host of physical changes that result in health issues of varying severity, from hot flashes and urogenital atrophy, to conditions of greater severity such as osteoporosis.[1–3]

2   **3**

In patients with osteoporosis, bones become fragile, causing an increased risk of fractures, even during everyday activities. Bone loss begins soon after menopause, but it is often several years until effects such as loss of height and kyphosis become evident, along with increased risk of bone fractures as a result of low bone mass or osteopenia. Of special concern are fractures of the hip and spine, which may require hospitalization and major surgery and may result in prolonged or permanent disability, deformity, or even death. The seriousness of this condition is underscored by its high incidence, with one out of every two women over age fifty experiencing fractures due to osteoporosis in their lifetime.[4–6]

Likewise, invasive breast cancer occurs when malignant cells break through the milk duct or lobule and invade surrounding fatty tissue, eventually spreading to the lymphatic system and bloodstream and thus to other parts of the body. As women reach menopause age, their risk of being diagnosed with invasive breast cancer is significantly higher than for younger women, with 66% of invasive breast cancers found in women above age 55.[4,5,7,8]

Estrogen replacement therapy (ERT) demonstrated effectiveness in reducing the frequency and severity of some menopause-related conditions and, therefore, became the treatment of choice for women with such symptoms. Unfortunately, patient compliance was compromised by the undesirable side effects associated with ERT, and its utility and widespread use were significantly limited after the publication of wide-scale studies showing that ERT increases the risk of breast cancer, endometrial cancer, and thrombosis.[9]

It is now believed that natural hormones **2** and **3** act via their interaction with intracellular estrogen receptors (ERs), of which two types have been described (ERα and ERβ). Upon binding hormone, the ERs modulate transcription of target genes on different tissues, resulting in the overall physiological effects. Given their wide distribution and rich pharmacology, ERs soon emerged as attractive targets for the development of therapies for a host of unmet medical needs of varying seriousness, including osteoporosis and breast cancer. The last few decades have seen an explosion in the development and launch of many steroidal and nonsteroidal ER modulators (see Table 1), which, in turn, has helped increase further understanding of ER physiology.[1–3]

Table 1. Representative Estrogen Receptor Modulators[1,10-21]

Molecule	Trade Names	Pharmacology	Indication	Company	Status
clomiphene (3)	Clomid Serophene Milophene	Partial Agonist	Anovulation	Sanofi Aventis	Approved 1967 (US)
tamoxifen (4)	Nolvadex Istubal Valodex	Mixed Agonist/ Antagonist	Breast cancer	AstraZeneca	Approved 1977 (US)
toremifene (5)	Fareston	Mixed Agonist/ Antagonist	Metastatic breast cancer	GTX Inc	Approved 1997 (US)
afimoxifene (6)	TamoGel	Mixed Agonist/ Antagonist	Cyclical mastalgia	Ascend Therapeutics	Under development
ormeloxifene (7)	Centron Saheli Sevista	Mixed Agonist/ Antagonist	Contraception	CDRI, India	Approved 1991 (India)
lasofoxifene (8)	Oporia Fablyn	Mixed Agonist/ Antagonist	Osteoporosis/ Vaginal atrophy	Pfizer	Pre-registered
fulvestrant (9)	Faslodex	Full Antagonist	Metastatic breast cancer	AstraZeneca	Approved 2002 (US)
Raloxifene (1)	Evista	Mixed Agonist/ Antagonist	Osteoporosis prevention/tre atment/ Invasive breast cancer risk reduction	Eli Lilly	Approved 1997/1999/2 007 (US)
Bazedoxifene (10)	Viviant Conbriza	Mixed Agonist/ Antagonist	Osteoporosis	Wyeth/Ligand	Approved 2009 (EU)

The first ER modulators studied clinically were clomiphene (4) and tamoxifen (5), which exemplify the triphenylethylene (TPE) structural class. TPEs were originally investigated as contraceptives, but molecules such as tamoxifen (5) and toremifene (6) were found to display strong antagonism of estrogen in mammary tissue, leading to their development for the treatment of breast cancer. A more recent member of this structural class, afimoxifene (7) is currently under development as a transdermal gel formulation for the treatment of cyclical mastalgia.[1,13]

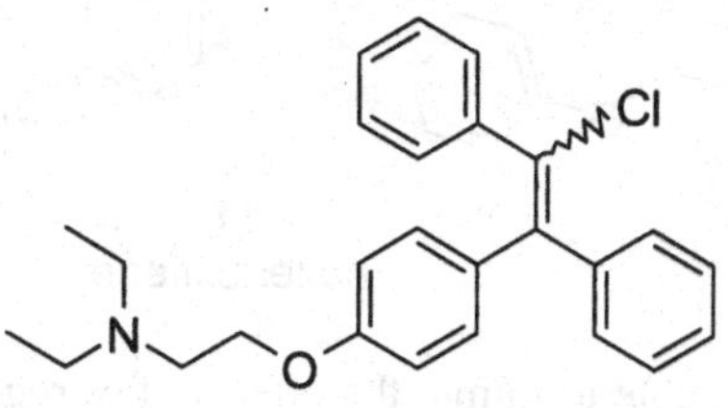

4
clomiphene

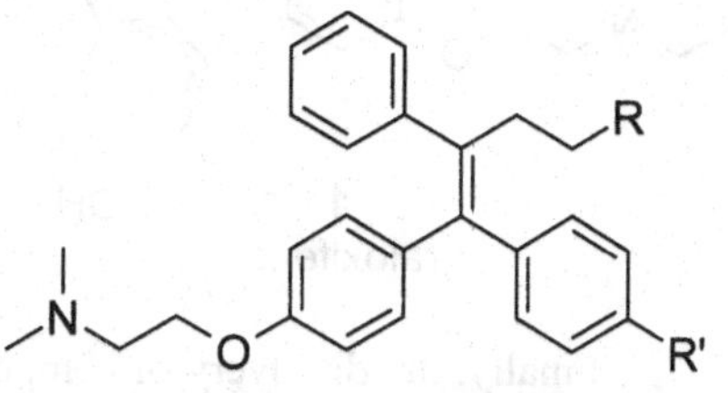

R = R' = H   tamoxifen (5)
R = Cl, R' = H toremifene (6)
R = H, R' = OH afimoxifene (7)

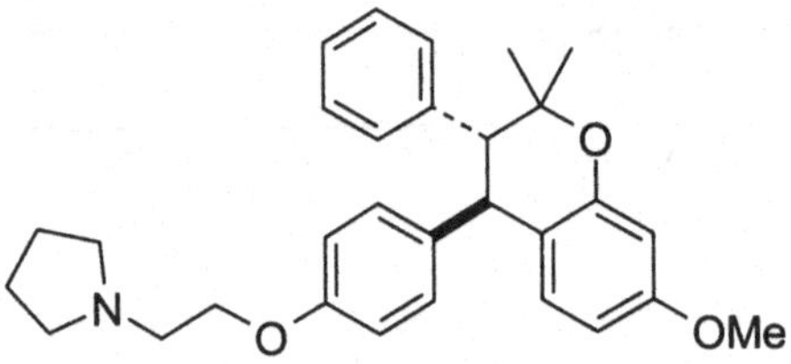

8
ormeloxifene

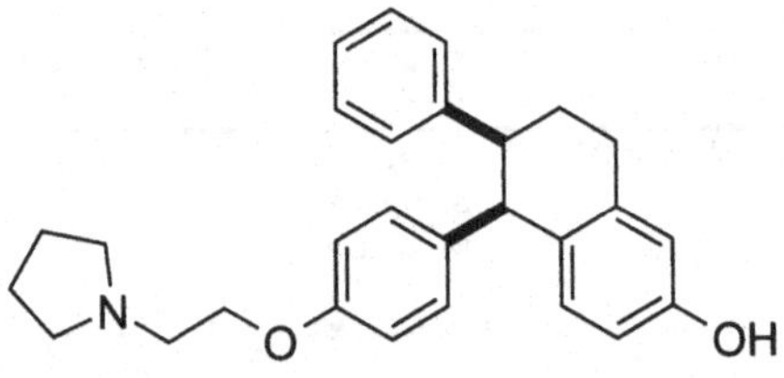

9
lasofoxifene

Another structural class of ER modulators is exemplified by ormeloxifene (**8**) and lasofoxifene (**9**). These molecules can be considered as conformational variants of the TPE scaffold. Ormeloxifene (**8**) was launched in India as a contraceptive in the early 1990s, but its effects on bone homeostasis were reported early on, hinting at other possible applications.[17] Indeed, lasofoxifene (**9**) is currently undergoing evaluation as an osteoporosis therapy.[18]

Early exploration of derivatives of the natural estrogens failed to provide the necessary tissue selectivity, but an example of successful application of the steroid-like scaffold was realized with fulvestrant (**10**), which has been used in breast cancer therapy since the early 2000s. Fulvestrant (**10**) is a full ER antagonist that displays no agonistic effects, and it works both by down-regulating and by degrading the ER.[1,19,20]

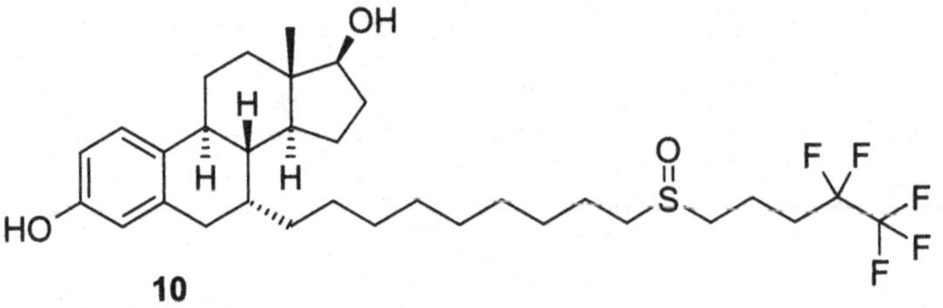

10
fulvestrant

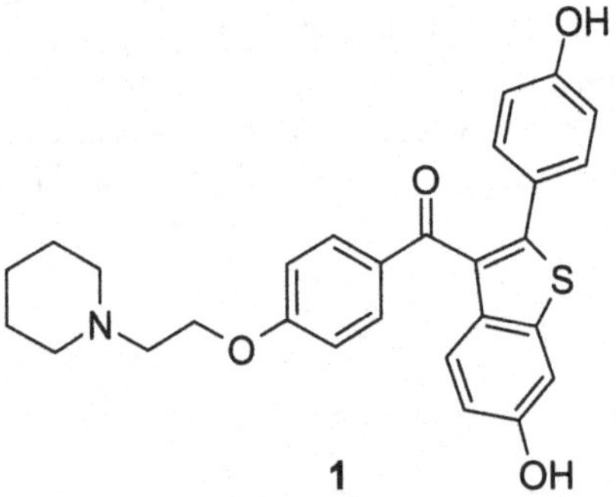

1
raloxifene

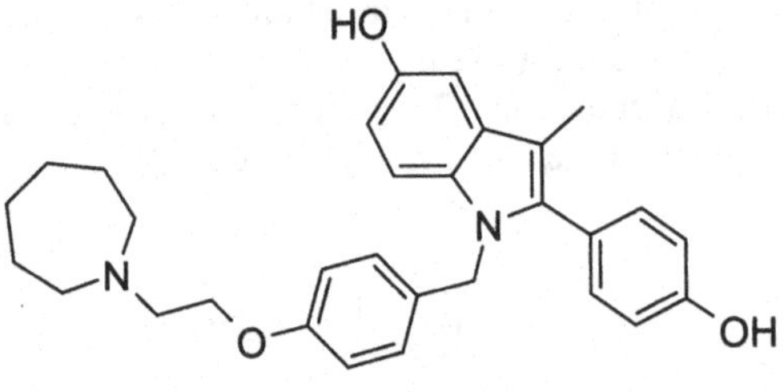

11
bazedoxifene

Finally, the discovery of compounds being able to mimic the effects of estrogen in skeletal and cardiovascular systems while producing virtually complete antagonism in breast and uterine tissues led to the coining of the term selective estrogen receptor modulator (SERM), of which raloxifene (**1**) is a representative and exemplifies the benzothiophene structural class. The discovery of raloxifene (**1**) spurred further investigation around the benzothiophene core, but it also encouraged development of related scaffolds.[1-5,21] Thus indole-based SERM bazedoxifene (**11**) obtained approval in

Europe in 2009 for the prevention and treatment of postmenopausal osteoporosis, and is currently under evaluation for approval in the United States.[11]

In addition to the molecules included in Table 1, research in the field of ER modulators has given rise to other novel scaffolds, which may produce additional SERMs to address unmet medical needs in the not-so-distant future.[3]

20.2   Mechanism of Action[1-3,22-29]

The mechanism of action of ER modulators spans the full fan of possibilities, from full agonists in all tissues, such as the natural endogenous estrogens, to pure antagonists such as fulvestrant (3), in addition to a wide variety of molecules that display agonism in some tissues while they behave as full antagonists in others (Table 1). The subset of ER modulators that lack uterine stimulation while they display agonism in bone tissue and cardiovascular system are defined as true SERMs. As a representative SERM, raloxifene (1) displays this beneficial tissue selectivity, which is the origin of its utility for reducing the risk of invasive breast cancer and prevention and treatment of osteoporosis in postmenopausal women.

The origin of the tissue selectivity displayed by SERMs is complex. Since description of the two receptor subtypes, ERα and ERβ, some selectivity has been attributed to the ligand's differential binding affinity for each subtype. A second factor is related to the existence of a host of possible receptor conformations that are induced by ligand binding, as demonstrated by crystal structures of the estrogen receptor bound to estrogen and raloxifene (1). Each conformational state interacts uniquely with target gene promoters and co-regulator proteins, which could be co-activators (leading to agonism) or co-repressors (leading to antagonism). A third complicating factor is derived from variations in the ratio of co-activators to co-repressors within the cell. Thus SERM pharmacology depends not only on the interaction of the ligand with the receptor, but also on the presence and relative amounts of gene promoters and co-regulators within the different cell types.

As an example of this complexity, tamoxifen (5) acts as an antagonist in breast and agonist in uterus, due to a higher concentration of co-activators in the uterus as compared to breast. In contrast, raloxifene (1) is an antagonist in both tissues, despite the higher concentration of co-activators relative to co-repressors in uterus, due to the fact that the receptor–raloxifene complex is able to recruit co-repressor proteins more strongly.

20.3   Pharmacokinetics and Drug Metabolism[21,30-32]

Table 2 offers a summary of clinical parameters for several representative ER modulators, showcasing similarities and differences across scaffolds. Thus the benzothiophene molecule raloxifene (1) and both TPEs tamoxifen (4) and toremifene (6) provide oral bioavailability, while the steroid-like molecule fulvestrant (9) requires administration via intramuscular injection. On the other hand, all three oral treatments require daily dosing, in contrast with monthly administration for the injectable, consistent with the relative values of elimination half-life.

As far as protein binding and elimination route all scaffolds seem comparable, whereas all molecules with the exception of raloxifene (1) have significant interaction with the CYP system. Finally, raloxifene (1) has a significantly higher volume of distribution resulting in activity on a number of organ systems, and it is the only molecule of the set to display a dual indication for both osteoporosis and breast cancer risk reduction.

Table 2. ADME Parameters for Selected Estrogen Receptor Modulators[21,30-32]

	Raloxifene (1)	Tamoxifen (4)	Toremifene (6)	Fulvestrant (9)
Indication(s)	Breast cancer risk reduction Osteoporosis	Breast cancer	Breast cancer	Breast cancer
Structural Class	Benzothiophene	TPE	TPE	Steroid-like
Route of Administration	PO	PO	PO	IM
Number of Doses	1/day	1/day	1/day	1/mo
Dose Amount (mg)	60	20 to 40	60	250
%F (Oral)	2	30 to 100	≈ 100	low
%Protein Binding	> 95	> 95	≥ 99	99
Apparent Volume of Distribution (L/Kg)	2348	50 to 60	8.3	3 to 5
CYPs involved in metabolism	not involved	3A4, 2C9, 2D6	3A4, others	3A4
Primary Elimination Route	Fecal	Fecal	Fecal	Fecal
Elimination Half-Life (days)	1.15	5 to 7	5	40[a]
Known Metabolites	Glucuronides	N-Desmethyl 4-Hydroxy	N-Desmethyl Deaminohydroxy 4-Hydroxy Didesmethyl	17-Ketone 3-Ketone Glucuronides Sulfates Sulfone

a Not true elimination half-life: with long-acting IM injection, absorption rate and not elimination determines the half-life.

20.4   Efficacy and Safety

The approvals of raloxifene (1) in 1997 and 1999 for the prevention and treatment of postmenopausal osteoporosis ushered a new era of drug development in this field.[4] In addition, the effect of raloxifene (1) on breast cancer was a secondary endpoint, and breast cancer risk reduction in postmenopausal women was demonstrated in the large clinical trials MORE, CORE, and RUTH.[33-38] These results culminated in the 2007 approval of raloxifene (1) for reduction in risk of invasive breast cancer in postmenopausal women with osteoporosis and for reduction in risk of invasive breast cancer in postmenopausal women at high risk for invasive breast cancer (United States only).[21,38,39]

Therapeutic efficacy in the treatment and prevention of bone loss in postmenopausal patients has been defined in terms of changes in bone mineral density (BMD) as well as fracture incidence. Although the occurrence of fracture is the clinical manifestation of osteoporosis, this disease is operationally defined by the WHO as a BMD equal or greater than 2.5 standard deviations below the young adult reference mean. In this regard, BMD has been shown to be an accurate predictor of some fractures, and it can be used as a diagnostic tool for at-risk individuals.[4]

The MORE study (Multiple Outcomes of Raloxifene Evaluation) was a multicenter, randomized, double-blind design which followed 7,705 postmenopausal women with osteoporosis for a total of 4 years.[5] Treatment with 60 or 120 mg raloxifene, with calcium and vitamin D_3 supplementation, resulted in a 2–3% increase in BMD versus placebo, along with reduced risk of vertebral fractures,[40,41] and a 71% reduction for the risk of invasive breast cancer versus placebo, with 1.1% absolute risk reduction.[36] A subset of patients from the MORE trial elected to enroll in a subsequent 4-year trial. Thus the CORE study (Continuing Outcomes Relevant to Evista) demonstrated that BMD increases were maintained after 7 years of use of raloxifene[42] as well as a 56% reduction for the risk of invasive breast cancer, with 0.9% absolute risk reduction.[37] An additional smaller study on osteoporosis prevention with 30, 60, and 150 mg raloxifene showed increased lumbar spine BMD of 2.6% versus placebo.[43]

Further demonstration of the utility of raloxifene in the prevention of breast cancer came from the STAR trial (Study of Tamoxifen and Raloxifene), which followed more than 19,000 post-menopausal women for 5 years. This was a prospective, double-blind, randomized trial comparing 20 mg/day tamoxifen versus 60 mg/day raloxifene, which resulted in the demonstration that incidence rates of invasive breast cancer for both treatments were comparable.[39]

Finally, the RUTH study (Raloxifene Use for the Heart)[38] was a placebo-controlled clinical trial following over 10,000 postmenopausal women with coronary heart disease (CHD) or with multiple risk factors for CHD.[44–46] This trial demonstrated 44% reduced incidence of invasive breast cancer versus placebo, with 0.6% absolute risk reduction,[47] thus confirming the findings from MORE and CORE, while also demonstrating that raloxifene did not increase or decrease risk for coronary events or stroke. However, there was an increase in stroke mortality and incidence of venous thromboembolic events (VTEs) as compared to placebo, already seen in MORE, which resulted in a recommendation that raloxifene should not be used for the prevention or reduction of the risk of cardiovascular disease.[21]

Other than the limitations in the use of raloxifene for the patient population studied by RUTH, raloxifene proved generally well tolerated at dosages up to 150 mg/day, as compared both with placebo and hormone replacing therapy (HRT). Adverse events reported were hot flashes, leg cramps, swelling, flu-like symptoms, joint pain, and sweating. Finally, no clinically important changes in hematological, renal or hepatic laboratory variables were observed during clinical trials, or any increased risk of uterine cancer, ovarian cancer or endometrial hyperplasia.[4,5]

20.5   Syntheses

The first reported synthesis of the raloxifene scaffold was based on the coupling of carboxylic acid fragment **12** with benzothiophene nucleus **13**. The main carbon-carbon bond formation was thus achieved via nucleophilic attach of the benzothiophene through its C(3) position on the activated carbonyl of **12**. The final step in the synthesis was the cleavage of both methyl ethers to unmask the requisite hydroxyls.[48,49]

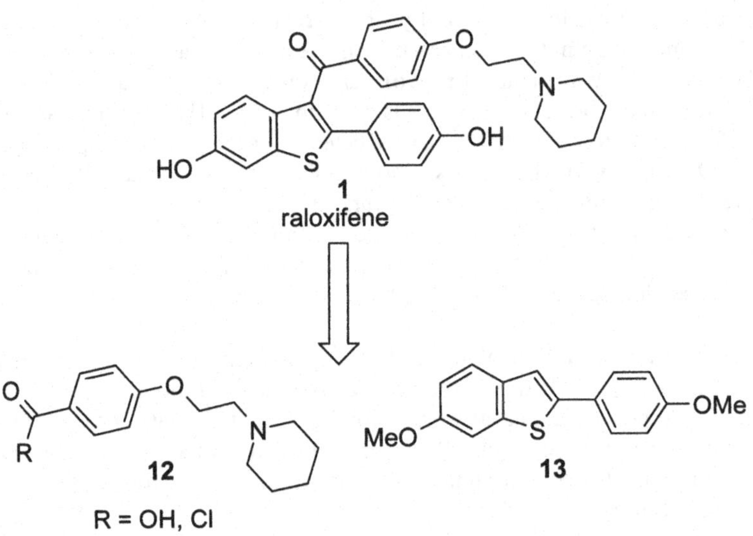

1
raloxifene

R = OH, Cl

In this case, carboxylic acid fragment **12a** was prepared from benzoate **14** and amine **15** in 82% overall yield. Likewise, benzothiophene **13** was generated from thiophenol **16** and bromoketone **17** in 55% overall yield.

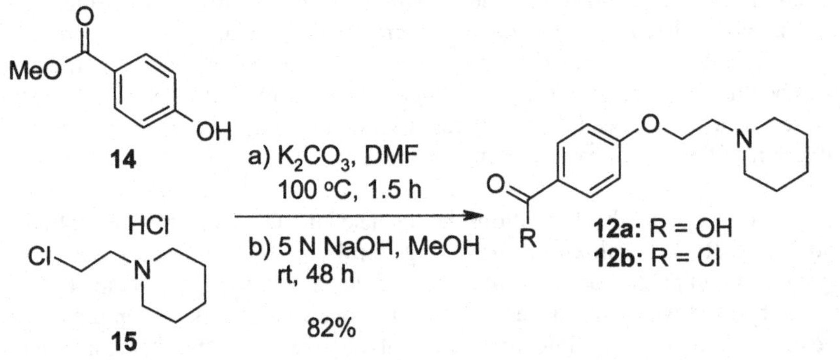

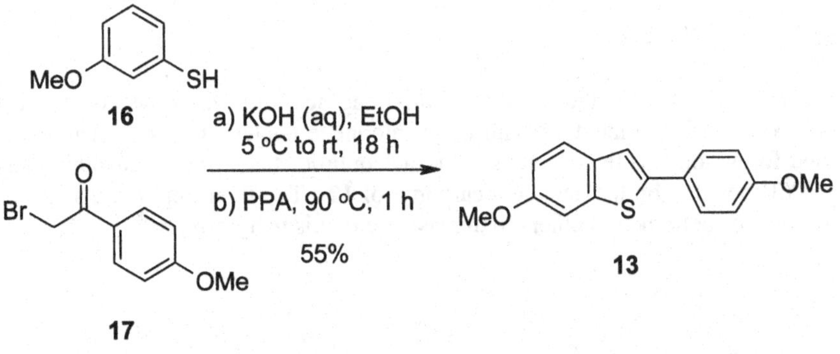

With the requisite materials **12a** and **13** on hand, the crucial coupling took place via activation of **12a** as its acid chloride **12b**, followed by treatment with benzothiophene **13** to afford the *bis*-methylated raloxifene precursor **18**. Cleavage of both methyl ethers with aluminum trichloride afforded the final compound raloxifene (**1**). Subsequent improvement of this synthetic route led to a one-pot acylation–demethylation procedure. Thus treatment of acid **12a** with thionyl chloride followed by direct addition of **13** in the presence of aluminum trichloride followed by work up with ethanethiol provided the desired final product raloxifene (**1**) as its hydrochloride salt.

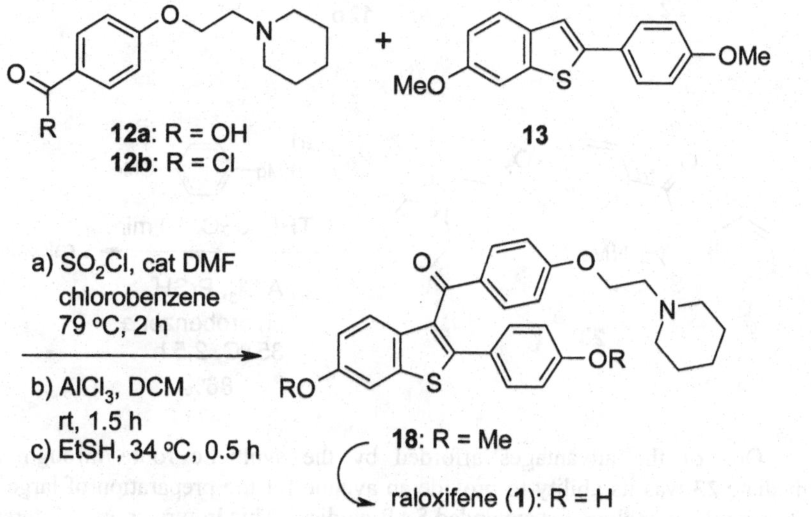

12a: R = OH
12b: R = Cl

13

a) SO$_2$Cl, cat DMF
   chlorobenzene
   79 °C, 2 h

b) AlCl$_3$, DCM
   rt, 1.5 h
c) EtSH, 34 °C, 0.5 h

18: R = Me

raloxifene (**1**): R = H

A second-generation approach improved the yield of the coupling step by strategically placing an electron-donating group at the C(2) position of the benzothiophene to boost its nucleophilicity.[50] Thus aldehyde **19** was converted to its cyanohydrin **20** under standard conditions, followed by a three-step conversion to thioacetamide **21** and acid-catalyzed cyclization to generate thiophene **22**. Coupling with acid chloride **12b** afforded intermediate **23**, which set the stage for introduction of the requisite C(2) phenyl moiety via attack by bromo-(4-methoxyphenyl) magnesium. It is interesting that only 1,4-addition to the enone moiety in **23** was observed, even in the presence of an excess of Grignard reagent.

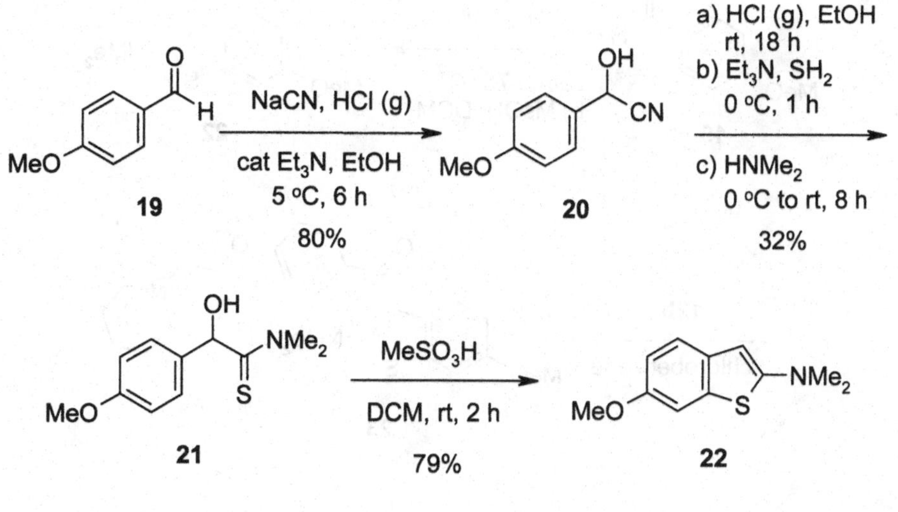

19

NaCN, HCl (g)

cat Et$_3$N, EtOH
5 °C, 6 h
80%

20

a) HCl (g), EtOH
   rt, 18 h
b) Et$_3$N, SH$_2$
   0 °C, 1 h

c) HNMe$_2$
   0 °C to rt, 8 h
32%

21

MeSO$_3$H

DCM, rt, 2 h
79%

22

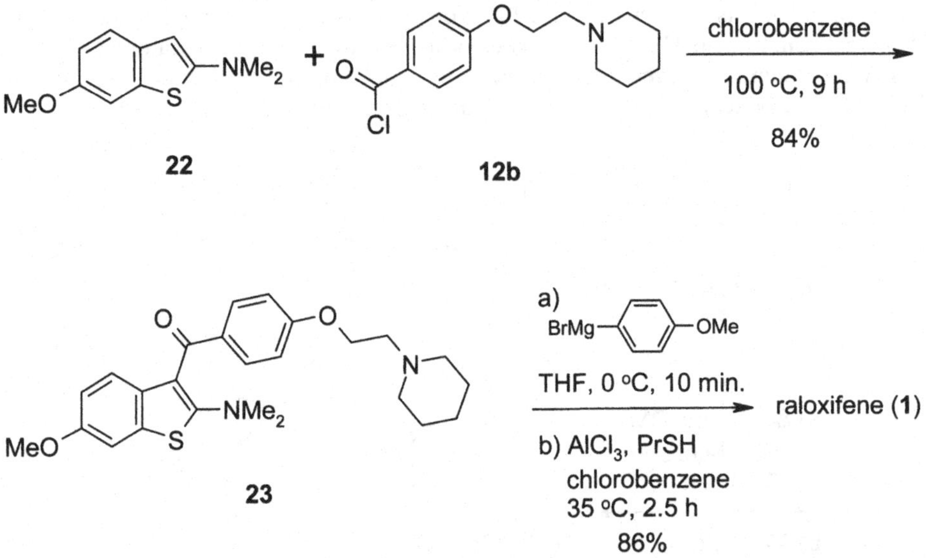

22 + **12b** → chlorobenzene, 100 °C, 9 h, 84%

23

a) BrMg—⟨⟩—OMe, THF, 0 °C, 10 min.

b) AlCl$_3$, PrSH, chlorobenzene, 35 °C, 2.5 h, 86% → raloxifene (**1**)

One of the advantages afforded by the synthetic route through versatile intermediate **23** was its ability to provide an avenue for the preparation of large number of C(3) raloxifene analogs for expanded SAR studies. This in turn spurred interest in the development of an improved route to **22** starting from aldehyde **19**. Therefore, condensation of **19** with *N,N*-dimethylthioformamide followed by an acid-catalyzed cyclization-aromatization sequence generated the desired benzothiophene **22**, which was converted to intermediate **23** as described before.[51,52] Treatment of **23** with a variety of Grignard reagents provided an entry to raloxifene analogs **24**. In addition, reaction with the Grignard reagent derived from bromide **A** produced **25a**, which afforded **26b** upon standard alcohol inversion. Finally, demethylation of **25a** and **26a** afforded raloxifene analogs **25b** and **26b**, which allowed the probing of the effect of geometric position of the *p*-hydroxyl on the biological activity of raloxifene (**1**).[51]

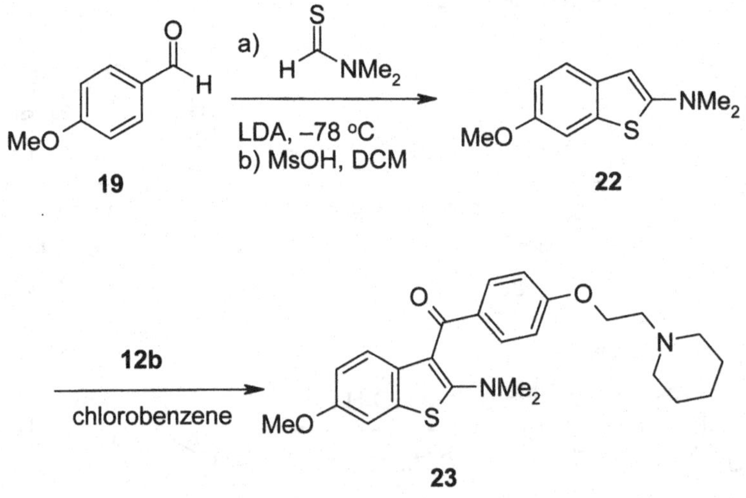

19

a) H—C(=S)—NMe$_2$, LDA, −78 °C
b) MsOH, DCM → **22**

12b, chlorobenzene → **23**

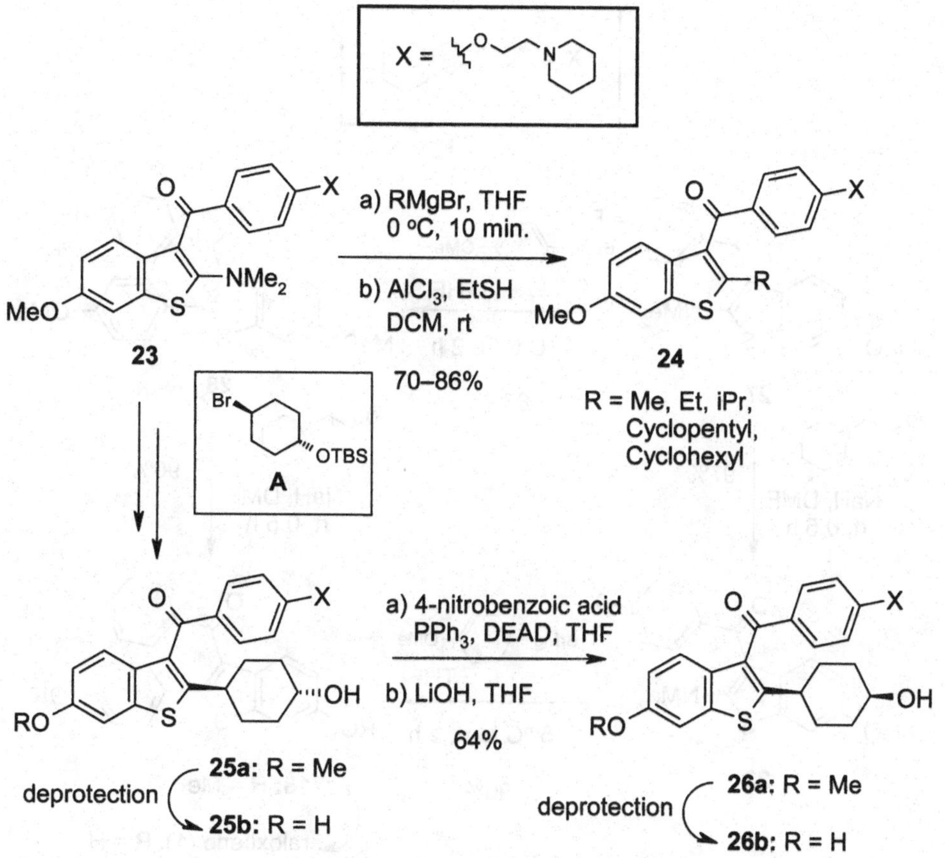

X = [structure: O-ethyl linked to piperidine nitrogen]

23 → 24

a) RMgBr, THF
0 °C, 10 min.

b) AlCl₃, EtSH
DCM, rt

70–86%

R = Me, Et, iPr,
Cyclopentyl,
Cyclohexyl

A [structure: Br-cyclohexyl-OTBS]

25a: R = Me
deprotection
25b: R = H

a) 4-nitrobenzoic acid
PPh₃, DEAD, THF

b) LiOH, THF

64%

26a: R = Me
deprotection
26b: R = H

A third-generation approach succeeded in introducing yet more flexibility into the synthesis by using difunctionalized intermediate **27**, which incorporated two synthetic handles for analog preparation. Further elaboration of this molecule could be achieved via the combination of Michael addition–elimination and S_NAr coupling methodologies, irrespective of the order.[53] This concept was illustrated by the completion of two concise syntheses of raloxifene (**1**) from intermediate **27**, obtained from aminobenzothiophene **22** in 70% yield. In the first synthesis, **27** was treated with *p*-methoxyphenylmagnesium bromide to provide adduct **28**, which was subsequently alkylated with hydroxyethylpiperidine to afford raloxifene precursor **18** in good overall yield. In contrast, the second synthesis was initiated with the alkylation of **27** to known intermediate **23**, which was readily converted to **18** in excellent overall yield. In both cases, the synthesis was completed by demethylation of **18** to afford raloxifene (**1**).

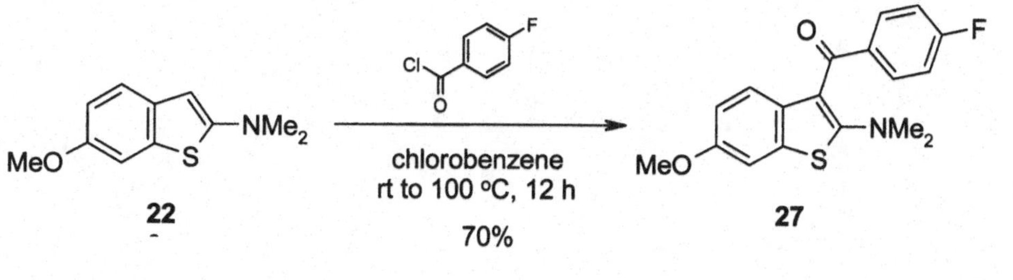

22 → 27

chlorobenzene
rt to 100 °C, 12 h

70%

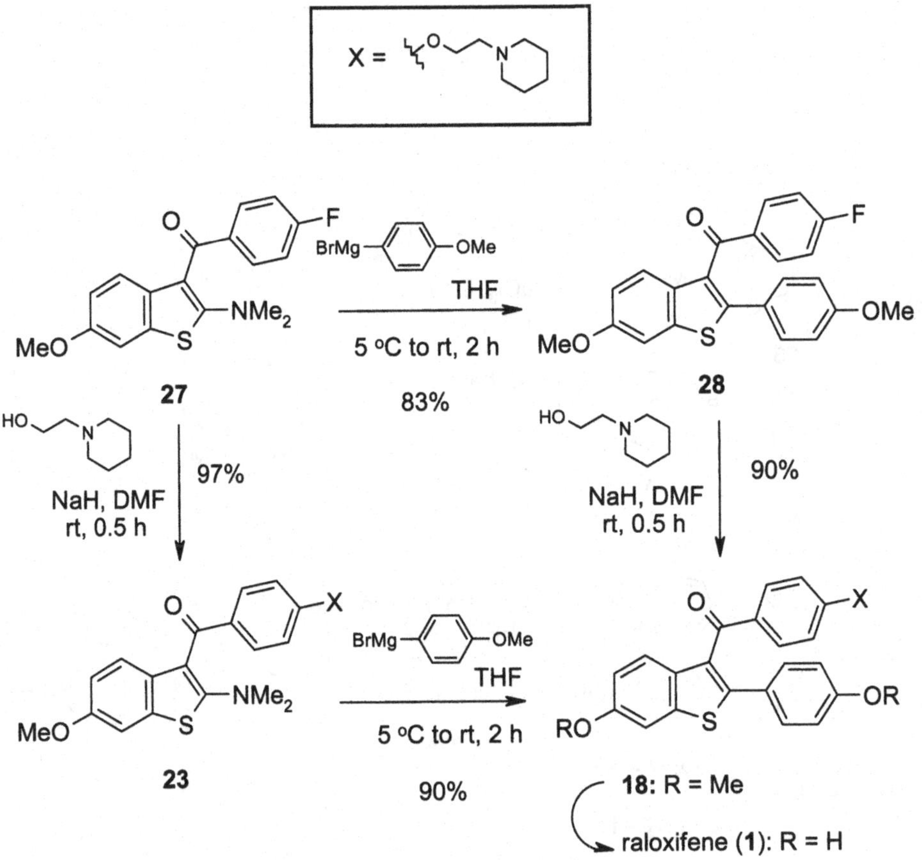

$X = $ (structure)

27   83%   **28**

23   90%   **18**: R = Me

raloxifene (**1**): R = H

Development of just-described S_NAr approach to introduce the requisite hydroxyethylpiperidine moiety late in the synthesis provided not only a rapid approach to raloxifene (**1**) but also a versatile entry into additional analog scaffolds **29**.[54] Thus advanced known intermediate **13** was acylated under standard conditions to provide key S_NAr substrates **28**, which reacted with a the series of O,S and N anion nucleophiles to generate the adducts **29**. Cleavage of both methyl ethers in **28** followed by application of the S_NAr methodology provided raloxifene (**1**) in overall 70% yield.

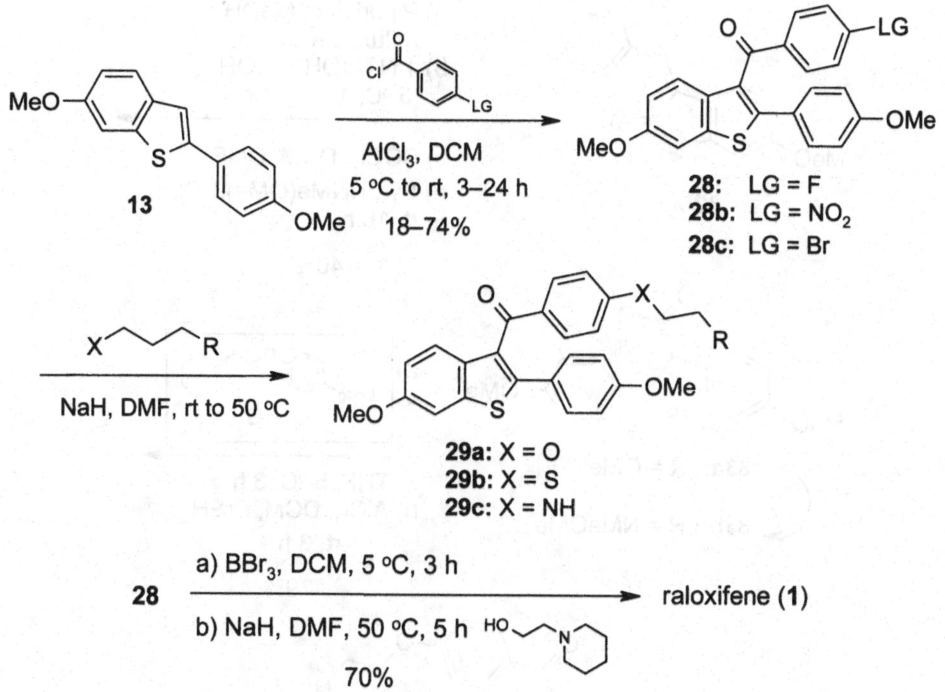

A synthesis of *trans*-2,3-dihydroraloxifene (**30**) used a base-catalyzed benzylidine–thiolactone rearrangement to install the requisite aryl moiety at C(2), instead of the previously used Grignard or Stille methodology.[55] Thus aminobenzothiophene **22**, itself a material employed in the synthesis of raloxifene (**1**), was converted into ketone **31** under basic catalysis in 80% yield. Condensation with *p*-anisaldehyde afforded a 75% yield of benzylidene thiolactone **32** as a mixture of geometrical isomers, which were subsequently treated with piperidine in methanol under reflux to afford the rearranged product **33a**. A three-step sequence afforded a 40% yield of the corresponding Weinreb amide **33b**, which was treated with the Grignard reagent **B** to afford the desired coupling product. Final cleavage of both methyl ethers provided the desired analog *trans*-2,3-dihydroraloxifene (**30**) in 79% yield.

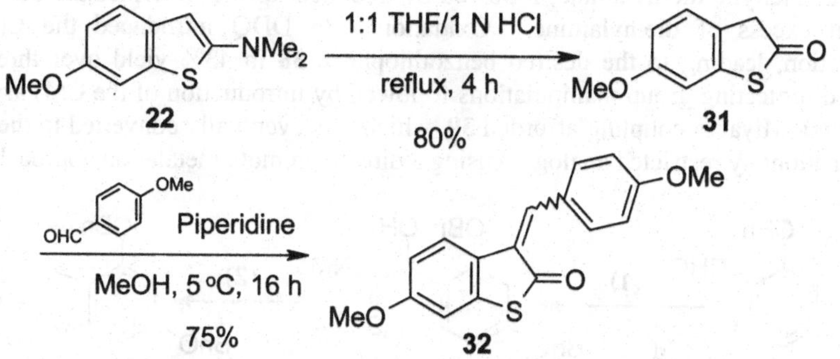

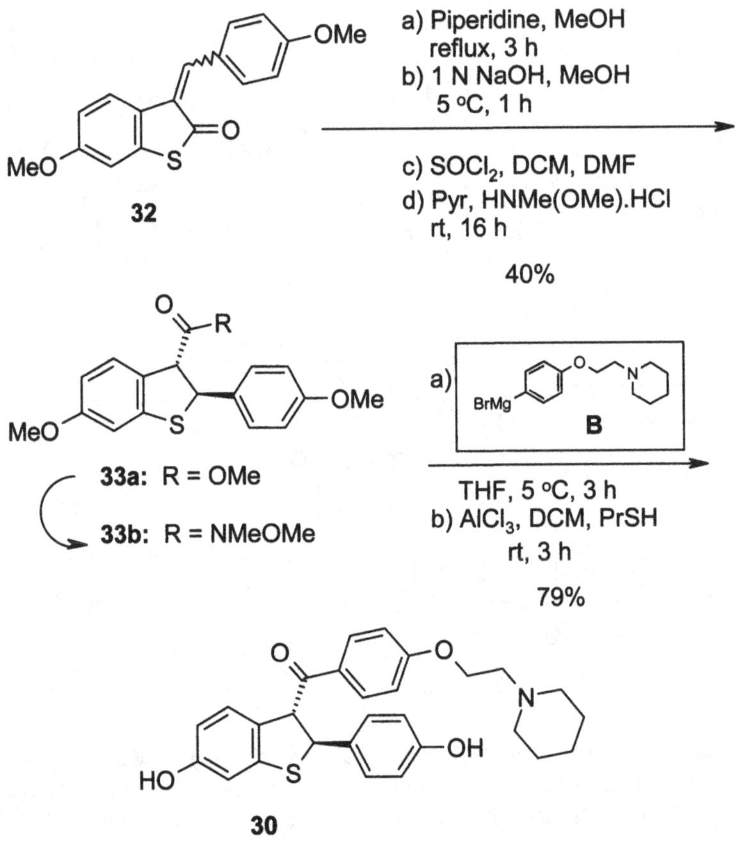
OMe
MeO
S
O
32
a) Piperidine, MeOH
reflux, 3 h
b) 1 N NaOH, MeOH
5 °C, 1 h
c) SOCl2, DCM, DMF
d) Pyr, HNMe(OMe).HCl
rt, 16 h
40%
R
OMe
33a: R = OMe
33b: R = NMeOMe
a)
BrMg
B
THF, 5 °C, 3 h
b) AlCl3, DCM, PrSH
rt, 3 h
79%
HO
OH
N
30

The benzylidine–thiolactone rearrangement was also applied to the synthesis of advanced intermediates for conformationally restricted raloxifene analogs.[56] Thus readily available aldehyde **34** was treated with the lithium anion of *N,N*-dimethylthioformamide to generate hydroxythioacetamide **35** in 41% yield, which gave rise to aminobenzothiophene **36** via dehydrative carbocationic cyclization promoted by methanesulfonic acid. Installation of the C(2) ketone to provide intermediate **37** in 59% yield set the stage for condensation with *p*-methoxyphenyl aldehyde in refluxing acetic acid with catalytic diethylamine, followed by treatment of the intermediate benzylidene with an excess of diethylamine. Oxidation with DDQ introduced the C(2)–C(3) unsaturation, leading to the desired benzothiophene **38** in 43% yield over three steps. Standard protecting group manipulations followed by introduction of the C(4) aryl group via Suzuki–Miyaura coupling afforded **39**, which was eventually converted to the desired conformationally restricted analog **40** using a directed remote metalation approach.

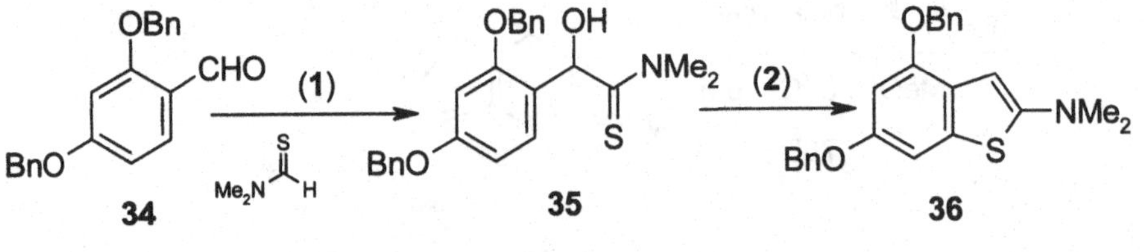
OBn
CHO
BnO
34
(1)
Me2N
H
S
OBn OH
NMe2
S
BnO
35
(2)
OBn
NMe2
S
BnO
36

(**1**) THF, –78 °C to rt (41%); (**2**) MeSO$_3$H, DCM, 0 °C to rt

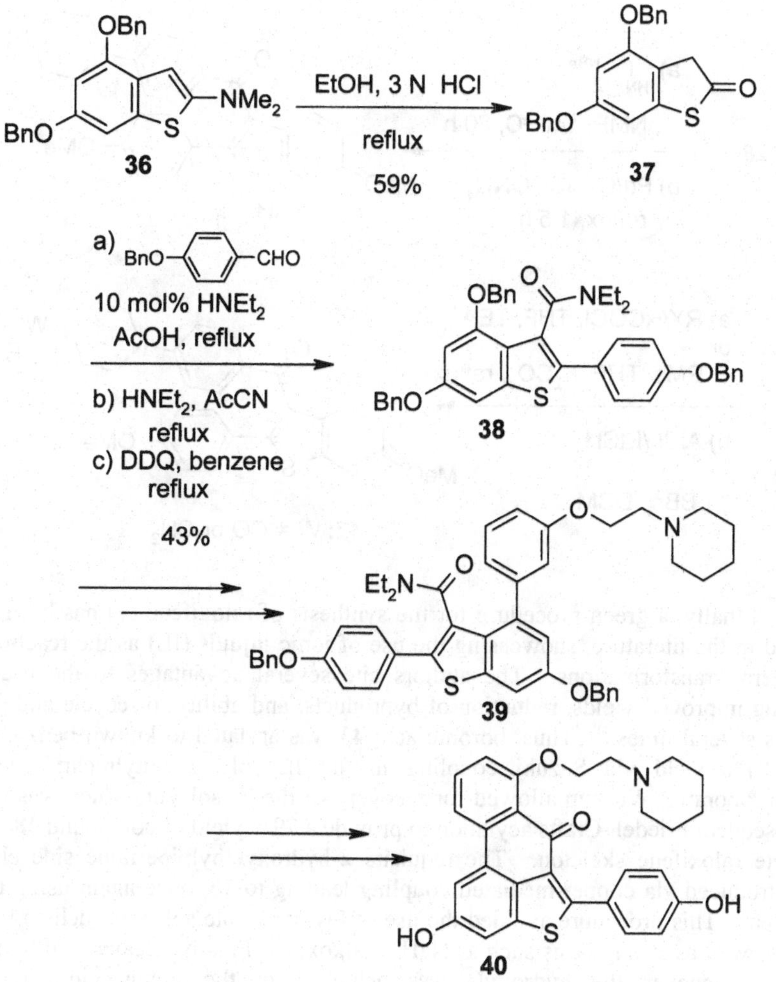

Raloxifene derivatives bearing a piperazine side chain were synthesized from known benzothiophene **13**.[57] Thus standard Friedel–Crafts acylation conditions yielded known compound **28**, possessing the necessary fluorine synthetic handle to allow introduction of the piperazine moiety. Subsequent removal of the benzyl protecting group gave rise to the necessary intermediate **41**, which was then further derivatized via alkylation or acylation conditions. Final cleavage of the methyl ethers with $AlCl_3$ or BBr_3 provided the final analogs **42**, which were tested for their affinity and selectivity in binding to ERα.

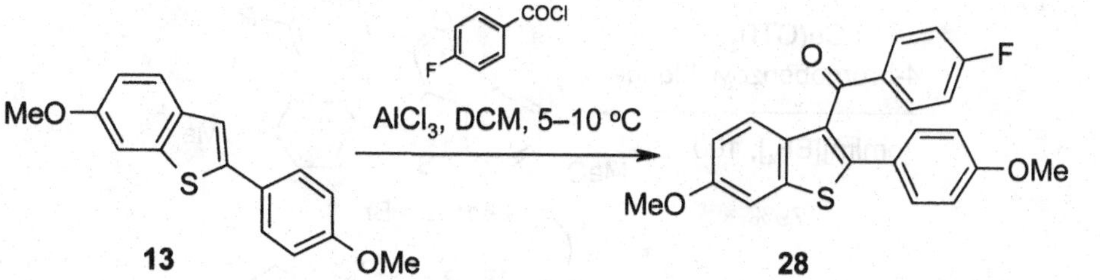

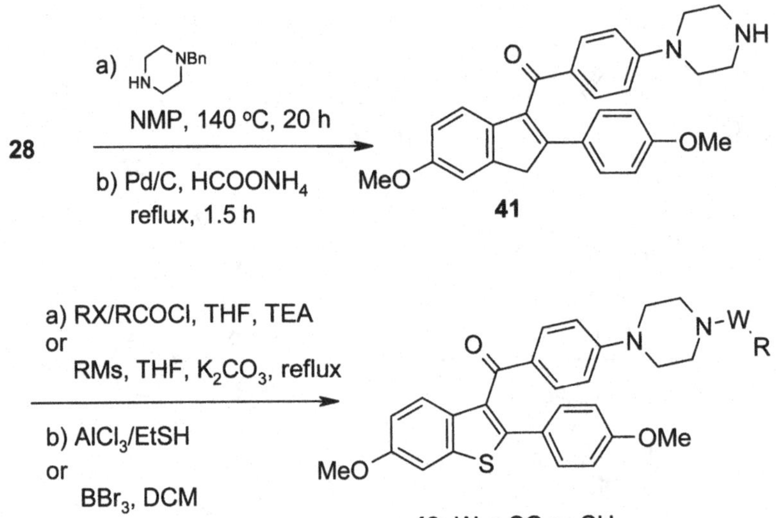

28 → **41**

a) [structure: NBn piperazine]
NMP, 140 °C, 20 h

b) Pd/C, HCOONH$_4$
reflux, 1.5 h

---

a) RX/RCOCl, THF, TEA
or
RMs, THF, K$_2$CO$_3$, reflux

b) AlCl$_3$/EtSH
or
BBr$_3$, DCM

42: W = CO or CH$_2$

Finally, a green procedure for the synthesis of raloxifene (**1**) has been recently reported in the literature, showcasing the use of ionic liquids (IL) as the reaction media for several transformations. The authors cite several advantages to the use of ILs, including improved yields, reduction of byproducts, and ability to recycle and reuse the solvents several times.[58] Thus, boronic acid **43** was arylated to known benzothiophene **13** in 81% yield via Suzuki coupling in the IL solvent butylmehtylimidazolium tetrafluoroborate. Workup allowed for recovery of the IL solvent, which was reused in the subsequent Friedel–Crafts acylation to provide a 79% yield of compound **28c** with the complete raloxifene skeleton. The requisite 2-hydroxyethylpiperidine side chain was then introduced via copper-mediated coupling leading to **18**, once again using the same IL solvent. This procedure avoided the use of less-desirable solvents such as DMSO or DMF, as well as strong basis such as NaH or alkoxide. Finally, cleavage of both methyl ethers to uncover the hydroxyls was achieved in the Lewis-acidic IL solvent trimethylammonium aluminum chloride, giving rise to raloxifene (**1**) in 43% yield over two steps.

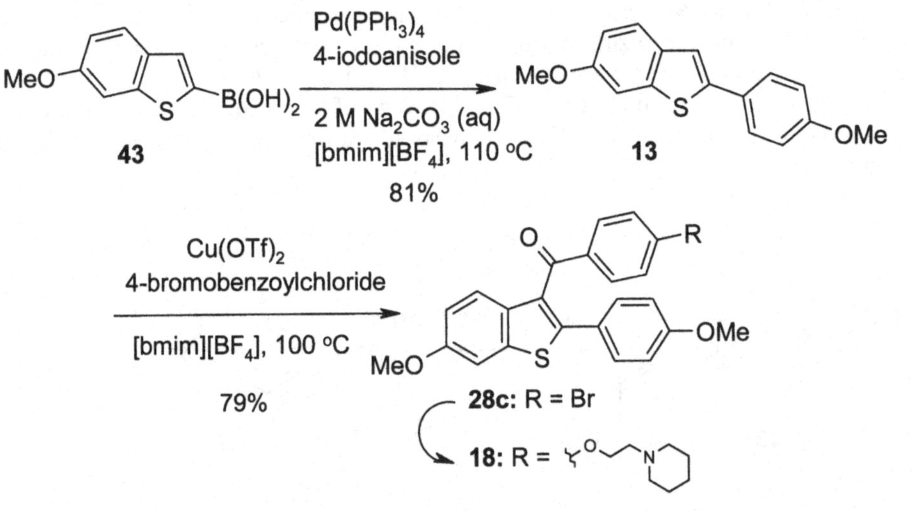

43

Pd(PPh$_3$)$_4$
4-iodoanisole

2 M Na$_2$CO$_3$ (aq)
[bmim][BF$_4$], 110 °C
81%

13

Cu(OTf)$_2$
4-bromobenzoylchloride

[bmim][BF$_4$], 100 °C
79%

28c: R = Br

18: R = [structure: piperidine side chain]

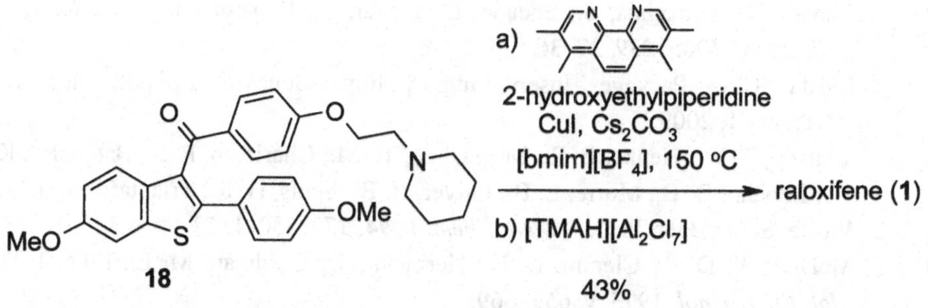

a)

2-hydroxyethylpiperidine
CuI, Cs_2CO_3
[bmim][BF_4], 150 °C
$\xrightarrow{\hspace{2cm}}$ raloxifene (1)

b) [TMAH][Al_2Cl_7]

43%

20.6   References

1   Grese, T. A.; Dodge, J. A. *Ann. Rep. Med. Chem.* **1996**, *31*, 181–190.

2   Jordan, V. C. *J. Med. Chem.* **2003**, *46*, 884–908 (Part I) and 1081–1111 (Part II).

3   Dodge, J. A.; Richardson, T.I *Ann. Rep. Med. Chem.* **2007**, *42*, 147–160.

4   Clemmett, D.; Spencer, C. M. *Drugs* **2000**, *60*, 379–411.

5   Evista information, www.evista.com, and healthcare professional area, www.evista.com/hcp/index.jsp, accessed December 1, 2009.

6   National Osteoporosis Foundation, www.nof.org/osteoporosis/index, accessed December 1, 2009.

7   National Breast Cancer Foundation, www.nationalbreastcancer.org, accessed December 1, 2009.

8   Breast cancer disease advocacy, www.breastcancer.org, accessed December 1, 2009.

9   Principal results from the Women's Health Initiative randomized controlled clinical trials: Rossouw, J. E; Anderson, G. L.; Prentice, R. L.; LaCroix, A. Z.; Kooperberg, C.; Stefanick, M. L.; Jackson, R. D.; Beresford, S. A.; Howard, B. V.; Johnson, K, C; Kotchen, J. M.; Ockene, J. *JAMA* **2002**, *288*, 321–333.

10   Thomson Micromedex Health Care Series database, www.thomsonhc.com, accessed December 1, 2009.

11   Drugs@FDA database, www.accessdata.fda.gov/Scripts/cder/DrugsatFDA, accessed December 1, 2009.

12   European Medicines Agency, www.emea.europa.eu, accessed December 1, 2009.

13   Nolvadex information on AstraZeneca: www.astrazeneca.com/medicines, accessed December 1, 2009.

14   Tamoxifen, Jordan, V. C. *Br. J. Pharmacol.* **2006**, *147*, 269–276.

15   Fareston information, www.fareston.com

16   Mansel, R.; Goyal, A.; Nestour, E. L.; Masini-Etévé, V.; O'Connell, K. *Breast Cancer Res. Treat.* **2007**, *106*, 389–397.

17   Singh, M. *Med. Res. Rev.* **2001**, *21*, 302–347.

18   Gennari, L.; Merlotti, D.; Martini, G.; Nuti, R. *Exp. Opin. Investig. Drugs* **2006**, *15*, 1091–1103.

19   Faslodex information, www.astrazeneca.com/medicines, accessed December 1, 2009.

20   Kansra, S.; Yamagata, S.; Sneade, L.; Foster, L.; Ben-Jonathan, N. *Mol. Cell Endocrinol.* **2005**, *239*, 27–36.

21   Evista U.S. Package Insert, http://pi.lilly.com/us/evista-pi.pdf, accessed December 1, 2009.

22   Willson, T. W.; Henke, B. R.; Momtahen, T. M.; Charifson, P. S.; Batchelor, K. W.; Lubahn, D. B.; Morre, L. B.; Oliver, B. B.; Sauls, H. R.; Triantafillou, J. A.; Wolfe, S. G.; Baer, P. G. *J. Med. Chem.* **1994**, *37*, 1550–1552.

23   McDonnell, D. P.; Clemm, D. L.; Hermann, T.; Goldman, M. E.; Pike, J. W. *Mol. Endocrinol.* **1995**, *9*, 659–669.

24   Katzenellenbogen, J. A.; O'Malley, B. W.; Katzenellenbogen, B. S. *Mol. Endocrinol.* **1996**, *10*, 119–131.

25   Grese, T. A.; Sluka, J. P.; Bryant, H. U.; Cullinan, G. J.; Glasebrook, A. L.; Jones, C. D.; Matsumoto, K.; Palkowitz, A. D.; Sato, M.; Termine, J. D.; Winter, M. A.; Yang, N. N.; Dodge, J. A. *Proc. Natl. Acad. Sci.* **1997**, *94*, 14105–14110.

26   Shang, Y.; Brown, M. *Science* **2002**, *295*, 2465–24688.

27   Riggs, B. L.; Hartmann, L. C. *N. Engl. J. Med.* **2003**, *348*, 618–629.

28   Wallace, O. B.; Richardson, T. I.; Dodge, J. A. *Cur. Topics Med. Chem.* **2003**, *3*, 1663–1682.

29   Smith, C. L.; O'Malley, B. W. *Endocr. Rev.* **2004**, *25*, 45–71.

30   Nolvadex, Fareston and Faslodex U.S. Package Inserts, http://dailymed.nlm.nih.gov/dailymed/about.cfm, accessed December 1, 2009.

31   Morello, K. C.; Wurz, G. T.; DeGregorio, M. W. *Clin. Pharmacolkinet.* **2003**, *42*, 361–372.

32   Robertson, J. F. R.; Harrison, M. *Br. J. Cancer* **2004**, *90* (suppl. 1), S7–S10.

33   Cummings, S. R.; Eckert, S.; Krueger, K. A.; Grady, D.; Powles, T. J.; Cauley, J. A.; Norton, L.; Nickelsen, T.; Bjarnason, N. H.; Morrow, M.; Lippman, M. E.; Black, D.; Glusman, J. E.; Costa, A.; Jordan, V. C. *JAMA* **1999**, *281*, 2189–2197.

34   Cauley, J. A.; Norton, L.; Lippman, M. E.; Eckert, S.; Krueger, K. A.; Purdie, D. W.; Farrerons, J.; Karasik, A.; Mellstrom, D.; Ng, K. W.; Stepan, J. J.; Powles, T. J.; Morrow, M.; Costa, A.; Silfen, S. L.; Walls, E. L.; Schmitt, H.; Muchmore, D. B.; Jordan, V. C. *Breast Cancer Res. Treat.* **2001**, *65*, 125–134.

35   Martino, S.; Cauley, J. A.; Barrett-Connor, E.; Powles, T. J.; Mershon, J.; Disch, D.; Secrest, R. J.; Cummings, S. R. *J. Natl. Cancer Inst.* **2004**, *96*, 1751–1761.

36   Data on file, Lilly Research Laboratories (EVI20070730).

37   Data on file, Lilly Research Laboratories (EVI20070913).

38   Barrett-Connor, E.; Mosca, L.; Collins, P.; Geiger, M. J.; Grady, D.; Komitzer, M.; McNabb, M. A.; Wenger, N. K. *N. Engl. J. Med.* **2006**, *355*, 125–137.

39   Vogel, V. G.; Costantino, J. P.; Wickerham, D. L.; Cronin, W. M.; Cecchini, R. S.; Atkins, J. N.; Bevers, T. B.; Fehrenbacher, L.; Pajon, E. R. Jr.; Wade, J. L. III; Robidoux, A.; Margolese, R. G.; James, J.; Lippman, S. M.; Runowicz, C. D.; Ganz, P. A.; Reis, S. E.; McCaskill-Stevens, W.; Ford, L. G.; Jordan, V. C.; Wolmark, N. *JAMA* **2006**, *295*, 2727–2741.

40   Ettinger, B.; Black, D. M.; Mitlak, B. H.; Knickerbocker, R. K.; Nickelsen, T.; Genant, H. K.; Christiansen, C.; Delmas, P. D.; Zanchetta, J. R.; Stakkestad, J.;

Glüer, C. C.; Krueger, K.; Cohen, F. J.; Eckert, S.; Ensrud, K. E.; Avioli, L. V.; Lips, P.; Cummings, S. R. *JAMA* **1999**, *282*, 637–645.

41 Delmas, P. D.; Ensrud, K. E.; Adachi, J. D.; Harper, K. D.; Sarkar, S.; Gennari, C.; Reginster, J.-Y.; Pols, H. A. P.; Recker, R. R.; Harris, S. T.; Wu, W.; Genant, H. K.; Black, D. M.; Eastell, R. *J. Clin. Endocrinol. Metab.* **2002**, *87*, 3609–3617.

42 Siris, E. S.; Harris, S. T.; Eastell, R.; Zanchetta, J. R.; Goemaere, S.; Diez-Perez, A.; Stock, J. L.; Song, J.; Qu, Y.; Kulkami, P. M.; Siddhanti, S. R.; Wong, M.; Cummings, S. R. *J. Bone Miner. Res.* **2005**, *20*, 1514–1524.

43 Johnston, C. C. Jr.; Bjarnason, N. H.; Cohen, F. J.; Shah, A.; Lindsay, R.; Mitlak, B. H.; Huster, W.; Draper, M. W.; Harper, K. D.; Heath, H., III; Gennari, C.; Christiansen, C.; Arnaud, C. D.; Delmas, P. D. *Arch. Intern. Med.* **2000**, *160*, 3444–3450.

44 Wenger, N. K.; Barrett-Connor, E.; Collins, P.; Grady, D.; Komitzer, M.; Mosca, L.; Sashegyi, A.; Baygani, S. K.; Anderson, P. W.; Moscarelli, E. *Am. J. Cardiol.* **2002**, *90*, 1204–1210.

45 Barrett-Connor, E.; Grady, D.; Sashegyi, A.; Anderson, P. W.; Cox, D. A.; Hoszowski, K.; Rautaharju, P.; Harper, K. D. *JAMA* **2002**, *287*, 847–857.

46 Martino, S.; Disch, D.; Dowsett, S. A.; Keech, C. A.; Mershon, J. L. *Curr. Med. Res. Opin.* **2005**, *21*, 1441–1452.

47 Data on file, Lilly Research Laboratories (EVI20070914).

48 Jones, C. D.; Suárez, T. U.S. Patent 4,133,814, **1979**.

49 Jones, C. D.; Jevnikar, M. G.; Pike, A. J.; Peters, M. K.; Black, L. J.; Thompson, A. R.; Falcone, J. F.; Clemens, J. A. *J. Med. Chem.* **1984**, *27*, 1057–1066.

50 Godfrey, A. G. U.S. Pat. 5,420,349, **1995**.

51 Grese, T. A.; Cho, S.; Bryant, H. U.; Cole, H. W.; Glasebrook, A. L.; Magee, D. E.; Phillips, D. L.; Rowley, E. R.; Short, L. L. *Bioorg. Med. Chem. Lett.* **1996**, *6*, 201–206

52 Grese, T. A.; Cho, S.; Finley, D. R.; Godfrey, A. G.; Jones, C. D.; Lugar, C. W.; Martin, M. J .; Matsumoto, K.; Pennington, L. D.; Winter, M. A.; Adrian, M. D.; Cole, H. W.; Magee, D. E.; Phillips, D. L.; Rowley, E. R.; Short, L. L.; Glasebrook, A. L.; Bryant, H. U. *J. Med. Chem.* **1997**, *40*, 146–167

53 Bradley, D. A.; Godfrey, A. G.; Schmid, C. R. *Tetrahedron Lett.* **1999**, *40*, 5155–5159

54 Schmid, C. R.; Sluka, J. P.; Duke, K. M. *Tetrahedron Lett.* **1999**, *40*, 675–678

55 Schmid, C. R.; Glasebrook, A. L.; Misner, J. W.; Stephenson, G. A. *Bioorg. Med. Chem. Lett.* **1999**, *9*, 1137–1140.

56 Kalinin, A. V.; Reed, M. A.; Norman, B. H.; Snieckus, V. *J. Org. Chem.* **2003**, *68*, 5992–5999.

57 Yang, C.; Xu, G.; Li, J.; Wu, X.; Liu, B.; Yan, X.; Wang, M.; Xie, Y. *Bioorg. Med. Chem. Lett.* **2005**, *15*, 1505–1507.

58 Shinde, P. S.; Shinde, S. S.; Renge, A. S.; Patil, G. H.; Rode, A. B.; Pawar, R. R. *Lett. Org. Chem.* **2009**, *6*, 8–10.

21

Latanoprost (Xalatan): A Prostanoid FP Agonist for Glaucoma

Sajiv K. Nair and Kevin E. Henegar

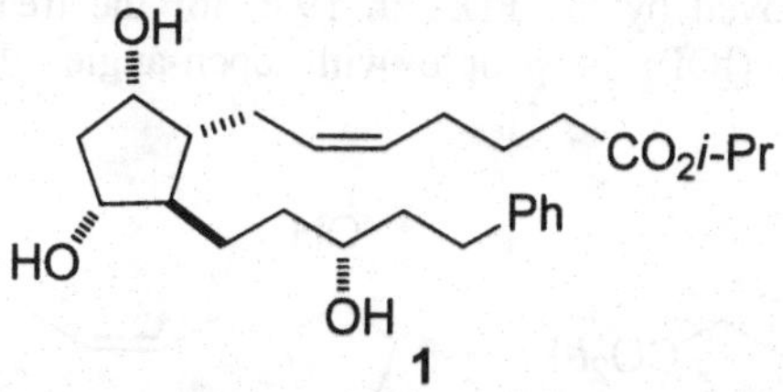

USAN: Latanoprost
Trade name: Xalatan®
Pfizer
Launched: 1996

21.1   Introduction

Glaucoma is one of the leading causes of irreversible blindness worldwide. Primary open-angle glaucoma, the most common form of glaucoma, is characterized by decreased outflow of aqueous humor through the conventional drainage pathway that results in elevated intraocular pressure (IOP). Glaucoma exhibits a characteristic intraocular pressure sensitive optic neuropathy leading to progressive loss of visual field. Thus reduction in IOP to within the normal range may prevent additional optic nerve damage and preserve remaining vision.[1] Current treatments of glaucoma are almost universally aimed at lowering IOP either through the chronic administration of topical drugs or through surgical methods.

Prostaglandins (PGs) are products derived from the cyclooxygenase-mediated metabolism of arachidonic acid to PGG/H_2, leading to the generation of five primary bioactive prostanoids: PGE_2, $PGF_{2\alpha}$, PGI_2, TxA_2, and PGD_2. These prostaglandins exert their effects by interacting with specific distinct G-protein coupled receptors designated EP, FP, IP, TP, and DP, respectively.[2] $PGF_{2\alpha}$ (2) and its iso-propyl ester $PGF_{2\alpha}$-IE (3) were found to lower intraocular pressure (IOP) in normotensive healthy volunteers and in open-angle glaucoma patients respectively. However, due to the poor corneal permeability of $PGF_{2\alpha}$ and the adverse side effects such as ocular irritation and conjuctival hyperemia seen with both $PGF_{2\alpha}$ and $PGF_{2\alpha}$-IE, these molecules were limited in their therapeutic potential.[3,4]

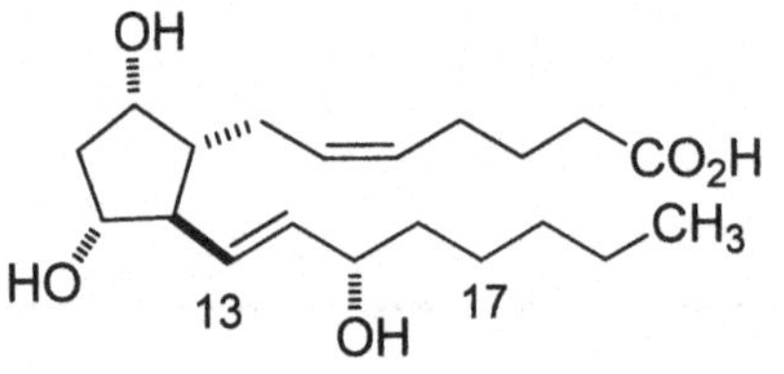

2

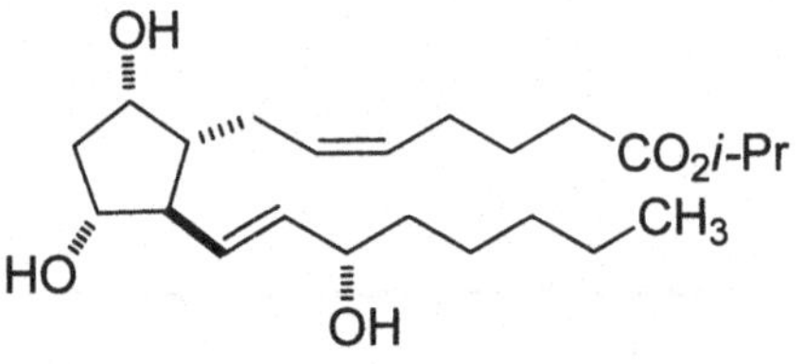

3

Medicinal chemistry efforts at Kabi-Pharmacia AB directed at addressing these issues found that the introduction of a phenyl ring at the C17 position on the omega chain of PG2F$_{2\alpha}$ and the saturation of the C13–C14 double bond, were beneficial from a pharmacological and safety perspective.[5,6] The use of the iso-propyl ester prodrug addressed the corneal penetration issue and this hindered ester was also stable to aqueous hydrolysis at pH 6–7. Thus a new class of phenyl-substituted prostaglandins was discovered, and the initial candidate evaluated for IOP-lowering studies was PhXA34 (13,14-dihydro-15[R,S]-17-phenyl-18,19,20-trinor-PGF$_{2\alpha}$-IE,   **4**).[6]   Although   this compound was therapeutically effective,[7] it was a mixture of the 15R- and S-epimers. Further work revealed that the 15R-epimer was more efficacious than the corresponding S epimer and this led to the discovery of PhAX41 (13,14-dihydro-15[R]-17-phenyl-18,19,20-trinor-PGF$_{2\alpha}$-IE, latanoprost, **1**), the key ingredient of the commercial product Xalatan.[8] Xalatan was approved by the FDA in 1996 for the treatment of high eye pressure/intraocular pressure (IOP) in people with open-angle glaucoma or ocular hypertension.

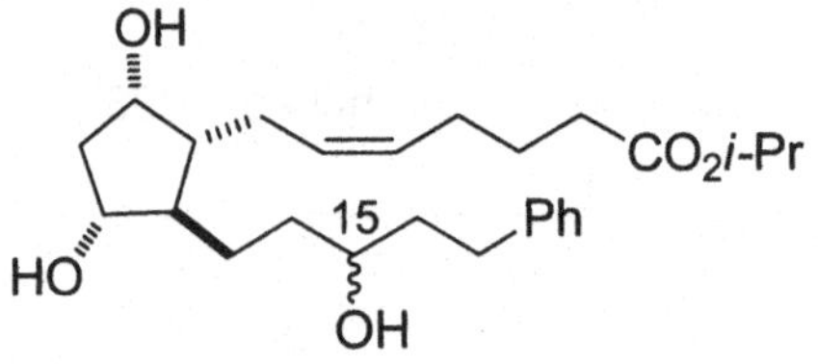

4

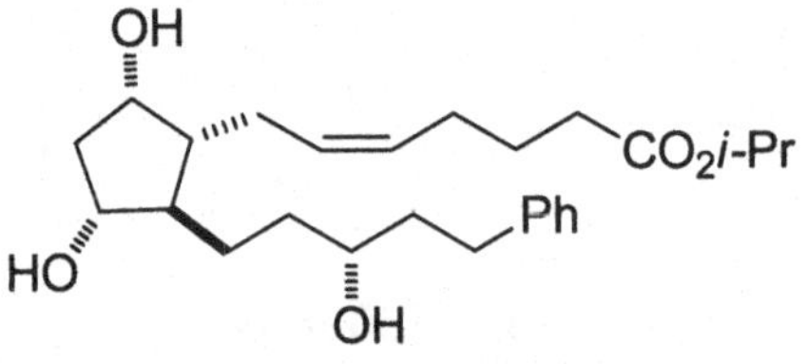

1

Latanoprost (**1**) is an iso-propyl ester prodrug that is hydrolyzed to the parent acid by esterases present in the cornea. The parent acid of latanoprost (**1**) is a selective FP-receptor agonist and reduces intraocular pressure by increasing the uveosceral outflow of aqueous humor.[9] Xalatan is supplied as a sterile, isotonic, buffered aqueous solution of 0.005% latanoprost (**1**) and is administered topically to the eye. The product also contains the preservative benzalkonium chloride (0.02%), and the pH of the solution is 6.7.[10] In Europe, latanoprost (**1**) is also sold as a combination with the maleate salt of timolol, a β1 and β2 (nonselective) adrenergic receptor blocking agent. The combined effect of this product named Xalacom results in additional intraocular pressure reduction compared to either component administered alone separately. The collective revenue for Xalatan/Xalacom was $1.7 billion in the year 2008, according to Pfizer's annual report.

21.2   Syntheses

21.2.1   Medicinal Chemistry Synthesis

The synthesis of latanoprost (**1**) commences with the lactone **5** derived from the Corey lactone (Scheme 1).[11,12] The early introduction of the *p*-phenylbenzoyl protecting group[12] in this lactone affords crystalline intermediates in the synthesis that are easy to handle and purify. In the next step, the primary alcohol functionality in the lactone **5** was oxidized using Pfitzner-Moffat conditions (DCC, DMSO and phosphoric acid) to yield the aldehyde **6**. The crude aldehyde was then treated with dimethyl (2-oxo-4-phenyl)-phosphonate under Wadsworth-Emmons conditions (NaH, DME) to afford the enone **7** as a white crystalline solid in 59% yield.

The introduction of the key C15 stereogenic center was achieved via the reduction of the α,β-unsaturated ketone **7** using lithium tri-*sec*-butylborohydride (*L*-Selectride) at –120/–130 °C to give a 7:3 epimeric mixture of the 15*S*-isomer **8a** and 15*R*-isomer **8b**. The epimers were separable by flash column chromatography on silica gel, and this purification step provided the 15*S*-isomer **8a** as a solid in 52% yield. A lower selectivity was obtained when this reduction was carried out using sodium borohydride in the presence of cerium chloride heptahydrate in methanol (Luche reduction) at room temperature. The double bond in the *trans*-allylic alcohol **8a** was then reduced to the saturated alcohol **9** under hydrogen atmosphere using Pd/C as a catalyst in the presence of sodium hydroxide or sodium nitrite.[13] The deprotection of the *p*-phenylbenzoyl group was achieved using potassium carbonate in methanol, and the lactone moiety was subsequently reduced to the lactol **10** using DIBAL-H.

The C1 acid–containing side chain was introduced using a Wittig reaction with (4-carboxybutyl)triphenylphosphonium bromide (CTP-bromide) and potassium *tert*-butoxide to give the olefin containing acid **11**. The crude acid was then treated with *iso*-propyl iodide and DBU in acetone to furnish PhAX41 (13,14-dihydro-15[*R*]-17-phenyl-18,19,20-trinor-PGF$_{2\alpha}$ isopropyl ester, latanoprost, **1**).[5,6,14] The stereochemistry at C15 is now designated *R* because of a reversal of substituent priorities relative to PGF$_{2\alpha}$ (**2**) and the early intermediate **8a**, that contain a double bond at the α-position to the stereocenter.

DCC, DMSO, H$_3$PO$_4$, DME

r.t., 2.5 h

NaH, DME, 0 °C to r.t.
59% over 2 steps

P = C(O)C$_6$H$_4$-*p*-Ph

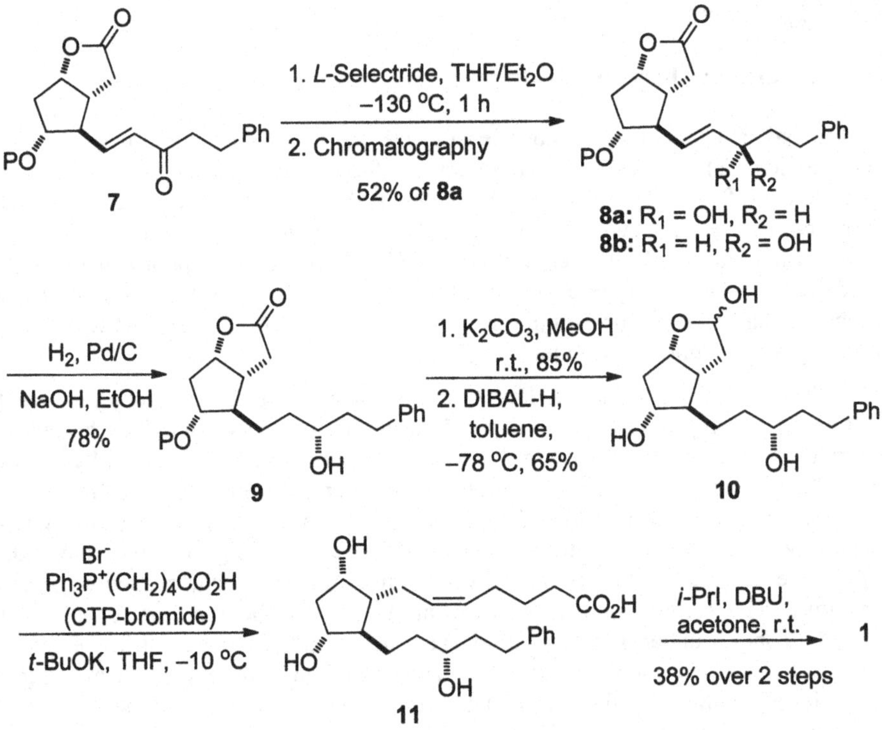

Scheme 1

21.2.2 Kabi-Pharmacia Commercial Process

The original Kabi-Pharmacia commercial process for the synthesis of latanoprost is similar to the Medicinal Chemistry synthesis and is shown in the Scheme 2.[5,15] Some modifications were made for commercial scale processing, but the general synthesis is basically the same.[16] Even with modifications for scale up, this process is limited by low yields and the need for multiple chromatographies, especially for the generation of intermediate **8** with adequate C15 diastereomeric purity. The instability of the deprotected lactol **10** complicates processing and the Wittig reaction with CTP-bromide (5-carboxybutyltriphenyphosphonium bromide) on the unprotected lactol **10** generates large amounts of 5-diphenylphosphinylpentanoic acid (P-acid) formed by ylide hydrolysis, which can be difficult to remove. Two chromatographies are required for final product isolation.

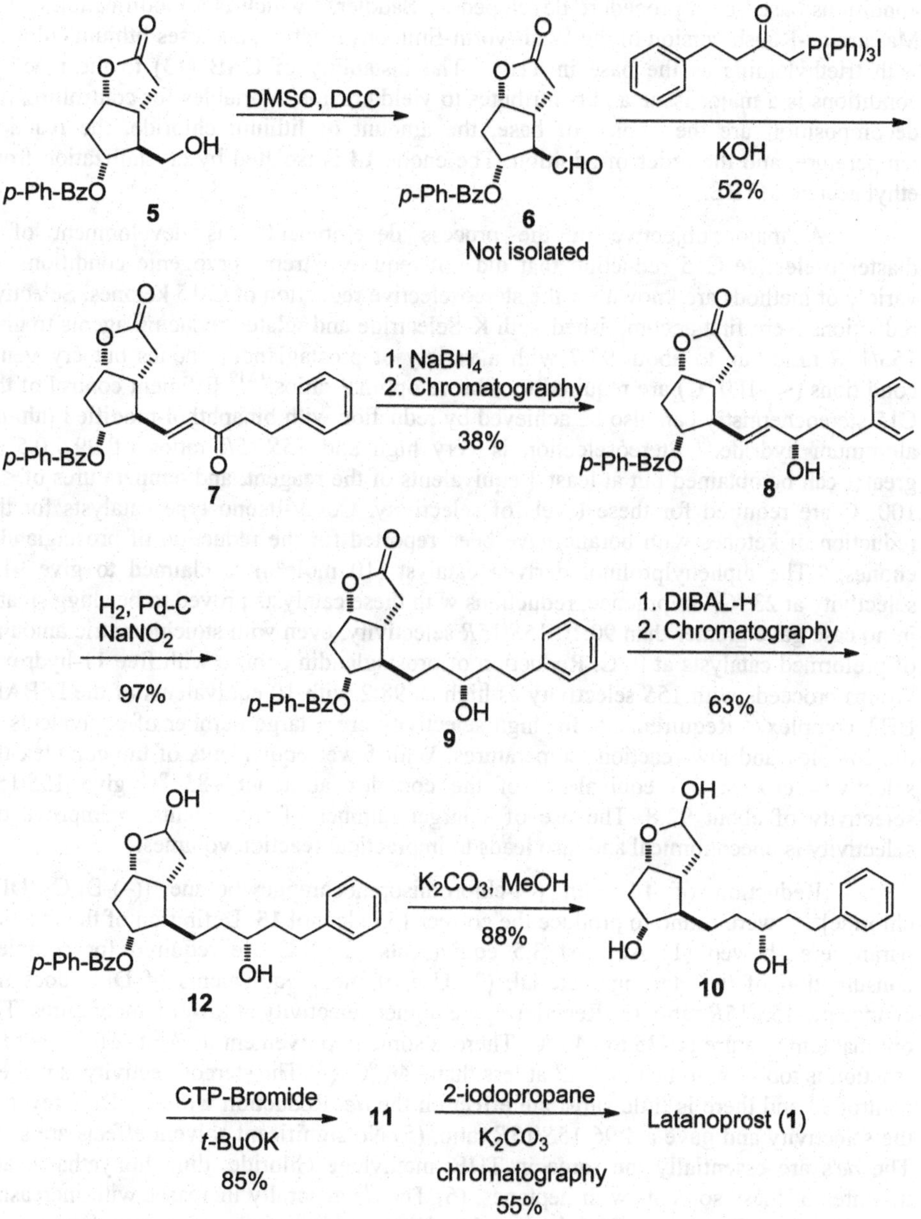

Scheme 2

21.2.3 Pfizer Commercial Process

The Pfizer process to latanoprost starts with the Corey aldehyde benzoate (**13**, CAB), available as a relatively stable crystalline solid via the historic Upjohn prostaglandin synthesis (Scheme 3).[17] The reaction of the phosphonate with CAB (**13**) is done using conditions based on a procedure developed by Saddler,[18] which is a modification of the Masamune-Roush version of the Wadsworth-Emmons reaction, and uses lithium chloride with triethylamine as the base in THF. The instability of CAB (13) to the reaction conditions is a major issue and contributes to yield erosion. Variables for controlling the decomposition are the choice of base, the amount of lithium chloride, the reaction temperature, and the order of addition. The enone 14 is isolated by crystallization from ethyl acetate/MTBE.

A major objective of the process development was development of a diastereoselective C15 reduction that did not require extreme cryogenic conditions. A variety of methods are known for the stereoselective reduction of C-15 ketones. Selective reductions were first accomplished with K-Selectride and related reducing agents to give 15S/15R ratios up to about 93:7 with a variety of prostaglandin enones but cryogenic conditions (< −100 °C) are required to achieve optimal ratios.[12,19] Efficient control of the C15 stereochemistry can also be achieved by reduction with binaphthol-modified lithium aluminum hydride.[20] Stereoselection is very high and 15S/15R ratios of 99.5:0.5 or greater can be obtained but at least 3 equivalents of the reagent, and temperatures of < −100 °C are required for these levels of selectivity. Corey-Itsuno type catalysts for the reduction of ketones with borane have been reported for the reduction of prostaglandin enones.[21] The diphenylprolinol derived catalyst (10 mole%) is claimed to give 91:9 selectivity at 23 °C. In practice, reductions with these catalysts proved to be sluggish and in no case gave greater than 90:10 15S/15R selectivity, even with stoichiometric amounts of preformed catalysts at 0 °C. Reduction of prostaglandin enones with free 11-hydroxyl groups proceeds with 15S selectivity as high as 98:2 with 10 equivalents of the DIBAL-BHT complex.[22] Requirements for high selectivity are a large number of equivalents of the complex and low reaction temperatures. With fewer equivalents of the complex the selectivity erodes; 3.5 equivalents of the complex at about −85 °C give 15S/15R selectivity of about 92:8. The use of a larger number of equivalents to improve the selectivity is uneconomical and also leads to impractical reaction volumes.

Reduction of 14 with (−)-chlorodiisopinocampheylborane [(−)-DPC, DIP-chloride][23,24] were found to produce the correct 15S alcohol 15. Definition of the reaction parameters showed (1) At least 3.5 equivalents of DPC are required for complete consumption of the starting material, (2) Use of more equivalents of DPC does not change the 15S/15R ratio, (3) Reactions give higher selectivity at lower temperatures. The optimal temperature is −35 to −40 °C. There is some improvement in de at −50 °C, but the reaction is too slow to be practical at less than −40 °C, (4) The stereoselectivity is reagent controlled and there is little substrate effect on the de. Reduction with (+)-DPC reverses the selectivity and gave a 4:96 15S/15R ratio, (5) No significant solvent effects are seen. The de's are essentially the same in THF, methylene chloride, dimethoxyethane, and mixtures of these solvents with heptanes, (6) The de generally increases with increasing concentration. The excess DPC is quenched by the addition of acetone and then sodium bicarbonate to neutralize the HCl produced. Diethanolamine quench is also effective but is less practical on scale than acetone. Both quenches generate the same boron-containing products and pinene. The pinene and other nonpolar products are removed by partitioning between heptane and acetonitrile. Alternatively, they can be removed later in the processing after hydrolysis of the benzoate.

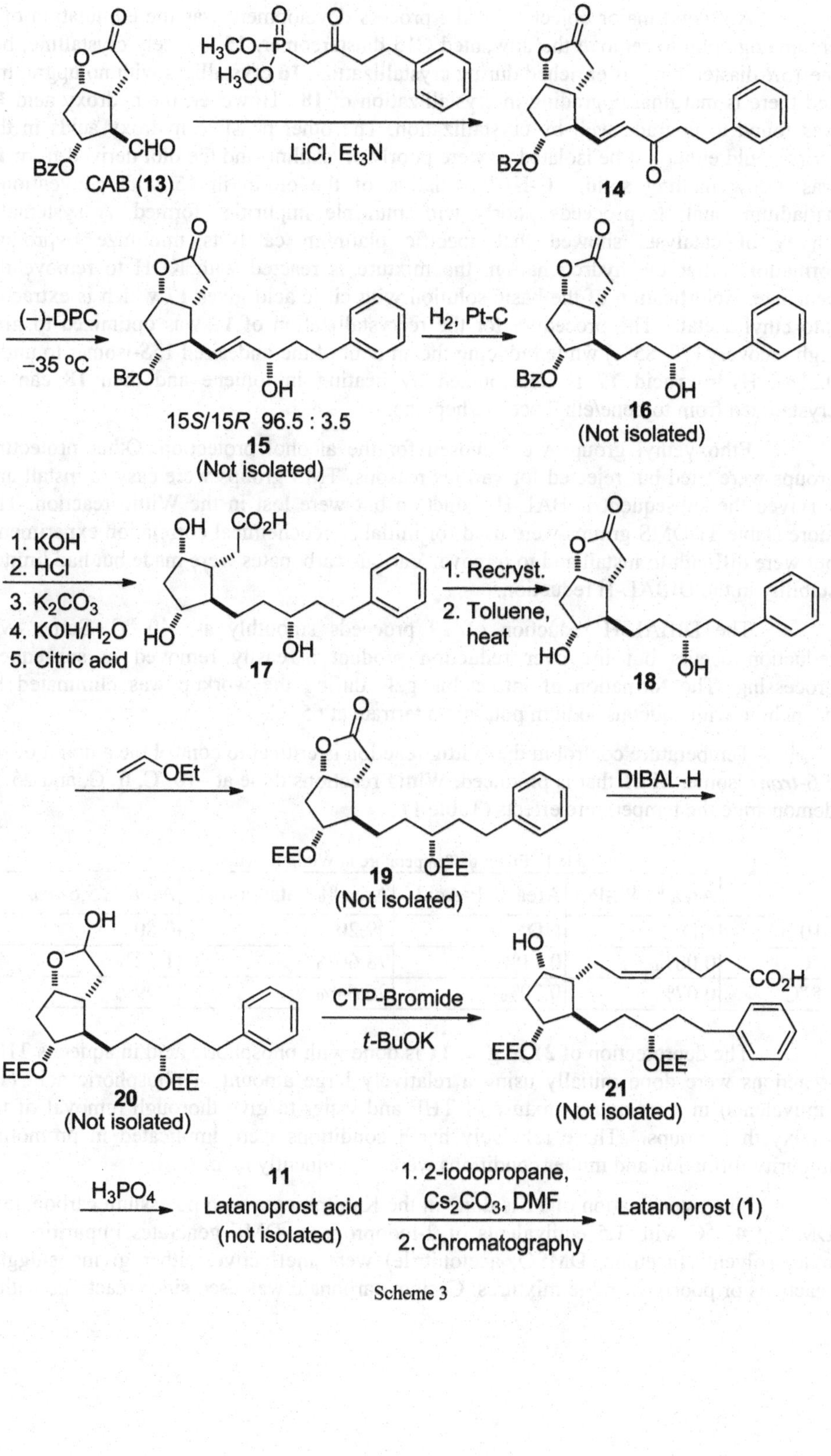

Scheme 3

A second major objective of the process development was the elimination of a chromatography to remove the unwanted C15 diastereomer. **15** is nicely crystalline, but the 15*R*-diastereomer is enriched during crystallization. **16** crystallizes with no upgrading and there is marginal upgrading in crystallization of **18**. However, the hydroxy acid **17** was found to upgrade well by crystallization. The other possible hydroxy acids in the series could either not be isolated or were poorly crystalline and the diol derived from **15** was a low melting solid.   C-Hydrogenation of the olefin in **15** with conventional palladium catalysts proceeds poorly with multiple impurities formed. A systematic survey of catalysts showed that specific platinum catalysts minimize byproduct formation. After the hydrogenation, the mixture is reacted with KOH to remove the benzoate. Acidification of the basic solution with citric acid gives **17** which is extracted into ethyl acetate. The procedure for the recrystallization of **17** was optimized to give high recovery (79–85%) while reducing the amount of the undesired 15*S*-isomer to under 0.25%. Hydroxyacid **17** is relactonized by heating in toluene and then **18** can be crystallized from toluene/ethyl acetate/heptane.

Ethoxyethyl groups were chosen for the alcohol protection. Other protecting groups were tried but rejected for various reasons. TMS groups were easy to install and survived the subsequent DIBAL-H reduction but were lost in the Wittig reaction. The more stable TBDMS groups were used for initial stereochemical correlation experiments but were difficult to install and to remove. Various carbonates were made but had limited stability in the DIBAL-H reduction.

The DIBAL-H reduction of **19** proceeds smoothly at –40 °C. Some over reduction occurs but the over reduction product is easily removed in subsequent processing. The formation of intractable gels during the workup was eliminated by quenching with aqueous sodium potassium tartrate at 65 °C.

Temperature control in the Wittig reaction is critical to control the amount of the 5,6-*trans* isomer of **21** that is produced. Wittig reactions done at –10 °C, 0 °C, and 25 °C demonstrate the temperature effects (Table 1).

Table 1. Effect of Temperature in Wittig Reaction

	Area % Ph$_3$PO	Area % P-Acid	Area % Latanoprost	Area % 5,6-*trans*
–10 °C	ND	ND	99.20	0.80
0 °C	0.08%	0.11%	98.60%	1.12%
25 °C	0.07%	0.30%	96.64%	2.99%

The deprotection of **21** to give **11** is done with phosphoric acid in aqueous THF. Reactions were done initially using a relatively large amount of phosphoric acid (1.8 equivalents) in a refluxing mixture of THF and water to give thorough removal of the ethoxyethyl groups. These relatively harsh conditions were implicated in promoting impurity formation and milder conditions were subsequently employed.

The esterification of **11** is done in the Kabi process with potassium carbonate in DMF at 45 °C with 1.5 equivalents of 2-iodopropane. DMF generates impurities, but other solvents (acetone, DMSO, acetonitrile) were ineffective, either giving sluggish reactions or poorly stirrable mixtures. Cesium carbonate was used since reactions with it were faster and cleaner than reactions with potassium carbonate. Finally, latanoprost (1), which is noncrystalline, is purified by a single chromatography using either MTBE or methylene chloride/IPA.

21.3   References

1.   Tsai, J. C.; Kanner, E. M. *Expert. Opin. Emerging Drugs* **2005**, *10*, 109.
2.   (a) Burk, R. M. *Ann. Rep. Med. Chem.* **2008**, *43*, 293, (b) Breyer, R. M.; Bagdassarian, C. K.; Myers, S. A.; Breyer, M. D. *Annu. Rev. Pharmacol. Toxicol.* **2001**, *41*, 661.
3.   (a) Giuffre, G. *Graefes Arch. Clin. Exp. Ophthalmol.* **1985**, *222*, 139, (b) Lee, P.-Y.; Shao, H.; Xu, L.; Qu, C.-K. *Invest. Ophthalmol. Vis. Sci.* **1988**, *29*, 1474.
4.   (a) Villumsen, J.; Alm, A.; Söderström, M. *Br. J. Ophthalmol.* **1989**, *73*, 975, (b) Bito, L. Z. *Arch. Ophthalmol.* **1987**, *105*, 1036.
5.   Resul, B.; Stjernschantz, J.; No, K.; Liljebris, C.; Selén, G.; Astin, M.; Karlsson, M.; Bito, L. Z. *J. Med. Chem.* **1993**, *36*, 243.
6.   Stjernschantz, J.; Resul, B. *Drugs Fut.* **1992**, *17*, 691.
7.   (a) Camras, C. B.; Schumer, R. A.; Marsk, A.; Lustgarten, J. S.; Serle, J. B.; Stjernschantz, J.; Bito, L. Z.; Podos, S. M. *Arch. Ophthalmol.* **1992**, *110*, 1733, (b) Villumsen, J.; Alm, A. *Br. J. Ophthalmol.* **1992**, *76*, 214.
8.   Bito, L. Z.; Stjernschantz, J.; Resul, B.; Miranda, O. C.; Basu, S. *J. Lipid Mediat.* **1993**, *6*, 535.
9.   Toris, C. B.; Camras, C. B.; Yablonski, M. E. *Ophthalmol.* **1993**, *100*, 1297.
10.   Bandyopadhyay, P. in *Prodrugs: Challenges and Rewards Part 1*, Stella, V. J., Borchardt, R. T., Hageman M. J., Oliyai, R., Maag, H., Tilley, J. W., eds. Springer: New York, **2007**; Part 5, pp. 581–588.
11.   Corey, E. J.; Weinshenker, N. M.; Schaaf, T. K.; Huber, W. *J. Am. Chem. Soc.* **1969**, *91*, 5675.
12.   Corey, E. J.; Albonico, S. M.; Koelliker, U.; Schaaf, T. K.; Varma, R. K. *J. Am. Chem. Soc.* **1971**, *93*, 1491.
13.   (a) Dart, M. C.; Henbest, H. B. *Nature* **1959**, *183*, 817, (b) Dart, M. C.; Henbest, H. B. *J. Chem. Soc.* **1960**, 3563.
14.   Collins, P. W.; Djuric, S. W. *Chem. Rev.* **1993**, *93*, 1533.
15.   Ivanics, J.; Szabo, T.; Hermecz, I.; Dalmadi, G.; Ivanics, J.; Bahram, R. U.S. Pat. 5466833, **1995**.
16.   (a) Resul, B. WO 9202496, **1992**. (b) Stjernschantz, J. W.; Resul, B. WO 9002553, **1990**.
17.   Kelly, R. C.; Van Rheenen, V. *Tetrahedron Lett.* **1976**, 1067–1070.
18.   Saddler, J. C.; Symonds, J. U.S. Pat. 5079371, **1992**.
19.   Corey, E. J.; Becker, K. B.; Varma, R. K. *J. Am. Chem. Soc.* **1972**, *94*, 8616–8618.
20.   Noyori, R.; Tomino, I.; Nishizawa, M. *J. Am. Chem. Soc.* **1979**, *101*, 5843–5844.
21.   Corey, E. J.; Bakshi, R. K.; Shibata, S.; Chen, C. P.; Singh, V. *J. Am. Chem. Soc.* **1987**, *109*, 7925–7926.

22.   Iguchi, S.; Nakai, H.; Hayashi, M.; Yamamoto, H. *J. Org. Chem.* **1979**, *44*, 1363–1364.

23.   (a) Brown, H. C.; Chandrasekharan, J.; Ramachandran, P. V. *J. Am. Chem. Soc.* **1988**, *110*, 1539–1546, (b) Cha, J. S.; Kim, E. J.; Kwon, O. O.; Kim, J. M. *Bull. Korean Chem. Soc.* **1994**, *15*, 1033–1034.

24.   Henegar, K. E. U.S. Pat. 6689901, **2004**.

Index

Printed in the United States
By Bookmasters